Particle Technology Forum 2017

Core Programming Area at the 2017 AIChE Annual Meeting

Minneapolis, Minnesota, USA
29 October – 3 November 2017

ISBN: 978-1-5108-5792-6

Printed from e-media with permission by:

Curran Associates, Inc.
57 Morehouse Lane
Red Hook, NY 12571

Some format issues inherent in the e-media version may also appear in this print version.

Printed by Curran Associates, Inc. (2018)

AIChE
120 Wall Street, FL 23
New York, NY 10005-4020

Phone: (800) 242-4363
Fax: (203) 775-5177

www.aiche.org

Additional copies of this publication are available from:

Curran Associates, Inc.
57 Morehouse Lane
Red Hook, NY 12571 USA
Phone: 845-758-0400
Fax: 845-758-2633
Email: curran@proceedings.com
Web: www.proceedings.com

TABLE OF CONTENTS

Some abstracts in this book refer to images that were not made available by the publisher

501271 Single Drop Impact on Heterogeneous Powder Beds through Granule Formation Mechanisms, Drop Penetration Time, and Granule Morphology and Structure

Abstract:

Wet granulation is a process by which agglomerates are formed using liquid to bind particles together. It is important in particulate applications like pharmaceuticals, detergents, agricultural chemicals and catalysts, with the purpose to achieve particle size enlargement, create dosage uniformity and improve the powder's flow properties. Specifically, the interaction between liquid and heterogeneous mixtures, where the powders have different wettabilities, has a lot of relevance in the pharmaceutical industry, where active ingredients are often relatively hydrophobic. Within granulation, single drop granule formation is of large interest, by revealing the fundamentals of the interaction between liquid and powders. Parameters like drop penetration time and granule formation mechanisms have been studied in single drop granulation to better quantify the drop impact and subsequent granule formation process. Studying the granule morphology and content uniformity can also help understanding the structure and property of granules, thus improving their quality. Therefore, it is novel to combine drop penetration time, granule formation mechanism, and granule characterization together to achieve a comprehensive map of the granulation process.

In this work, single drop impact of liquid on a static powder bed was studied to investigate the granule formation mechanism, droplet penetration time, granule content uniformity, as well as the characterization of granules (morphology and internal structure). Water was used as the liquid binder and two pharmaceutical powders, microcrystalline cellulose (MCC) and acetaminophen (APAP), were mixed to make heterogeneous powder beds of varying compositions. The complete drop impact and penetration was recorded with a high-speed camera. Three granule formation mechanisms identified previously [1] have occurred: Spreading, Tunneling, and Crater formation. At a low impact height, Spreading occurred for mixtures with an APAP amount of less than 25%, while there is a transitional regime from Spreading to Tunneling when the APAP amount increased from 25% to 50%. Tunneling dominates from 50% to 100% of APAP proportion. Crater formation occurred at a high impact height for the materials that exhibit Spreading at low impact height. The granule morphology and the internal structure were characterized by a prism method with image analysis [1] and micro-CT, respectively. The content uniformity was characterized via UV-vis spectrometry by measuring the UV absorbance of the APAP dissolved in ethanol. It is believed that the mean particle size of the powder bed is the predominant factor in influencing the granule formation mechanism, drop penetration time, and granule properties.

[1] Emady, H. N., Kayrak-Talay, D., Schwerin, W. C., & Litster, J. D. (2011). Granule formation mechanisms and morphology from single drop impact on powder beds. Powder technology, 212(1), 69-79

489513 Development of an Inline Measurement Tool for Particle Size and â €“Shape Analysis during the Granulation Process

Abstract:

Development of an Inline Measurement Tool for Particle Size and –Shape Analysis during the Granulation Process

Maria Niesing, Technical University Dortmund

Dominik Weis, Technical University Kaiserslautern

Sergiy Antonyuk, Technical University Kaiserslautern

Markus Thommes, Technical University Dortmund

Inline measurement tools provide an efficient way of monitoring industrial processes. Generated data during the process can prove to be valuable for process optimization, and processing times can be shortened by implementing endpoint-detection. Especially for pharmaceutical applications, close process surveillance is necessary to ensure quality by design. Therefore, several process analytical tools for particle size measurement of granules have been developed. However, the quality of a granule is not only determined by the particle size, but also by the particle shape. With most commercially available tools for inline particle size measurement, no information about the particle shape is gained[1].

A common technique to gain uniform, spherical granules for pharmaceutical use is extrusion/spheronisation. Here, extrudate strands are generated by wet extrusion. The wet extrudates are then rounded in a spheronizer, a machine which consists of a cylindrical jacket and a rotating plate at the bottom. Through breakage, attrition and agglomeration, the long extrudate strands are formed into spherical particles[2]. Previous works show that so far, no commercially available process analytical tool can provide inline particle size and shape measurement of moving particles in a spheronizer. The aim of this study is the development of a method for the inline particle size and shape determination for the spheronization process. The measurement is facilitated by the round shape of the particles and the narrow size- and shape distribution[3].

An optical method was preferred to get information about the particle shape. A script for image analysis was written in Python 2.7, using the libraries Pillow 3.2.0, Scipy 0.17.1, Numpy 1.11.1, Scikit-image 0.12.3 and Mahotas 1.4.1. The equivalent diameter was used for size determination of each particle. For the shape description, the maximum feret diameter, its perpendicular feret diameter and the minimum feret diameter was calculated. The aspect ratio, calculated by dividing the maximum feret diameter by the minimum feret diameter, was used to describe the sphericity. The calculation of the aspect ratio by the maximum feret diameter and its perpendicular feret diameter is also possible.

To validate the correct function of the script, granules of different shapes and sizes were generated by extrusion/spheronization of a powder mixture (20% MCC, 80% Lactose) with different amounts of water, which served as granulation fluid. Images of the dried particles where taken under a microscope on dark background. These images were analyzed with the python script and with the open source software ImageJ.

The inline measurement was achieved by placing a High-Speed-Camera (HighSpeedStar 3G, LaVision, Göttingen, Germany) in front of the glass cylinder of a spheronizer(Ø 120 mm). The friction plate was rotating with a peripheral speed of 2 m/s. To ensure that no particle was contained in consecutive images, the Camera took 60 images per second, each with an exposure time of 1/4000 s., LED lights were placed beside, over and under the spheronizer, illuminating the particles from a sharp angle and enhancing particle-background contrast Granules from the same batches that were measured under the microscope were now measured inline in the spheronizer.

Comparison of the results from the python script and ImageJ showed that the equivalent diameter, the maximum feret diameter and the minimum feret diameters did not differ between the two methods. ImageJ is by default not able to detect the feret diameter that lies orthogonal to the maximum diameter. The python script was able to analyze a picture of 2592 x 1944 pixels in less than six seconds on a standard office computer, while the semiautomatic analysis with ImageJ took about two minutes per image.

The equivalent diameter measured inline was systematically smaller than the equivalent diameter that was obtained by the offline measurement. This was expected, since the inline measurement technique tends to ignore particle areas that are shadowed by other particles. Therefore, the outer edges of a particle are often cut off. The amount of deviation between the inline and offline measurement depends on the particle size and shape. Since the minimum feret diameter is affected stronger than the maximum feret diameter, the calculated aspect ratio is too high. However, the inline measurement still gives valuable information about trends in the change of particle size and shape.

The results show that inline measurement of particle size and shape during extrusion/spheronization is possible. However, improvement concerning the measurement of the particle shape is needed.

1. A.F.T. Silva, A. Burggraeve, Q. Denon, P.V.d. Meeren, N. Sandler, T.V.D. Kerkhof, M. Hellings, C. Vervaet, J.P. Remone, J.A. Lopes, and T.D. Beer, Particle sizing measurements in pharmaceutical applications: Comparison of in-process methods versus off-line methods. European Journal of Pharmaceutics and Biopharmaceutics, 2013. 85(3): p. 12.

2. M.D. Köster, Spheronization Process - Particle Kinematics and Pellet Formation Mechanisms, in Institut für pharmazeutische Technologie und Biopharmazie. 2012, HHU Düsseldorf. p. 108.

3. L. Hellen and J. Yliruusi, Process variables of instant granulator and spheroniser: III. Shape and shape distributions of pellets International Journal of Pharmaceutics, 1993. 96.

491868 Surface Velocity Measurements of a Powder Bed in a High Shear Wet Granulator Using High Speed Video Analysis

Abstract:

Current scale up strategies for wet granulation methods from lab scale (5 L) to plant scale (300+ L) normally involve maintaining a constant impeller tip speed and operating at a Froude number which ensures the granulator is operating in the roping regime. However, for highly soluble drugs, powder wetting and water distribution are the most important parameters. Therefore an alternative approach is to improve water distribution through the implementation of spraying apparatus and then to subsequently scale up using the spray flux density. The spray flux density is a dimensionless parameter which can be used to maintain equivalent liquid distribution across different scales of granulator. An important value used for calculating this parameter is the powder surface velocity, which is significantly lower than the tip speed of the impeller or the velocity of the bulk powder. This work tested the application of a high speed video camera running at up to 750 frames per second to film the movement of a large fraction of the area of the powder bed. A range of bowl sizes was studied, with various bowl fill levels using a lactose based placebo blend. To analyse the data, a combination of ImageJ with the TrackMate plug-in and Matlab were used. This allowed the determination of powder flow properties as a function of impeller speed and bowl fill in great detail. This approach was able to accurately measure powder bed velocity, using thousands of tracked particles. As a result, the transition from the bumping to the roping regime, with increasing impeller speed, was clearly seen as a distinct plateau in the data. It also revealed that the speed at which the transition between regimes occurred increased with increasing bowl fill. This suggests that the Froude number could be modified for use with wet granulators to take bowl fill into account. Further to this the video data was able to reveal the oscillatory nature of the velocity of the powder bed caused but the periodic movement of the impeller blades under the bed. It was also established from this information that powder movement varied distinctly as a function of powder bed location, both radially and with relative impeller position. The results were also compared against DEM simulations and other measurement approaches to demonstrate the accuracy and repeatability of the method. This work has demonstrated that the high speed video camera is a useful tool for monitoring the movement of powder in a wet granulator and the data can be used to assist in scale up. Moreover, careful analysis of the wealth of data provided will help to adjust system parameters as needed.

502471 DEM Investigation on the Dynamics of Singlet-Doublet Collisions of Cohesive Particles

Abstract:

Particulate and multiphase flows can be impacted significantly by interparticle cohesion such as the van der Waals force, liquid-bridging and electrostatics. Research toward the establishment of a general theory for cohesive particles is an active area of research. For relatively rapid flows, such a theory is comprised of a kinetic-theory-balances coupled to the population balance . Population balances track the evolution of agglomerate size distributions via source and sink terms, which are one of the most challenging aspects of closing such theories. Specifically, developing robust closures for these terms requires a criterion to differentiate between different outcomes (daughter distribution) of collisions among cohesive particles and/or agglomerates. While the criterion is well understood for binary collisions, it remains less explored for collisions involving agglomerates. In this work, we study the dynamics of a cohesive particle (singlet) impacting an agglomerate of two particles (doublet) via DEM simulations. Compared with binary collisions where two singlets either agglomerate or bounce depending on their impact velocity, four collision outcomes are observed in single-doublet collisions, including agglomeration, bounce, transfer and breakage. The occurrence of each outcome is affected by the pre-collisional configurations with four degrees of freedom. Sweeping the parameter space of the pre-collisional configurations reveals oscillatory transitions between different collision outcomes with increasing impact velocity. Detailed kinematic analysis suggests the oscillatory transitions are caused by the multiple, consecutive collisions between the singlet and the two particles in the doublet. By summarizing the DEM results in non-dimensional form, we propose a general method to estimate the transition criteria of collision outcomes, which can be applied to derive the closures for the population balances in the continuum theory to predict the behavior of granular and multiphase flows of cohesive particles.

INTERVAL TYPE-2 FUZZY PREDICTIVE MODELLING FOR A HIGH SHEAR GRANULATION PROCESS

Wafa' H. Alalaween[1], Mahdi Mahfouf[1] and Agba D. Salman[2]

[1] Department of Automatic Control and Systems Engineering, University of Sheffield, UK
[2] Department of Chemical and Biological Engineering, University of Sheffield, UK
Email: whalalaween1@sheffield.ac.uk

Nowadays, there is a strong need for a further step to be taken towards the development of a cost-effective product that meets the stringent regulations that are being imposed on most industries. Such a step is very important but represents a non-trivial task, particularly for those industries where, for instance granulation is defined as being a unit operation of the production process; this being due to the complex nature of the granulation process itself. Consequently, developing a fast, more accurate, transparent and cost-effective predictive model is a target that researchers in both industry and academia strive to achieve. Various approaches have hitherto been utilized to model the process such as population balance or neural networks based models [1-2]. However, these modelling approaches cannot effectively deal with the uncertainties present in both the inputs and the outputs of the granulation process systematically. Such uncertainties may be due to measurement errors or constraints, or simply to the heterogeneous distribution of both porosity and binder content during the granulation process. In this research, a type-2 fuzzy modelling paradigm has been implemented to predict the properties of the granules produced by a high shear granulation process and also to absorb any uncertainties that surround the process. This choice was motivated by the fact that fuzzy logic can deal with uncertainties more naturally and is considered to be a universal approximator. In addition, such a model has been utilized to extract meaningful information to describe the process linguistically in a simple way that can be understood by users. In order to take account of any unmodelled stochastic behaviour of the process, and also to improve the prediction performance, a Gaussian mixture model has been used. Experimental results show that the proposed predictive model was successful in predicting the properties of the granules accurately with the added advantage of the modelling framework being interpretable.

1. Introduction

The process of high shear granulation (HSG) is typically used in a wide range of industries to control and obtain the particle properties required, such as strength, flowability and homogeneity [3]. Modelling a wet granulation process has received a considerable attention, where the main aim is not only to predict the properties of the granules but also to understand the process and its mechanisms. The modelling approaches that have been developed to represent the granulation process can be classified as either physical-based or data-based models. One of the most widely used physical-based models is a population balance model (PBM), where the evolution of particles is followed with time [1]. Recently, various compartmental models that integrate PBM and computational fluid dynamics have been developed [4].

Intelligent systems associated with data-based models, which are simply based on intensive computations, have also been implemented to model the HSG process [2]. Many paradigms such as neural networks with different structures and topologies and fuzzy logic systems have been utilized to predict the properties of the granules produced by using different granulation equipment and materials [5]. Furthermore, models that integrate both physical-based models and data-based ones have also been reported in the open literature [6]. The idea behind such an integration is to circumvent some potential limitations, which may rise when each model is implemented in isolation [6]. Some of these models focus on providing a better understanding of the process, whereas others focus on achieving highly predictable capabilities. However, none of these models is capable of dealing with uncertainties; such uncertainties may be due to the heterogeneous distribution of the binder content and porosity or due to the measurement methods or apparatus [2].

In this research, which is an extension of a previous study by the same authors [7], a modelling framework that integrates interval type-2 fuzzy logic systems (IT2FLS) and a Gussian mixture model (GMM) has been utilized to model the HSG process. First, an IT2FLS model has been implemented to map the granulation inputs to the properties of the granules by extracting meaningful features from the collected data. Such a model has the ability to handle the uncertainties in both the inputs and the outputs. However, this model, as most of the other models do, assumes that the error is normally distributed. Such an assumption is not valid in most cases. Therefore, GMM has been developed to characterise the model prediction errors and thus to improve the modelling performance.

2. Interval Type-2 Fuzzy Logic Systems

2.1 Interval Type-2 Fuzzy Logic Systems: Theory

Generally, data-driven intelligent systems based models include, but are not limited to, artificial neural networks, fuzzy systems and evolutionary genetic algorithms [8]. In addition to predicting the process output accurately, developing a model that provides useful information about the process under investigation is essential in many cases, especially, when such information can be used by users to understand and consequently to control the process. Fuzzy logic based models can extract accurate, transparent and consequently attractive architectures for a complex process [9]. Both type-1 fuzzy logic systems (T1FLS) and type-2 fuzzy logic systems (T2FLS) characterise the process by linguistic IF-THEN rules. The distinction is associated with the nature of the fuzzy sets of the antecedents and the consequent of these rules. In T2FLS, at least one of the fuzzy sets is a type-2 fuzzy set, where the membership function is itself a fuzzy set. Therefore, the extra degree of freedom allows T2FLS to handle the uncertainties more accurately compared to its counterpart T1FLS. However, T2FLS models, generally, are more computationally expensive compared to T1FLS ones [9]. For simplicity, an interval type-2 fuzzy logic systems (IT2FLS) model has been utilized in this research paper. Figure 1 depicts the structure of IT2FLS.

As shown in Figure 1, the crisp inputs are fuzzified into fuzzy input sets using an interval type-2 set to produce the lower and upper membership functions $\left[\underline{\mu}_{\tilde{A}_j^i}, \overline{\mu}_{\tilde{A}_j^i}\right]$, where $\tilde{A}_j^i$ is the ith interval type-2 fuzzy set for the jth variable. By using a Gaussian function with an uncertain mean, the membership function can be defined as follows:

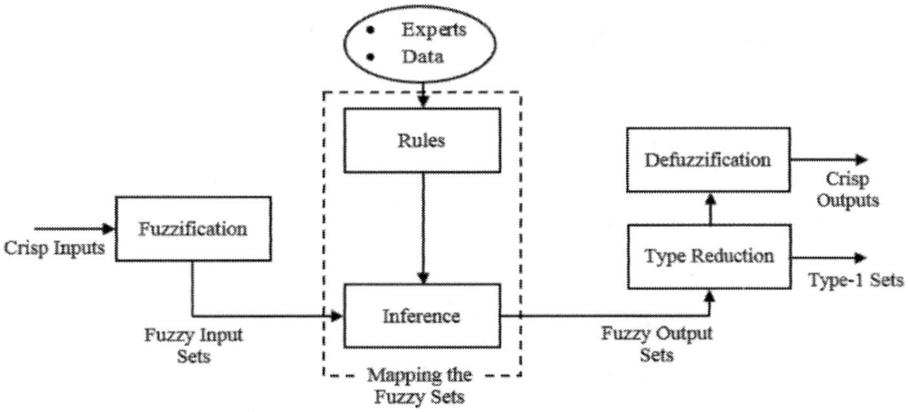

Figure 1. The Structure of IT2FLS.

$$\mu_j^i(x_j) = \exp\left[-\frac{1}{2}\left(\frac{x_j - m_j^i}{\sigma_j^i}\right)\right], \quad m_j^i \in \left[m_{j1}^i, m_{j2}^i\right] \tag{1}$$

where m_j^i and σ_j^i are the mean and the standard deviation, respectively. The region bounded by the upper and the lower membership function is called the footprint of uncertainty [9]. The fuzzification step is then followed by estimating the firing strength of the rules which are usually expressed as follows:

Rule$_i$: IF x_1 is $\tilde{A}_1^i$ and x_2 is $\tilde{A}_2^i$ … and x_n is $\tilde{A}_n^i$, THEN y is $\tilde{G}^i$

where $\tilde{G}^i$ is the output interval type-2 fuzzy set for the ith rule. This fuzzy set is then reduced into type-1 fuzzy set by using the Karnik-Mendel (KM) algorithm [10]. Finally, the output is defuzzified by the use of simple average calculation.

2.2 Interval Type-2 Fuzzy Logic Systems: Results and Discussion

Calcium Carbonate (D_{50}=85μm) was granulated by adding Polyethylene Glycol (PEG 1000), in the liquid state, using the high shear Eirich mixer (1 Litre vertical axis granulator with a top-driven impeller, Maschinenfabrik Gustav Eirich GmbH & Co KG, Germany). Four input parameters were considered, namely, impeller shape, impeller speed, granulation time and L/S ratio. The size of the granules produced was measured [2].

To develop an IT2FLS model for the HSG process, the collected data (108 experiments) were divided into two sets; 90 and 18 data points were used to train and to validate the model, respectively. In order to take account of the input measurement uncertainties, some of the aforementioned input variables were modelled as type-2 non-singleton (*i.e.* fuzzy numbers) whereas the impeller shape was considered as a crisp variable. The number of rules that corresponds to the minimum value of the root mean square error (RMSE) was selected. The steepest descent algorithm was utilized with an adaptive back-propagation network to tune the parameters of the model. For D_{50} and by using 6 rules, Figure 2 shows a simple fuzzy rule which would read as follows:

***Rule*: *IF* the impeller type is Type II and impeller speed is medium and the granulation time is small and the L/S ratio is small, *THEN* the D_{50} is small.**

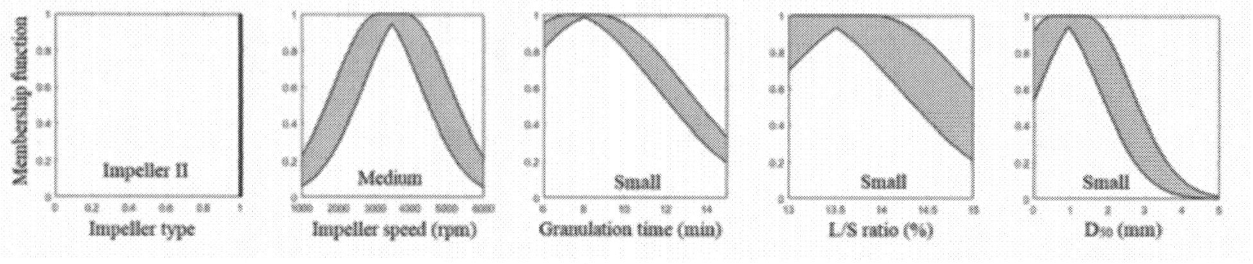

Figure 2. An example of the fuzzy rule for the D_{50}.

The model performances for the training and the testing data sets for D_{50} are shown in Figure 3, where the root mean square error values are RMSE (training, testing) = [0.26,0.31]. It can be seen that most of the predictions fall within a 90% confidence interval. Figure 4 shows an example of the predicted and the measured size distributions.

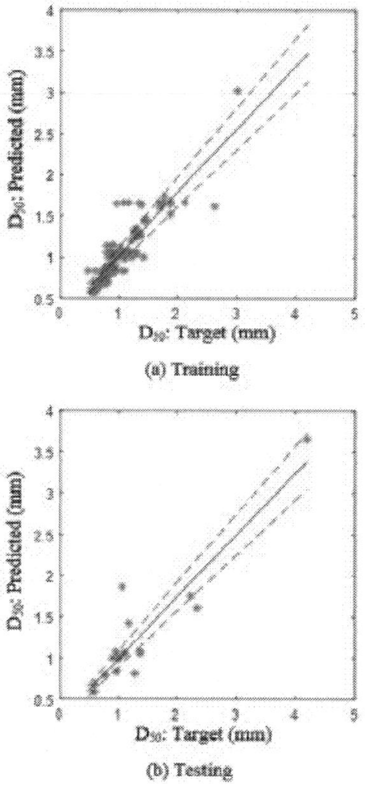

Figure 3. The IT2FLS model for D_{50}: (a) training and (b) testing (with 10% bands).

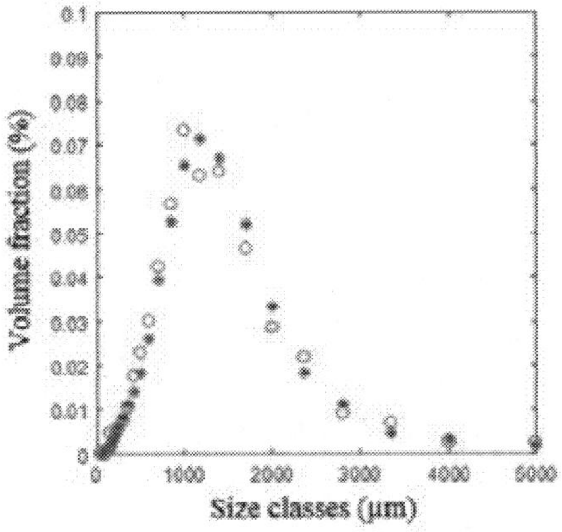

Figure 4. The IT2FLS model: The predicted and the experimental size distribution (using impeller type II, speed=6000rpm, L/S ratio (w/w)=15% and granulation time=15min).

The overall performance for the IT2FLS model (0.74) is superior to that of the IT1FLS model presented in [7], where the coefficient of determination (R^2) was approximately 0.73. The improvement percentage might be considered as insignificant, in other words, one might think it was not worth taking account the process uncertainties which increased the computational effort. However, it should be clearly stated that in the previous study [7] the impeller type was considered as a non-singleton variable, where it should have been modelled as a crisp one instead. This would consequently reduce the degree of freedom for the model, which might lead to a decrease in the modelling performance. Furthermore, including a constraint in the optimization algorithm to ensure that the variable is dealt with as a crisp one may lead to performance deterioration. By implementing the IT1FLS model and considering the impeller type variable as a crisp one, the model performance is (R^2=0.69), which when compared with the one of the IT2FLS model shows the effectiveness and the efficiency of the IT2FLS model to handle such uncertainties which may be included in the granulation process.

3. Gaussian Mixture Model

3.1 Gaussian Mixture Model: Theory

Generally, most of the predictive models are based on the assumption that the errors are normally distributed. Such an implicit assumption may lead to losing useful information in the errors. Therefore, a plethora of different algorithms has been proposed previously to extract the information hidden behind the errors residuals [11-12]. One of these algorithms is the Gaussian mixture model (GMM), which has already been utilized efficiently to refine a model performance by characterising the resultant errors [11]. However, using these algorithms, including GMM, to refine the performance of the fuzzy logic systems may change the nature of the rules that are usually extracted from collected data to provide a simple understanding of a process. In this study, GMM has been used to modify the rules based on the extra features that have been extracted from the error residuals.

The GMM is simply a linear combination of Gaussian components, which can be expressed as follows:

$$p(x)=\sum_{j=1}^{J}\pi_j \mathrm{N}\left(x \setminus \mu_j, \Sigma_j\right) \quad (2)$$

where x is data used to characterize the error, μ_j and Σ_j are the mean and the covariance of the jth component, respectively. Finally, π_j is the mixing coefficient. The optimal parameters for each component can usually be defined by maximizing the log likelihood function [8]. However, finding these parameters is not an easy task, thus the Expectation Maximization (EM) algorithm has been utilized to derive the analytical solution [13]. Once the optimal parameters are defined, the mean is added to the upper and lower mean values of each rules to compensate for the bias [11].

3.1 Gaussian Mixture Model: Results and Discussion

To improve the performance of the IT2FLS model, the error residuals were characterized. Since the model considered the main effects of the investigated variables, two input variables out of four were included in GMM. The combination of these variables that corresponds to the maximum error compensation is selected. The number of the Gaussian components has been optimised based on the EM algorithm.

Figure 5 shows an example of the predicted and the measured size distributions after bias compensation. This shows a good performance, where the predicted results are closer to the experimental ones. It is worth mentioning that the overall improvement achieved by implementing GMM is 5.3% in R^2. The overall modelling performance proved the effectiveness of the proposed modelling framework in anticipating the complex behaviour of the granulation process. Additionally, a simple understanding of the process was obtained and retained during the model refinement process.

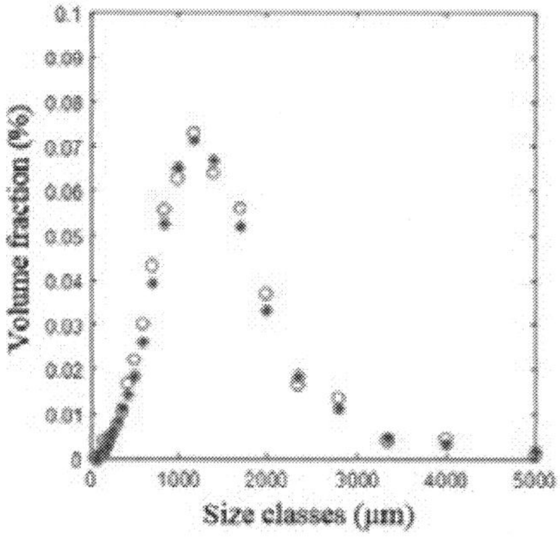

Figure 5. The modelling framework: The predicted and the experimental size distribution (using impeller type II, speed=6000rpm, L/S ratio (w/w)=15% and granulation time=15min).

4. Conclusions

In this paper, a new modelling framework integrating IT2FLS and GMM was developed to model the HSG process. First, IT2FLS was implemented to model the HSG and to extract linguistic rules that one can easily understand. Then, GMM was utilized to refine the rules in such a way that compensates for any potential bias. Such a modelling architecture successfully predicted the size distribution of the granules produced by the high shear mixer. Moreover, it can be used to control and optimize the process systematically and consequently to reduce the recycling ratio and waste in the related industries.

Acknowledgment

The Authors would like to thank Maschinenfabrik Gustav Eirich GmbH & Co KG (Germany) for providing the Eirich mixer that has been used for the experimental work.

References

[1] C. Sanders, A. Willemse, A. Salman, M. Hounslow, Development of a predictive agglomeration model, Powder Technology, 138 (2003) 18-24.

[2] W.H. AlAlaween, M. Mahfouf, A.D. Salman, Predictive modelling of the granulation process using a systems-engineering approach, Powder Technology, 302 (2016) 265–274.

[3] G. Reynolds, J. Fu, Y. Cheong, M. Hounslow, A. Salman, Breakage in granulation: A review, Chemical Engineering Science 60 (2005) 3969-3992.

[4] X Yu, M.J. Hounslow, G.K. Reynolds, A. Rasmuson, I.N. Bjorn, P.J. Abrahamsson, A compartmental CFD-PBM model of high shear wet granulation. AIChE Journal.2016.

[5] R. Mansa, R. Bridson, R. Greenwood, H. Barker, J. Seville, Using intelligent software to predict the effects of formulation and processing parameters on roller compaction, Powder Technology 181 (2008) 217-225.

[6] W.H. AlAlaween, M. Mahfouf, A.D. Salman, Integrating the physics with data analytics for the hybrid modelling of the granulation process, submitted to AIChE Journal.

[7] W.H. AlAlaween, M. Mahfouf, A.D. Salman, Data-driven deterministic and stochastic modelling of the wet granulation process, The Joint IFPRI Robert Pfeffer Symposium & UK Particle Technology Forum, Guilford, United Kingdom, 2016.

[8] C. Bishop, Pattern recognition and machine learning, Springer, New York, 2006.

[9] J.M. Mendel, Uncertain rule-based fuzzy logic systems: introduction and new directions, Prentice Hall, 2001.

[10] N.N. Karnik, J.M. Mendel, Centroid of a type-2 fuzzy set, Inform Sci 132 (2001) 195-220.

[11] Y. Y. Yang, M. Mahfouf, G. Panoutsos, Probabilistic Characterisation of Model Error Using Gaussian Mixture Model– with Application to Charpy Impact Energy Prediction for Alloy Steel, Control Engineering Practice 20 (2012) 82-92.

[12] J. A. Mauricio, Computing and using residuals in time series models, Computational Statistics & Data Analysis 52 (2008) 1746-1763.

[13] G. J. McLachlan, T. Krishnan, The EM Algorithm and Extensions, John Wiley and Sons Inc., Hoboken, New Jersey, 2008.

499163 Experimental Investigation and Modelling of the Deformation Behavior of Pharmaceutical Pellets As a Basis for DEM Simulations of the Spheronization Process

Abstract:

Spherical granules with a narrow size distribution and a regular shape are highly interesting for pharmaceutical applications. To produce such pellets, a combined extrusion and spheronization process is widely used. A spheronizer consists of a stationary cylindrical wall and a rotating disk with a structured surface. The rounding of the cylindrical extrudates is influenced by various overlapping mechanisms. An important factor affecting those mechanisms is the particle dynamics in the spheronizer, which depends on the process conditions and the pellets properties. The simulation of the particle dynamics with the Discrete Element Method (DEM) is a useful tool to get detailed information of what happens during the pellet rounding. In addition to particle kinematics and collision characteristics, other effects like segregation and mixing can be analyzed.

For the prediction of real particle interactions, the contact behavior of the simulated pharmaceutical pellets needs to be modelled with a suitable model. For this purpose, the deformation behavior was estimated experimentally in single particle uniaxial compression tests with the Texture Analyser (TA.XTplus, Stable Micro Systems, UK). The obtained force-displacement curves were used to develop a contact model which describes the particle interactions in the simulation. In the quasistatic compression tests, viscous effects cannot be investigated. Therefore additionally impact tests at different impact velocities were performed in addition to the compression tests as shown in [1] and considered in the model.

For the experiments pellets were produced with the combined extrusion and spheronization process. A mixture consisting of microcrystalline cellulose (Vivapur 102, JRS Pharma, Germany) with a mass fraction $w_i = 0.2$ g/g and a-lactose monohydrate (Granulac 200, Meggle, Germany) with $w_i = 0.8$ g/g was extruded (Micro 27 GL-28D, Leistritz, Germany) with water at 300 rpm. Afterwards, 450 g of the extrudates were spheronized in a lab scale spheronizer (RM300, Schlüter, Germany) with a rotational speed of the friction plate of 750 rpm for 5 minutes.

The spheronized pellets have a high moisture content of about 40%. Since the moisture has a significant impact on the deformation and breakage behavior, drying of the pellets during the compression tests must be avoided. For that reason, the compression tests were performed in a climatic chamber, where the temperature and the moisture can be controlled. During the spheronization process, the pellets undergo multiple loadings. To study the cyclic deformation behavior cyclic compression tests of pellets at different force levels and stress rates were carried out.

In our previous works [2, 3], the force-displacement curve were approximated with linear relations according to the model of Walton&Braun [4, 5]. In [6] different approaches to model plastic material behavior based on the work of Tomas and Luding can be found. In this contribution, a new contact model is presented. The force-displacement relationship during loading and reloading is still modelled using a linear function, however the unloading stage is

approximated with a power law to further increase the accordance of model and experiment. Moreover, the change in stiffness for reloading because of plastic deformation and consolidation in the contact area is considered, using the flattening of the particle. The force displacement curves of the linear and the new model are shown in figure 1 with a characteristic experimental result.

Figure 1: Force displacement curves of the linear and the new model compared with a characteristic experimental result

DEM simulations of the particle dynamics in a spheronization process were performed using the developed contact model and compared against simulations based on the linear contact model regarding computing time and deviation of the results. The collision rates and forces of the pellet interaction with other pellets and the apparatus walls were in particular focus. Distributions of the collision rates and the time averaged collision forces in the spheronizer were obtained.

References:

[1] Antonyuk, S., Heinrich, S., Tomas, J., Deen, N.G., van Buijtenen, M.S. and J.A.M. Kuipers: Energy absorption during compression and impact of dry elastic-plastic spherical granules, Granular Matter 1(12), 2010, 15-47.

[2] Weis, D., Antonyuk, S.; Thäte, D.; Thommes, M.: DEM Simulations of Particle Dynamics in a Spheronization Process to describe the Pelletization Mechanisms. IV International Conference on Particle-Based Methods, Proceedings, ISBN 978-84-944244-7-2, S. 567-578

[3] Weis, D., Antonyuk, S., Thommes M.: Manufacturing of Pharmaceutical Pellets by Spheronization - DEM Simulation of Particle Kinematics. PARTEC 2016, International Congress for Particle Technology, Proceedings

[4] Walton, I. OR, and R. L. Braun: Stress calculations for assemblies of inelastic speres in uniform shear. Acta mechanica 63(1-4), 1986, 73-86.

[5] Walton, O. R., Braun, R. L.: Viscosity, granular-temperature, and stress calculations for shearing assemblies of inelastic, frictional disks. Journal of Rheology 30(5), 1986, 949-980.

[6] Tykhoniuka, R. et al.: Ultrafine cohesive powders: From interparticle contacts to continuum behavior. Chemical Engineering Science 62, 2007, 2843 – 2864.

494045 A Mechanistic Model for Granule Breakage in Twin Screw Granulation

Abstract:

Wet granulation is a particle size enlargement process commonly used in solids handling industries to alleviate issues related to fine powder handling such as powder segregation, dust hazards, and poor flowability. Twin Screw Granulation provides several advantages over other methods of wet granulation, such as compact size, flexible design, flexible throughput, and robust performance, and is thus becoming increasingly common in several industries.

Granule breakage has been shown to be an important rate process in controlling the formation of large agglomerates as well as providing good liquid distribution. This presentation describes a mechanistic model for granule breakage in Twin Screw Granulation. Breakage-isolating experiments were performed at various screw speeds and powder feed rates using model granules having a wide range of dynamic yield strengths. The breakage probability and daughter size distribution were measured as functions of the formulation and process parameters. The process parameter dependence was quantified using the dimensionless powder feed number. Breakage probability was observed to be a strong function of the dynamic yield strength and the powder feed number. Breakage rate and daughter size distribution models were proposed based on fundamental understanding of the breakage mechanisms in Twin Screw Granulation.

496517 Experimental Investigation of a Dry Slag Granulation Unit Using a Rotating Disc

Abstract:

Crude steel production has increased significantly over the past decade. Blast furnace slag is one of the major by-products from the steel industries. The molten slag exits the blast furnace at 1500-1700 K. At present, this molten slag from the blast furnace is quenched with water. Upon cooling, the molten slag solidifies to form a glassy phase that can be used as an alternate to Portland cement. The major drawbacks in the existing methodology of treating BF slag are (i) Heat from BF slag is not recovered (ii) consumption of water for quenching is very high.

There is significant scope to improve the energy efficiency of the steel manufacturing sector by recovering high quality energy from the blast furnace slag and reducing the consumption of water using the dry slag granulation technology. In this work, we have performed an experimental analysis on a dry slag granulation unit using rosin-paraffin wax mixture (liquid mixture in 4:1 ratio based on weight) as an analogue to BF slag. A spinning disc was employed to atomize the molten liquid mixture in the form of fine droplets under the action of centrifugal force. The focus is on analysing the effect of temperature of rosin-paraffin wax mixture and the material of the spinning disc on average particle size. Aluminium, glass and wood are chosen as the disc material. The range of operating conditions in terms of dimensionless numbers is analysed to demarcate the particle forming regime and the fibre forming regime. The effect of diameter and rotational speed of the disc and molten liquid flow rate on the average diameter of the granules formed is studied for the chosen materials. Scale up details of the existing setup will also be highlighted based on a correlation relating the dimensionless numbers and particle diameter.

497316 Sustained Supersaturation of Erlotinib Sdd Ternary Amorphous Systems

Abstract:

As more low-solubility drug candidates enter the pharmaceutical development pipeline, the need for solubilization-enhancing technologies is ever greater. In an amorphous dispersion, the solubility of a drug is enhanced by kinetically trapping the molecules into the higher-energy amorphous state, often realizing a 5-50x enhancement in solubility relative to the crystalline state. Spray dried dispersions (SDDs), containing amorphous drug and a concentration-enhancing matrix polymer are a scalable and mature technology which takes advantage of the amorphous drug solubility enhancement. Typically, an SDD must meet numerous performance metrics including rapid dissolution rate, sustainment of supersaturation, physical stability, chemical stability and manufacturability. We have successfully formulated over 1,000 low-solubility compounds as SDDs over the past two decades.

Despite this success, for some compounds it can be challenging to meet all of the performance requirements using only one excipient in the SDD system. In these cases, a multi-component amorphous system approach may warranted, in which two excipients are utilized, each tailored to separately achieve specific performance metrics. In this presentation, we discuss the case study of erlotinib, a rapidly-crystallizing compound which requires both fast dissolution and sustained supersaturation. In this case, erlotinib SDD is spray dried with a fast-dissolving matrix polymer (Eudragit L100) and formulated with a crystallization-inhibiting polymer (HPMC) to maximize both dissolution rate and sustainment. The combination of Eudragit L100 and HPMC succeeds by increasing the in vitro test AUC over ten-fold compared to binary SDD formulations.

488954 Connecting the Product-Process-Performance Interplay for Improved Understanding in Development of a Spray Dried Dispersion Drug Product

Abstract:

The pharmaceutical industry has faced a shift in the physicochemical properties of pre-clinical drug candidates rendering them less soluble and bioavailable. Spray drying has become an important unit operation in the pharmaceutical industry to produce spray dried amorphous dispersions (SDD) which can improve apparent solubility versus the crystalline drug as well as dissolution rate, thus improvingthe bioavailability of poorly soluble active pharmaceutical ingredients (API). The development of drug products consisting of spray dried solid dispersions is complex as it involves the need for strengthened understanding of the solid dispersion and drug product formulation attributes as they relate to enhancement of drug dissolution rate and maintenance thereof in the drug product. In contrast to standard drug particle erosion as main dissolution mechanism in conventional crystalline API formulation, solid dispersions can release drug in various types of species ranging from free drug, nano-particles, bile salt micelles, precipitated drug, and so on[1]. The characteristic dissolution mechanisms are impacted by API physiochemical properties and formulation as well as the process conditions utilized to make them.

This work will cover: 1. different mechanisms of dissolution and species formed for various APIs, 2. how the spray drying process influences dissolution and amorphous product stability, 3. utility of novel spectroscopic and other characterization techniques to study dispersion quality and performance, 4. influence of dispersion and drug product formulation properties on dissolution and tablet properties, 5. end to end considerations for ensuring how key quality and performance attributes are designed in and maintained in the drug product.

Spray dried dispersion particles and powders were manufactured on various scales ranging from single particle levitation, ProCepT 4M8, GEA mobile minor and GEA PSD-2 spray dryers. Multiple process conditions resulting in a range of drying rates were studied and will be presented. The resulting powders will be characterized by standard techniques such as modulated differential scanning calorimetry (mDSC) and x-ray powder diffraction (XRPD) to evaluate bulk amorphous dispersion uniformity. In several cases, additional techniques such as focused ion beam scanning electron calorimetry (FIB-SEM), x-ray photoelectron spectroscopy (XPS), and energy dispersive x-ray spectroscopy (EDX) were utilized to look a particle level homogeneity. Drug product formulations were made by standard tumble blenders, dry granulated, lubricated and compressed into tablets. Discriminating dissolution methodologies were used to study drug release from the dispersions and tablet formulations. Environmental stressing was conducted to look at stability and impact on drug release from tablets.

The choice of polymers and surfactants that make up the SDD composition was shown to have a significant impact on the amorphous quality, particle level uniformity, resulting phase stability and dissolution behavior. The choice of solvent composition and spray drying conditions was shown to have a modest to profound impact on these quality attributes. This was shown through dissolution studies coupled with human PK data as well as phase uniformity issues

discovered immediately following manufacture or after post-process stressing. The API has inherent physio-chemical properties that tend to give rise to characteristic speciation behavior during dissolution of SDD powder. The rate of release of overall drug into these entities was negatively impacted as the dispersion load in the tablet increases, tablet void fraction exceeded critical levels, and tablet disintegrant system was not properly designed.

In conclusion, spray dried dispersion based drug products require end to end study and understanding of the API behavior, formulation and process components, characterization methodology for amorphous quality and dissolution methodology and environmental stresses to be successful. Discriminating dissolution methodologies can help understand the specific mechanisms of dissolution enhancement and novel characterization methods can be useful in characterizing and ensuring maintenance of amorphous quality in the SDD and final drug product.

References:

1: D.T Friesen et al. Hydroxypropyl Methylcellulose Acetate Succinate-Based

Spray-Dried Dispersions: An Overview. Molecular Pharmaceutics. 2008. VOL. 5, NO. 6, pg. 1003â€"1019

495763 Strategically-Designed Polymer Synthons As Pharmaceutical Excipients in Oral Drug Delivery

Abstract:

Synthetic polymer excipients, equipped with tunable structures and properties, can enable more bioavailable, inexpensive, and safe oral drug formulations when rationally designed to solubilize active pharmaceutical ingredients in amorphous solid dispersions. However, the underlying polymer-drug interactions in these materials to control drug supersaturation generation and maintenance currently remain poorly understood. We present a new approach to discover specialized excipients for a highly-crystalline drugs, driven by molecular evolution-inspired ideas to rapidly construct polymer synthons (macromolecules with structural subunits correlating to desired noncovalent interactions, analogous to the retrosynthetic analysis taught in organic chemistry) using high-throughput synthesis and screening protocols. In this manner, the molar mass and chemical composition state spaces were sampled to identify an optimal excipient synthon among 70 copolymers to solubilize phenytoin. The leading system contained N-isopropylacrylamide (NIPAm, responsible for inhibiting phenytoin crystallization) and N,N-dimethylacrylamide (DMA, providing polymer hydrophilicity), and polymer-drug interactions were revealed using NOESY and DOSY NMR spectroscopy. Complementary all-atom molecular dynamic simulations corroborated experimental evidence of NIPAm synthons strongly binding to phenytoin. When administered to rats, we observed a record 23-fold increase in area-under-the-curve bioavailability over neat drug with excellent in vitro-in vivo correlations. Altogether, these discoveries demonstrate how existing synthesis tools in the polymer chemistry landscape can accelerate the translation of new drug candidates (or even failed compounds) into more effective and accessible therapeutic treatments.

497295 Pill Burden Reduction through Engineering the Mechanical Properties of Spray-Dried Amorphous Solid Dispersions

Abstract:

Drug candidates with poor oral bioavailability due to low aqueous solubility and dissolution rate are common in the pharmaceutical industry. One of the most mature and well understood techniques to improve oral absorption of drug candidates is to formulate the drug as a spray-dried amorphous solid dispersion (SDD). Given SDDs are diluted with excipients to, primarily, stabilize the amorphous form of the drug, a common challenge in SDD oral delivery of tablets is pill burden. One promising solution to reduce pill burden is engineering the SDD particles to replace the use of tablet fillers, which primarily attenuate undesirable mechanical and powder flow properties of crystalline drugs. Given that the mechanical and powder flow properties of SDDs can be significantly tuned during the spray drying process, the use of fillers can be substantially reduced. However, tuning the particle properties of SDDs is a hierarchical multidimensional optimization problem, and for that reason, a clear relationship between particle properties and physical and chemical state stability of the drug and in vivo performance is generally established prior to optimization of mechanical and powder flow properties. There have been several recent reviews on this subject, but there lacks the direct connection between spray-drying process conditions and their impact on final tablet formulation [1-2].

The aim of the current study is to establish relationships between spray drying process parameters and the resulting physical and mechanical properties of the SDD particles, and secondly, to use these relationships to formulate high SDD loaded (greater than 70%) immediate release tablets. Felodipine spray-dried with PVP VA64 was used as the model SDD system. The powder mechanical properties measured include compression profiles, SEM imaging, modified in-die compression analysis [3], BET surface analysis, and Hiestand indices.

Results showed that, for a given SDD formulation, specific surface area of the powder was controllable by tuning the particle size, particle wall thickness, and to a lesser extent, particle morphology through drying kinetics and atomization during spray drying. This resulted in SDDs that spanned a 10-fold difference in specific surface area. The resulting mechanical properties of these high surface area SDDs included significant increases in tensile strength and bonding index (BI) when compressed, due to larger bonding area between particles [4].

The material properties were strongly dependent on the drug loading, with higher drug loading SDDs having diminished compactibility compared to their placebo counterparts. These results suggest that, counter to selecting the highest drug loading SDD with acceptable physical stability and in vivo performance, a lower drug loading SDD formulation could allow for less filler in the final drug product, resulting in an overall higher loading within the tablet. Consequently, this also improves the physical stability robustness.

In addition, it was observed that, when compressed as the neat material, the compacts had varying degrees of anisotropy, which resulted in greater axial elasticity. High elastic recovery decreases the bonding area between particles, reducing the strength of the tablets and increasing the risk for capping and lamination. Therefore, decreasing the degree of anisotropy could reduce the risk of these tablet defects and should be considered when engineering SDD particles for integration into tablet formulations [5].

Overall, the powder properties resulted in various immediate release tablet formulations with as high as 80% concentration of SDD. It was demonstrated that the SDD could span the tabletability range of typical fillers depending on how it was spray dried, allowing it to â€œdo the jobâ€ of different types of fillers. The study ultimately concludes that rationally selecting spray-drying conditions to target specific particle properties, such as high surface area, high compactibility, and isotropic geometries, allows tablet formulations that can successfully decrease pill burden.

[1] Patel, B.B. et al. Revealing facts behind spray dried solid dispersion technology used for solubility enhancement. Saudi Pharm. J. 2015, 23 (4), 352-365.

[2] Singh, A. and Van den Mooter, G. Spray drying formulation of amorphous solid dispersions. Adv. Drug Deliv. Rev. 2015.

[3] Katz, J. M., Roopwani, R., Buckner, I. S. A Material-Sparing Method for Assessment of Powder Deformation Characteristics Using Data Collected During a Single Compression-Decompression Cycle. J. Pharm. Sci. 2013, 102 (10), 3687-3693.

[4] Osei-Yeboah, F., Chang, S., and Sun, C.C. A critical Examination of the Phenomenon of Bonding Area - Bonding Strength Interplay in Powder Tableting. Pharm Res. 2016, 33, 1126-1132.

[5] Sun, C.C. Microstructure of Tablet â€“ Pharmaceutical Significance, Assessment, and Engineering. Pharm. Res. 2017, 34, 918-928.

500057 Reprocessing of Spray Dried Dispersions: Minimizing Resources in Process Development and Dealing with the Possible Failure

Abstract:

Reprocessing of spray dried dispersions: minimizing resources in process development and dealing with the possible failure

T. Porfirio[1,2*], I. Duarte[1], R. Ferreira[1], B. Henriques[1], J. Vicente[1], V. Semião[2]

[1] Hovione Farmaciência SA, Sete Casas, 2674-506 Loures, Portugal; *tporfirio@hovione.com

[2]LAETA, IDMEC, Mechanical Engineering Department, Instituto Superior Técnico, Universidade de Lisboa, Av. Rovisco Pais 1, 1049-001 Lisbon, Portugal

Spray drying is a well-established manufacturing process to produce amorphous solid dispersions to improve the bioavailability of poorly soluble drugs. However, the scale-up of the process is often considered a resource-intensive challenge requiring large quantities of expensive drugs [1]. Therefore, the possibility of reprocessing spray dried dispersions (SDD) is very appealing since it minimizes the material's requirements during the process development. Furthermore if the feasibility of the reprocessing steps is demonstrated during development it can be applied during commercial routine as risk mitigation strategy.

However, the reprocessing can have significant potential to adversely affect the identity, strength, quality, purity or other attribute of the drug and excipients [2]. The reprocessing should be performed on the cases where there is an improvement in one or more quality attributes without adverse effect on the others. The drive of the present work is to understand the effect of successive reprocessing steps on chemical and physical properties of spray dried solid dispersions using different stabilizing polymers commonly used in the production of amorphous solid dispersions.

Three different systems were produced and reprocessed (as shown in Figure 1) using itraconazole (ITZ) as model drug and HPMC E3, HPMCAS-MG and Copovidone as different stabilizing polymers. After each spray drying step a sample was collected for analytical characterization to assess the physicochemical attributes of spray dried material viz. particle size, bulk density and content of residual solvents, assay, and dissolution profile. The remaining spray dried product was reprocessed to prepare the feed solution of the subsequent spray drying step. This study carried out two reprocessing in order to understand when the material begins to present modifications in its properties.

Figure 1 – Scheme of the adopted method

All the tests were performed in a lab-spray dryer equipped with a two-fluid nozzle and using the same drying conditions (such as outlet temperature, drying gas flowrate, feed flowrate and atomization flowrate). A solvent mixture of dichloromethane and methanol was used to dissolve the material in a solid concentration of 5 wt.%.

Placebo tests with polymer-only were also performed to assess the physiochemical properties, such as molecular weight, monomer content and heavy metal levels of the stabilizing polymers. The same conditions were used as in the tests with active substance.

The use of different formulations and placebos tests will clarify the role of the polymer type on the SDD reprocessing.

Reference:

[1] Hallow, D., Mudryk, B., Braem, A., Tabora, J., Lyngberg, O. & Tummala, S., 2010. An Example of Utilizing Mechanistic and Empirical Modeling in Quality by Design. Journal of Pharmaceutical Innovation, 5(4), pp.193–203.
[2] Guidance for Industry - Drug Substance Chemistry, Manufacturing, and Controls Information, U.S. Department of Health and Human Services – Food and Drug Administration

494745 *In silico* Development of Amorphous Solid Dispersions for Optimal Performance and Stability

Abstract:

The focus of the present work lies on screening formulations for amorphous solid dispersions (ASDs) with a computational tool which allows a comprehensive set of virtual experiments to be performed without API expenditure and with a fast turnover and consequently to explore the design space in full prior to manufacturing prototypes in the laboratory.

ASDs of poorly water soluble active pharmaceutical ingredients (APIs) in hydrophilic polymeric matrices are a commonly used strategy to increase the solubility and dissolution rate of oral-dosage forms [1]. However, successful development of amorphous solid dispersions for optimal performance and stability depends on both the formulation (e.g. drug load, polymer system and surfactants) and on the manufacturing technology/process conditions. For example, adequate solvent mixtures used in the process offer the possibility of influencing particle morphology and drying rate [2] and have been reported to offer advantages in the solid state drug-polymer miscibility and on the stability of the ASDs [3,4]. On the other hand, excessively complex formulations can lead to undesirable phase separation which can compromise physical stability [5]. However, optimizing the formulation (e.g. drug load, polymer system and solubilizers) and the process (e.g. drying kinetics and droplet/particle size) requires extensive laboratory work, which is often compromised by time and cost constraints. Therefore the ability to perform virtual experiments is highly advantageous and cost-effective.

In this presentation we cover the mathematical modeling of the various concurrent physical phenomena that a solution of API and polymer (and sometimes other excipients) undergo prior to forming an amorphous solid dispersion, such as atomization, co-solvent evaporative dynamics, propensity for phase separation and shell formation [6,7]. We show a set of case studies of *in silico* screening benchmarked against data from solvent casting and spray drying experiments and also show a case-study where an excessively complex ASD formulation led to phase separation and compromised the stability of the amorphous intermediate.

Figure 1: Sketch of various stages throughout the droplet drying history and the concomitant propensity for phase separation and crystalization if the formulation and/or process are not adequately designed.

References

[1] Perrie Y., Rades T. "Pharmaceutics - Drug Delivery and Targeting", Pharmaceutical Press, 2010.

[2] Wulsten, E., Kiekens, F, van Dycke, F., Voorspoels, J. & Lee, G., "Levitated single-droplet drying: Case study with itraconazole dried in binary organic solvent mixtures", Int. J. Pharmaceutics, 2009, 378(1-2) pp. 116-121

[3] Janssens, S, Anné, M., Rombaut, P. & Van den Mooter, G., "Spray drying from complex solvent systems broadens the applicability of Kollicoat IR as a carrier in the formulation of solid dispersions", Eur. J. Pharm. Sci. 2009, 237, pp. 241-248.

[4] Paudel, A. & Van den Mooter, G., "Influence of Solvent Composition on the Miscibility and Physical

Stability of Naproxen/PVP K 25 Solid Dispersions Prepared by Cosolvent Spray-Drying", Pharm. Res. 2012, 29, pp. 251-270.

[5] Janssens, S., Nagels, S., de Armas, H.N., D'Autry, W., Van Schepdael, A. & Van den Mooter, G., "Formulation and characterization of ternary solid dispersions made up of Itraconazole and two excipients, TPGS 1000 and PVPVA 64, that were selected based on a supersaturation screening study", Eur. J. Pharm. Biopharm, 69:1, 2008

[6] Duarte, Í., Santos, J. L, Pinto, J. G & Temtem, M., "Screening methodologies for the development of spray-dried amorphous solid dispersions", Pharm. Res., 2015, 32 pp. 222 – 237, 2015.

[7] Prudic, A., Ji, Y. & Sadowski, G., "Thermodynamic Phase Behavior of API/Polymer Solid Dispersions", Mol. Pharmaceutics, 2014, 11(7), pp. 2294-2304

GHENT UNIVERSITY

AIChE

Applying Inline Spatial Filter Velocimetry to Improve Process Robustness of a Continuous Twin-Screw Wet Granulator

Niels Nicolaï

Laboratory of Pharmaceutical Process Analytical Technology

BIOMATH

Outline

- » **Introduction**
  - › Continuous Drug Product Manufacturing
  - › Process Robustness
- » **System description**
  - › ConsiGma system
  - › Spatial filter velocimetry
- » **Measurement optimisation**
  - › Response, Sensitivity, Accuracy and Precision
  - › Sample size
  - › Response time
- » **Applications**
  - › Offline vs. Inline measurements
  - › Mathematical modeling
- » **Conclusions and Future Perspectives**

LP PAT

BIOMATH

Continuous drug product manufacturing

» **Domains of interest**

- › Optimisation of the mechanical design of each unit operation
- › Understanding the influence of raw material attributes and process parameters
- › Derivation of mathematical models for simulation
- › Integration of real-time instrumentation and multivariate data analysis
- › Design of advanced process control strategies

LP PAT

BIOMATH

Continuous drug product manufacturing

» Domains of interest

- › Optimisation of the mechanical design of each unit operation

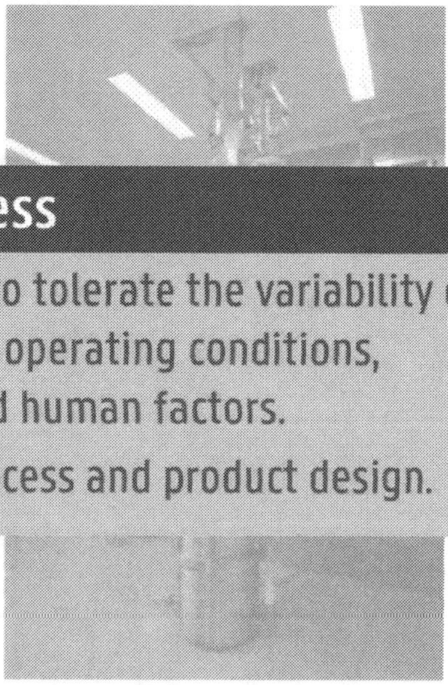

Robust process

The ability of a manufacturing process to tolerate the variability of raw materials, process equipment, operating conditions, environmental conditions and human factors.

Robustness is an attribute of both process and product design.

analysis

- › Design of advanced process control strategies

LP PAT

BIOMATH

ConsiGma™ WG System

LP PAT

BIOMATH

Continuous twin-screw wet granulation

LP PAT

BIOMATH

Spatial filter velocimetry

Parsum IPP70 inline size analyser

(Measurement principle: SFV)

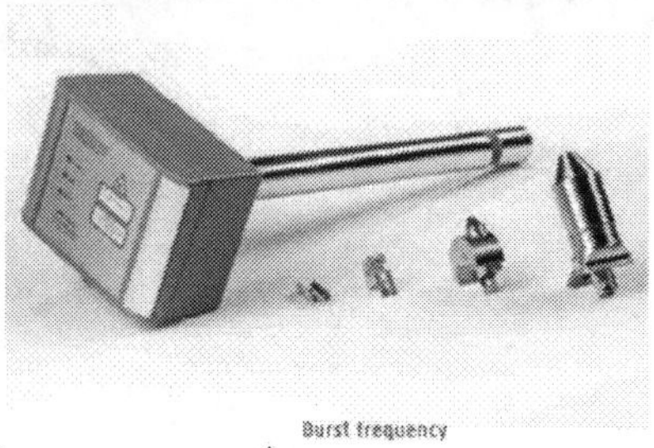

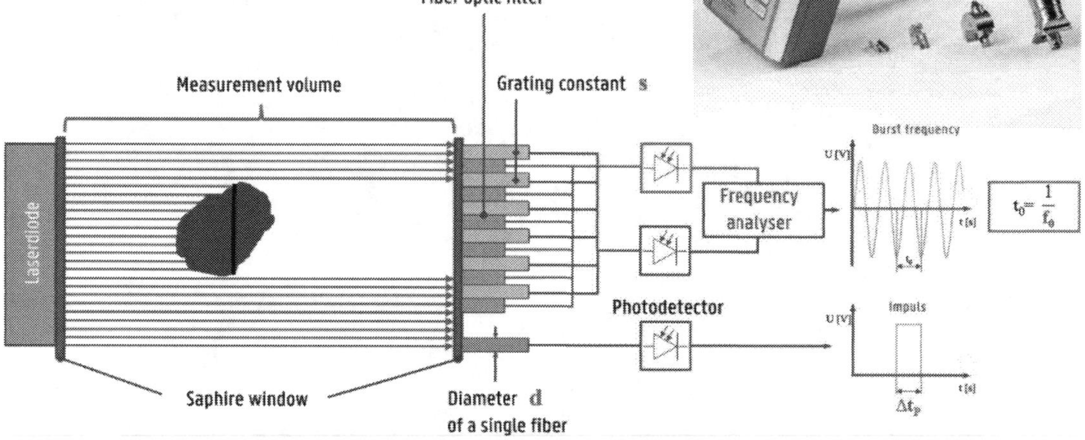

Particle velocity: $v_P = f_0 \cdot s$

Chord length: $x_P = \Delta t_P \cdot v_P - d$

LP PAT

BIOMATH

Process-instrument interface

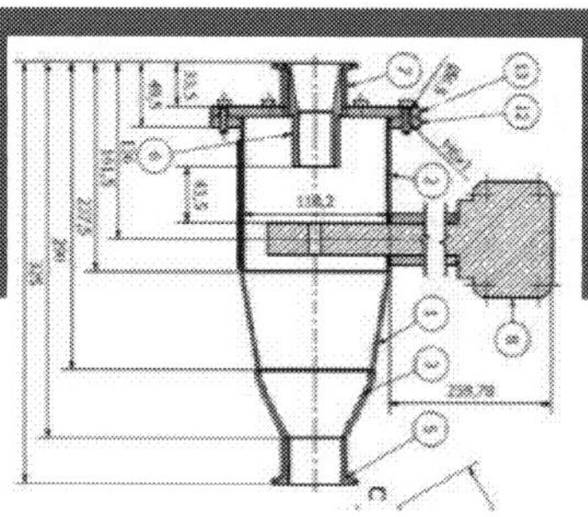

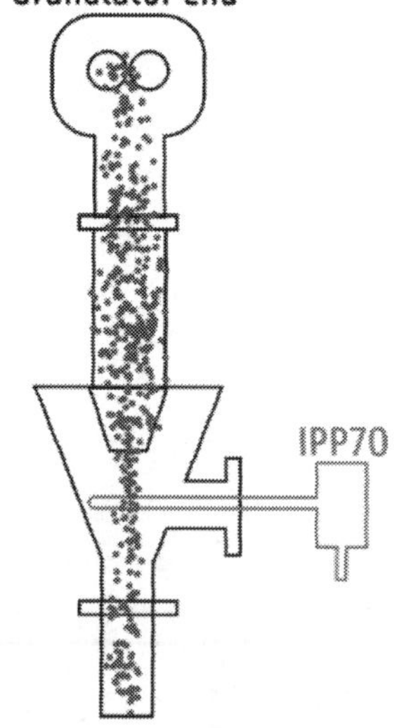

LP PAT

BIOMATH

Measurement optimisation

INPUT signal to liquid addition pumps of the wet granulation unit

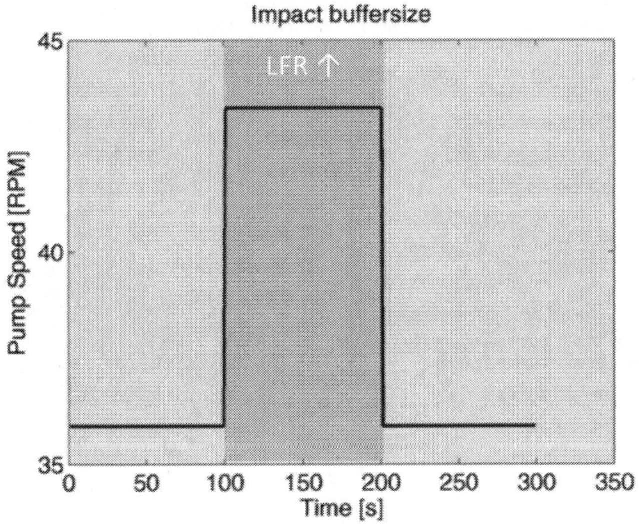

1. Instrument performance
2. Particle size

LP PAT

BIOMATH

Measurement optimisation

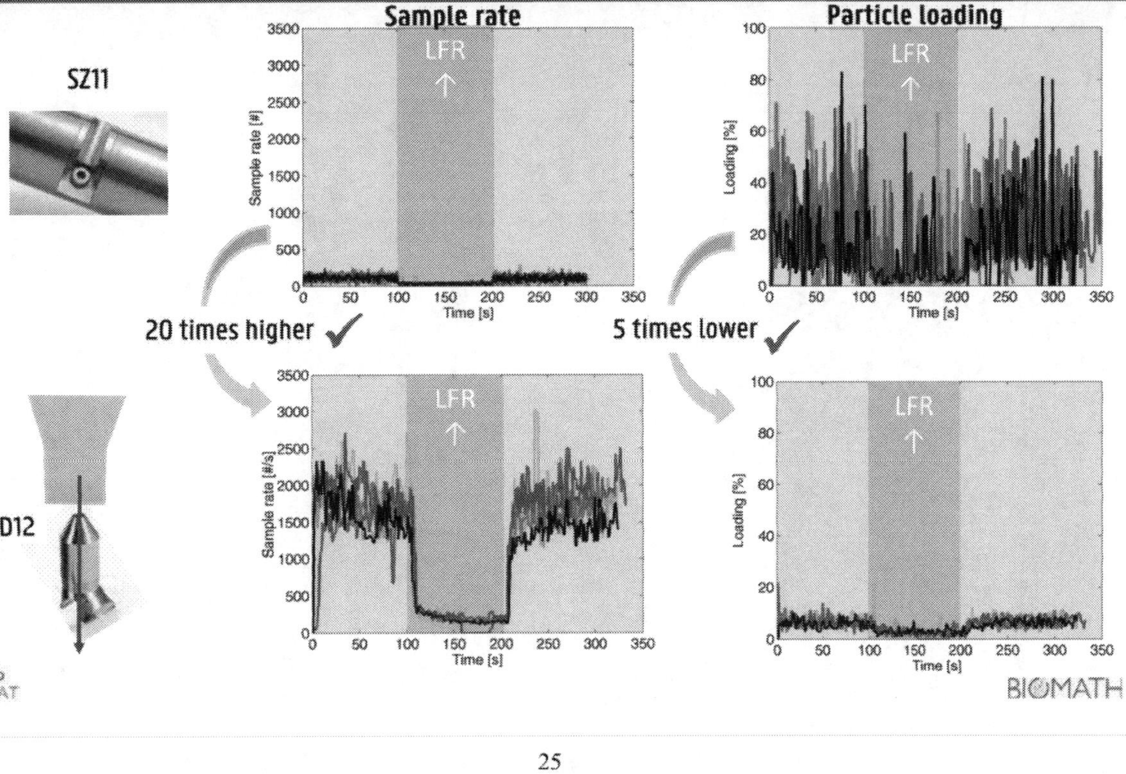

LP PAT

BIOMATH

Response

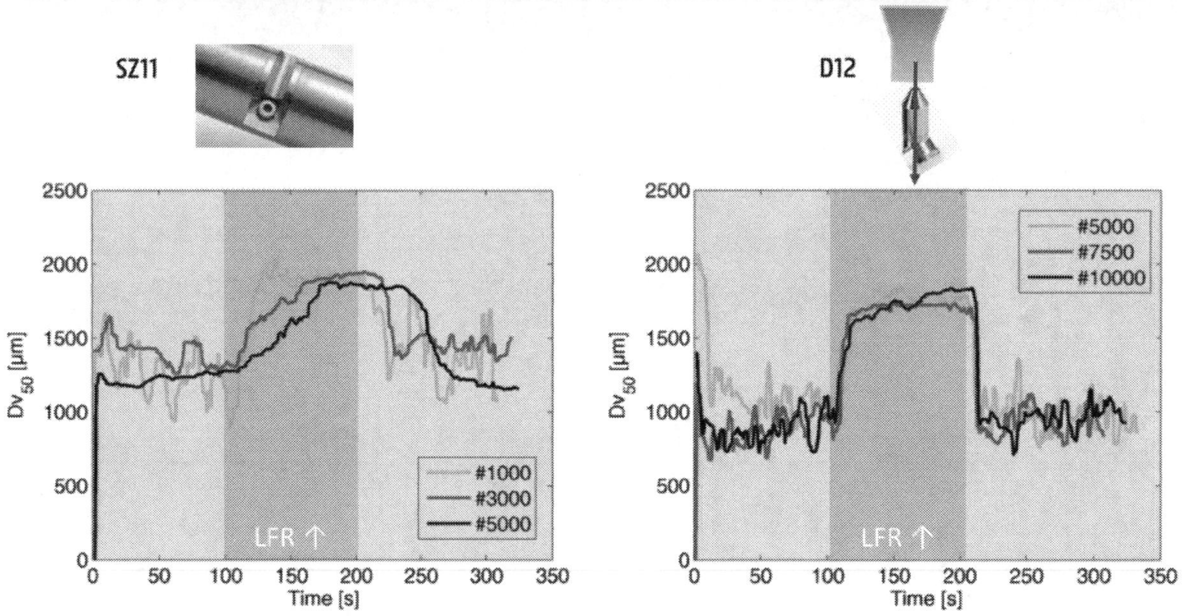

LP PAT

BIOMATH

Sensitivity

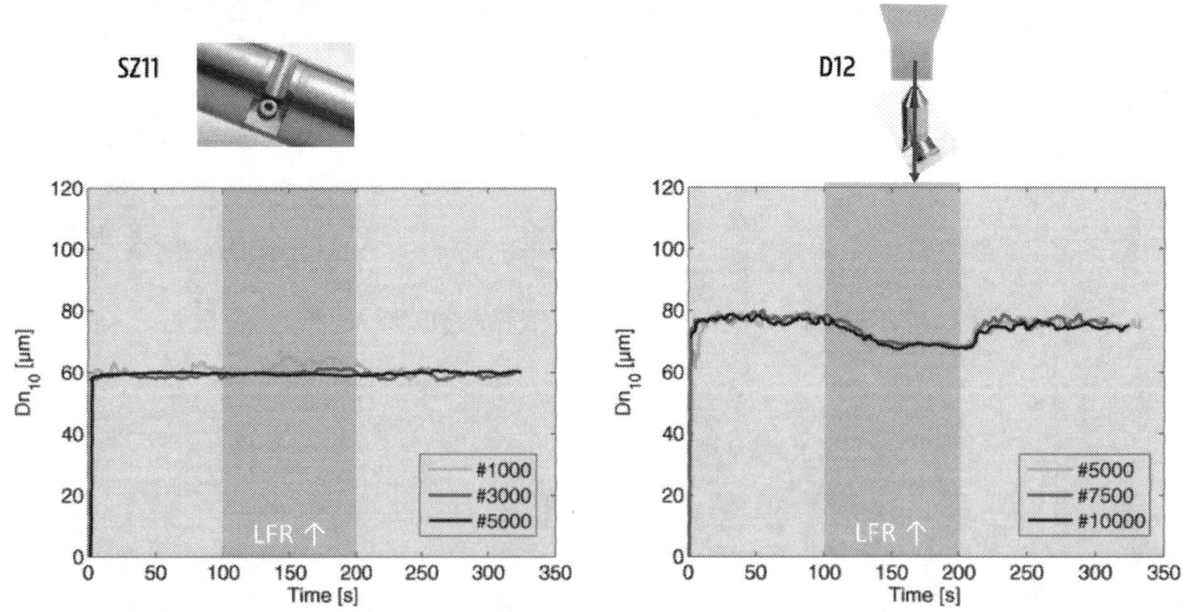

LP PAT

BIOMATH

Accuracy

SZ11

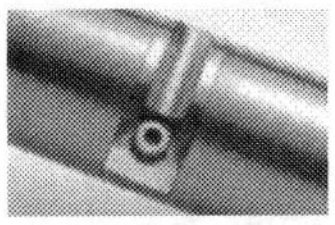

D12

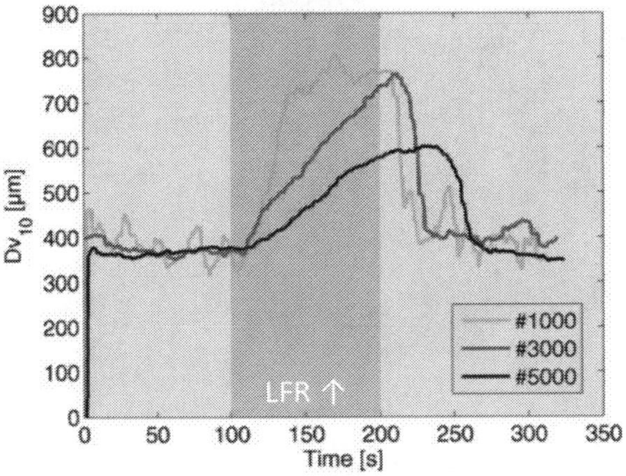

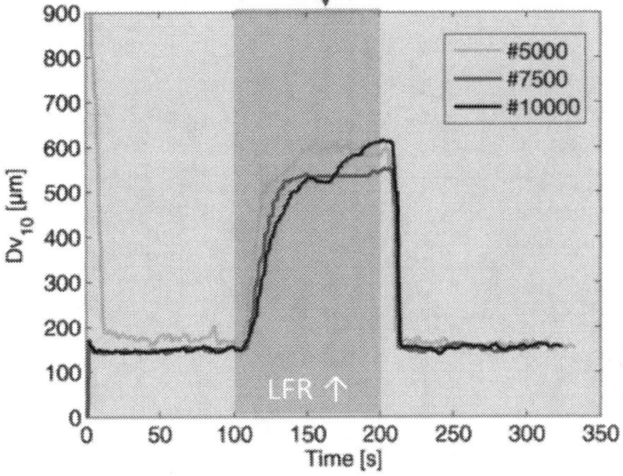

LP PAT

BIOMATH

Precision

SZ11

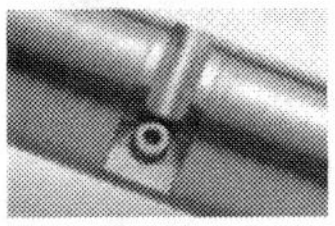

D12

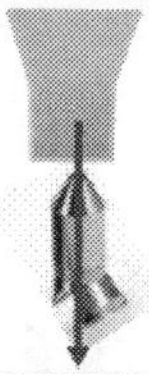

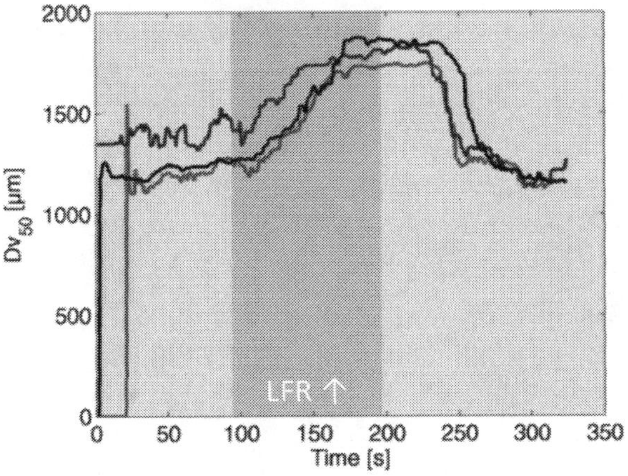

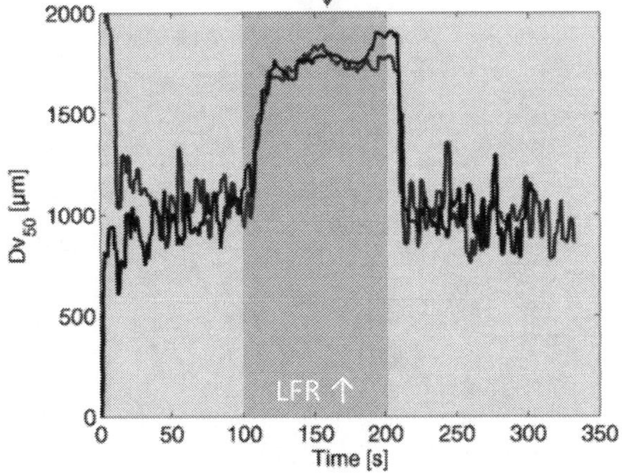

LP PAT

BIOMATH

Sample size - Measured

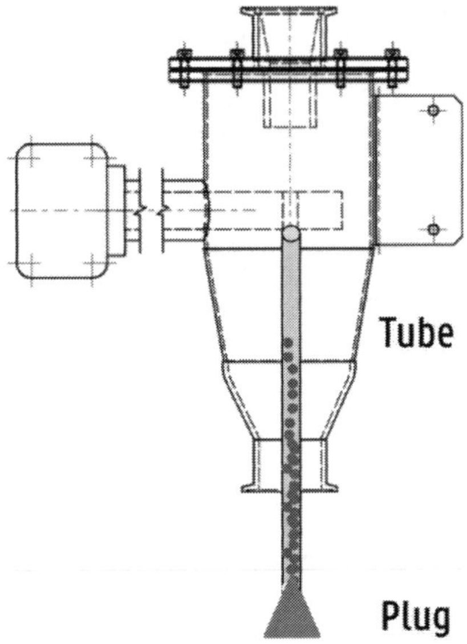

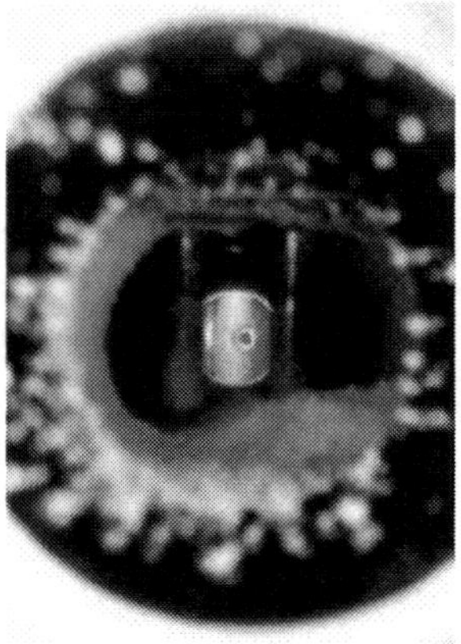

Sample size - Measured

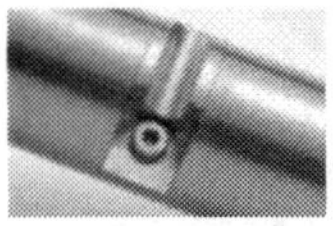

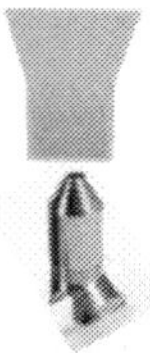

	SZ11	
Sample 1	4.10	%w/w
Sample 2	5.10	%w/w
Sample 3	5.10	%w/w
Average	4.53	%w/w
	0.91	**kg/h**
	6.28	**g/Distr.**

D12 + No Air + Cone	
9.60	%w/w
9.30	%w/w
9.10	%w/w
9.33	%w/w
1.87	**kg/h**
12.90	**g/Distr.**

D12 + Air + Cone	
9.30	%w/w
10.30	%w/w
10.10	%w/w
9.90	%w/w
1.98	**kg/h**
13.66	**g/Distr.**

Sample size - Calculated

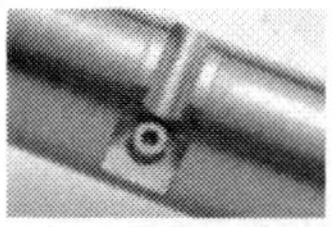

SZ11		
ϕ inlet	0.042	m
ϕ slit	0.010	m
r1 inlet	0.021	m
r2 slit	0.005	m
Ratio $r1^2/r2^2$	5.67	%w/w
Calc.	**1.13**	**kg/h**
Meas.	**0.91**	**kg/h**

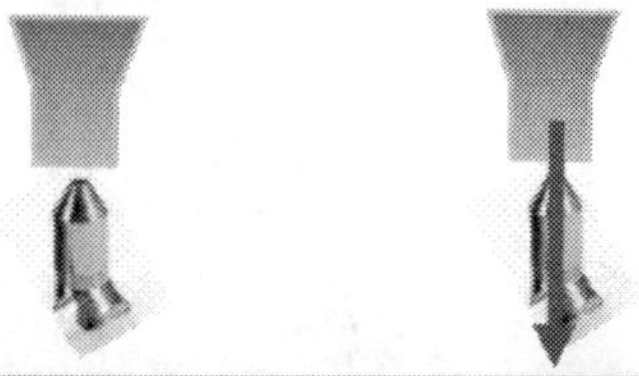

D12 + Cone		
ϕ inlet	0.027	m
ϕ slit	0.007	m
r1 inlet	0.014	m
r2 slit	0.004	m
Ratio $r1^2/r2^2$	6.67	%w/w
Calc.	**1.33**	**kg/h**
Meas. no air	**1.87**	**kg/h**
Meas. air	**1.98**	**kg/h**

LP PAT

BIOMATH

Response time

INPUT

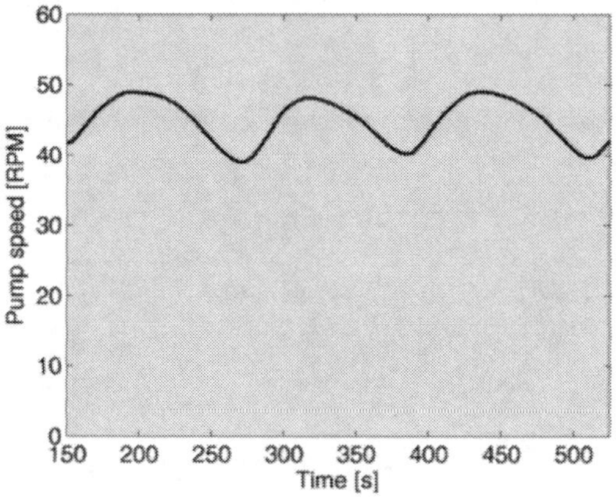

OUTPUT

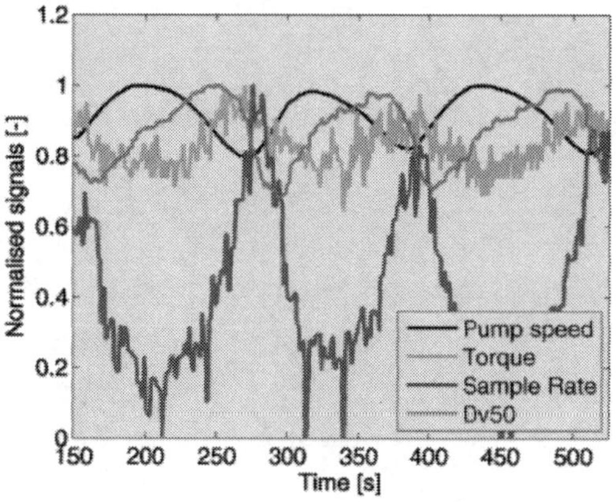

Residence time granulator << Lag time measurement

LP PAT

BIOMATH

Response time

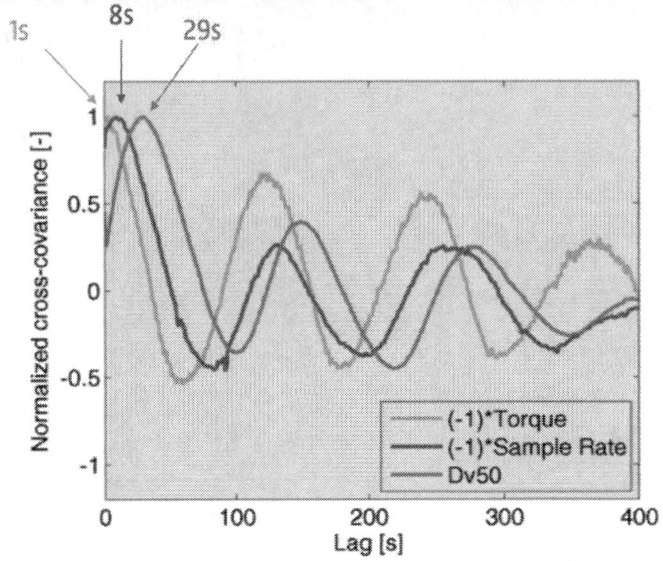

Additional machine information should be taken into account

LP PAT

BIOMATH

Offline vs. inline comparison

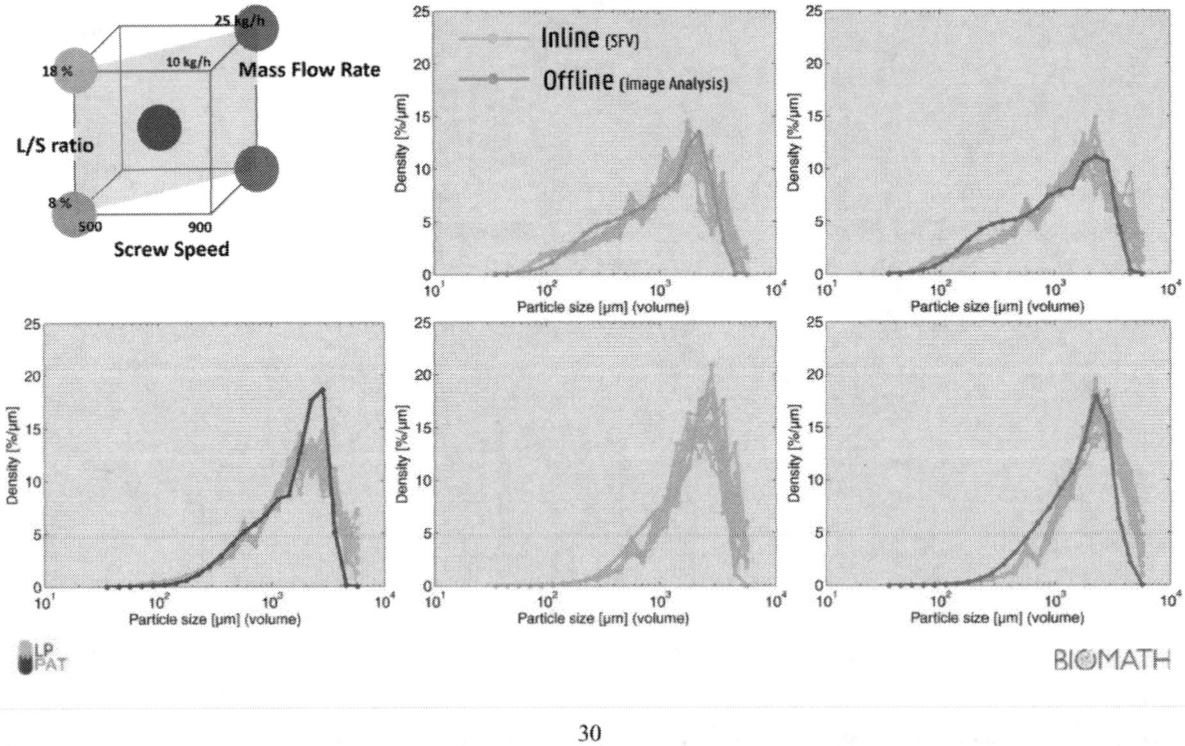

LP PAT

BIOMATH

Modeling granule size distribution

Population balance model (PBM)

$$\frac{\partial n(t,x)}{\partial t} = \frac{Q_{in}}{\tilde{V}} n_{in}(x) - \frac{Q_{out}}{\tilde{V}} n_{out}(x)$$

GSD balance

Aggregation term

$$+\frac{1}{2}\int_0^x \underbrace{\beta(t, x-\varepsilon, \varepsilon)}_{\text{aggregation rate}} n(t, x-\varepsilon)\, n(t,\varepsilon)\, d\varepsilon$$

$$-n(t,x)\int_0^\infty \underbrace{\beta(t,x,\varepsilon)}_{\text{aggregation rate}} n(t,\varepsilon)\, d\varepsilon$$

Breakage term

$$+\int_0^\infty \underbrace{b(x,\varepsilon)}_{\text{breakage fun.}} \underbrace{S(\varepsilon)}_{\text{selection rate}} n(t,\varepsilon)\, d\varepsilon$$

$$-\underbrace{S(x)}_{\text{selection rate}} n(t,x)$$

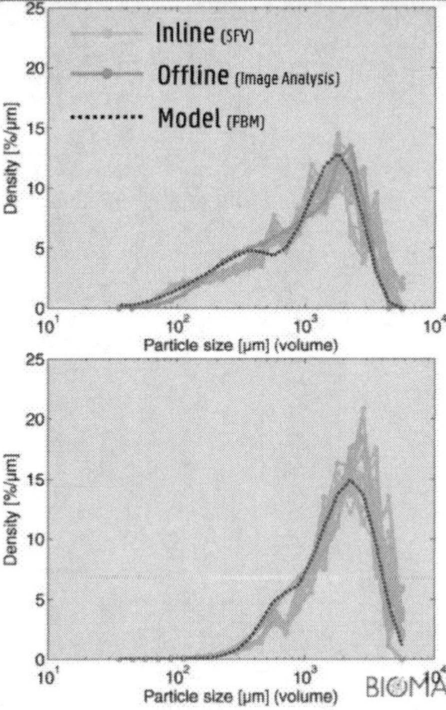

Daan Van Hauwermeiren
223f: Development and Calibration
223g: Validation 1
223h: Validation 2
Monday, October 30, 2017
Room 200H at 4:45 pm

17 AIChE Annual Meeting, Minneapolis, MN

LP PAT

BIOMATH

Conclusions

» Successful implementation of a real-time granule size analyser directly after the continuous twin screw granulator.

» The interface between the process and the instrument is crucial.

» Dispersion of the granules is required to decrease the loading

» Up to 9%w/w of the continuous granule stream is sampled.

» Offline and inline size analysis are in good agreement.

LP PAT

BIOMATH

Future perspectives

Model-based control of the granule size

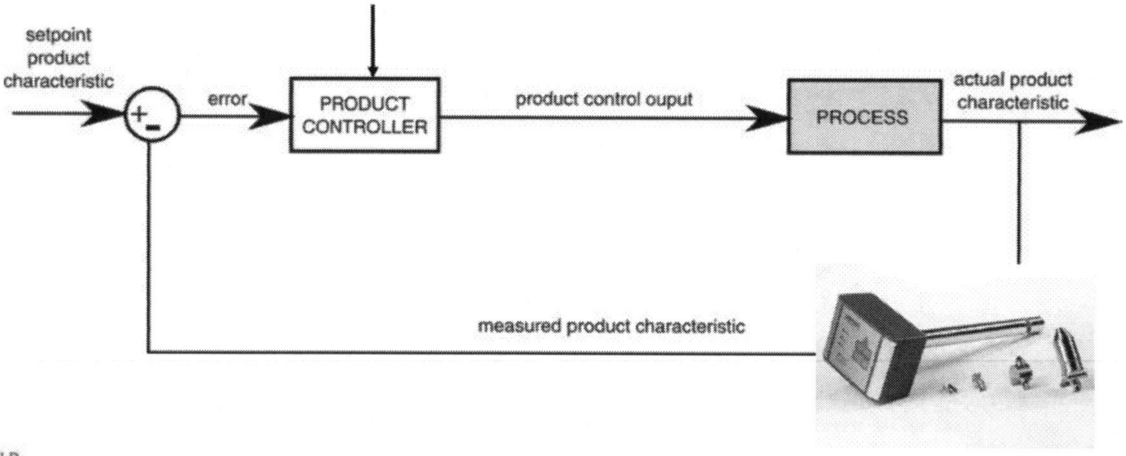

LP PAT

BIOMATH

GHENT UNIVERSITY

Acknowledgement

Prof. Thomas De Beer

Laboratory of Pharmaceutical Process Analytical Technology,
Dept. of Pharmaceutical Analysis, Ghent University, Belgium

Prof. Ingmar Nopens

BIOMATH - Model-based analysis and optimization of bioprocesses,
Dept. of Mathematical Modelling, Statistics and Bio-informatics, Ghent University, Belgium

Prof. Krist V. Gernaey

PROSYS Research Centre, Dept. of Chemical and Biochemical Engineering, Technical University of Denmark

Janssen Pharmaceutica

Pharmaceutical Companies of Johnson & Johnson

2017 AIChE® ANNUAL MEETING

17AIChE Annual Meeting, Minneapolis, MN
OCTOBER 29 - NOVEMBER 3, 2017
MINNEAPOLIS, MN

491673 The Use of Flow Aids to Prevent Caking of Bulk Solid Materials

Abstract:

Flow aids have long been used to take a cohesive material that has significant instantaneous strength and reduce the particle-particle interaction to decrease the bulk unconfined yield strength. Some feel that this method works because the inter-particle forces are generally caused by van Der Waals attractive forces and the flow aids separate the particles, thereby decreasing the attractive forces and reducing the bulk strength. Not all mechanisms of strength are due to van Der Waals forces. Material strength increase can be the result of capillary bond formation and crystal growth between particles. The questions to be addressed by this paper are:

- What is the role of flow aids in situations where strength increase is primarily due to crystal growth between adjacent particles (i.e. caking due to recrystallization)?
- What is the optimal size of flow aid that will prevent this type of caking?
- Should the flow aid be hydrophobic or hydroscopic?
- What is the optimal coverage of flow aid for these situations?

This paper will address caking prevention in a systematic manner, incorporating a particle scale model of the caking phenomena and using bulk scale caking tests to build a bridge between the bulk strength and inter-particle forces for caking materials.

492278 Novel Starch -Salt Composites; Their Strength, Composition and Scale upÂ Investigated and Characterized

Abstract:

Novel bio-based composite materials attract a lot of attention at the moment. In this work, we have looked in combinations of starch, salt and water which under the right heating conditions form a very strong composite material which could be used for building and design purposes. The work presented will reveal the difference in strength properties using different starches. It will also give more insight in the working principle of forming the composites on a molecular level, with the focus on the types of salts used and the strength properties explained.

Next to this scaling up of the composite formation process will be shown and the issues surrounding this scaling up.

Again it shows that a simple salt starch mixture can form remarkable strong materials which can be used in different application fields.

497986 Derivative Peak Fitting of Differential Diffuse Reflectance for Compositional Analysis of Multiphase Semiconductor, P25 TiO_2

Abstract:

The ability to characterize changes in polymorph composition and analyze the compositional of variations of multi-phase semiconductors thoroughly and quickly is important to catalyst manufacturers and users alike. Diffuse-reflectance UV-Visible spectroscopy derivative peak fitting (DPR) is a technique that can measure both concentration of a multi-phase semiconductor as well as the band gap energies of each semiconductor in the material. Degussa P25 TiO_2 is a widely-used semiconductor catalyst in both industry and academia. As-received P25 samples were subjected to grinding, sieving, and calcination treatment, and a combination of these operations. As-received and as-prepared samples were analyzed via DPR, X-ray Diffraction (XRD), and Raman Spectroscopy to quantify the percentage of anatase and rutile polymorphs present in the sample. Compositional measurements from DPR were consistent with XRD and Raman analysis. The effect of treatment conditions on the phase composition of P25 was measured. The onset of rutilization upon calcination of as-received P25 was 873 K. However, grinding and sieving (between 53 Î¼m and 45 Î¼m) P25 lowered the onset of rutilization during calcination, and had a three-fold increase in rutilization when calcined at 773 K as compared to as-received P25, as determined via in situ diffuse-reflectance UV-visible spectroscopy during calcination.Â

499915 Real-Time, in-Line Assessment of Flow Performance for Compressible and Anisotropic Bulk Solids

Abstract:

Feeding and handling problems have been recognized as a major contributor to low production of renewable cellulosic ethanol in the US, which achieved only 7% of nameplate capacity in 2016. Current methods that assess material feeding and handling properties require multiple hours to complete, making them ill-suited for quality assurance and control (QA/QC) in production environments. This talk presents a method, based upon a custom hopper with adaptable geometry, to assess feeding and handling behavior of materials in a production or receiving line in real-time, so that those properties can be used to either sort material or adjust preprocessing conditions to ensure that incoming material meet feeding and handling specifications. Flow properties are determined by continually adjusting the size of the outlet in a plane-flow hopper to maintain a constant fill level in the hopper. The angle of the hopper walls can also be adjusted to tune the geometry to different materials. Tests conducted at approximately 1,000 lbs per hour indicate that this method readily assesses the flow performance differences of corn stover and pine wood due to changes in particle size from 3 to 13 mm and due to changes in moisture content from 10 to 20%. Comparison of this method with shear tests using an automated Schulze ring shear indicate that this method is more sensitive for determining differences in flowability performance than traditional shear tests for corn stover and pine wood. Test results, including wall pressure measurements, are compared to simulations and indicate that this method shows promise for in-line quantitative estimation of the flow properties of compressible and anisotropic materials, including bulk density, cohesion, and internal angle of friction.

488043 Drag Force Flow Sensors for in-Line Monitoring of Powder Processes

Abstract:

In-line Process Analytical Technology (PAT) tools can produce data which allows for real-time monitoring and control of material without the need to disrupt the manufacturing process. As well as providing minimal disruption, in-line measurements have several advantages when compared with off-line and at line techniques.Â Most importantly the data represents a greater proportion of the in-process material and the measurements are made under process conditions â€" this is especially important for powder and granular systems whose behaviour is dependent on localised stress and strain rates.

Several PAT tools exist for in-line measurement, typically based on particle size analysis or spectrographic techniques, however, there are some limitations associated with these methods. The application of a Drag Force Flow (DFF) sensor overcomes these limitations and provides robust, highly sensitive and high (temporal) resolution in-line measurement of the local flow forces within process equipment. A DFF sensor is a thin, hollow pin, whose deflection in the flow is measured by two optical strain gauges affixed to the inner surface of the pin. The sensor is calibrated to measure DFF, this is then converted to a Force Pulse Magnitude (FPM) which is influenced by fundamental parameters of the material such as density and shear viscosity. In addition to force, the DFF sensor also outputs temperature information.

The fast measurement rate (500 Hz), sensitivity (0.4 Pa to 100 MPa) and stainless steel construction mean that the DFF can be used in a diverse range of applications from High Shear Wet Granulation through to mixing and blending and flow in powder transport systems

To evaluate the applicability of the DFF sensor, the flow behaviour of several types of granular material was investigated during a simple blending process. The DFF probe was able to detect clear and repeatable differences between samples and identify variations in in-process behaviour.

In a second study, the ability of the DFF sensor to track the addition of a liquid binder and consequently inform on blend uniformity was investigated. A short chain soluble carbohydrate was mixed with two binders with notably different viscosities (Binder A = ~0.001 Pa.s and Binder B = 100 Pa.s) at 1% w.w. and 3% w.w concentration. Clear differences were observed in the FPM profiles for the four blends. Both blends containing Binder A exhibited a sharp increase in FPM followed by a decrease until a steady state was achieved, suggesting uniform mixing. The 3% formulation required approximately 60 s less to reach steady state. Binder B (3%) generated a similar profile however the peak in the FPM was notably broader suggesting significant differences in the dispersion mode for the two binders.

The DFF sensor was demonstrated to be a highly sensitive device, capable of detecting minor variations in powder and granulate properties. As such, this technique methodology may be suitable for real-time in-line continuous monitoring of typical pharmaceutical unit operations such as blending, mixing and High Shear Wet Granulation (detection of end point/blend uniformity) and in powder transport systems.

493088 Surface Energy and Its Effect on Interparticle Interaction during Particle Flow

Abstract:

Surface Energy and its Effect on Interparticle Interaction during Particle Flow

C.G. Jange[1]; R. P. K. Ambrose[2]

[1]Graduate Research Assistant; [2]Assistant Professor, Department of Agricultural and Biological Engineering, Purdue University, West Lafayette, Indiana, USA.

The aim of this study is to understand the mechanism that drives the surface intermolecular interaction effects on dry cohesive powder flowability. Silica beads of 125 μm, 180 μm and 250 μm were coated with chemical components to mimic the intermolecular interaction of surface components present in common food products. The average Sauter mean diameter ($d_{3,2}$) of coated particles were 143.32 ± 8.92 for 125 μm, 176.21± 6.05 for 180 μm and 257.23 ± 2.57 for 250 μm, respectively. As per dynamic flow properties, the stability index for all samples ranged from 0.9 to 1.1, indicating stable powder flow. No significant sensitivity to change in flow rate was reported. An increment in specific energy was observed for all surface modified samples as compared to the control group, except for particles coated with molasses. Lower basic flow energy values were observed for particles coated with butter and molasses as compared to the control specimens. The surface modified samples presented higher compressibility percentages as compared to control systems. The angle of internal friction (AIF) also varied with surface composition and applied pressure. To investigate the influence of surface composition over kinematic performance, dispersive and polar contributions of the surface free energy of the solids were obtained using a surface energy analyser (SEA) . Heptane, octane, nonane, and decane were used as probe molecules based on an infinite dilution method. With the completion of this study, a granular Bond number based relationship will be developed to establish a relation between surface composition and powder flow.

495432 Novel Computational Strategy for Particle Charge Determination

Abstract:

Electrostatically charged granular matters are important in a wide variety of contexts, ranging from particulate matter (PM) pollution, to industrial handling of pharmaceutical products. Various electrostatic properties such as the total charge of a cluster of particles, and macroscopic electric fields of granular systems can be easily quantified. However, an efficient tool to measure individual charges has yet to be developed to better understand how charged particles interact. In this work, we build a novel computational strategy to obtain charges of individual particles by combining proper force fields, which are used to simulate particle trajectories, and an optimization technique that evolves the particlesâ€™ charges. The force fields include the dielectric polarization due to the nontrivial particle-particle charge mismatch and particle-environment dielectric contrast. We apply the strategy to simple systems: charged particles in vacuum environment, where only electrostatic interactions determine inter-particle forces. The extensibility of the strategy to more complex systems will be discussed.

496469 Novel Applications of Light Scattering Techniques for on-Line Characterization of Nano-Particles

Abstract:

Novel Applications of Light Scattering Techniques for On-Line Characterization of Nano-Particles

Abstract.

As the nature of active pharmaceutical ingredients changes and more and more biologics are present in the drug development pipelines, nanoparticles like liposomes and nanoemulsions are emerging as useful vehicles for drug delivery. Nanoemulsions, consisting of dispersed colloidal droplets, typically around 50- 600 nm in size, within an immiscible continuous phase, are used routinely to encapsulate drugs that are poorly soluble or susceptible to hydrolysis and/or enzymatic breakdown. For these drugs nanosized micelles act as protective delivery vesicles *in vivo*, ensuring successful drug delivery to the target site.

In pharmaceutical nanoemulsions droplet size typically has a defining influence on *in vivo* behavior thereby impacting the pharmacological profile of a drug and its clinical performance. Droplet size often influences factors such as absorbance, bioavailability, diffusion rate and stability, and as result tends to be a critical to quality parameter for the drug product. Understanding how formulation composition and processing parameters impact droplet size is therefore a critical step in the development of a robust and well-controlled nanoemulsification process.

The use of active ingredients that are only sparingly soluble in solid dosage forms also creates a need to disperse particles to a very fine size. Increasingly the optimal particle size for an active ingredient is in the nanometer, rather than the micrometer range. Nanomilling or micronization processes are applied to grind the particles, in suspension, and achieve the necessary comminution. Controlling particle size during the milling process is essential to ensure a safe and efficacious drug.

DLS is a technique used routinely to measure all types of nanoparticles across the size range 0.3 nm to 10 μm. The principle that underpins it is a relatively simple one, indeed simplicity is a key attraction. Small particles in a dispersion or solution are subject to Brownian motion which is driven by collisions with the solvent molecules. A DLS instrument determines the rate of diffusion of particles moving under Brownian motion, from measurements of fluctuations in scattered light intensity. The resulting diffusion coefficient is used to calculate particle size through application of the Stokes-Einstein equation:

Where D = Diffusion coefficient, k = Boltzmann's constant, T = absolute temperature, h = viscosity, and D_H = hydrodynamic radius.

This relationship shows clearly how size can be determined from diffusion speed, provided that the temperature and continuous phase viscosity of the sample are known however, there are additional features of modern instrumentation that further enhance the industrial usefulness of the technique. For example, Non-Invasive Back Scattering technology (NIBS) makes it possible to achieve high sensitivity across a broad size range and at high measurement concentrations. It is also possible to apply DLS to measure a sample in flow, providing that the flow is laminar and of a limited rate.

The conventional way to use DLS in process development or monitoring is to manually extract a sample and carry out a measurement in the laboratory to determine particle size. The endpoint of the process is marked by particle size reaching a target value and this can be confirmed by offline sampling and analysis. However this approach introduces a significant lag between an action and any assessment of the impact with the results often obtained only once the batch is complete. This does not allow for an assessment of the trajectory of the process, as it progresses, or provide an opportunity to tune processing parameters to achieve a better result.

For example, in the production of a pharmaceutical nanoemulsion the endpoint of processing is marked by particle size reaching a stable value and the top end of the droplet population dropping below an upper limit. This limit is often guided by the need to achieve certain performance characteristics in subsequent processing steps, such as sterile filtration using a 0.2 μm filter, for example. Ensuring that the droplet size has been sufficiently reduced while minimizing over-processing requires multiple manual analyses over the course of a batch or during continuous production. This is a time and labor intensive approach, both at the pilot scale, when the focus is to scope and optimize processing conditions, and during commercial production, where precise monitoring is linked directly with confidence in quality and variable cost management. The need to submit in-process samples to an analytical lab for analysis results in the loss of valuable time and inhibits the ability to respond to process deviations.

The use of on-line particle characterization tools, in contrast, enables rapid, time relevant process monitoring. Real-time data make it possible to quickly see the impact of changing a process parameter and efficiently establish conditions that will deliver a successful outcome. Online measurement therefore brings clarity and reduces development efforts during the scoping and scale-up studies associated with QbD, as well as enabling efficient process monitoring and control at the commercial scale.

The case studies described in this presentation will discuss the comparability of manual and automatic analysis and results achieved from evaluating various formulations with emphasis on practical value of utilizing Process Analytical Technologies.

500077 Core-Shell Nanostructured Anodes for Lithium Ion Batteries

Abstract:

In the last ten years there have been great advances in the production technologies of Lithium Ion batteries, making them one of the most popular sources for portable computing and telecommunication equipment. In particular, a lot of work was done on the electrochemical properties of electrode Materials for Lithium Ion Batteries using Graphene (G) and Graphene Oxide (GO) with silicon, however, these works suffer from several drawbacks associated with the breakage of the Solid Electrolyte Interface (SEI) during repeated cycling. In this work we developed a new method that avoids breakage of the SEI at the silicon surface. We first created novel core shell/hollow nanocarriers made with GO (*) and then incorporated silicon as the high capacity electrochemically active material which has the potential to greatly increase the capacity of these materials.

The ultrathin graphene oxide membranes encapsulate the nanoparticles with controlled void space, allowing for the expansion and contraction of the silicon as it is alloyed with lithium during cycling of the battery. Under the correct processing conditions the graphene oxide was almost completely templated from the aqueous phase to form ultrathin membranes about the silicon nanoparticles. We used a modified emulsion precipitation method to encapsulate silicon nanoparticles. Because this is fundamentally a directed self-assembly process, it was possible to achieve control over the size of the resulting droplets of the emulsion. It is a directed self-assembled process in the sense, that although the membrane of GO spontaneously forms upon the interface between the oil phase and aqueous phase, energy input (e.g., sonication, rotor-stator, high pressure homogenization, or energetic stirring) is required to finely divide the phases and create the interfacial area. The graphene oxide provided high stability to the emulsions, on the order of minutes or hours. The emulsions could be stabilized down to graphene oxide concentrations as small as 40 ppm. Increasing the graphene oxide concentrations above 40 ppm enhances the stability of the emulsions.

We used different size of graphene oxide sheets to develop the process for different size emulsion particles. By purposely using relatively large graphene oxide sheet (relative to the size of the particles) we ensured that the graphene oxide did not become overly hydrophilic. The reason for this is because GO sheets tend to have carboxylic acid groups in the edges, so the smaller the length, the more edges they have and the more hydrophilic the GO sheets become. In order to have the GO sheets attract to each other, that situation is not favorable because it affects the stacking and plating of the sheets. By increasing the sheet size, the sheets become larger in length and therefore the number of edges per unit area decreases, and hence, the hydrophilic groups decrease so the sheets become more hydrophobic and attract to each other forming thicker membranes (plating). We present the analysis of the graphene oxide sheet size distribution and mean sheet sizes use in the synthesis of the core-shell particles. The composite structures produced by this method were thermally reduced to convert GO to a graphitic structure capable of providing high electrical conductivity as well as high lithium ion conductivity. We solidified the composite to effectively trap the electrochemically active silicon nanoparticles within the matrix, preventing segregation of the active components from the conductive matrix. Our results show anodes cycled at very high charge and discharge rates of 3A/g are possible. We varied the percent of silicon inside the graphene oxide structures, and we report capacities starting at 2000 mA-hr/g. We present the results of both thermally reduced and chemically reduced graphene oxide based anodes. Our

particles were characterized via laser diffraction, zeta potential, FE-SEM, TEM, BET, and AFM.

(*) Langmuir March 2017. Kurt B. Smith, Maria S. Tomassone, â€œUltrathin Hollow Graphene Oxide Membranes for Use as Nanoparticle Carriersâ€

DOI: 10.1021/acs.langmuir.6b04583

499469 Connecting the *in-Operando* reduction Kinetics of Porous Nickel Oxide and Copper Oxide in Dilute H_2 from the Molecular- to Micron-Scales for Clean Energy Generation

Abstract:

The kinetics of porous nickel and copper oxide reduction has applications that range from chemical looping combustion to solid oxide fuel cells. While previous studies reported the structural changes in materials independently of the microstructural evolution, we can simultaneously probe the microstructural and structural changes from the molecular- to micron-scales using multi-scale X-ray scattering (Ultra Small Angle/ Small Angle/ Wide Angle X-ray Scattering (USAXS/SAXS/WAXS)). By combining multi-scale X-ray scattering with X-ray Absorption Spectroscopy (XAS), it is now possible to connect the chemically-specific short-range changes in materials with the structural and microstructural changes across longer spatial scales. This *in-operando* experimental methodology is applied to contrast the reduction kinetics of nickel and copper oxides in dilute H_2 (~3%) at temperatures as high as 500°C. The dynamic changes in the microstructure (i.e., porosity and particle size) of nickel and copper oxides as the materials are reduced are related to their corresponding structural changes.

496858 Non-Contact AFM Measurement of the Hamaker Constants for the Improvement of Trace Explosive Detection

Abstract:

The detection of homemade explosive devices in public areas, such as airports, is a primary concern for public safety. Trace amounts of explosive particles can contaminate the clothing of individuals in contact with the explosive materials during manufacturing and/or transportation of the device. This particle-surface interaction is controlled primarily by interfacial intermolecular forces. Of these, the van der Waals (vdW) force is of great importance since it is omnipresent. The Hamaker constant, *A*, is a quantitative measure of the vdW interaction and captures each material's compositional effects. Therefore, in order to improve the efficiency of detection checkpoints and better understand the adhesive nature of these materials, it is essential to determine the Hamaker constants for the explosives, the surfaces to which they adhere, and the swab materials used to collect the explosives from these surfaces during checkpoint screening.

Experimental attempts to determine *A* using an AFM are often plagued by deviations from expected behavior caused by issues inherent to the contact regime, such as surface roughness and deformation. Thus, we developed a method for estimating Hamaker constants from the non-contact approach regime of an AFM experiment (Fronczak et al., 2017, *Langmuir* **33**, 714-725). This method relies on a quasi-dynamic understanding of the cantilever tip's approach to contact, in which the inertial effects of the tip motion are accounted for when describing the trajectory of the tip as it approaches the substrate. The method was tested experimentally using silica, alumina and polystyrene substrates, and was demonstrated to yield estimates of *A* for these materials that were in very good agreement with previously published Lifshitz calculations.

This new method, as with other approaches, relies heavily on the accuracy of the geometric model governing the interaction between the AFM tip and the substrate. Therefore, original validation efforts were focused on remaining as true to the shape of cantilever tip as possible. A complex model of a truncated pyramid with a spherical cap was thus used in the first experimental interpretation. Although this geometry can be confirmed and the dimensions estimated via scanning electron microscopy (SEM), even high-resolution SEM analysis of the tip cannot provide sufficient detail to allow appropriately precise determination of the tip's geometry. In addition, the numerical methods that are required to evaluate the Hamaker constants with high accuracy for complex tip geometry can quickly become prohibitively complex.

Therefore, we proposed an adaptation of the current method in which the geometric complexity of the cantilever tip was eliminated from the determination of the Hamaker constant by describing the tip as an 'effective' perfect sphere, thereby capturing all the geometric effects in the single dimension of this sphere. We began by measuring the interaction between the tip and well-defined surfaces for which the values for *A* are documented in the literature. After the forces against these well-known surfaces were measured, the interactions were modeled by assuming that our effective spherical tip had an effective radius of R_{eff}. When the dimensions of this effective shape were properly captured, the Hamaker constants for the tip-surface interactions can be correctly calculated in all cases where this cantilever is applied. In this way, the dimensions of the effective tip shape were "calibrated" based on the attractive forces of the well-characterized ideal surfaces. Since an optimization procedure was already required to fit the geometric parameters for the pyramidal tip model, this amendment merely applies the same optimization principle to a simpler spherical tip model. Once we generated the tip's R_{eff} we modeled the tip's approach to contact using just this one geometric parameter. Then, this was used to determine the unknown *A* of a desired material.

We first demonstrated the practicality and accuracy of this updated method by comparing the results with both the original pyramid model and Lifshitz approximations (when available) for flat substrates composed of silica, polystyrene, highly ordered pyrolytic graphite (HOPG), sapphire (α-Al3O2), Plexiglas (PMMA), and acrylonitrile butadiene styrene (ABS). After calibrating the tip against these materials, we then measured the Hamaker constant of the previously undetermined trinitrotoluene (TNT) and several other commercial swabs used at airport security checkpoints. Since the dielectric response functions for an appropriate range of the electromagnetic spectrum is not available for these materials, the Hamaker constants could not be compared against predictions from the Lifshitz theory. The results of our measurements do agree, however, with those previously published in the literature, which were obtained from surface energy based experiments such as contact angle measurements and inverse gas chromatography. Finally, by using a geometric averaging function, the strength of the interaction between TNT and these swab materials are inferred, key information that is needed as input for generating improved detection protocols.

Proceeding for AIChE Annual Meeting 2017 *Paper ID# 485506*

High Resolution Characterization of Colloid Size Distributions by Field Flow Fractionation

Tianying Jiang, Erick Soto-Cantu, Charlie Chan
3M Company, Maplewood, MN, USA

Abstract

High resolution characterization of particle size distribution in colloidal dispersions has remained a great challenge for long time, making it difficult to understand suspensions containing particles with size covering a wide range, which have unique optical, mechanical and optical properties. Previous analytical tools have weaknesses of low resolution, statistical non-reproducibility, or difficulty in data analysis. With motivation in exploring characterization method of multi-modal size distribution, Field Flow Fractionation (FFF) was used to separate colloidal particles first before analyze the size by Multi-Angle Light Scattering (MALS) and dynamic light scattering (DLS). Effects of particle size range, cross flow rate setting, composition of particle, eluent selection, membrane type were searched to build up universal understanding of the principles of designing test method for a variety of experimental systems in FFF.

Introduction

Colloidal dispersions are suspensions containing small particle with size ~1nm to ~10μm, which are dispersed in a certain continuum phase. This type of materials have high application potential in industrial products and processes where high solid concentration and larger interfacial areas are required.[1] It is important and critical to control particle size in colloidal dispersions, as which can significantly influence a variety of properties of the materials, including optical properties, rheology, shelf-life.[2] Traditional particle size characterization methods for colloidal dispersions applied widely in industry include Dynamic Light Scattering (in batch mode), Analytical Centrifugation, and Nanoparticle Tracking Analysis.[3] However, all these methods have weakness when applied to characterize the colloidal dispersion particle size distribution with high resolution. Dynamic Light Scattering (in batch mode) could only obtain intensity weighted particle size distribution, which does not offer high enough resolution to resolve multi-modal particle size distribution, and smaller size portion cannot be detected due to low scattering intensity. Analytical centrifugation can separate particles with multimodal size distributions, but it requires sufficient density difference between particle and solvent to complete experiment in a fair amount of time, and it is not applicable to analyze colloidal dispersions with heterogeneous chemical composition. Nanoparticle Tracking Analysis is not statistically reliable due to the limited number of particles detected and the resulting poor reproducibility. Thus it is important to introduce a technique which can physically separate the colloidal dispersion containing multi-modal size distributions and analyze the particle size online. Asymetrix Flow Field Flow Fraction (FFF) was introduced inspired by this motivation, in which the separation of the samples is achieved in a liquid zone of several hundred micro meter height contained within a channel with semipermeable membrane lining the bottom surface.[4] In the channel, the particle fractionation has been completed in three steps: focus, focus/injection, and elution. Following that, Online Multi-Angel Light Scattering (MALS) and Dynamic Light Scattering (DLS) are applied to analyze the particle size in the eluent. In the following sections, experimental approaches, principles of validating test methods, effects of different parameters on separation will be introduced to show how high resolution characterization of particle size distribution is performed in FFF.

Instrument and Experiment

The separation was controlled by the FFF separation module (from Wyatt, under the trade designation "ECLIPSE® DUALTEC"). Following the separation, the analyses were completed by connected light scattering detectors MALS (from Wyatt, under the trade designation "DAWN HELEOS II") and/or DLS characterization (from Wyatt, under the trade designation "WYATTQELS").

Results and Discussion

Methodology Validation: Stability vs. Resolution

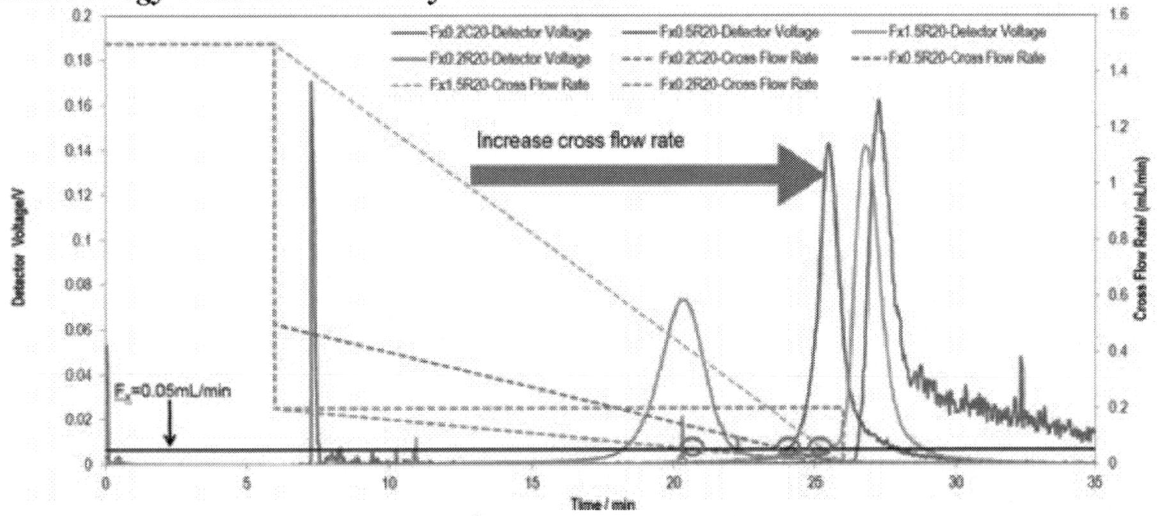

Figure 1. Detector voltage and cross flow rate vs. time for elution of 60 nm PSL standard (Eluent: 800ppm SDS, Membrane: 10kDa PES, Injection Volume: 10μL, Channel Flow Rate: 1mL/min, Concentration: 50ppm).

FFF retention equation can be expressed as [5]

$$t_r = \frac{\pi \eta d w^2}{2kT} \bullet \frac{\dot{V}_{cross}}{\dot{V}_{channel}} \qquad \textbf{Equation 1}$$

where t_r is the retention time, w is the channel thickness, d is the Stokes diameter, η is the eluent viscosity, k is the Boltzmann constant, T is the absolute temperature, $\dot{V}_{cross}$ is the cross flow rate and $\dot{V}_{channel}$ is the channel flow rate. So shown in this equation, we can more efficiently separate two particle populations by increasing channel thickness, decreasing channel flow rate and/or increasing cross flow. Usually the channel thickness, eluent type and channel flow rate are fixed, thus the cross flow rate variance can change the fractionating power. In principle, large and constant cross flow rate can improve the resolution.

However, a key challenge remained for nanoparticle characterization is that aggregation could occur and influence the accuracy of the analysis results. A high cross flow rate can sometimes cause aggregation and make it difficult to load more particles. It is necessary to find a balance point where we can keep good stability of the particle s and still pursue efficiency in separating particles. As shown in Figure 1, different cross flow rate methods are used to separate the particles. Fx0.2C20 (Cross flow rate is 0.2mL/min, and kept constant for 20 min); Fx0.2R20 (Cross flow rate is initiated as 0.2mL/min, and ramping down to 0 in 20 min), et al. It is obvious that Fx0.2C20, Fx0.5R20 and Fx1.5R20 did make elution occurred after the cross flow stops. Therefore we do need to go with Fx0.2R20 to make sure of particle stability during the analysis and get the particle geometric radius distribution. This is a key point we need to keep in mind in the following exploration and test method development.

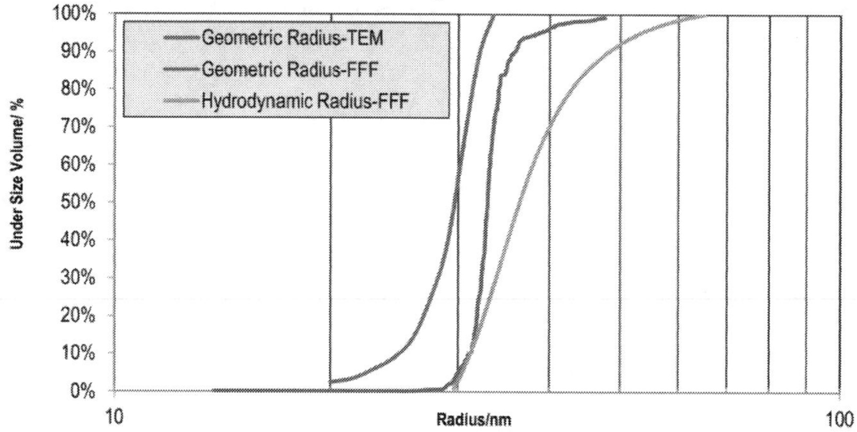

Figure 2. Comparison of FFF-MALS, FFF-DLS with TEM on 60 nm PSL standard.

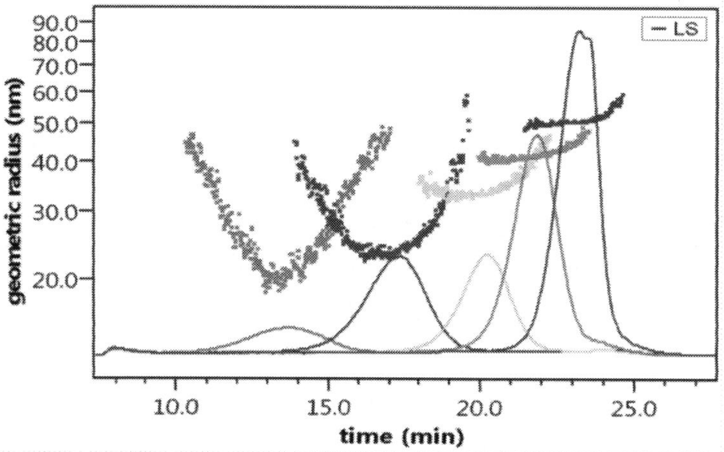

Figure 3 FFF elution results of different PSL nanosphere standards (channel flow rate: 1.0mL/min, Fx0.2R20) (Red: 20nm, Blue: 40nm, Green: 60nm, Fuchsia: 80nm, Brown: 100nm)

To further confirm the validation and advantage of using FFF in particle size analysis, TEM results have been used for particle size distribution analysis of these 60 nm PSL spherical standards. Converting the number weighted distribution to volume weighted distribution based on the fact that all these standard particles have spherical shape. The results are compared to FFF analysis results (both geometric radius from MALS and hydrodynamic radius from DLS) as in

Figure 2. Just by comparing the geometric radius, it is obvious that FFF-MALS has only <10% shift on Dv50 compared to TEM results despite that they are complete different techniques and TEM results are converted from directly measured number weighted distributions, confirming the accuracy of FFF in detecting particle size distribution.

Effects of particle size on elution

Table 1. Results of FFF-DLS and Batch DLS.

	FFF+DLS Result		Batch DLS Result	
	Rh(Q)w	Rh(Q)z (nm)	Z-ave Rh(Q)	V-ave Rh(Q)
20nm	15.8	22.6	9.7	11.6
40nm	23.0	25.7	19.8	21.9
60nm	33.8	36.0	33.8	36.7
80nm	39.8	40.8	42.3	45.2
100nm	48.9	50.9	53.4	55.6

PSL sphere standards are used to study the effects of particle size on elution. As shown in Equation 1, larger Stokes diameter will result in longer retention time with all other parameters fixed. The results in Figure 3 suggest this tendency. With the exactly same test method, larger particles were eluted later, which suggests the necessity of adjusting test method to achieve the best resolution in separation. Hydrodynamic sizes are compared to Batch DLS results in Table 1, with the similarity proving the particle stability.

Effects of cross flow rate setting on separation resolution for different particle size ranges

To study the effects of different parameters on separation resolution, we purposely introduce the definition of resolution on FFF. We note here that it is universal to use separation resolution to describe particle analysis efficiency, but a rough reference to judge the test method. Below in Equation 2 resolution R_s is defined:

$$R_s = \frac{\Delta t_R}{2(\delta t_{R1} + \delta t_{R2})} \qquad \textbf{Equation 2}$$

Where the key parameters are defined in Figure 4. The value of R_s simply indicates the degree of the baseline separation.

As we discussed above, setting ramping cross flow rate could raise up the stability of particle and increase the loading capacity although this sacrificed the resolution. So we keep using ramped cross flow rate in this study and the cross flow rate mentioned below will be the value at the very beginning unless indicated differently. In Figure 5 (A), the results of R_s has been illustrated as a function of initial cross flow rate (F_x/ (mL/min)) and (R/ min) for different 1:2 size ratio binary mixtures. Obviously it is complex and there is no obvious rule to develop test method. As shown more obviously in Figure 5 (B), the same test method will work very differently for binary mixtures with size in different ranges, thus it is necessary to choose the most efficient test method. For example, Fx0.2R20 works better than Fx0.1R40 for 40nm/80nm mixture, but the former works a lot worse than the latter for 100nm/200nm mixture and 200nm/400nm mixture.

So here we have the main conclusions (1) Cross flow in elution mode is important in separation; (2) One method does not work for all different size ranges; (3) Larger particle needs smaller cross flow rate.

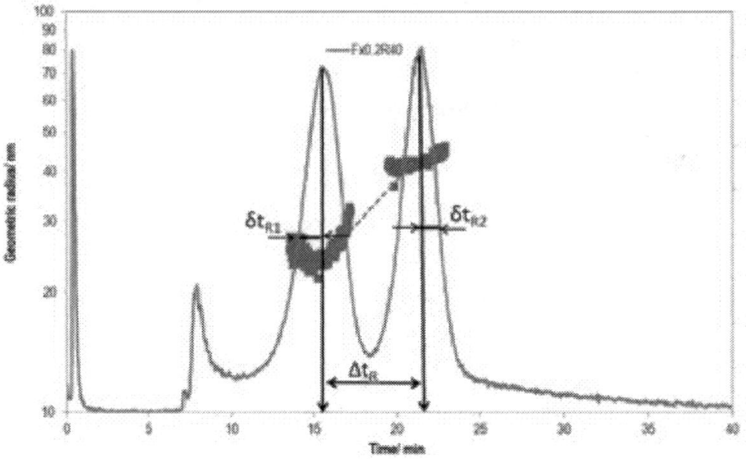

Figure 4. FFF separation result of a traditional nanoparticle dispersion with bimodal distribution, with key parameters used for resolution definition illustrated in the figure.

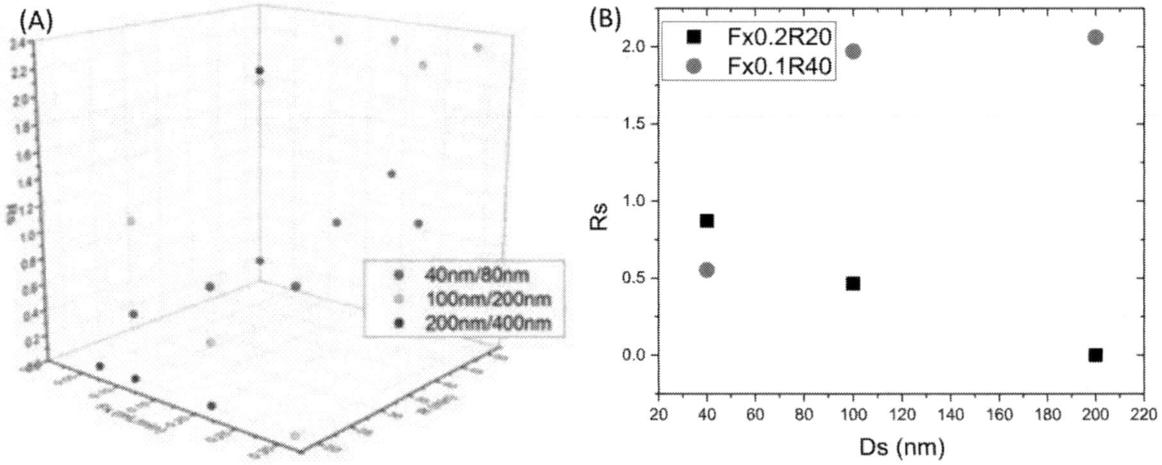

Figure 5. (A) R_s as a function of F_x and R for three different 1:2 binary mixtures, (B) R_s as a function of D_s at two different cross flow rate settings.

Effects of selection of membrane type

We kept using surfactant SDS and PES membrane for the study of PSL sphere above, which did not result in too much trouble in keeping particle stability. However with increasing particle size, very long elution time occurs near the end point of cross flow ramping to zero. For D=400 nm + 600 nm mixture, stability can no longer be guaranteed as shown in Figure 6. A fairly amount of particles are eluted after the cross flow is decreased to zero.

This decline in particle stability and this increase of particle elution time with increase of particle size could be a consequence of enhanced particle-membrane interaction. In addition to relevant steric interaction, the particle-membrane interaction can be composed of two parts: the electrostatic interaction and the van der Waals attraction, the latter can be expressed as a function of particle size R, Hamaker coefficient A, and membrane-particle surface separation distance D. So the aggregation and instability of the particles due to size increase can be explained as a consequence of enhanced van der Waals attraction. Since we cannot change the particle size and its average distance from the membrane, we can only manipulate the Hamaker coefficient in this situation, which is an intrinsic material property determined by chemical composition of membrane, chemical composition of particle and the continuum phase as estimated in the last equation in Equation 3, where n_i (i=1 for particle, 2 for membrane, 3 for continuum phase) stands for the refractive index and ε_i (i=1 for particle, 2 for membrane, 3 for continuum phase) stands for the dielectric constant.

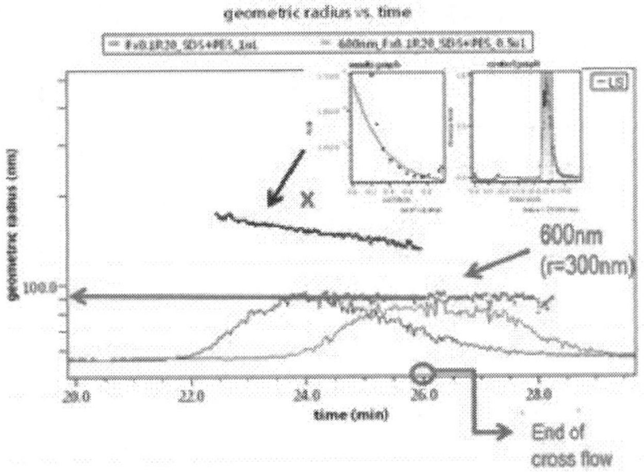

Figure 6. Light scattering intensity and geometric radius as a function of time for 400 nm+ 600 nm PSL sphere mixtures (blue) and 600nm PSL (red) at Fx0.1R20.

$$W(D) = W_{ele}(D) + W_{vdW}(D)$$

$$W_{vdW}(D) = -\frac{AR}{6D}$$

$$A = \frac{3}{4}kT\frac{(\varepsilon_1-\varepsilon_3)(\varepsilon_2-\varepsilon_3)}{(\varepsilon_1+\varepsilon_3)(\varepsilon_2+\varepsilon_3)} + \frac{3h\nu_e}{8\sqrt{2}}\frac{(n_1^2-n_3^2)(n_2^2-n_3^2)}{(n_1^2+n_3^2)^{1/2}(n_2^2+n_3^2)^{1/2}[(n_1^2+n_3^2)^{1/2}+(n_2^2+n_3^2)^{1/2}]}$$ **Equation 3**

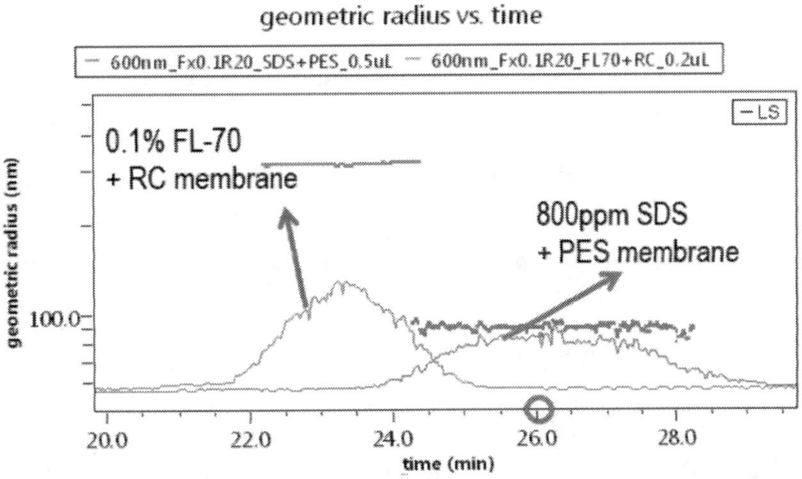

Figure 7. Scattering intensity and geometric radius as a function of time for 600 nm PSL particle comparing two different membrane+ surfactant settings.

It is obvious that we might reduce the van der Waals attraction for PSL sphere dispersed in water when we use Regenerated Cellulose (RC) membrane instead of PES membrane. Therefore we moved from the PES membrane + SDS surfactant setting to RC membrane + FL-

70 surfactant setting to study 600 nm PSL sphere using the exactly same test method, with the results compared in Figure 7. Obviously with all other settings kept, RC membrane + FL-70 surfactant resulted in shorter elution time and reasonable particle size information from the scattering data. Therefore, this mode could protect particles from aggregating, and could be used for further exploration of cross flow rate settings to improve the separation resolution.
In Table 2, we summarized values of A for some most commonly used experimental systems.

Table 2. Summary of Hamaker coefficient for most commonly studied systems in this study.

Particle	Eluent	Membrane	A/kT
Polystyrene	water	PES	3.67
Silica	water	PES	2.24
Polystyrene	water	Cellulose	2.58

Effects of chemical composition of particles

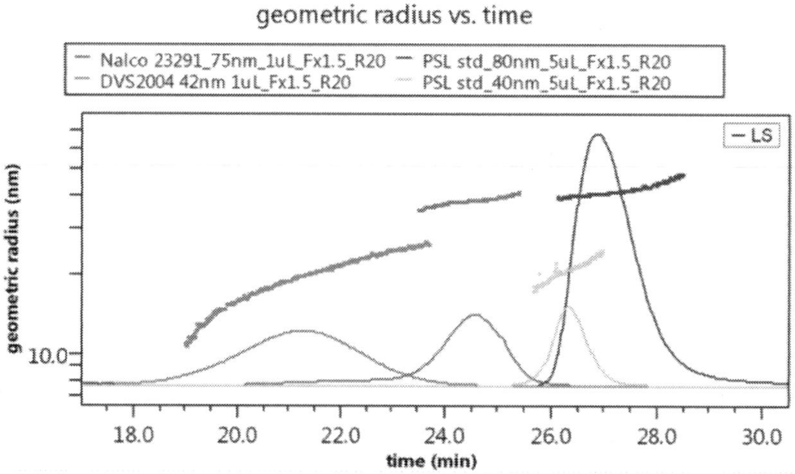

Figure 8. Comparing FFF-MALS results for silica particles and PSL particle in the similar size range at the same cross flow setting (10kDa PES membrane, Fx1.5R20).

As suggested by Equation 1, ideally with size and all the experimental approached fixed, we should expect the elution at a certain time regardless of the materials of the particles. However, we did discuss and find the effects of particle-membrane interaction which could make this problem more difficult. As we summarized in Table 2 about Hamaker coefficient for different experimental systems, silica particles have weaker van der Waals attraction with the PES membrane than PSL particles when the two have similar size. Figure 8 illustrate the results of 40 nm PSL spherical particles, 80 nm PSL spherical particles, 42 nm silica spherical particles, 75 nm silica spherical particles. Apparently with the same experimental settings, it is possible to separate 42 nm +75 nm Nalco silica particle, while it is impossible to separate 40nm PSL and 80 nm PSL at this same condition. Considering the long elution time, it is necessary to bring down the cross flow rate to efficiently separate PSL particles.

Effects of eluent viscosity: beyond aqueous systems

Using organic solvent as the eluent can be very different from studying aqueous system even the particle has the same chemical composition. Isopropanol as the solvent has a significantly larger viscosity (1.96 cP) than water. Using channel flow setting 1 mL/min as in aqueous system created a large pressure, thus it is necessary to decrease the channel flow to 0.5 mL/min and use the cross flow setting Fx0.5R30 to achieve the separation as shown in Figure 9.

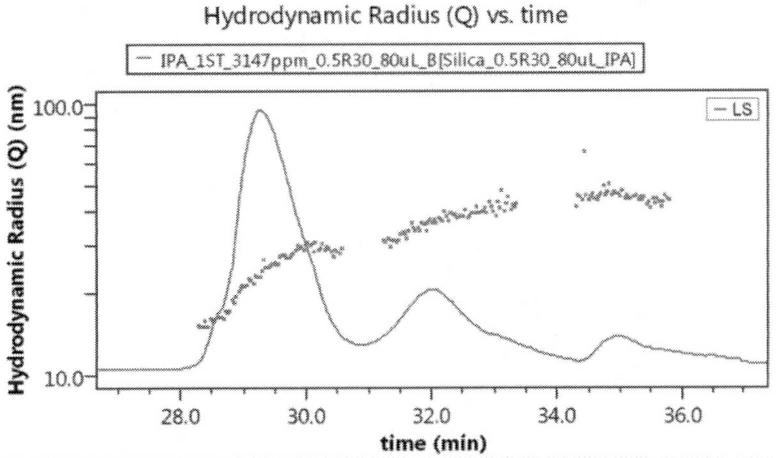

Figure 9. FFF-DLS result of IPA-ST using Fx0.5R30 (Fc=0.5 mL/min, using 10kDa PES membrane).

Conclusion

This study focused on advancing particle size distribution analysis capabilities by using Field Flow Fractionation. It added important extension in nanoparticle characterization where high resolution study of multi-modal distributions is desired. This could not be achieved by previously existent nanoparticle analysis techniques including batch Dynamic Light Scattering, Nanoparticle Tracking Analysis, Analytical Centrifugation. This study could be extended to study submicron particles with different chemical compositions. Test method development for an unknown samples follows the rules of optimizing the usage of membrane type, eluent type, injection volume/concentration, channel flow rate, cross flow rate and detectors, to achieve the an appropriate condition for separation.

References

1. W. B. Russel, D. A. Saville and W. R. Schowalter (1989), Colloidal Dispersions, New York: Cambridge University Press.
2. A. E. Edris (2012), "Formulation and Shelf Life Stability of Water-Borne Lecithin Nanoparticles for Potential Application in Dietary Supplements Field," J Diet Suppl, vol. 9, no. 3, pp. 211-222.
3. H. G. Merkus (2009), Particle Size Measurements, Spriner.
4. M. Baalousha, B. Stople and J. R. Lead (2011), "Flow field-flow fractionation for the analysis and characterization of natural colloids and manufactured nanoparticles in environmental systems: A critical review," J. Chromatogra. A, vol. 1218, no. 7, pp. 4078-4103.
5. G. Yohannes, M. Jussila, K. Hartonen and M. L. Riekkola (2011), "Asymmetrical flow field-flow fractionation technique for separation and characterization of biopolymers and bioparticles," J. Chromatogr. A, vol. 1218, no. 27, pp. 4104-4116.

501501 Monitoring Magnetic Nanoparticle Synthesis Progress Using a Benchtop Magnetic Particle Relaxometer

Abstract:

Current tracer imaging techniques like positron emission tomography (PET) and singe photon emission computed tomography (SPECT) are valuable in imaging molecular level features but they utilize radioactive tracers. Magnetic Particle Imaging (MPI)[1], which makes use of biocompatible and non-radioactive iron oxide nanoparticles as tracers has been shown to have comparable resolution and applications in real-time cardiovascular[2] and cancer imaging[3], cell labeling and tracking[4], and magnetic fluid hyperthermia[5]. In MPI the image is obtained by localizing the signal induced by the magnetic particles using a magnetic field gradient with a field free point (FFP). The resolution of the image is influenced by the particleâ€™s response to time varying magnetic fields, which can be assessed using a Magnetic Particle Relaxometer (MPR)[6] or Magnetic Particle Spectrometer (MPS),[7] which lacks the spatial imaging capability of an MPI scanner. This equipment provides a quick performance assessment of newly synthesized tracers. It can also be employed to explore interesting phenomena like tracking of particle growth during synthesis, evaluating particle aggregation, measuring changes in viscosity, studying particle interactions with tissue, and biomarker detection for disease diagnosis.

In this work we make use of MPR/MPS to study particle response by observing changes in the acquired signal strength and its full-width-at-half-maximum (FWHM), which is related to the achievable resolution in MPI. These can be affected by change in size of particles in an Extended LaMer semi-batch thermal decomposition synthesis[8] that allows controlled growth of particles over time or changes in the environment surrounding the particles. In our preliminary work we were able to observe a drop in the FWHM, thus indicating an increase in expected MPI resolution, as the particle size increased with time during synthesis, without the need of time-consuming operation step as required in other characterization tools like SQUID magnetometry and transmission electron microscopy (TEM).

The custom-built MPR is inexpensive, sensitive, and small in size, thus showing a promise of being widely adopted in laboratory and commercial settings. Also, with the tests requiring less than a minute and a tiny amount of sample, we envision the MPR to serve as an efficient benchtop particle analyzer and a complementary characterization tool.

[1] B. Gleich and J. Weizenecker, Nature **435,** 1214 (2005).

[2] J. Weizenecker, B. Gleich, J. Rahmer, H. Dahnke, and J. Borgert, Physics in Medicine and Biology **54,** L1 (2009).

[3] E. Y. Yu, M. Bishop, B. Zheng, R. M. Ferguson, A. P. Khandhar, S. J. Kemp, K. M. Krishnan, P. W. Goodwill, and S. M. Conolly, Nano Letters **17,** 1648 (2017).

[4] B. Zheng, M. P. von See, E. Yu, B. Gunel, K. Lu, T. Vazin, D. V. Schaffer, P. W. Goodwill, and S. M. Conolly, Theranostics **6,** 291 (2016).

[5] D. W. Hensley, Z. W. Tay, R. Dhavalikar, B. Zheng, P. Goodwill, C. Rinaldi, and S. Conolly, Physics in Medicine and Biology **62,** 3483 (2017).

[6] P. W. Goodwill, E. U. Saritas, L. R. Croft, T. N. Kim, K. M. Krishnan, D. V. Schaffer, and S. M. Conolly, Advanced Materials **24,** 3870 (2012).

[7] N. Garraud, R. Dhavalikar, L. Maldonado-Camargo, D. P. Arnold, and C. Rinaldi, AIP Advances **7** (2017).

[8] M. Unni, A. M. Uhl, S. Savliwala, B. H. Savitzky, R. Dhavalikar, N. Garraud, D. P. Arnold, L. F. Kourkoutis, J. S. Andrew, and C. Rinaldi, ACS Nano (2017).

502510 Contact Strength Between Powder Particles and Strength of Powder Compacts

Abstract:

Micromechanical models for the compaction behaviour of powders are based on understanding of the compressive and tensile stresses transmitted between particles in contact during compaction. The strength of the contacts between particles determines the strength of powder compacts.

Firstly, the tensile and compressive strength of powder compacts are characterised as functions of the relative density of the compacts for a range of pharmaceutical powders and model materials consisting of spherical particles. Using micromechanical models we relate the compact strength to the strength of contacts between particles. In effect the contact strength is inferred from macroscopic behaviour. We discuss the use of this framework as a means of specifying particle behaviour requirements to produce compacts of prescribed strength.

502472 Extraction of Parameters for a Square-Force Cohesion Model from Defluidization Experiments

Abstract:

Interparticle cohesion originating from van der Waals force and capillary force significantly impacts the flow behavior of particulate and multiphase systems. Though commonly seen in nature and industry, cohesive particles remain poorly understood, triggering continued interest from physicists and engineers. An accurate inter-particle cohesion model is key for improved understanding, though the development and validation of such models remains largely elusive. A key obstacle is the extreme sensitivity of the cohesive force to separation distance, and thus surface roughness. The precise measurement of surface roughness is time-consuming, requires specialized training, and involves expensive equipment such as atomic force microscopes (AFM). Such model development is often not conducive to applying to a wide range of materials and/or in an industrial setting. In previous work [LaMarche et al., Towards a universal description of cohesive-particle flows. (submitted)], we have shown that the behavior of cohesive-particle systems can be collapsed for either a dimensionless cohesive force or energy, depending on particle concentration. Accordingly, we propose a simplified cohesion model, the â€œsquare forceâ€ model, with only two parameters that capture both cohesive force and energy of more rigorous treatments. Perhaps most importantly, these two particle-particle (micro) parameters can be extracted directly from bulk (macro) measurements, namely defluidization curves. Specifically the incorporation of the extracted parameters into DEM simulations reproduces the results based on more rigorous cohesion models applied to both riser (gas-solid, dynamic and dilute) and hopper (granular, quasi-static and dense) systems. Access to such a practical method for developing reliable particle-particle cohesion models is a key step in bridging microscopic models to their continuum counterparts â€“ e.g., population balances.

489926 Quantitative Simulation of Tablet Compression and Hardness Test Processes Using Discrete Element Method

Abstract:

Keyword: material properties, compressibility, tensile strength, discrete element method

In the formulation development of solid dose drug product (DP), understanding on the influence of physical properties of drug substance (DS) and excipients is critical on quality control of DP. While empirical correlation has been widely applied in qualitative estimation of DP quality, potential risk to fail DP specifications still exists, especially in technical transfer and manufacture scale-up activities. It may result in significant waste of DS or serious quality issue on DP. In order to avoid risk of failure and save DS, modeling using Discrete Element Method (DEM) has been interest of formulation development and DP manufacture in decades. The main challenge was how to develop quantitative correlation so that DEM particles were accurately representative of behavior of pharmaceutical DS and excipients in various unit operations.

In this work, the development of quantitative powder compressibility platform was described. A powder flowability platform developed in the previous studies [1, 2] was applied to guarantee that the flow behavior of DEM particles was representative of pharmaceutical powder. Material properties such as particle size distribution, true density, and flowability were collected and applied into simulations. Detailed compression profile of pharmaceutical powder was collected using ESH Tablet Compaction Simulator by Huxley Bertram. Tablet hardness data was collected using HT-300 hardness tester by Key International and tensile strength was calculated based on tablet shape. Development of the compressibility platform was performed using the commercial DEM software Star CCM+® v11.06 by Siemens and with processors of dual CPUs, Intel® Xeon® E52640 @ 2.50 GHz. The Walton Braun Hysteretic contact model and the Parallel Bonds Model (PBM) were involved to model plasticity of pharmaceutical powder and rigidity of compacted tablet, respectively. After development and optimization, the platform was applicable to simulate one compression module and one hardness test module within 48 hrs with setting of material properties similar to pharmaceutical powders.

Based on the developed platform, influence of different DEM parameters on powder compressibility was estimated. Results showed that the setting of particle size distribution and particle flowability had little influence on compression profile. On the other hand, compression profile was significantly impacted by the setting of particle porosity, Youngâ€™s modulus, and stiffness ratio, while tablet hardness was mainly determined by the setting of tensile strength in PBM. Quantitative correlation was then developed between DEM particles and example pharmaceutical powders by adjusting DEM particle material properties. It was observed that simulation matched experimental data well. Several case studies were made to demonstrate capability of the platform in different scenarios such as binary system, punch adhesion, shape of tooling etc. The studies showed potential benefits of the proposed simulation method in the pharmaceutical powder compression processes, such as save of DS and risk reduction of batch failure, especially when technical transfer and scale-up of solid dose manufacture were involved.

1. Gao Y, Quantitative Investigation on Segregation Phenomenon Using Discrete Element Method in Large Scale Pharmaceutical Manufacture Process, AIChE Annual Meeting, San Francisco CA, Nov 2016, paper 788e.

2. Gao Y, Chalasani S, Mittal B, Computational Modeling of Low Shear Blending Process Using Discrete Element Method, AIChE Annual Meeting, Salt Lake City UT, Nov 2015, paper 412e.

491333 Simulating Convex Polyhedral Particles Utilizing the Discrete Element Method on GPU

Abstract:

Simulations of granular flows are an effective tool to gain a deeper understanding and subsequently, optimization of processes such as fluidized beds, mixing, powder transport, etc. The discrete element method (DEM) allows simulating these kinds of processes. Each particle is treated individually and its motion is described by Newtonâ€™s equation of motion. A soft-sphere approach is used, where colliding particles are allowed to slightly overlap. This overlap results in a repulsive force.

The particle shape plays a crucial role in modelling the process as realistically as possible. There are several approaches for modelling the shapes of more complex particles. The multisphere approach clusters several spheres to one rigid particle [1]. This approach is very flexible and can describe various shapes. Biconvex tablets can be represented as the overlap of three spheres. Superquadrics represent many shapes, such as ellipsoids, cylinder- and box-like particles [2]. To increase the accuracy, particles can be modelled as convex polyhedrons [3]. This approach allows simulating sharply-edged non-spherical particles. Although it is computationally expensive, it is the only way to model complex shapes realistically.

While the collision of spheres is easy to handle, more complex shapes are challenging. This work shows an accurate contact algorithm for arbitrarily shaped convex polyhedrons. The algorithm has been implemented in the GPU based DEM software XPS, allowing simulation of a very large number of particles [4].

References

[1] Ketterhagen W. Modeling the motion and orientation of various pharmaceutical tablet shapes in a film coating pan using DEM. International Journal of Pharmaceutics, Volume 409, Issues 1-2. May 2011, S. 137-149.

[2] Podlozhnyuk A., Pirker S., Kloss C. Efficient implementation of superquadric particles in Discrete Element Method within an open-source framework. Computational Particle Mechanics. 2016, S. 1-18.

[3] Govender N., Wilke D.N., Kok S., Els R. Development of a convex polyhedral discrete element simulation framework for NVIDIA Kepler based GPUs. Journal of Computational and Applied Mathematics, Volume 270. November 2014, S. 386-400.

[4] Jajcevic D., Siegmann E., Radeke C., Khinast J. Large-scale CFD-DEM simulations of fluidized granular systems. Chemical Engineering Science, Volume 98. 2013, S. 298-310.

498511 DEM Simulation Studies on the Effect of Particle Size and Morphology on Mixing and Fluid Content Uniformity in a Double Cone Impregnator

Abstract:

Impregnation of active metals onto a porous support is an important step in the preparation of heterogeneous catalysts. In a typical impregnation process, metal solutions are sprayed over a particulate bed in a mixing vessel until the pore volume is reached. The inter-particle variability of the impregnated liquids inside the particles and metal content may significantly affect the activity and selectivity of the resulting catalyst. Current scale-up practices lead to poor fluid distribution and inhomogeneity in metal content. The aim of this work is to understand the dynamic behavior of the particles under the spray nozzle, which is essential for a desired content uniformity in the impregnation process.

Discrete Element Method (DEM) simulations coupled with an algorithm for the transfer of fluid to and between particles was used to investigate the impregnation process. The fluid transfer algorithm has been developed in our previous work [Powder Technology, 221, 57-69, 2012], and has been validated by geometrically equivalent experiments. In this work, using this DEM model, the effects of rotational speed, particle size and particle morphology were explored in order to achieve the best mixing and content uniformity in the particle bed. Mixing analysis and liquid distributions were used to investigate the propagation of the fluid throughout the particle bed. The mixing in the axial direction was found to be a function of the number of revolutions, and was independent on the rotational speed of the vessel. The content uniformity was characterized by the relative standard deviation of the fluid content from all the particles in the impregnated particle bed and was found to be better at lower rotational speeds. We also found that fluid content uniformity depends strongly on the rotational speed of the vessel, contrary to what is found in non-impregnated systems.

Since catalyst support particles come in a wide variety of sizes and shapes, it is with great interest that we investigate the behavior of different catalyst particles during the impregnation process. Both spherical and cylindrical particles were considered to determine the effect of particle size and particle morphology on the impregnated particle bed. Spherical particles were examined at various diameters, ranging from 2.5mm to 10mm; cylindrical particles were selected to examine various particle sizes and aspect ratios. The results show that the mixing time for cylinders is always shorter than that for spheres. This is consistent with work by Guo and Wassgren et al [Journal of Fluid Mechanics, 713, 1-26, 2012] where they found that cylinders have smaller fluctuating velocities and larger particle collision frequency, so they take less time to mix. In addition the mixing time in the axial direction for both spheres and cylinders followed an exponential function of the surface area to volume ratio (S/V) of the particle. Likewise the impregnation times as a function of S/V in the entire particle bed has the same trends than for the axial mixing times as a function of S/V. We found a power law correlation between axial mixing and fluid content uniformity suggesting that axial mixing in a double cone vessel is a dominating factor in achieving the fluid content uniformity in the particle bed.

502419 CFD-DEM Simulations and Uncertainty Quantification (UQ) of Horizontal Jets in Gas-Solid Fluidized Bed

Abstract:

A recent experiment conducted at PSRI of opposing horizontal air jets impinging into a semi-circular fluidized bed of 6 mm plastic beads is studied here computationally. The experimental particle count is ~60,000, making it amenable to direct comparison with discrete-particle simulations. We use MFiX CFD-DEM to simulate this problem. The gas phase is treated as compressible due to the high velocity of the jets. All uncertainties in model inputs are considered other than those relating to the geometry of the jets, although the uncertainty in the area is considered in the jet velocity. An extensive parameter identification and ranking table (PIRT) is constructed. We use a local sensitivity analysis to rank the impact of all model input uncertainties on the system response quantities (SRQs), i.e., the model output variables of interest in this problem. Four SRQs are considered: the time averaged mean- and standard deviation of the bed pressure drop and the left- and right-side jet penetration depths (which were calculated experimentally from a high speed video imaging analysis). Somewhat surprisingly, the PIRT reveals that the superficial distributor velocity is, by far, the most important parameter controlling the jet penetration depts. After eliminating low impact uncertainties, we directly propagate the remaining uncertainties through the model to determine the uncertainties in the output SRQs. The results show that the CFD-DEM-based uncertainty quantification (UQ) bounds the jet penetration depths, though it is worth noting that the predicted bounds are significantly wider than the experimental data. Results for the bed pressure drop does not bound the experimental data, showing a slight bias, namely a 10 â€“ 15% overestimation. These comparisons and their implications on CFD-DEM+UQ analysis will be discussed.

THE INSTITUTE FOR CLEAN AND SECURE ENERGY

•155 South 1452 East Room 380 •Salt Lake City, Utah 84112 •1-801-585-1233

Particle Fluid Dynamic Code Validation with Hydrodynamics

Matthew A. Hamilton, Zach Reinking#, Kevin J. Whitty, and JoAnn S. Lighty

2017 AIChE Annual Meeting

Minneapolis, Minnesota

Oct 29th – Nov 3rd

INSTITUTE FOR Clean and Secure Energy
THE UNIVERSITY OF UTAH

University of Utah Plan

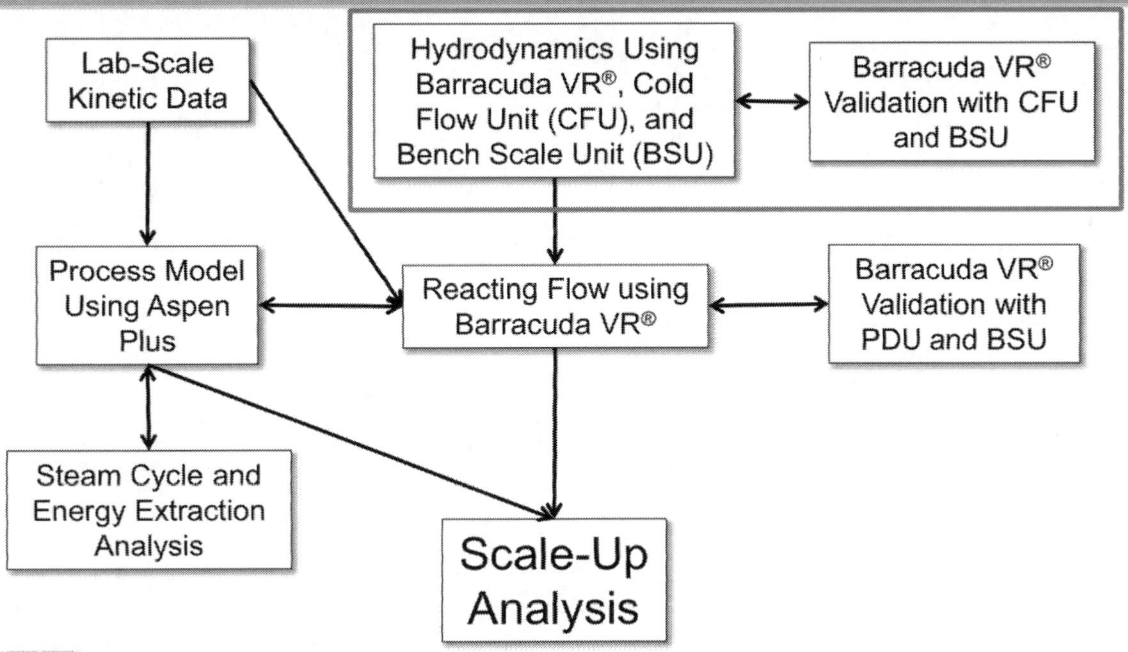

Objectives

- Show initial validation of Barracuda VR® hydrodynamics for dual circulating fluidized system
  - Pressure profile
  - Global circulation rate
  - Bed mass
- Discuss initial design of uncertainty quantification of Barracuda-VR® simulations

Technology Background: Chemical Looping Combustion

- CLC achieves *in situ* air separation by using a metal to transport oxygen from *air reactor* to *fuel reactor*
- Fuel (e.g. natural gas, coal) fed to fuel reactor is indirectly combusted by oxygen on oxidized metal
- Metal returns to reduced state in fuel reactor and "loops" back to air reactor
- Overall balance same as for conventional combustion
- Economic evaluations indicate CLC yields lowest COE of any CO_2-capture technology

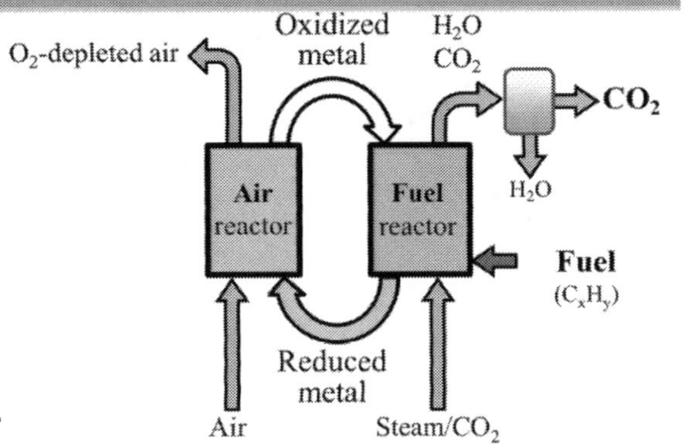

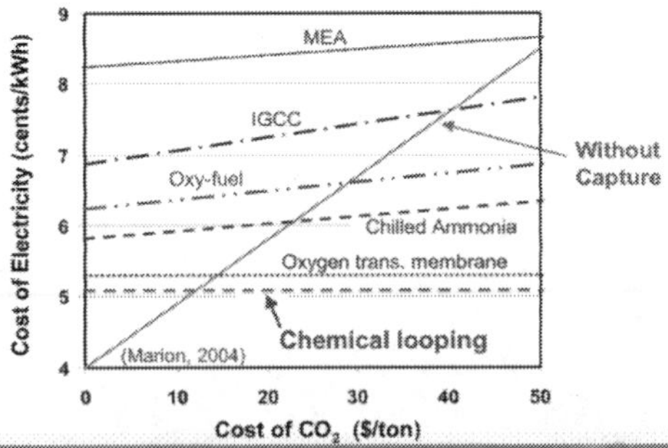

Technology Background: Chemical Looping with Oxygen Uncoupling (CLOU)

$Cu_2O(s) + \frac{1}{2} O_2(g) \rightleftarrows 2\,CuO(s)$

- Copper is one of few metals for which oxidation equilibrium (Cu_2O/CuO) lies within CLC operating temperatures.
- Cu_2O is oxidized in air reactor
- CuO spontaneously releases O_2 in fuel reactor due to low O_2 partial pressure
- Released O_2 reacts with solid coal char, converting more than 50x faster than with non-CLOU oxygen carriers

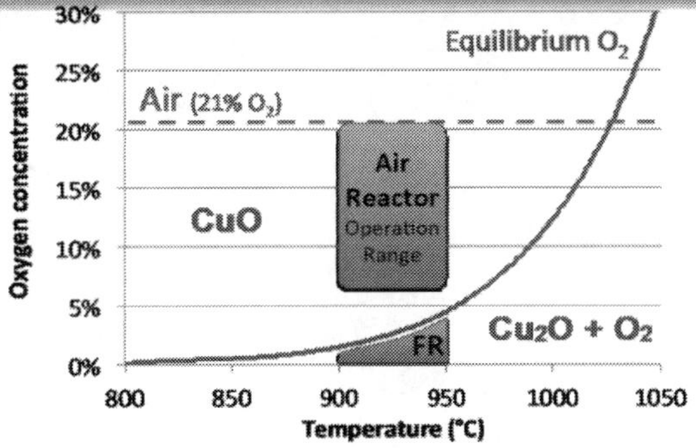

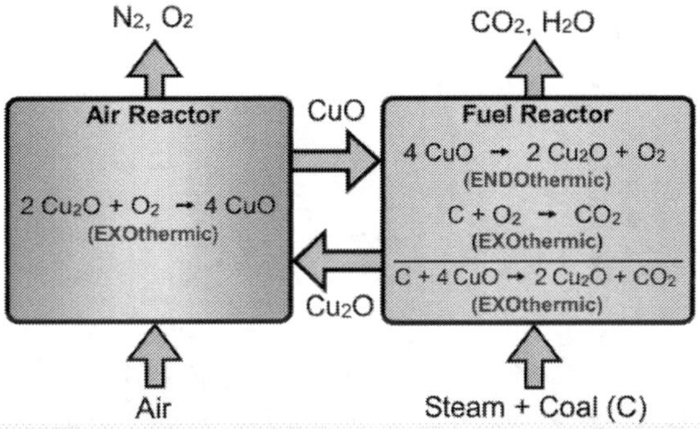

THE INSTITUTE FOR CLEAN AND SECURE ENERGY

Cold-Flow Unit

- Scaled from 200 kW process development unit
  - Dual circulating fluidized bed (CFB) design
  - Air reactor designed as a fast-fluidized bed
  - Fuel reactor designed as a turbulent-fluidized bed
- Glicksman's scaling parameters
  - Geometric $f_{geometric} = 0.577$
  - Time scaling $f_{geometric}^{1/2} = f_{time} = 0.76$

Part	Value
AR & FR Inner Diameter	15 cm (6 in)
AR Height	3.58 m (141 in)
FR Height	3.25 m (128 in)
AR & FR Cyclone Diameter	30 cm (11.75 in)
AR & FR Cyclone Inlet	7.3 cm x 14.6 cm (2.125 in x 5.75 in) (w x h)
LS1-3 Diameter	3.2 cm (1.25 in)

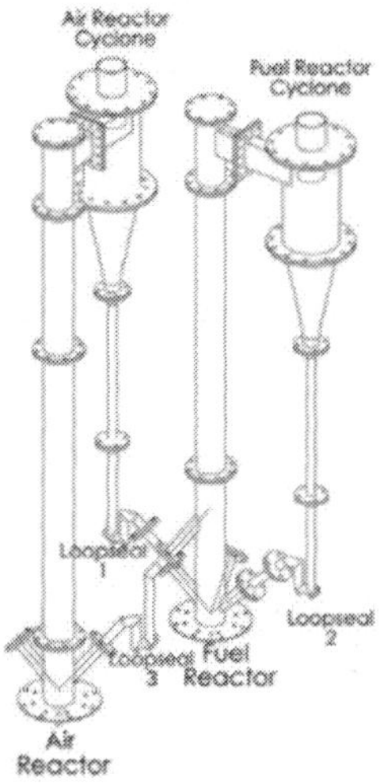

Computer model design of cold-flow unit

THE INSTITUTE FOR CLEAN AND SECURE ENERGY 6

Bed Particle Parameters

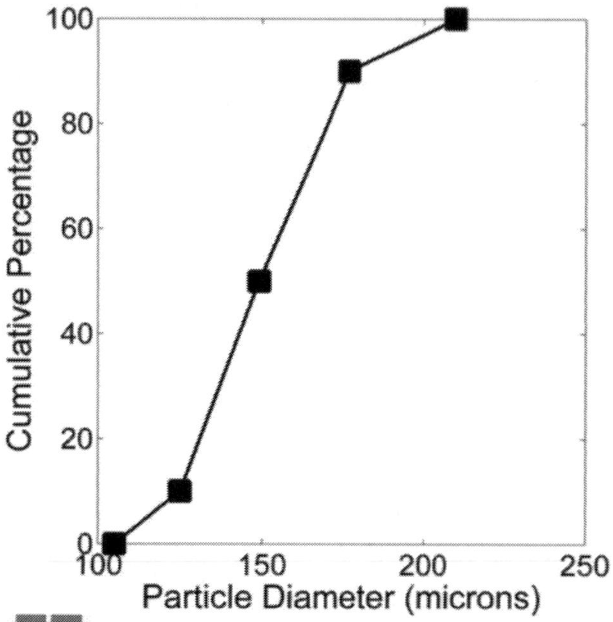

Parameter	Unit	CFU Value
Particle Density	$\frac{kg}{m^3}$	3900
Average Diameter	μm	149
Sphericity	-	0.85
Close Pack Solid Volume Fraction	$\frac{m^3_{solids}}{m^3_{volume}}$	0.63
Reactors Mass Inventory	kg (lbs)	40 (88)

- Real Cells: ≈100,000
- Particles: 6.62 x 10^9
- Particle Clouds: 4.60 x 10^5
- Exit Pressure: 82.5 kPa
- Isothermal Temperature: 300 K

Circulation Rate

- Trend captured
- Experimental circulation rate below the simulated circulation rate
- Simulated choked flow around 22 kg/min this is not reached with experiments

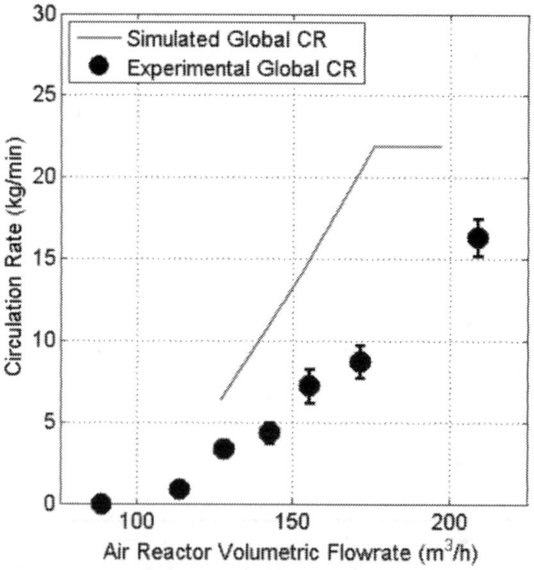

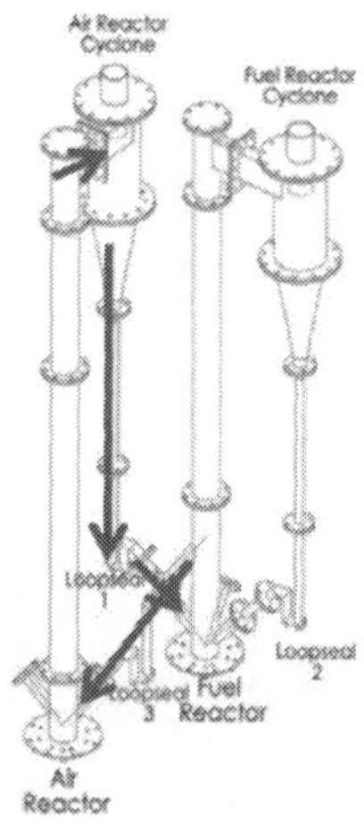

Reactor Bed Mass

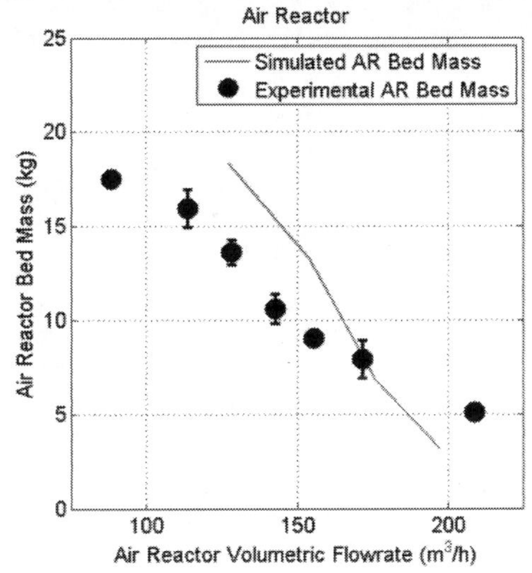

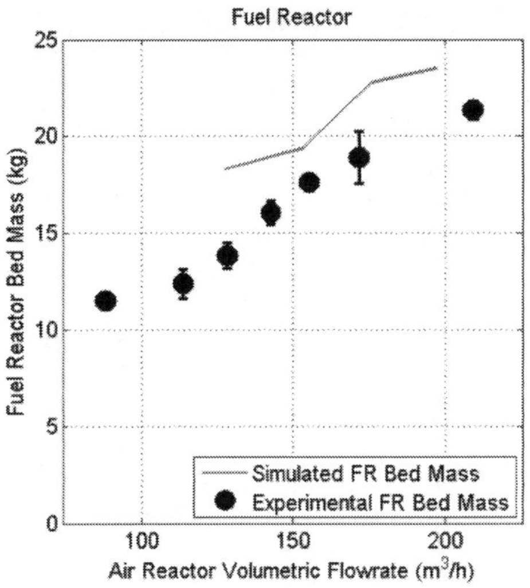

- Trends have been captured
- Air reactor simulated higher then lower than bed mass
- Same order of magnitude
- Experimental does not show bed mass jump

Pressure

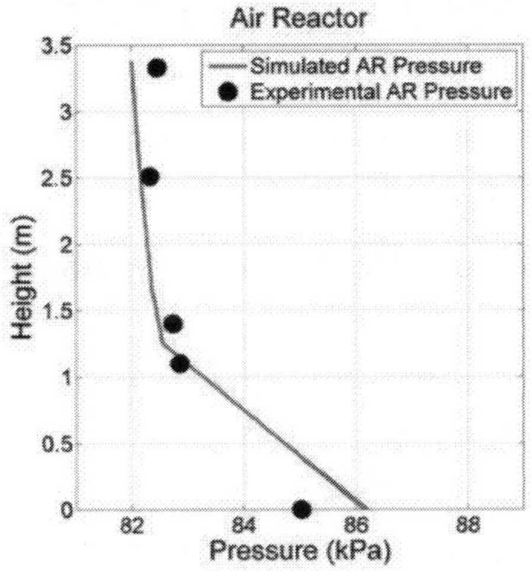

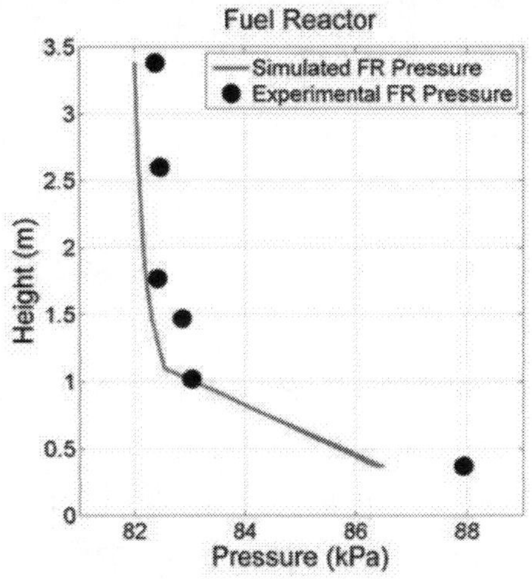

- Trend is captured
- Simulation over predicts the pressure at the bottom of the air reactor and under predicts at the bottom of the fuel reactor
- Simulation under predicts the pressure at the upper portion of both reactors

Conclusions and Future Work

- Out of the box Barracuda VR® can reasonably represent a dual circulating fluidized bed system
- Drag models have a major effect on the system and more work is needed to quantify the effect and determine an appropriate drag model
- Use of uncertainty quantification will decrease the risk of scaling up the process
- Further work is going to focus on finding an appropriate drag model with drag multiplier

THE INSTITUTE FOR CLEAN AND SECURE ENERGY 11

Acknowledgements

CPFD LLC.

This material is based upon work supported by the University of Wyoming through the Clean Coal Technology Research Fund under grant number 1001541 and by the Department of Energy (DOE) under Award Number DE-FE0025076.

This material is based upon work while Dr. Lighty served at the National Science Foundation. Any opinions, findings, and conclusions expressed in this publication are those of the authors and do not necessarily reflect the views of the National Science Foundation.

THE INSTITUTE FOR CLEAN AND SECURE ENERGY 12

Momentum Retention

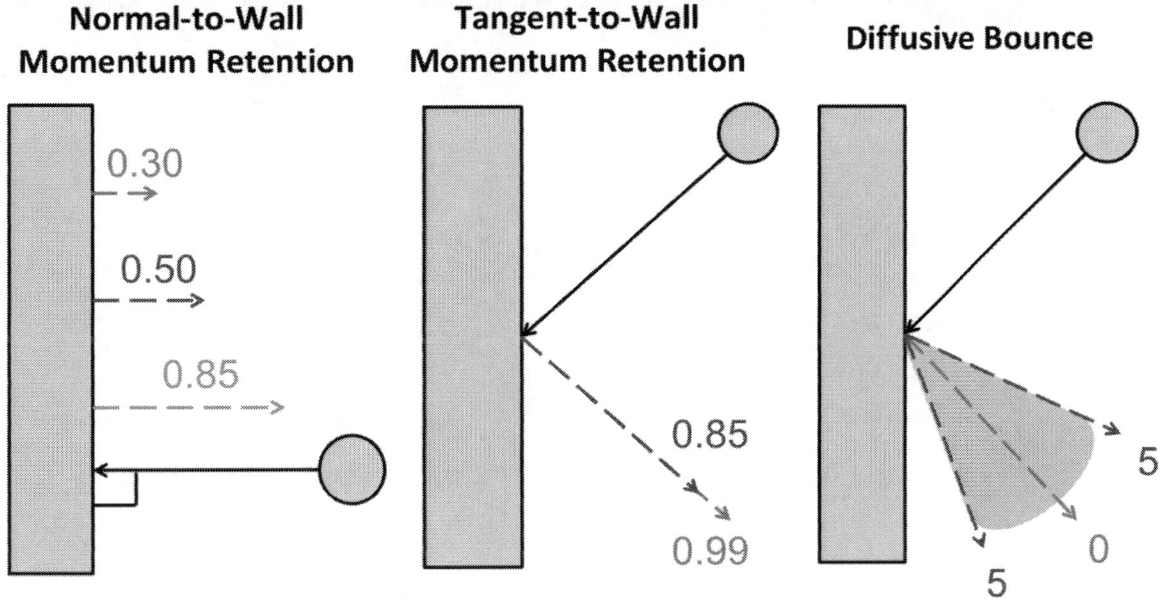

THE INSTITUTE FOR CLEAN AND SECURE ENERGY 13

Effect of Normal-to-Wall Momentum Retention Pressure

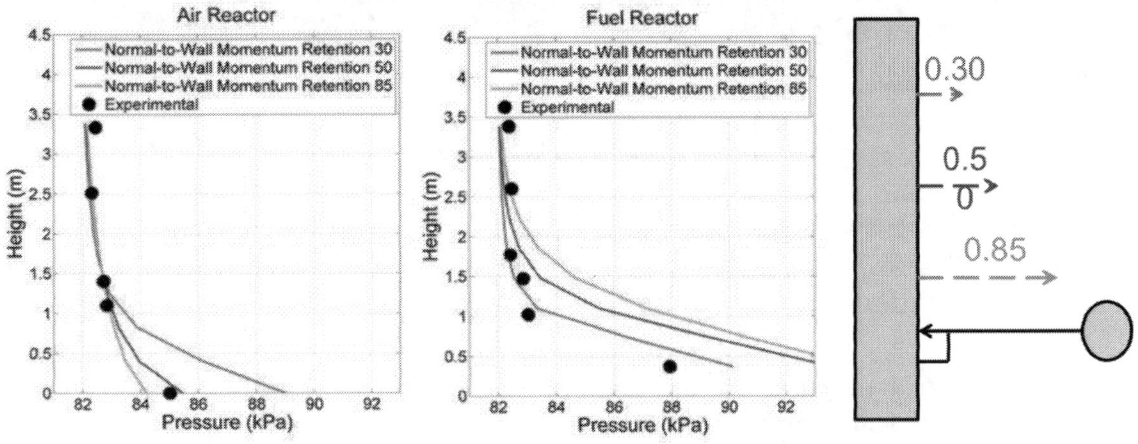

- Air reactor increasing the normal-to-wall momentum retention decreases the pressure
- Fuel reactor increasing the normal-to-wall momentum retention increases the pressure
- Increase in the FR bed mass with a decrease in the AR bed mass

THE INSTITUTE FOR CLEAN AND SECURE ENERGY 14

Effect of Normal-to-Wall Momentum Retention Particle Flux

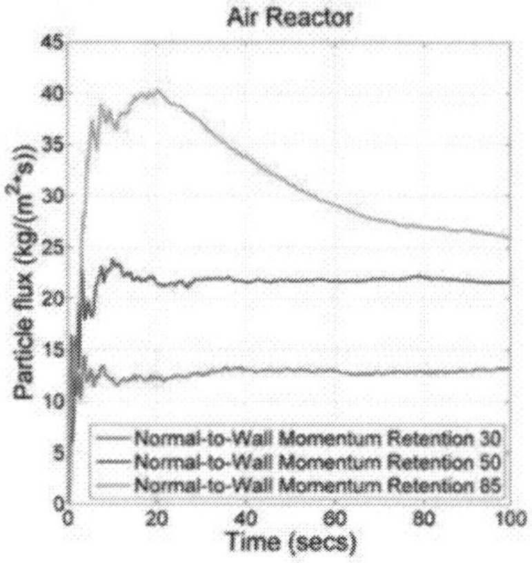

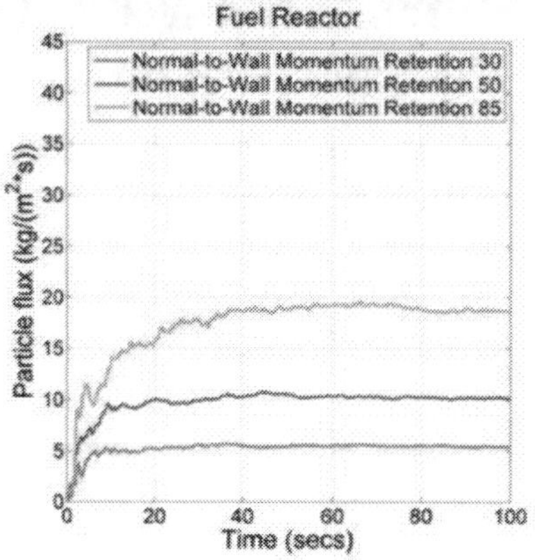

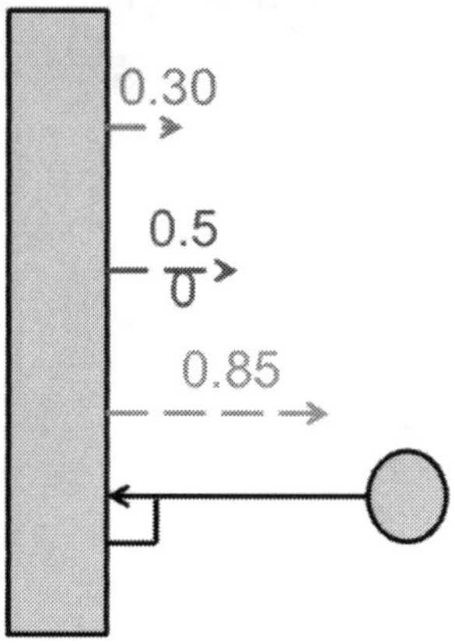

- Spike when normal-to-wall is at 0.85
- Decrease after peak due to bed mass loss
- Stable to both the 0.30 and 0.50 cases
- Fuel reactor cases are stable

Effects of Tangent-to-wall Momentum Retention and Diffusive Bounce

- These had minimal effects
- Pressure profiles
    - Air Reactor: Decrease tangent-to-wall momentum retention or diffusive bounce decreases pressure
    - Fuel Reactor: Decrease tangent-to-wall momentum retention or diffusive bounce increase pressure
- Circulation rate
    - Air Reactor: Decrease tangent-to-wall momentum retention or diffusive bounce increase circulation rate
    - Fuel Reactor: Decrease tangent-to-wall momentum retention or diffusive bounce increase circulation rate

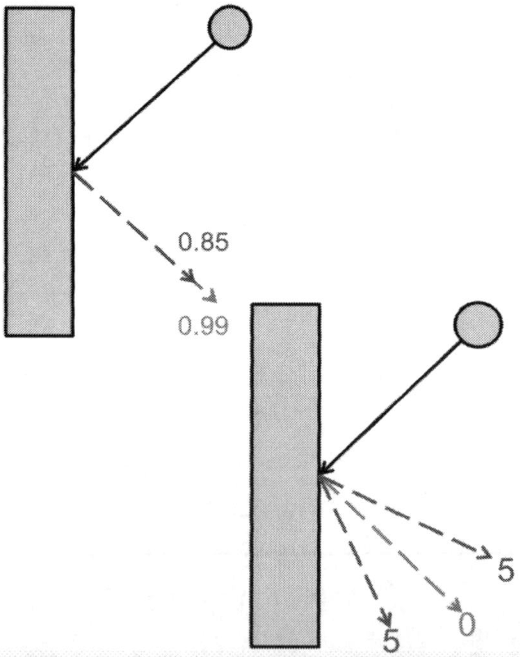

PROPPANT TRANSPORT EFFICIENCY SIMULATION IN A WELLBORE-FRACTURE SYSTEM

Dr. Jianxin Lu and Dr. Andrey Filippov, Halliburton, Houston, Texas, United States

Abstract

Controlling proppant transport in the field is essential to a successful hydraulic-fracturing operation. Proppant particles need to be delivered to the desired fracture locations to help prevent those fractures from closing. Flow conditions in the wellbore around each perforation connected to the fractures can vary, and proper delivery of proppants to each perforation is crucial. This presentation focuses on the development of a method for predicting particle flow rates into individual perforations, accounting for the non-Newtonian fluid property effect on particle transportation. Two dimensionless numbers were generated that dominate the transport efficiency in Newtonian and non-Newtonian fluids, and proppant transport operations can be designed or optimized using this dimensionless analysis.

Introduction

During hydraulic fracturing, proppant particles are transported downhole and into fractures. Proppant particles are usually fine, solid particles composed of sand or ceramic having a typical diameter range from 0.1 to 1 mm. Proppant particles are carried by hydraulic fracturing fluid and placed into fractures to help prevent fracture closure, which can improve production rates in fracture-stimulated reservoirs. The carrier fluids can be water or liquid hydrocarbon with additives to increase viscosity and reduce particle sedimentation resulting from gravity. The entire wellbore length can be 3000 to 6000 m (10,000 to 20,000 ft), the wellbore diameter is usually 10 to 20 cm, and the diameter of the perforation is approximately 1 to 2 cm.

Because flow conditions around each perforation connected to the fractures can differ, the amount of proppant carried to each fracture can vary. Particle inertia also needs to be taken into account. In cases of high flow rates and small perforation diameters, particles do not follow the flow streamlines, resulting in declining proppant delivery efficiency to fractures. These effects make designing proppant transport during a hydraulic fracturing process challenging.

Proppant Delivery Model

This work models an individual perforated exit from the wellbore. Figure 1 shows a T-shaped junction used to describe the wellbore-perforation geometry and the streamlines of the carrier liquid, with the color of the streamlines reflecting the velocity amplitude. A solid particle in this system might or might not follow liquid flowlines. Figure 2 shows a three-dimensional (3D) view and the mesh setting for the simulation.

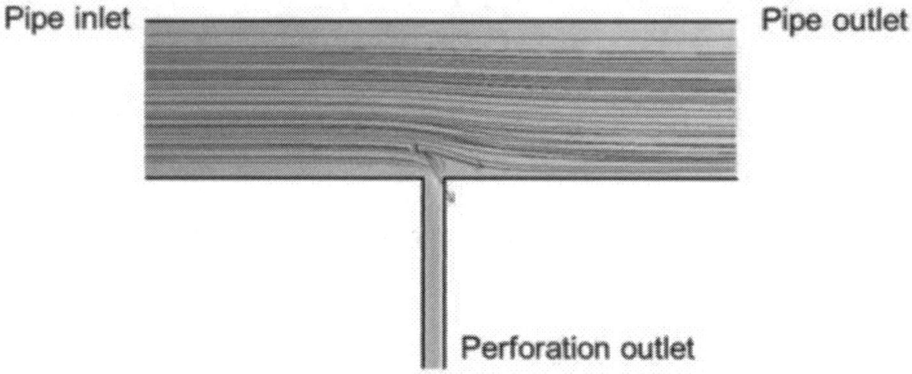

Figure 1. Flow schematic: proppant particle trajectories near a fracture exit.

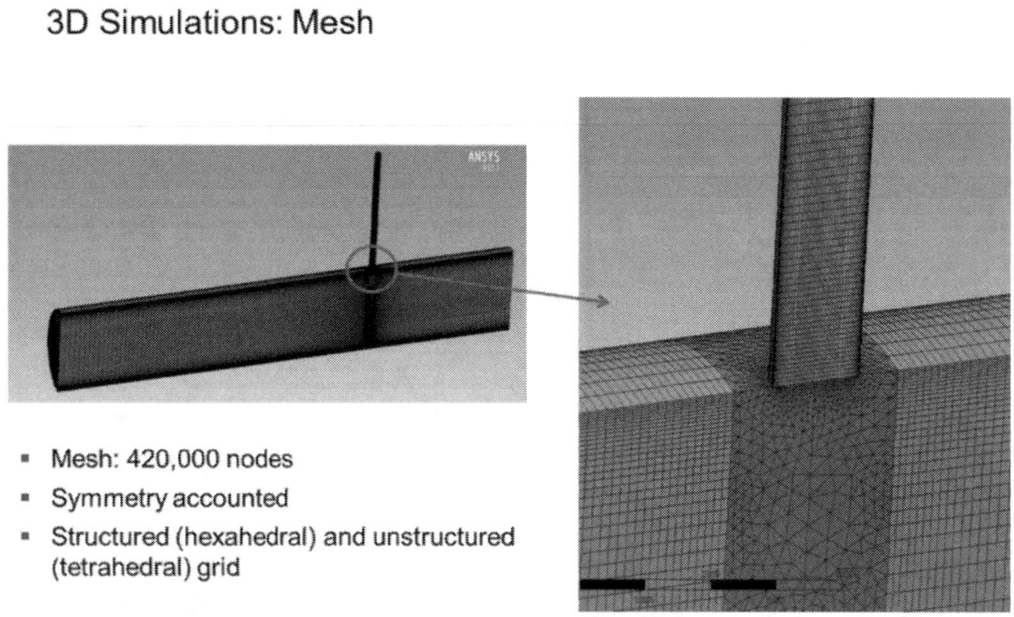

Figure 2. 3D computational fluid dynamics (CFD) simulation geometry and mesh settings.

Dimensionless Analysis

The interaction of the particle-liquid interphase was modeled using a specific drag-force correlation for non-Newtonian fluid. Analysis of particle-laden fluid flow near an individual perforation showed that the particle transport efficiency into the fracture (ratio of the particle concentration in the fracture to the particle concentration in the wellbore) depends on two dimensionless parameters: the particle Stokes number in a non-Newtonian fluid (St) and the ratio of the flow velocity in the wellbore to the flow velocity in the fracture channel (Λ). Using a detailed 3D CFD model of fluid-particle flow into the fracture (built with commercial software), the particle transport efficiency into the fracture was calculated for various combinations of parameters, shown in Table 1. A description of the efficiency follows; the derivation can be found in Ref. 1.

$$R = \frac{\alpha_{pw}}{\alpha_{pf}} \tag{1}$$

where α_{pw} and α_{pf} are the average particle mass concentration sufficiently far upstream in the wellbore and far downstream in the fracture, respectively. In an ideal case, when particles move along the carrier fluid's streamlines, $R = 1$. However, if particles' paths deviate from the liquid streamlines because of their inertia or the action of external forces, R is no longer equal to 1. In particular, in the case of the geometry shown in Figure 1, the effect of particle inertia is always negative, and ratio R is less than 1. The dimension analysis yields the following dimensionless parameters defining the local two-phase flow shown in Figure 1:

$$St = \frac{4}{3}\frac{\rho_p D_p}{D_f C_d \rho_l}; \qquad \Lambda = \frac{V_f}{V_w}; \tag{2}$$

where D_p is the particle diameter, ρ_p and ρ_l are the particle and fluid density, respectively, V_f and V_w are the average fluid speed in the fracture and wellbore, respectively, and D_f and D_w are the diameters of the fracture and wellbore, respectively. It is assumed that the fracture diameter is sufficiently small to make the gravity effect on the particle motion near the junction negligible.

In the case of non-Newtonian, power-law fluid, the viscosity depends on the liquid's shear rate and shear stress:

$$\tau = k\gamma^n; \qquad \gamma = \left(\frac{\partial u_l}{\partial y}\right); \quad \mu = \frac{\partial \tau}{\partial \gamma}, \tag{3}$$

where γ is the shear strain, τ is the shear stress, k is the viscosity index, and n is the power-law index. For power-law fluid flow near a spherical particle, the drag law can be defined as follows (Ref. 2 and Ref. 3):

$$C_d = \frac{24}{\mathrm{Re}_p}\left(1 + 0.418\mathrm{Re}_p^{2.35n/(2.42n+0.918)}\right) \tag{4}$$

Thus, the Stokes number for non-Newtonian power-law fluid can be defined as:

$$St = \frac{4}{3}\frac{\rho_p D_p}{h_f \rho_l} \Bigg/ \left[\frac{24}{\mathrm{Re}_p}\left(1 + 0.418\,\mathrm{Re}_p^{2.35n/(2.42n+0.918)}\right)\right]$$

$$\mathrm{Re}_p = \frac{\rho_l (u_p)^{2-n} D_p^n}{k} \tag{5}$$

The dependence of the collection efficiency on the parameters St and Λ was determined numerically by solving equations of particle and fluid motion with the geometry of the fracture entrance area shown in Figure 1 using the ANSYS® Fluent® commercial software package. Figures 3 and 4 show the dependence R calculated for various Stokes number values. As expected, the efficiency is close to 1 at low values of St but declines monotonously with increasing St.

Table 1. Parameter Set for Different Operating Conditions in the Simulations

Parameters	**Value Range**
Particle bulk density (kg/m^3)	1500 to 2500
Particle radius (m)	5×10^{-5} to 5×10^{-4}
Liquid velocity in wellbore (m/s)	2.38 to 3.332
Liquid viscosity (Pa·s)	0.001003 to 0.01
Perforation diameter (m)	0.0107 to 0.01605
Reynolds number (Re)	240,000 to 500,000
Diameter ratio between wellbore and perforation	9.5 to 14.3
Ratio of velocities in wellbore and perforation	0.1 to 50

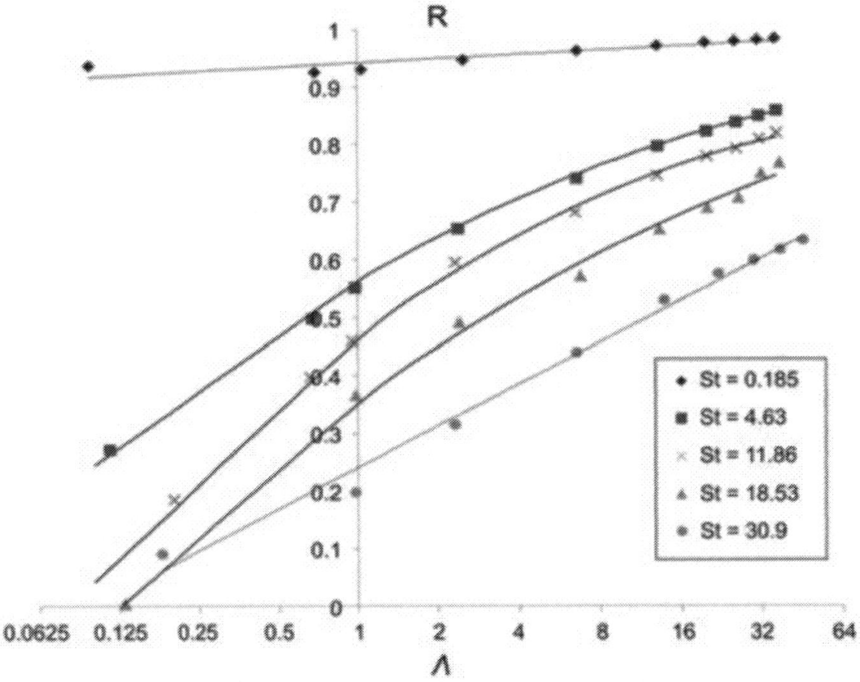

Figure 3. Dimensionless analysis for proppant transport efficiency in Newtonian fluids.

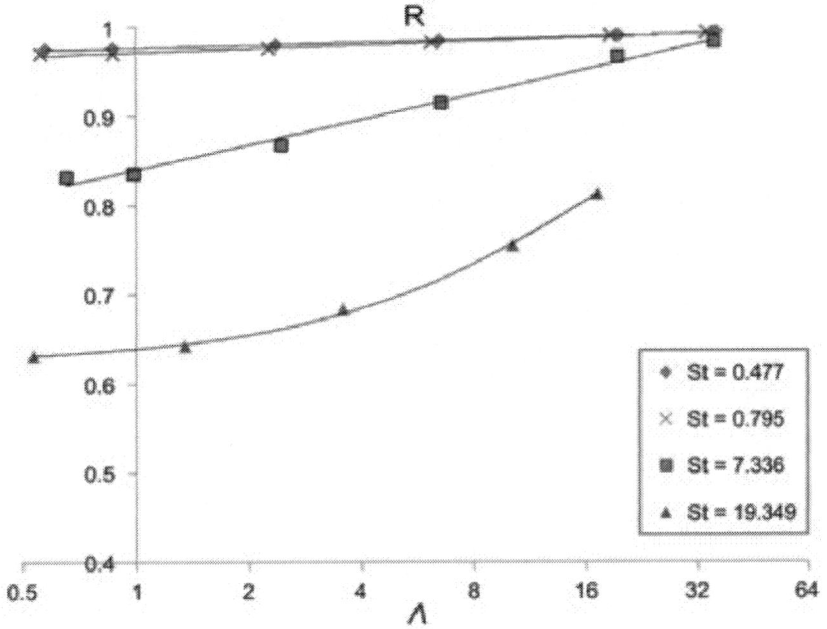

Figure 4. Dimensionless analysis for proppant transport efficiency in non-Newtonian fluids.

Discussion and Conclusion

The CFD simulation results showed that proppant particle distribution is more homogeneous in non-Newtonian fluid than in Newtonian fluid under the same operating conditions. CFD analysis of fluid and particle motion showed that because the flow conditions around each perforation connected to the fractures can differ, the amount of proppant carried to each fracture can vary. The particle inertia in non-Newtonian fluid was analyzed. The efficiency of particle delivery was calculated for a sequence of parameters, St and Λ, and the results were tabulated. In the case of relatively low particle-volume fractions, using such tables enables quick and accurate calculation of delivery efficiency in non-Newtonian fluid systems. The proppant-mass flow rate, which is the product of particle mass, particle concentration, and flow velocity in the fracture, can be then calculated for flow to the fracture.

References

1. Filippov, A., J. Lu, and F. Popa (2016), "Efficient Calculation of Proppant Transport in a Wellbore-Fracture System," presented at the SPE Annual Technical Conference and Exhibition, Dubai, UAE, 26–28 September, SPE-181606-MS.
2. Dhole, S.D., R.P. Chhabra, and V. Eswaran (2007), "Drag of a spherical bubble rising in power law fluids at intermediate Reynolds numbers," *Industrial & Engineering Chemistry Research*, 46 (3), pp. 939-946.
3. Tripathi, A., R.P. Chhabra, and T. Sundararajan (1994), "Power law fluid flow over spheroidal particles," *Industrial & Engineering Chemistry Research*, 33 (2), pp. 403-410.

Acknowledgements

The authors thank Halliburton for project support and permission to publish this article. They also express appreciation for suggestions and discussions from colleague Ron Dusterhoft and previous colleague F. Popa.

ANSYS® Fluent® is a registered trademark of Ansys Inc., Ann Arbor, Michigan.

Contact:

Jessie.Lu@halliburton.com

492291 Two-Compartment Modeling and Simulation Study of a Top Sprayed Fluidized Bed Granulator

Abstract:

Spray fluidized bed granulator (SFBG) is promising technique for the production of granules in various industries. In this study, a top spray fluidized bed granulator is being modeled and analyzed with the help of population balance equation (PBE). The mathematical model for SFBG is derived using the concept of compartment modeling. A top spray fluidized bed granulator is divided into two compartments, wet compartment and dry compartment. In the wet compartment, the aggregation is assumed whereas the breakage process is considered in dry compartment as these are the dominating processes in respective compartments. Further, a recently developed discretization ([1] and [2]) is modified and used to solve this model whose idea is based on the conserving the important properties of the system. Moreover, the numerical results of the integral moments derived by the discretization are validated against the newly developed exact results for various combinations of the aggregation and breakage kernels. The model is also tested for physical tractable kernels and the numerical results are also authenticated with the constant volume Monte Carlo [3]. The volume ratio between the wet and dry compartments and the rate of mass exchange between them are important variables and produce size distributions that are distinctly different from those in a single compartment granulator with simultaneous aggregation and breakup. Moreover, the dynamics of the two-compartment model exhibit a diverse range of behaviors that do not always reach steady state, and in some cases are unstable. We conclude that the proposed discretization scheme is capable of tracking the size distribution accurately and provides a robust numerical methodology to study the dynamics of two compartment granulators.

References:

[1] Kumar, J., Kaur, G. and Tsotsas, E. (2016) An accurate and efficient discrete formulation of aggregation population balance equation. *Kinetic and Related Models* 9(2), 373-391.

[2] Saha, J., Kumar, J., Bück, A. and Tsotsas, E. (2016) Finite volume approximations of breakage population balance equation. *Chemical Engineering Research and Design* 110, 114-122.

[3] Smith, M. and Matsoukas, T. (1998) Constant-number Monte Carlo simulation of population balances. *Chemical Engineering Science* 53(9), 1777-1786.

504313 Understanding Particulate Flow Behavior: Fluidized Bed, Pneumatic Conveying and Membrane Filtration Processes

Abstract:

A key focus of my group is on obtaining mechanistic understandings of particulate flow behavior in practical applications, like fluidized bed, pneumatic conveying and membrane filtration processes.

In fluidized bed systems, many of the probes used to understand hydrodynamics intrude into the environment they are measuring, although assumptions are typically asserted that the intrusive probes do not affect the data collected. This could be a poor assumption in some cases and conditions. In circulating fluidized bed risers, we found that intrusive fiber optic probe measurements consistently mis-predicted the solids concentration compared to the non-intrusive pressure drop measurements. The discrepancy was sensitive to superficial gas velocity, solid circulation rate, probe position and flow direction. Computational fluid dynamics simulations confirmed this, and indicated that particle momentum was lost at the leading edge of the probe and particles were spilling over to the probe tip.

Pneumatic conveying is a common operation in a wide range of chemical industries. A critical parameter for ensuring the efficient operation of pneumatic conveying systems is the minimum pickup velocity, U_{pu}, defined as the minimum superficial gas velocity required to initiate rolling or suspend a particle initially at rest. In view of the increasing prominence of nano-scale particles, we have attempted to bridge the knowledge gap between the nano- and micro-scale particles by investigating the U_{pu} of particle diameters traversing both scales. We found that the understanding available on micro-scale particles may not be directly applicable to nano-scale ones.

For membrane filtration processes, which is a key unit operation in water treatment, membrane fouling is the key obstacle that leads to higher energy costs and lower productivity. Solid-liquid fluidization, whereby the fluidized particles mechanically scour the membrane, is a promising energy-efficient means of mitigating membrane fouling. The optimal fluidization condition would give (i) good enhancement of permeate flux, (ii) uniform scouring across the membrane, and (iii) a modest power requirement. We have attempted to correlate the hydrodynamics of the fluidized particles with the extent of membrane fouling mitigation in order to understand and optimize such processes.

492272 CFD-DEM Simulation of the Heat Transport in a Wurster Coater

Abstract:

The Wurster process is widely applied to coat particles in the pharmaceutical industry. Beads are often coated with a functional film that controls the release of drug substances or masks a specific taste. Multiple films can be coated onto the pellets. The coating process generally takes place in the Wurster tube, in which the particles are also transported centrally upwards. Outside the Wurster tube the particles are moving in a downward direction. Thus, the particle flow resembles a torus shaped recirculation pattern.

The spray applied through the spray nozzle onto the particles and the solvent temperature is usually lower than the fluidization air temperature. The inlet temperature of the fluidization air is set in such a way, that the drying process of the sprayed particle is accelerated and to facilitate a good film forming temperature. Particles are sprayed in the Wurster tube. Drying starts immediately and the biggest part of the energy is absorbed in the Wurster tube. After a short while an equilibrium of the temperature inside the Wurster coater is reached. This equilibrium depends on the spray rate, liquid temperature and used solvent.

In this work a model to simulate the heat transport in a CFD-DEM simulation is presented. The eXtended Particle System XPS is used for the DEM simulation and is coupled to AVL Fire® for the CFD simulation[1]. Drag as well as the energy information are exchanged between the fluid and solid phase at every time step.

First the validity of the model is shown. A lab scale Wurster coater[2] is used to show the influence the spray rate and fluidization air flow rate has on the heat transfer. To model the influence of the different spray rate the particle temperature is set constant. The influence the fluidization air flow rate has on the heat transfer rate is shown. A correlation between fluidization air flow rate and temperature difference was discovered. At last the heat transfer in the Wurster coater at various scales going from the lab up to production scale sized processes will be shown.

References

1. Jajcevic D, Siegmann E, Radeke C, Khinast JG. Large-scale CFDâ€"DEM simulations of fluidized granular systems. *Chem Eng Sci*. 2013;98:298-310.

2. Liang L, Remmelgas J, van Wachem BGM, et al. Residence time distributions of different size particles in the spray zone of a Wurster fluid bed studied using DEM-CFD. *Powder Technol*. 2015;280:124-134.

493266 Development of a Rheological Model for Cohesive Granular Materials across Dense and Dilute Flow Regimes

Abstract:

Introduction:

Cohesive granular materials are ubiquitous in nature and industrial processes, and exhibit behaviors unseen in non-cohesive materials, including jamming at low packing fractions and agglomerate formation. To describe their flow behavior through a continuum model, one must first formulate a model for particle stress. Based on Discrete Element Method (DEM) simulation results, stress models for cohesive granular materials have been developed for the dense flow regime where the solid volume fraction is high (larger than ~0.4) [1-2]. For the dilute flow regime, many studies [1, 3-6] have found that inhomogeneous flow structures would emerge, rendering model development difficult. Thus, there is a need (1) to investigate the mechanism that leads to inhomogeneous flow structures at low dilute flow regime, and (2) to develop a comprehensive stress model for cohesive granular materials that covers both dense and dilute flow regimes. This talk will address both issues.

Methodology:

To study the aforementioned questions, we perform CFD (Computational Fluid Dynamics) - DEM simulations for gas-fluidization of frictional, cohesive particles in a fully periodic domain. By analyzing snapshots gathered from the simulations, quantities of interest in formulating a rheological model are determined. It is found that, unlike previous studies, this approach allows for examination of stresses at both dense and dilute flow regimes.

Results and Discussions:

For dense flow regime, it is found that stress results consistent with previous studies [1-2] are obtained, thus validating the approach. For dilute flow regime, it is found that pressure follows a behavior similar to van der Waals equation of state. Specifically, at low granular temperatures, pressure decreases with increasing solid volume fraction for a range of solid volume fractions, which is consistent with a previous analysis by Takada et al. [5]. This behavior explains the formation of inhomogeneous flow structures. We then propose a pressure model for this dilute flow regime. In the end, we bridge the models for both dense and dilute flow regimes to form a comprehensive pressure model for cohesive granular materials. For further studies, we seek to develop an analogous model for particle shear stress.

References:

[1] Y. Gu, S. Chialvo, and S. Sundaresan. *Phys. Rev. E* **90**, 032206 (2014)

[2] P.R. Rognon, J.N. Roux, M. Naaim, and F. Chevoir. *J. Fluid. Mech.* **596**, 21 (2017)

[3] E. Irani, P. Chaudhuri, and C. Heussinger. *Phys. Rev. L.* **112**, 188303 (2014)

[4] S. Takada, K. Saitoh, and H. Hayakawa. *Phys. Rev. E.* **90**, 062207 (2014)

[5] S. Takada, K. Saitoh, and H. Hayakawa. *Soft Matter* **11**, 6371 (2015)

[6] E. Irani, P. Chaudhuri, and C. Heussinger. *Phys. Rev. E.* **94**, 052608 (2016)

494130 CFD-DEM Simulations of Bubbling Fluidization: Global Sensitivity Analysis for the Identification and Validation of Critical Model Parameters

Abstract:

Bubbling fluidized beds are used extensively in the energy and chemical industries because of their excellent heat and mass transfer characteristics. The performance of these reactors is strongly dependent on particle-scale phenomena such as chemical conversion, clustering, cohesion and their interaction with reactor-scale transport. In this context, CFD-DEM simulations are extremely valuable because particles are individually tracked and their interactions are completely resolved. However, despite significant progress over the past decade, there is considerable uncertainty regarding the choice of particle-interaction parameters such as the stiffness, restitution and interphase drag force, most of which cannot be measured experimentally (under typical operating conditions of reactors). Therefore, they are often treated as *tunable* parameters in simulations and their impact (highly coupled and non-linear) on hydrodynamic predictions is largely unknown. In this study, an exhaustive global sensitivity analysis is performed and statistical tools are employed to (a) identify critical parameters in CFD-DEM simulations and (b) validate these with experimental data.

For the bubbling fluidization of Geldart B particles, a large parametric space is initially considered including model, operating and numerical parameters in CFD-DEM simulations. The set of insensitive parameters is first dicounted using MOAT (Morris One-At-Time) screening analysis [1] and thorough sensitivity analysis is then performed to quantify the dominant and coupled interactions of the remaining sensitive parameters. Numerical experiments are designed using Latin-hypercube based sampling, optimized to ensure sampling uniformity in the parameteric space. Critical parameters are identified using response surface methodology based on bubble statistics and solids circulation metrics as quantities of interest [2] and these parameters are validated using experimental measurements in a thin rectangular pulsating fluidized bed. All numerical CFD-DEM simulations are conducted using open-source software MFiX and bubble statistics are computed using MS3DATA [3]. This is one of the first large-scale parametric studies of CFD-DEM simulations and the developed framework will guide the selection and validation of critical sub-models in other applications of particle-gas flows.

REFERENCES

[1] F. Campolongo, J. Cariboni and A. Saltelli, An effective screening design for the sensitivity analysis of large models. Environmental Modelling and Software 22:1509-1518, 2007

[2] A. Bakshi, C. Altantzis, R.B. Bates and A.F. Ghoniem, Eulerianâ€"Eulerian simulation of dense solidâ€"gas cylindrical fluidized beds: Impact of wall boundary condition and drag model on fluidization. *Powder Technology* 277: 47â€"62, 2015

[3] A. Bakshi, C. Altantzis, R.B. Bates and A.F. Ghoniem, Multiphase-flow Statistics using 3D Detection and Tracking Algorithm (MS3DATA): Methodology and application to large-scale fluidized beds. *Chemical Engineering Journal*, 293: 355-364, 2016

494425 The Role of Particle Friction in the Stabilization of Pulsed Gas-Solid Fluidized Beds. from Surface Waves to Structured Bubble Nucleation

Abstract:

Many operations in the process industry and the energy sector, e.g., catalytic reactors and dryers, rely on good contact between solid and gas phases. Fluidized beds in their various designs represent the means to disperse a solid phase and promote the transport of mass and heat with a surrounding fluid. The level of mixing and the circulation of the solids become critical factors in determining the transport properties and so the overall performance of a bed.

However, optimization and, particularly, scale-up remain a challenge when designing gas-solid fluidized systems. The instabilities arising from the coupling between the transport of momentum in the gas and the particles and the nonlinear behavior emerging from highly dissipative particle-particle interactions give rise to macroscopic phenomena that are difficult to predict and are highly dependent on particle level properties. Spouting, channeling, jetting, bubbling or expansion are some of these, often associated to the characteristic features of classic fluidization regimes. Over the decades, much work has focused on understanding their hydrodynamics and how the transport properties are affected or could be modified, but not much is known about how more regular and predictable flow structures may occur. This has been the focus of some efforts in recent years; one such way to induce structure in fluidized beds, is by pulsating the gas flow. This gives rise to surface patterns in vibrating and fluidized shallow granular layers [1, 2], and dynamic bubble patterns in quasi-2D gas-solid fluidized beds under certain conditions, whereby bubbles rearrange into a triangular tessellation pattern with a characteristic bubble size and wavelength[3, 4] . This contribution investigates the role of frictional contacts in the formation of such dynamic patterns. It studies the transition from the generation of a surface wave in a shallow granular layer to the nucleation of bubbles in alternate sites, as the thickness increases. We report experimental data in a quasi-2D pulsating bed along numerical studies under different flow conditions and particle friction factors. A comparison between a continuous and discrete formulation is used to display the role of friction in inducing the bubble nucleation at a critical bed height and in stabilizing the subsequent pattern. Both Eulerian-Eulerian (TFM) and Eulerian-Lagrangian (CFD-DEM) models succeed in predicting the formation of a surface wave in a shallow fluidized granular layer, which appears dominated by hydrostatic effects. However, as the height increases, the solid circulation leads to locally dense regions where particles interact by sustained multi-particle contacts and the effect of friction becomes dominant. The inherent difficulty of continuum fluid mechanical formulations to deal with correlated particle velocities and the resulting anisotropic distribution of contacts explains why the commonly used closures for the solid stresses cannot reproduce the rearrangement of bubbles in dense systems. On the contrary, we will show how the Eulerian-Lagrangian formulation explicitly solves the granular rheology and can track the boundary of the areas locked in the plastic regime, rendering the correct bubble size, wavelength and sequence in the location of the nucleation sites, up to a quantitative agreement with the experiments.

[1] F. Melo, P. Umbanhowar, H.L. Swinney, Transition to parametric wave patterns in a vertically oscillated granular layer, Phys. Rev. Lett. 72 (1994) 172.

[2] J. Bougie, K. Duckert, Continuum simulations of shocks and patterns in vertically oscillated granular layers, Phys. Rev. E 83 (2011) 011303.

[3] M.-O. Coppens, J.R.van Ommen, Structuring chaotic fluidized beds, Chem. Eng. J. 96 (2003) 117-124.

[4] M.-O. Coppens, M.A. Regelink, C.M. van den Bleek, Pulsation induced transition from chaos to periodically ordered patterns in fluidised beds, Proceedings of the 4th World Congress on Particle Technology, Sydney, 2002.

494495 Critical Comparison of Electrostatic Effects on Hydrodynamics, Heat Transfer and Chaotic Analysis in a Bubbling Fluidized Bed with a Central Jet

Abstract:

Critical comparison of electrostatic effects on hydrodynamics, heat transfer and chaotic analysis in a bubbling fluidized bed with a central jet

Haotong Wang, Lungu Musango, Jingdai Wang, Yongrong Yang

Abstract: Gas-solid fluidized bed has a very high heat transfer and mass transfer efficiency and can handle a large number of particles, and thus has a wide range of applications in many industrial processes. In a gas-solid fluidized bed, the flow behavior exhibits randomness and structural instability due to the existence of bubble agitation, particle motion and gas-solid interaction which results in a typical nonlinear transient system. The study of gas-solid fluidized bed is mainly focused on the time-averaged parameters. Compared with the time-averaged parameters, the fluctuating parameters (pressure and porosity fluctuation, etc.) of the fluidized bed also contain the information of flow structure and transport behaviors. Therefore, the combination of time domain analysis and frequency domain analysis can be a more in-depth study of fluidized bed, especially for gas and solid flow, heat and mass transfer characteristics. In particular, electrostatic phenomenon is unavoidable in many industrial processes, such as ethylene polymerization, due to particle-particle and particle-wall collision and friction [1, 2]. The electrostatic and hydrodynamic effects are interdependent. After the particles are charged, parameters such as bubble size and particle velocity will also be affected, which further influences the heat and mass transfer ability in the fluidized bed reactor. Whereas changes in hydrodynamic properties, bubble size, particle velocity and phase fraction distribution, etc., also affect the electrostatic potential distribution of the fluidized bed.

In this paper, a comprehensive study of the electrostatic effect on bubble behavior, particle fluctuation velocity and heat transfer coefficient in the fluidized bed with a central jet has been evaluated by Eulerian-Eulerian two fluid model coupled with electrostatic model and energy model. The simulated voidage profiles at different positions, bubble detachment time and initial bubble diameter are compared with experiment results in the literature. The predicted bubble behaviors including bubble frequency and bubble numbers, and particle fluctuation parameters, including particle fluctuation velocity, granular temperature and Reynolds stress are analyzed in both charged and uncharged system. Electrostatic effect on two kinds of heat transfer coefficients, namely bubble to emulsion phase heat transfers based on the gas throughflow velocity and gas solid local heat transfer are quantitatively compared. Simulation results show that electrostatic charges decrease bubble numbers and granular temperature, whereas the averaged heat transfer coefficients are enhanced. A more chaotic behavior is observed by time series analysis and phase space reconstruction. Overall, the electrostatic effect on the hydrodynamic and heat transfer characteristics can be revealed.

Keywords: Fluidization, Bubble dynamics, Electrostatics, heat transfer

References

1 J.Y. Sun, Y. Yan, Characterization of flow intermittency and coherent structures in a gas-solid circulating fluidized bed through electrostatic sensing. *Ind Eng Chem Res*, 55 (2016):12133-12148.

2 G. Hendrickson, Electrostatics and gas phase fluidized bed polymerization reactor wall sheeting. *Chem Eng Sci*, 61 (2006): 1041-1064.

500604 Verification of Euler-Lagrange and Euler-Euler Simulations of Meso-Scale Gas-Solid Flows

Abstract:

Simulations of gas-solid flows will aid the modeling effort for many chemical engineering technologies, such as fluidized bed reactors. However, fully resolved simulations of gas-solid flows are too expensive to capture emerging dynamics involving clusters of particles. Reactor-scale modeling requires data from â€œmeso-scaleâ€ simulations that resolve these emerging dynamics, but model smaller scales. These simulations are challenging to perform, particularly in flows with a range of particle Knudsen numbers. Two classes of techniques are commonly used: Euler-Lagrange (EL) and Euler-Euler (EE). However, standard EL and EE simulations are incapable of converging under mesh refinement. Previous work has introduced methods that can, in principal, converge: Euler-Euler Anisotropic Gaussian (EE-AG) and volume-filtered Euler-Lagrange (VF-EL).

In this work, we study the convergence properties of three gas-solid simulation techniques: VF-EL, EE-AG, and Euler-Euler Two-fluid model (EE-TFM), a method that has been shown to yield unphysical shocks in dilute regions of flow. Using these three techniques, we perform simulations of two idealized problems: particles in frozen homogeneous isotropic turbulence (HIT), and cluster-induced turbulence (CIT). Statistical quantities from these simulations, such as the volume fraction distribution and the granular temperature, will inform macro-scale models of gas-solid flows.

We find, for HIT, that flow statistics from only the VF-EL simulations converge. In contrast, for CIT, both EE-AG and VF-EL simulations converge and approach similar values, while statistics from EE-TFM show no such convergence. We explore the mechanisms behind convergence in these simulation techniques. In VF-EL we also study the recovery of the particle density function and associated statistics. Here, we use kernel density estimation for reconstructing the particle density function and show the estimated statistics depend strongly on the kernel bandwidth. This work demonstrates the need for further verification of meso-scale simulation techniques.

502308 Fundamental Prediction of Agglomeration and Entrainment Rates for Cohesive Powders in a Riser Flow

Abstract:

Empirical models for the entrainment rates of non-cohesive powders in fluidized bed systems can vary by several orders of magnitude in their predictions for the same system ((Chew et al. 2015)). For cohesive particles, the formation and breakage of agglomerates results in further uncertainty to the entrainment rate. In this work, a fundamental approach is used for entrainment prediction. Specifically, the population balance is coupled with kinetic-theory-based balances. This new continuum framework is based on fundamental closures for the birth and death rates due to aggregation in the population balance. These novel closures take into account the effect of impact velocity (the granular temperature) on the outcome of a collision as agglomeration, rebound, or breakage. The agglomerate size distribution and entrainment rates predicted by the population balance model are compared to discrete element method (DEM) simulations of a riser flow as well as to experimental results.

Chew, J. W., A. Cahyadi, C. M. Hrenya, R. Karri and R. A. Cocco (2015). "Review of entrainment correlations in gasâ€"solid fluidization." Chemical Engineering Journal **260**: 152-171.

501016 Quantitative Validation and Analysis of Heat Transfer in a Rotary Drum Using Experiments and Simulations

Abstract:

Quantitative validation and analysis of heat transfer in a rotary drum using experiments and simulations.

Manogna Adepu, Shaohua Chen, Yang Jiao, Aytekin Gel, Heather Emady

Arizona State University, Tempe, AZ 85287

Heat transfer in granular processing and handling is commonly encountered in material manufacturing . Granular processing consumes roughly ten percent of the world's energy [1]. Approximately 1–3% of the world's energy is utilized in cement industries alone [2]. Since most of these processes are energy-intensive, the handling of granular materials is tightly connected to a high-energy demand. The flow properties of granules alone are governed by their intrinsic material properties (e.g., cohesion, the coefficient of friction and yield stress) and particle properties (e.g., size distribution, morphology, porosity, and roughness), which makes predicting the heat transfer behavior of dynamic granular processes even more complex. Over the last five decades, there has been a continued interest in understanding the role of system parameters and the mechanisms of heat transfer between granular media and the boundary surfaces. The thermal behavior of the granules may change dramatically during processing, introducing variability into the process and affecting final product quality and characteristics. It is, therefore, important to understand the heat transfer mechanism in process specific granular systems.

This work focuses on understanding the heat transfer mechanism in the granular bed inside a rotary drum, one of the most commonly used process equipment. Experiments are performed using silica particles to investigate the granular flow and heat transfer mechanism inside a stainless-steel rotary drum. All the modes of heat transfer are quantified under varying operation conditions to establish a strong understanding of the heat transport. For this, both experiments and CFD-DEM (using MFIX-DEM, an open source multi-solver suite) simulation techniques are used to analyze the thermal behavior. The experimental setup is designed to handle up to 1000^0 C, making it possible to study all modes of heat transfer via conduction, convection, and radiation. Omega thermocouples and a mid-wave infrared (MWIR) camera were used to record the temperature profile. The digital image analysis (DIA) and the particle image velocimetry (PIV) are combined with IR thermography (DIA/PIV/IR technique [3]) to obtain combined quantitative hydrodynamic and thermal data sets.

Reference:

[1]. Duran, Jacques. "Introduction." Sands, Powders, and Grains. Springer New York, 2000. 1-18.

[2]. Engin, T., Ari, V., 2005. Energy auditing and recovery for dry type cement rotary kiln systems – a case study. Energy Convers. Manag. 46, 551–562.

[3]. Patil, A. V, Peters, E. A. J. F., Sutkar, V. S., Deen, N. G., & Kuipers, J. A. M. (2015). A study of heat transfer in fluidized beds using an integrated DIA / PIV / IR technique. Chemical Engineering Journal, 259, 90–106.

498924 A Comparison of Numerical Optimization Methods for Cyclone Separators

Abstract:

Cyclone separators, more commonly known as cyclones, are widely used in the industry with multiple different purposes. The most common usage is in gas-solid separation. In order to find the best possible performance for this equipment, optimization procedures are conducted with mathematical models. Studies that employ optimization techniques to cyclones have demonstrated the existence of geometries with better characteristics than those with classical dimensions, such as the Lapple or Stairmand geometries. Although there are several mathematical optimization methods available in the literature, to the authorsâ€™ knowledge no comparison has been made to evaluate which optimization method is more adequate specifically for cyclone separators. This is important as some optimization methods will be significantly more efficient than others, depending on the topology of the objective function and on the computational cost of the model for the objective function utilized. As an example for cyclone separators, in cases where the computational cost of the mathematical model used is negligible (e.g. when using meta-models or simple empirical models), the number of objective function calls is not important as its impact on the total optimization time will most likely be negligible. On the other hand, if the model is costly such as in the computational fluid dynamics approach, the total time taken by the procedure will heavily depend on the function call efficiency of the optimization method. This paper therefore describes the comparison of different optimization methods for cyclone geometry optimization studies, evaluating mainly the number of objective function calls and the quality of the results. For this purpose, the semi-empirical model of Muschelknautz is chosen as mathematical cyclone model due to its reduced computational cost added to its good prediction and extrapolation capabilities. It is assumed that this model is sufficiently representative of the behavior that other good fidelity cyclone models will have with optimization methods. The six optimization methods compared include the derivative-free COBYLA, Nelder-Mead and Powell methods. Results show that, even with the relatively limited sample of six methods tested, the choice of optimization method has a big impact on the total computational time required by the complete procedure. Overall the COBYLA method is shown to be the indicated one for cyclone optimization studies, with consistent results and a low number of objective function evaluations required.

499046 Optimization of Cyclone Separators in Series Based on Computational Fluid Dynamics

Abstract:

Cyclone separators, simply known as cyclones, are commonly used equipment in the industry with the main purpose of gas-solid separation. There are also examples of them being used for other functions, such as heat transfer or as reactors. Cyclones are frequently employed in parallel or series configurations to better conform to process requirements. In serial configuration for gas-solid separation, the gas cleaned in one cyclone is fed into the next cyclone, to try to further collect the remaining particles. In order to find the best possible performance for this equipment, optimization procedures are conducted with mathematical models. Studies which employ optimization techniques to cyclones have demonstrated the existence of geometries with better characteristics than those with classical dimensions, such as the Lapple or Stairmand geometries. Although there are several geometry optimization studies of single cyclone configurations, to the best of our knowledge no paper explores multiple cyclones in series. In addition, in spite of the better predictions provided by phenomenological models such as those used in computational fluid dynamics (CFD), authors mostly use lower fidelity models due to their lower computational resources requirement. This paper therefore describes the optimization of three cyclones in series based on high fidelity CFD cyclone modeling. For this purpose, a methodology on how to solve this problem is evaluated, with the COBYLA optimization method and an Eulerian-Eulerian six-phase two-way gas-solid CFD approach, solved using an axisymmetric approach with an in-house developed code. The optimization of three cyclones in series had as objective the efficiency maximization. Results indicate that the optimized trios of cyclones outperform the classical high efficiency Stairmand trio of geometries with 30 times less solids emitted. We show that the optimization of cyclones in series delivers excellent results and is highly feasible in industrial time without compromising the fidelity of the mathematical cyclone model used.

492648 Application of Psychrometric Principles for Predicting the Bulk Density of Spray Dried Dispersions

Abstract:

A closed-loop spray drying process may be deconstructed into a thermodynamic cycle comprising evaporative cooling, sensible cooling, dehumidification, and sensible heating. Consequently, its outcomes may be predicted through application of psychrometric principles. Here, we present psychrometric analyses of the spray drying process of a pharmaceutical product at large pilot scale. In this approach, the thermophysical properties of the vapor and gas are modeled using the Redlich-Kwong equation of state. The gas-vapor mixture properties are evaluated assuming ideal behavior (Raoultâ€™s law), and the saturation vapor pressure of each mixture component is computed using the Clausius-Clapeyron relation. Vapor-liquid phase equilibria are approximated through the Van-Laar activity model. Through the resulting model, an algebraic, semi-empirical relation that describes factors governing droplet evaporation, namely the Sherwood, Reynolds, and Spalding thermal and mass numbers, is derived. We demonstrate that such a relation trends with the bulk density of the spray dried dispersion, and may provide key process insights beyond those achieved from a conventional empirical QbD approach.

493401 A Model-Guided Selection of Media Size in Wet Stirred Media Milling of Poorly Water-Soluble Drugs

Abstract:

Wet stirred media milling (WSMM) of poorly water-soluble drugs (BCS Class II) increases surface area of drug crystals significantly, thus enhancing their dissolution rate and bioavailability (Li et al., 2016). Although WSMM has proven to be a robust process for producing nanoparticle suspensions of a wide range of materials, selection of media (bead) size has been largely empirical, lacking fundamental rationale (Bilgili et al, 2016). This study aims to establish such rationale by investigating the impact of bead size at various stirrer speeds on the breakage kinetics of a BCS Class II drug. In previous studies (Bilgili and Afolabi, 2012; Afolabi et al., 2014), we used a variation of the microhydrodynamic model proposed by Eskin et al. (2005), where the impact of various formulation and process parameters on the breakage kinetics of BCS Class II drugs was investigated. In this study, we used this model to gain insight into the impact of bead size at various energetic conditions. To this end, stable suspensions of griseofulvin, a model BCS Class II drug, were prepared using hydroxypropyl cellulose and sodium dodecyl sulfate. The suspensions were milled at four different stirrer speeds using various sizes of yttrium-stabilized zirconia beads. Laser diffraction, SEM, and XRPD were used for characterization. Our experimental and modeling results suggest that there is an optimal bead size that achieves fastest breakage at each stirrer speed and that it shifts to a smaller size at higher speed. Calculated microhydrodynamic parameters reveal two counteracting effects of bead size: more bead–bead collisions with less energy/force upon a decrease in bead size. The optimal bead size exhibits a negative power-law correlation with either specific energy consumption or the microhydrodynamic parameters. Overall, this study has rationalized the use of smaller beads for more energetic wet media milling and provided an overarching explanation as to the use of smaller beads in wet stirred media mills than in low-energy mills such as ball mills including centrifugal or planetary ball mills. This study is expected to guide the design and optimization of wet media milling processes in terms of proper choice of bead size for optimal performance.

References:

Afolabi, A., Akinlabi, O., Bilgili, E., 2014. Impact of process parameters on the breakage kinetics of poorly water-soluble drugs during wet stirred media milling: A microhydrodynamic view. Eur. J. Pharm. Sci. 51, 75–86.

Bilgili, E., Afolabi, A., 2012. A combined microhydrodynamics–polymer adsorption analysis for elucidation of the roles of stabilizers in wet stirred media milling. Int. J. Pharm. 439, 193–206.

Bilgili, E., Capece, M., Afolabi, A., 2016. Modeling of Milling Processes via DEM, PBM, and Microhydrodynamics, in: Pandey, P., Bharadwaj, R. (Eds.), Predictive Modeling of Pharmaceutical Unit Operations. Elsevier, Amsterdam, Netherlands, pp. 159–203.

Eskin, D., Zhupanska, O., Hamey, R., Moudgil, B., Scarlett, B., 2005. Microhydrodynamic analysis of

nanogrinding in stirred media mills. AIChE J. 51, 1346–1358.

Li, M., Azad, M., Davé, R., Bilgili, E., 2016. Nanomilling of drugs for bioavailability enhancement: A holistic formulation–process perspective. Pharmaceutics 8, 17.

499098 Reduced Order-Discrete Element Method Modeling of Comilling for Efficient Integration into Continuous Process

Abstract:

Continuous pharmaceutical manufacturing is known for the advantages it offers for cost, efficiency and reduced 'time to market', specifically in the time of rising manufacturing, research and development costs, and competition from generics. The gradual shift from batch to continuous process along with introduction of US food and drug administration's (FDA) Quality by Design (QbD) initiative has provided the necessary impetus for research efforts to build quality into the product through a thorough understanding of the effect of variability in material properties and process conditions on product properties. The role of process modeling in this paradigm is apparent as effect of a disturbance in continuous process on the product quality needs to be well predicted [1].

Comilling unit in the dry and wet granulation routes in pharmaceutical manufacturing is used to break granules into a required size distribution. Particle size reduction in comill through the impact of a rotating impeller on granules and their eventual exit through a screen is a complex phenomenon. Past work on predicting the breakage has used Population balance models (PBM) where the kernel that represents the probability of granule breakage is represented semi-empirically through their dependence on impeller speed, granule size etc. A mechanistic understanding of the breakage phenomenon can be achieved through discrete element method (DEM) modeling of breakage in comill where particle-particle, particle-wall interactions are captured through underlying system of equations based on Newton's laws of motion. The DEM model proposed in [2] takes into account a threshold impact energy above which the granules break. A multi scale framework is established where cumulative energy of granules from DEM simulations are utilized in an energy based PBM kernel and an iterative algorithm is used for estimating material specific parameters in the kernel. However, the high computational expense of DEM simulations and the required cumbersome post processing limits its use in any predictive capacity for continuous manufacturing such as integration into flowsheet model. Reduced order modeling can bridge this gap [3].

Current work proposes a reduced order-discrete element method modeling to represent the multi-dimensional and time dependent energy data from DEM simulations. The methodology reduces the computational expense associated with the high fidelity DEM simulations while retaining accuracy of the mechanistic data [4]. DEM simulations of comilling are run at various flow rates and impeller speeds based on an experimental design and the energy distribution data from these simulations are used to build a reduced order model. An analytical relationship that captures dependence of energy of granules on size, time, flow rate and impeller speed is developed. This enables accurate representation of energy based kernel in PBM model. A comparison of various surrogate modeling techniques such as kriging, radial basis functions is also presented. The improvement achieved through this framework is drastic as the DEM simulations take approximately five days where as the reduced order DEM model incorporated into PBM takes few seconds. Through this work, a framework to represent the mechanistic information from DEM efficiently into PBM of a comill operation is established, thus enabling its incorporation into flowsheet modeling of DG and WG routes of continuous manufacturing. The incorporation of reduced order model into

a continuous WG route flowsheet model is also demonstrated.

References

[1] A. Rogers, A. Hashemi, M. Ierapetritou, Modeling of Particulate Processes for the Continuous Manufacture of Solid-Based Pharmaceutical Dosage Forms, Processes, 1 (2013) 67.

[2] N. Metta, Ierapetritou, M., Ramachandran, R., A combined experimental and computational approach using discrete element method for the development of a mechanistically motivated breakage kernel, AIChE annual meeting, San Francisco, CA, USA, 13-18 November 2016.

[3] D. Barrasso, A. Tamrakar, R. Ramachandran, Model Order Reduction of a Multi-scale PBM-DEM Description of a Wet Granulation Process via ANN, Procedia Engineering, 102 (2015) 1295-1304.

[4] A. Rogers, M.G. Ierapetritou, Discrete Element Reduced-Order Modeling of Dynamic Particulate Systems, Aiche Journal, 60 (2014) 3184-3194.

495870 A Unified Theory for the Solid Stresses in Particle Laden Flow

Abstract:

An improved model describing the solid stresses in particle laden flow is presented. The model is shown to be valid for slow and intermediate flows and is developed based on the theory originally proposed by Tardos et al. [Powder Technol. 131 (2003) 23]. The study is an extension of the work proposed by Makkawi et al. [Powder Technol. 163 (2006) 69], where a steady one-dimensional particle flow model was used to demonstrate the smooth merge of quasi-static flow with â€œliquid-likeâ€ behaviour. In this study, the concept is further developed and extended to develop a general three dimensional constitutive model. The model is applied to simulate a bubbling fluidised bed and it is validated using experimental data obtained by electrical capacitance tomography (ECT) measurements.

489330 Numerical Determination of Contact Laws for Compressible Particles

Abstract:

In the study of particulate matter, numerical simulation can play a role in efforts to gain insight into the influence of intermolecular forces, crystal structure, surface topology, particle material properties and surface chemistry on powder bulk properties by enabling controlled investigation of the effects of small-scale mechanisms on large-scale behaviour. The accuracy of discrete approaches to modelling the static and dynamic behaviour of granular media is dependent on their accuracy in representing particle-particle interactions, which may include plastic, viscoelastic, viscoplastic, adhesive and large deformation effects (see, for example, Tomas (1993)). Parameters for contact laws including these features may be found experimentally (see, for example, Pasha et al (2014)). However, normal contact relations for particles exhibiting compressible plastic behaviour derived from underlying smaller-scale mechanics are not found in the literature. We present work aimed at developing such relations based on parametric finite element (FE) studies of spherical particles in contact. This includes (1) A description of the finite element model and compressible plasticity models employed for the simulations, (2) nondimensionalisation of the problem, (3) derivation of an explicit normal force-displacement relation for contact of compressible spheres and its dependence on material parameters, (4) qualitative mapping of the spatial development of plastic and compression zones, (5) investigation of the sensitivity of the resulting contact law to the compression plasticity model used, and (6) introduction of adhesive effects in the FE contact law and a study on its effects on the resulting inter-particle contact law. It is intended that the results from this work will help improve the accuracy of discrete element and similar tools by providing normal contact laws dependent on properties that can be determined from experiment.

References

Tomas, P., 2003, Mechanics of Particle Adhesion, In Particles on Surfaces 8: Detection, Adhesion and Removal, pp. 183-229.

Pasha, M., Dogbe, S., Hare, C., Hassanpour, A., Ghadiri, M., 2014, A linear model of elasto-plastic and adhesive contact deformation, Granular Matter 16, pp. 151-162.

UF Herbert Wertheim College of Engineering
UNIVERSITY *of* FLORIDA

Determining Collisional Dissipation Rate for Elongated Rods with Friction

Kevin Buettner[1], Yu Guo[2], Jennifer Curtis[3]

[1] Department of Chemical Engineering, University of Florida, Gainesville FL
[2] Department of Engineering Mechanics, Zhejiang University, Hangzhou, China
[2] College of Engineering, University of California, Davis CA

POWERING THE NEW ENGINEER TO TRANSFORM THE FUTURE

Acknowledgements

- NASA STTR Phase II
- US-Israel BARD Fund

Simulating granular flows

- Gas-solid flows common in many industries
  - Pharmaceuticals, agriculture, energy, etc.
- Process design and scale-up can benefit from simulations
- CFD-DEM simulations are too computationally expensive
- Two-fluid simulations utilize spherical closure models

New models for two-fluid simulations

Particle Properties
- Shape
- Wet/dry
- Smooth/rough
- Breakage
- Flexible/rigid

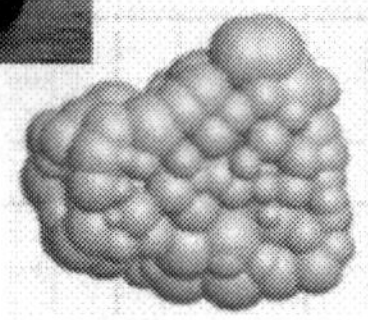

DEM Simulations
- Measure particle-particle interactions
- Vary properties

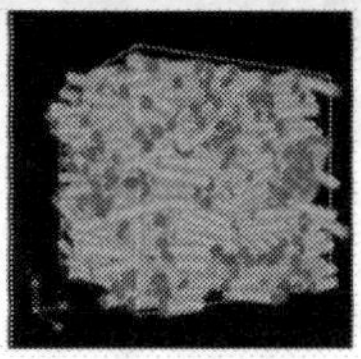

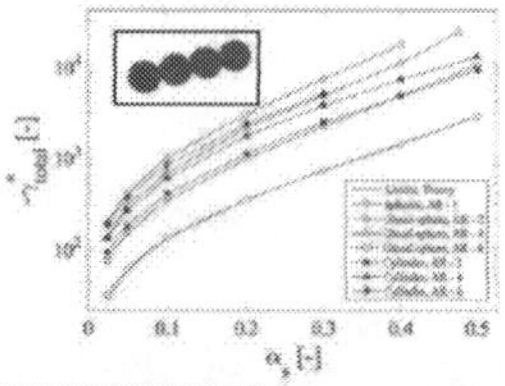

Functional Model
- Fitted equations
- Scaling relationships

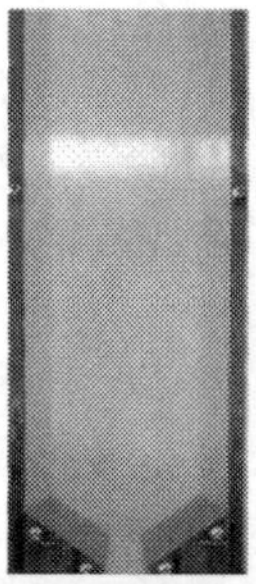

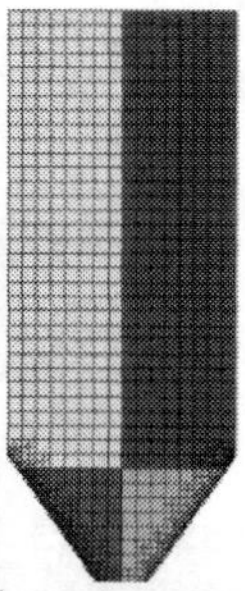

CFD Simulations
- Realistic domains
- Validation
- Prediction

Measuring dissipation rate

- **Homogeneous cooling system (HCS)** used in previous study[1] of dissipation rate (γ_s) for smooth cylinders
  - Fully periodic domain
  - Dissipation rate is quantified by decrease in velocity fluctuations over time

1 Buettner, K.E., Guo, Y., Curtis, J.S., n.d. Using the Discrete Element Method to develop Collisional Dissipation Rate Models that Incorporate Particle Shape. AIChE J. doi:10.1002/AIC.15933
2 Guo, Y., Wassgren, C., Ketterhagen, W., Hancock, B., James, B., Curtis, J., 2012. A numerical study of granular shear flows of rod-like particles using the discrete element method. J. Fluid Mech. 1–26. doi:10.1017/jfm.2012.423

Measuring dissipation rate

- **Shear flow (SF)** simulations used in previous studies[2] of solid stress for cylinders with complex properties
  - Dissipation rate is measured using the following equation:

$$\gamma_s = -\dot{S}\sigma_{xy}$$

Where $\dot{S}$ is the shear strain rate and σ_{xy} is the shear stress on the x-y plane

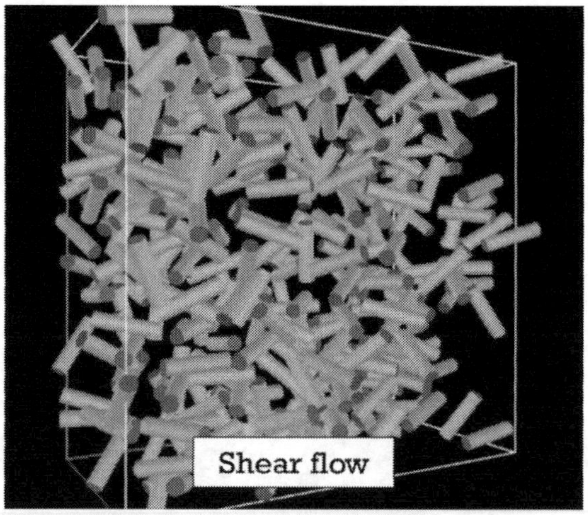

1 Buettner, K.E., Guo, Y., Curtis, J.S., n.d. Using the Discrete Element Method to develop Collisional Dissipation Rate Models that Incorporate Particle Shape. AIChE J. doi:10.1002/AIC.15933
2 Guo, Y., Wassgren, C., Ketterhagen, W., Hancock, B., James, B., Curtis, J., 2012. A numerical study of granular shear flows of rod-like particles using the discrete element method. J. Fluid Mech. 1–26. doi:10.1017/jfm.2012.423

Frictionless to rough particles

	Frictionless Spheres	**Rough Cylinders**
Degrees of freedom	• Translational (x, y, z) movement • No particle rotation • **Total DOF - 3**	• Translational (x, y, z) movement • Particle rotation in **3 directions** • **Total DOF - 6**
Granular temperature	• $\boldsymbol{\theta_{trans} = \frac{1}{3} < u_i'^2 >}$ $u_i' = u_i - < u_i >$	• $\theta_{trans} = \frac{1}{3} < u_i'^2 >$ • $\boldsymbol{\theta_{rot} = \frac{1}{3m} < I_{ij}\omega_i'\omega_j' >}$
Scaled dissipation rate (γ_{total}^*)	• $\theta_{total} = \theta_{trans}$ • $\gamma_{total}^* = \frac{\gamma_s d_p}{\theta_{total}^{\frac{3}{2}}}$	• $\theta_{total} = \theta_{trans} + \theta_{rot}$ • $\gamma_{total}^* = \frac{\gamma_s d_p}{\theta_{total}^{\frac{3}{2}}}$

Spherical comparison of HCS and SF

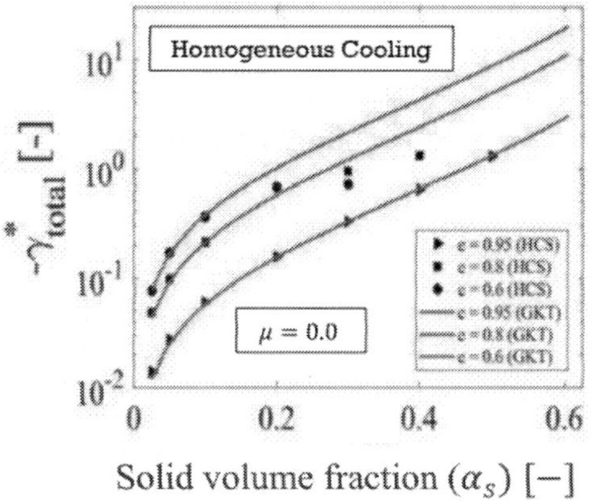

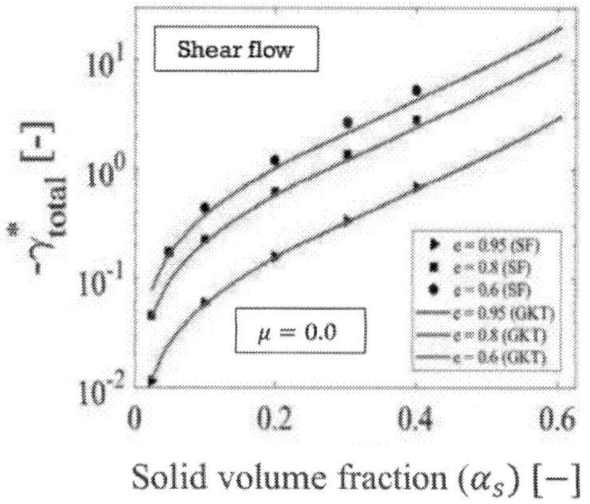

As α_s increases and e decreases, HCS simulations deviate from granular kinetic theory (GKT) predictions due to particle clustering

Herbert Wertheim College of Engineering

POWERING THE NEW ENGINEER TO TRANSFORM THE FUTURE

9

Shear flow alignment

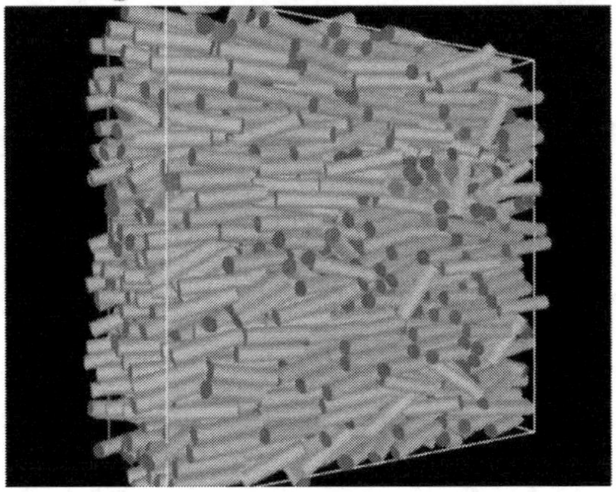

As α_s increases, elongated particles in SF simulations become aligned with the flow. We want to avoid this in our model development

Herbert Wertheim College of Engineering

POWERING THE NEW ENGINEER TO TRANSFORM THE FUTURE

10

Simulation design

- Considered true cylinder particles
  - Aspect ratio (AR) – **1, 2, 4, & 6**
  - Coefficient of restitution (e) – $[\mathbf{1.0, 0.6}]$
  - Coefficient of friction (μ) – **[0.01, 1.1]**
  - Solid volume fraction (α_s) – **0.05**
    - 100 particle simulations
- Radial distribution function (g_0) and α_s scale γ^*_{total}
  - Allows for focus on single α_s to represent all values of α_s

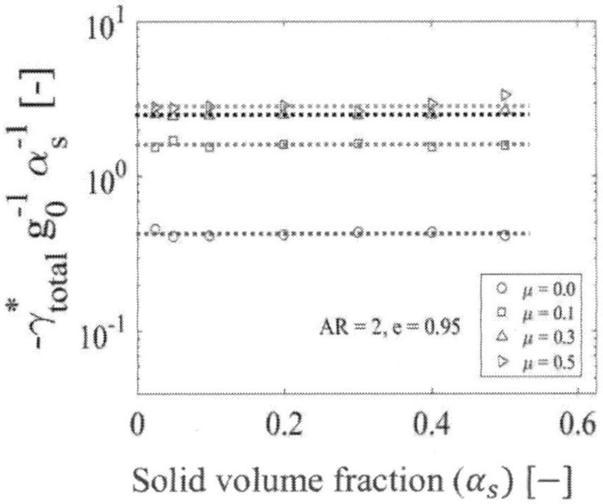

General HCS and SF results

Key points:

- As μ increases, $-\gamma^*_{total}$ increases
- γ^*_{total} plateaus around $\mu = 0.3$
- As e decreases, effect of μ diminishes
- Similar results seen for all AR

During these simulations, energy is lost due to collisions and friction

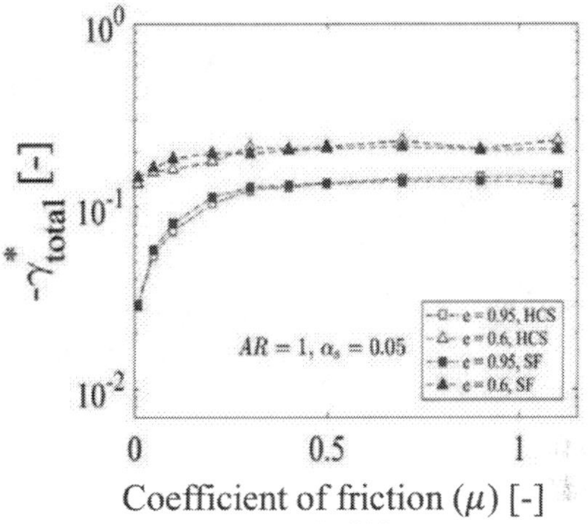

Modeling γ^*_{total}

- Previous work showed simulations of smooth cylinders where:
  - $\mu = 0.0$ & $e = xx$ ("Normal dissipation" dominated) [1]
- This work focuses on rough cylinders:
  - $\mu = xx$ & $e = 1.0$ ("Friction dissipation" dominated)
- Basic model for γ^* could be additive where the normal and friction dominated measurements are summed together
  - i.e. $\gamma^*_{total}(e, \mu) = \gamma^*_{normal}(e) + \gamma^*_{friction}(\mu)$
  - Already have model for γ^*_{normal} [1]

[1] Buettner, Kevin E., Yu Guo, and Jennifer S. Curtis. "Using the Discrete Element Method to Develop Collisional Dissipation Rate Models That Incorporate Particle Shape." *AIChE Journal*.

Will the additive approach work?

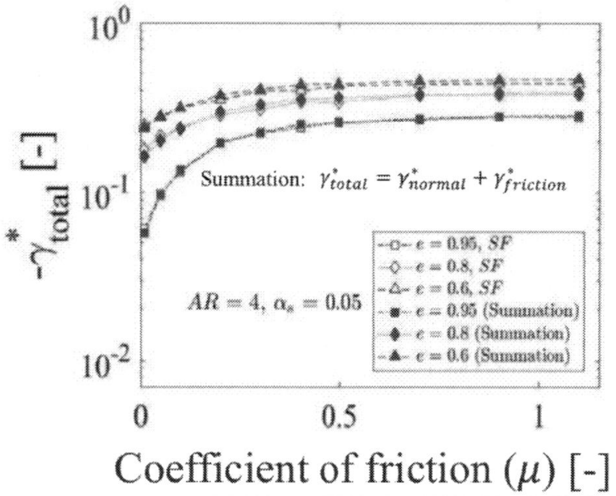

- Summation results (blue symbols) compared with shear flow results where simulanteously $e \neq 1.0$ and $\mu \neq 0.0$
- γ^*_{normal} values taken from previous true cylinder simulations where $\mu = 0.0$
- $\gamma^*_{friction}$ values taken from current set of simulations where $e = 1.0$

Key point:

- Agreement is good, can focus on model for when $e = 1.0$ and μ is varied

Model for $\gamma^*_{friction}$

- Accurate portrayal of γ^*_{total} vs. μ must include piecewise function
- Hence,

$$\gamma^*_{friction} = \frac{12\left(1 - e_{eff}^2\right)}{d_v} g_0 \alpha_s F_f(AR)$$

$$e_{eff} = \begin{cases} -0.59 \cdot \mu + 1, & \mu \leq 0.3 \\ 0.82, & \mu \geq 0.3 \end{cases}$$

$$F_f = 0.23 \cdot AR + 0.76$$

- Model is similar to spherical model, main changes are the inclusion of $e_{eff}(\mu)$ and $F_f(AR)$

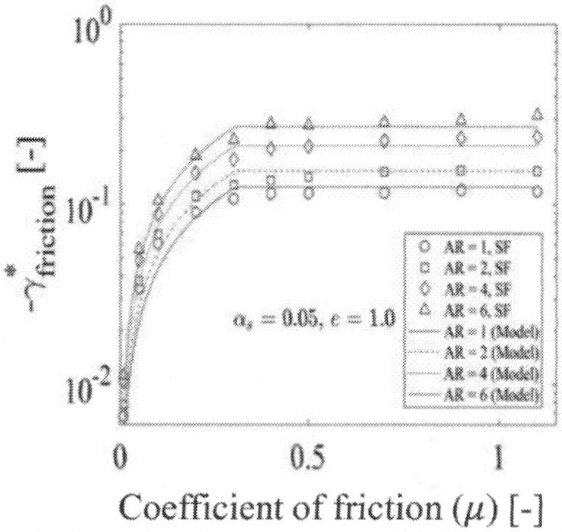

Combining $\gamma^*_{friction}$ and γ^*_{normal}

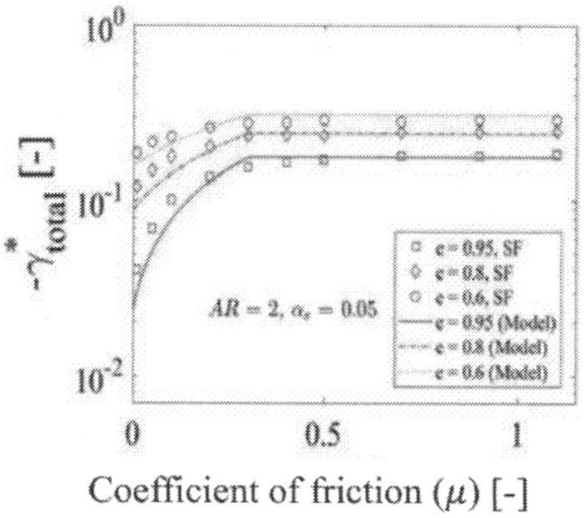

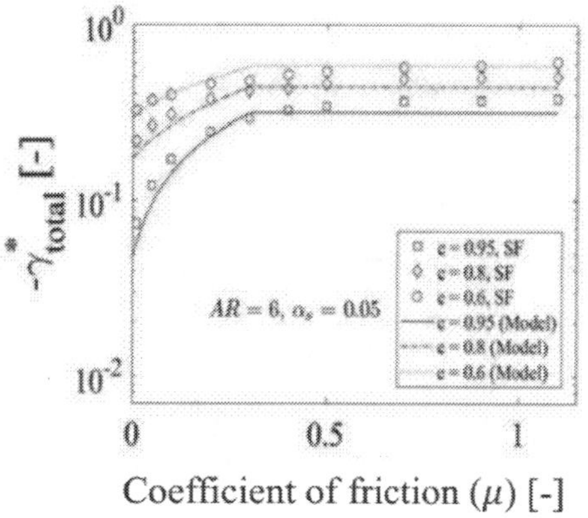

$$\gamma^*_{total}(e,\mu) = \gamma^*_{normal}(e) + \gamma^*_{friction}(\mu)$$

Implementation into two-fluid solver

- Granular temperature ratio ($R_\theta = \theta_{trans}/\theta_{rot}$) describes the equipartition between the translational and rotational DOF
- Previous study with smooth cylinders showed $R_\theta \approx 1$ for all conditions
  - θ_{rot} can be described in terms of θ_{trans}
  - Thus, there is no modification of the governing equations for smooth cylinders
- For frictional cylinders, R_θ varies as μ varies and is different for HCS and SF simulations

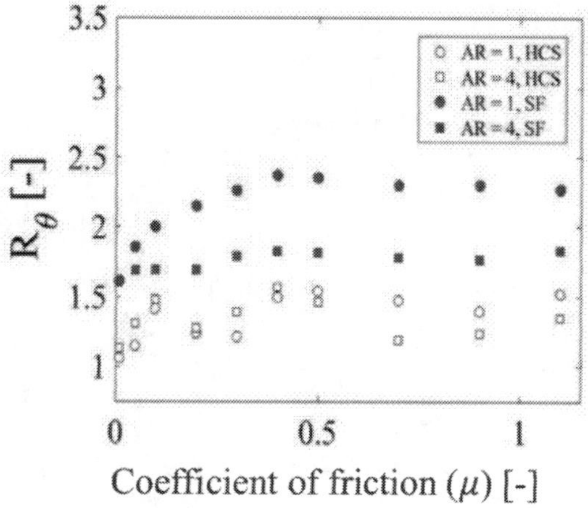

Future work

- Look into possible modifications of kinetic theory governing equations for implementation into two-fluid solver
- Investigate the effect of particle alignment on γ^*_{total} with friction
- Examine effect of particle flexibility on γ^*_{total}
- Perform validation studies to examine efficacy of non-spherical kinetic theory models

Questions?

Particle Properties
- Shape
- Wet/dry
- Smooth/rough
- Breakage
- Flexible/rigid

DEM Simulations
- Measure particle-particle interactions
- Vary properties

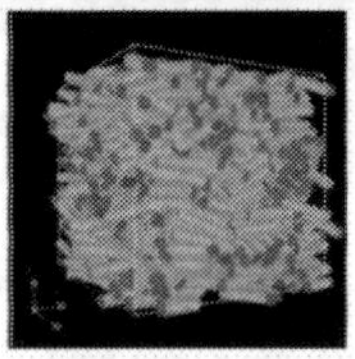

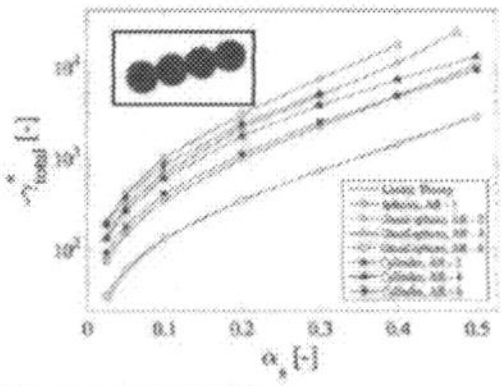

Functional Model
- Fitted equations
- Scaling relationships

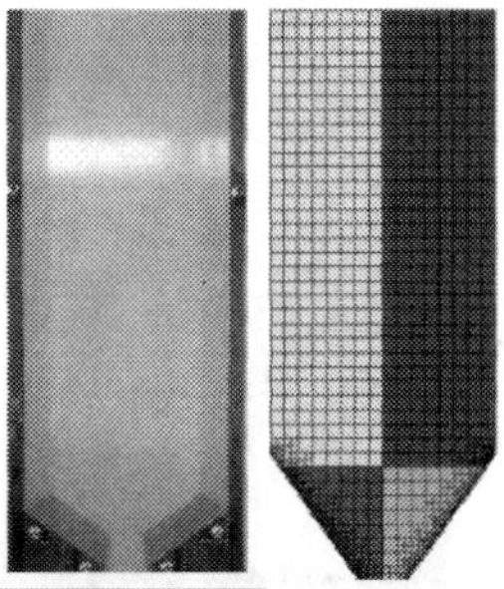

CFD Simulations
- Realistic domains
- Validation
- Prediction

UF Herbert Wertheim College of Engineering

POWERING THE NEW ENGINEER TO TRANSFORM THE FUTURE

Does α_s scaling hold up?

- Goal was to create model for beds of randomly oriented particles
- Hence, model is compared with results from HCS simulations which stay randomly oriented
- Agreement between HCS results and model for different values of μ is quite good for all AR and α_s

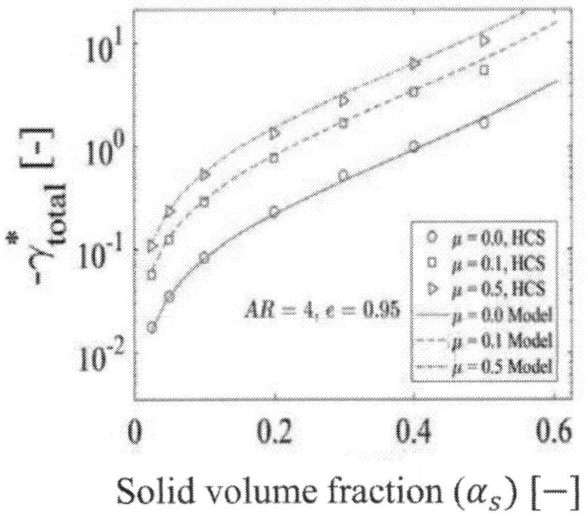

494166 An Integrated Workflow for Numerical Generation and Meshing of Packed Beds of Non-Spherical Particles

Abstract:

An integrated workflow for numerical generation and meshing of packed beds of non-spherical particles

Behnam Partopour, Anthony G. Dixon

Department of Chemical Engineering, Worcester Polytechnic Institute,

Worcester, MA, USA, 01609

Packed beds are widely used in different areas of the chemical industry such as reaction engineering and separation processes, and researchers from different disciplines such as chemical and mechanical engineering and material science are interested in studying them. Therefore, along with experimental methods, the numerical generation of these beds is extensively investigated. Unlike packed beds of spherical particles, the contacts between non-spherical particles could include lines and surfaces. Therefore, collision detection and subsequent meshing in those areas can computationally be challenging.

In this work we introduce the automated packed bed generator (PBG) package for spherical and non-spherical particles based on the Bullet physics library. The library includes a robust collision detection module which uses the impulse based collision detection method[1]. The package gets the user input for parameters such as shape, dimensions and number of particles, as well as friction and restitution factors, etc., and returns an STL file of the constructed geometry along with the bed properties (e.g. void fraction and angle distributions).

The constructed packed bed STL file then is imported to meshing software and a 3-dimensional mesh is generated based on the shrink-wrap method. The bed properties for packed beds of different shapes of particles are validated against existing experimental and computational data in the literature. The meshed geometries are imported to CFD software, and momentum and heat transfer simulations are carried out. Velocity and temperature profiles are validated against the experimental data, and previous CFD simulations, respectively. The package has shown to have two important advantages over similar tools such as the Discrete Element Method (DEM). First, the PBG simulation time is significantly lower than DEM tools (~1 hour for a packed bed of 1500 particles). Second, Bullet physics uses direct mesh representations of the original objects to detect the collisions and contact points while DEM tools use multi-spherical approximate shapes to detect the contact points and then replace them by the original representations of the particles (e.g. cylinders or rings). Several studies have raised concerns about the accuracy as well as computational cost of such methods[2, 3]. Finally, for the first time the generated and meshed geometries are successfully used for

resolved particle fixed bed CFD simulations.

References

1- Pytlos M, Gilbert M, Smith CC. Modelling granular soil behaviour using a physics engine. Géotechnique Letters 2015; 5:243–249

2- Marigo M, Stitt EH. Discrete element method (DEM) for industrial applications: comments on calibration and validation for the modelling of cylindrical pellets, KONA Powder & Particle J 2015; 32:236-252.

3- Guo Y, Wassgren C, Ketterhagen W, Hancock B, Curtis J. Some computational considerations associated with discrete element modeling of cylindrical particles. Powder Technol 2012; 228;193–198.

499835 Flow of Granular Materials in a Bladed Mixer: Effect of Particle and Process Parameters on Impeller Torque

Abstract:

Impeller torque is an important parameter that is measured to indicate the status of a system and to control the progress of processes such as mixing and granulation in the pharmaceutical industry. In this study, experiments were performed using monodisperse, spherical glass beads flowing in a cylindrical bladed mixer agitated by an impeller. Experimental measurements of the torque that the impeller blades exerted on the granular bed were made using a data acquisition and recording device. Computational simulations of spheres flowing in a four-bladed mixer were also carried out by means of the discrete element method (DEM). The effects of various material properties and impeller blade design configurations on the torque were investigated as a function of the impeller blade rotational speed. It was demonstrated that the time-averaged torque values exerted on the granular system were dependent on the particle size and friction coefficient, the fill height of the glass beads above the blades, and the moisture content in the system. It was also shown that average torque depended upon the number of impeller blades, their design configuration, and their position in the granular bed. It was also found that the experimentally measured impeller torque was largely different in the cases of shallow and deep bed granular systems. A scale-up relationship for the deep granular bed was also developed. It was proposed that time-averaged torque in deep bed granular systems scaled with the square of the normalized height of the material above the blades.

497687 Study of Spontaneous Structure Formation in Granular Systems Using DEM Framework

Abstract:

Spontaneous cluster and pattern formation can happen in granular systems with multiple particles. Such phenomena have been observed in nature as well as in experiments. Studying these phenomena can help us understand the physics of these systems and build and validate models that can then be used in process equipment design. In this work, we investigate the spontaneous organization of granular particles in a cylindrical vessel by using discrete element method (DEM) simulations. Two different types of motion of the cylindrical vessel has been studied. In the first type of motion, the cylindrical vessel follows the horizontal circular motion, i.e., the distance between any point in the cylinder and the center of motion is fixed during the entire process. In the second type of motion, the cylindrical vessel follows the motion of an orbital shaker without self-rotation. This type of swirling motion is generated by superposing two sinusoidal vibrations in two perpendicular directions respectively. When the cylinder vessel contains a group of granular particles follows the first type of motion, the particles form a steady, symmetric structure on the side wall of the cylindrical vessel. However, if the cylindrical vessel follows the second type of motion, the particles form an asymmetric structure and becomes dynamic in nature. We examine the results of our simulations and compare to existing published experimental data to gain a better understanding of this phenomenon.

500496 Assessment of the Effect of Dairy Powder Properties on Lean-Phase Pneumatic Transport: A CFD-DEM Modelling Approach

Abstract:

The design of pneumatic transport for dairy powders depends on multiple parameters: several of them are associated with powder characteristics - such as powder type (e.g., fat-filled milk powders or skim milk powders), particle size distribution, bulk density, flowability. et cetera - with operating variables - such as solids loading, operating pressure, velocity, and pressure drop - and with the lay-out of the transport line. Essentially, the dynamics of the pneumatic conveying system is dominated by the dynamics of the fluid-particle interaction, the particle-particle interactions and the particle-wall interactions which all are affected by the above system parameters.

In this study, a multi-scale CFD-DEM model was developed and validated with data from a 50 mm diameter pneumatic transport rig based on conveying of cohesive fat-filled milk powder fines. The impact of conveying parameters and the properties of the powder fines on pressure drop and blockage boundary in the rig was analyzed. Pressure drop and local particle volume fractios were measured with the aid of a differential pressure gauge and an optical fibre probe respectively.

The multi-scale model is achieved by dynamically coupling computational fluid dynamics (CFD) for modelling the fluid flow field, and discrete element method (DEM) for modelling the behaviour of the powder particles, referred to as CFD-DEM. The validated CFD-DEM model shows a reasonable accuracy in predicting the system pressure drop and particle volume fraction. The validated model can be used to solve pneumatic transport problems associated with highly cohesive dairy powders.

497968 Advances in Voidage Reconstruction Schemes for the Simulation of Dense Gas-Particle Flows

Abstract:

Euler-Lagrange (EL) models of dense fluid-particle suspensions are often treated as the "gold standard" for simulating flow in these suspensions. When using so-called particle-unresolved (PU) EL models, one relies on (i) tracking every particle and (ii) using spatially-averaged transport equations to describe fluid flow. In the context of such PU-EL models of dense particulate systems it is necessary to accurately reconstruct the local voidage at the particle position. This is since the voidage is of central importance for estimating exchange coefficients, e.g., for the drag force or the rate of interphase heat transfer. Preliminary work showed that standard approaches to reconstruct the voidage at the particle position often lead to results that are highly grid sensitive. This is the case even though we use advanced smoothing, mapping and interpolation schemes.[1] Especially systems involving polydisperse granular materials in which the particle volume fraction can surpass 0.8, i.e., an extremely stiff particle-fluid coupling is present, are problematic.

Similar as Li et al.[2] we start our study with a naïve analysis of the total force acting on a fixed bed of particles bound by regions void of particles. In these situations fluid flow is aligned with the voidage gradient. As expected, upon coarsening we find that the grid both force and heat transfer rates are significantly underpredicted. Thus, a positive drag correction is required for these systems due to the latent underprediction of voidage gradients when using a finite grid size. Surprisingly, this underprediction is more extreme for PU-EL-based simulations compared to predictions by an Euler-Euler approach. Consequently, we work towards an improved voidage reconstruction algorithm for PU-EL that compensates this underprediction, both for systems with finite and infinitely large voidage gradients. In order to generalize our ideas, we combine the idea of an angle-dependent correction to the drag coefficient[2] with the newly proposed voidage reconstruction scheme. By considering the closure for the modified drag coefficient proposed by Radl and Sundaresan,[3] we then benchmark the combined approach for a variety of systems including fixed and fluidized beds.

Finally we put our developments into perspective by a comparison with the recent study of Li et al.[2] (which considered Euler-Euler models), as well as the progressive analysis provided by Schneiderbauer.[4] While the latter study already laid the foundation for a positive drag correction, it relied on an oversimplified picture for the correlation of voidage and gas-phase velocity fluctuations. The study of Li et al.[2] relied on pre-tabulated correction factors, which lacks of generality and is valid for Euler-Euler-based simulations only. We conclude our study by documenting key advantages of our advanced voidage reconstruction algorithm, which appears to most significantly improve predictions when studying polydisperse fluid-particle suspensions.

References

1. Radl, S., Gonzalez, B., Goniva, C. & Pirker, S. State of the Art in Mapping Schemes for Dilute and Dense Euler-Lagrange Simulations. *10th Int. Conf. CFD Oil Gas, Metall. Process Ind.* 1–9 (2014).

2. Li, T., Wang, L., Rogers, W., Zhou, G. & Ge, W. An Approach for Drag Correction Based on the Local Heterogeneity for Gas–Solid Flows. *AIChE J.* **63,** 1203–1212 (2017).

3. Radl, S. & Sundaresan, S. A drag model for filtered Euler-Lagrange simulations of clustered gas-particle suspensions. *Chem. Eng. Sci.* **117,** 416–425 (2014).

4. Schneiderbauer, S. A Spatially-Averaged Two-Fluid Model for Dense Large-Scale Gas-Solid Flows. *AIChE J.* **in press,** 1–19 (2017).

Hydrodynamics of Sound Assisted Fluidization of Rigid-microsized Powder

Cherntongchai, P.*

Department of Industrial Chemistry, Faculty of Science, Chiang Mai University, Chiang Mai, Thailand

Email: Parimanan@gmail.com

Outline:

- **Introduction**
- **Objectives**
- **Research novelty**
- **Theory**
- **Experiments**
- **Results and discussion**
- **Conclusions**
- **Acknowledgment**
- **Reference**

Introduction: Sound assisted fluidization

modification of conventional fluidization by sound vibrational force

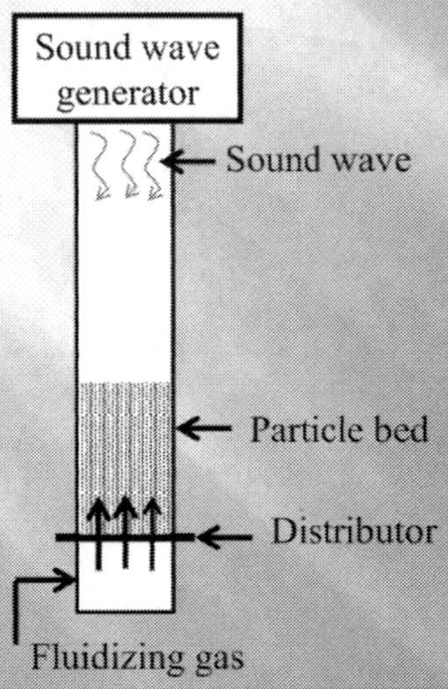

Effects:
- Better fluidization quality
- Smaller agglomerate size
- Particle elutriation reduction
- Hydrodynamics behaviors

Theory:
- Standing wave characteristics
- Force balance principle

Objectives:

- To investigate and explain influence of sound frequency on minimum fluidization velocity of Geldart's group A powder, using standing wave characteristics.

Research novelty:

- Rationale for the influence of sound frequency on fluidization hydrodynamics of Geldart's group A powder, using tanding wave characteristics.

Theory: Standing wave characteristics [1]

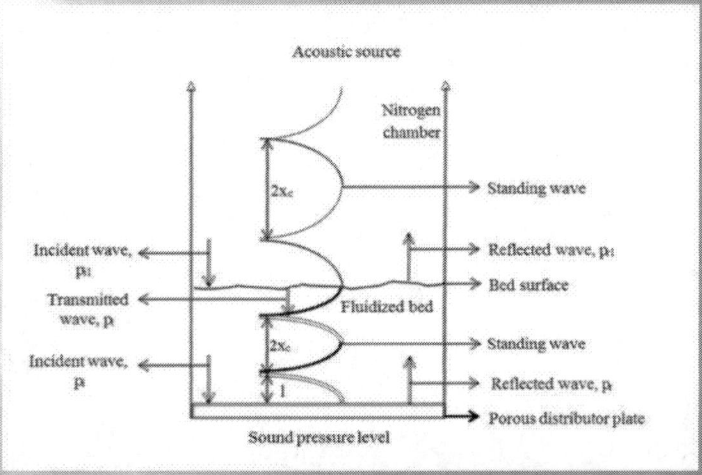

Figure 1 Standing wave pattern in fluidization media

Theory: Standing wave characteristics [1]

- Sound propagation was an adiabatic one-dimensional process and a wave equation in terms of pressure is;

$$\frac{\partial^2 p}{\partial x^2} = \frac{1}{\varepsilon^2}\frac{\partial^2 p}{\partial t^2} \qquad (1)$$

- A solution of Eq. (1) is;

$$p(x,t) = Ae^{-j(\omega t - kx)} \qquad (2)$$

- A Sum of incident and reflected waves within the fluidized bed in form of total pressure.

$$p(x,t) = Ae^{-j(\omega t - kx)} + Be^{-j(\omega t - kx + \theta)} \qquad (3)$$

- Pressure distribution becomes;

$$|p(x,t)| = P_{fs}\sqrt{\frac{\cos^2 kx}{\cos^2 kh}} \qquad (4)$$

Theory: Standing wave characteristics [1]

- At critical state, $|p(x,t)|$ was very large which occurred at;

$$kh = (2n-1)\frac{\pi}{2}, \text{for } n = 1,2... \tag{5}$$

- Value of $|p(x,t)|$ can be interchanged using Eqs. (6) and (7).

$$P_{rms} = \frac{|p(x,t)|}{\sqrt{2}} = \frac{P}{\sqrt{2}} \tag{6}$$

$$SPL = -20\log\left[\frac{p_{rms}}{p_{ref}}\right] \tag{7}$$

Experiments: Experimental setup

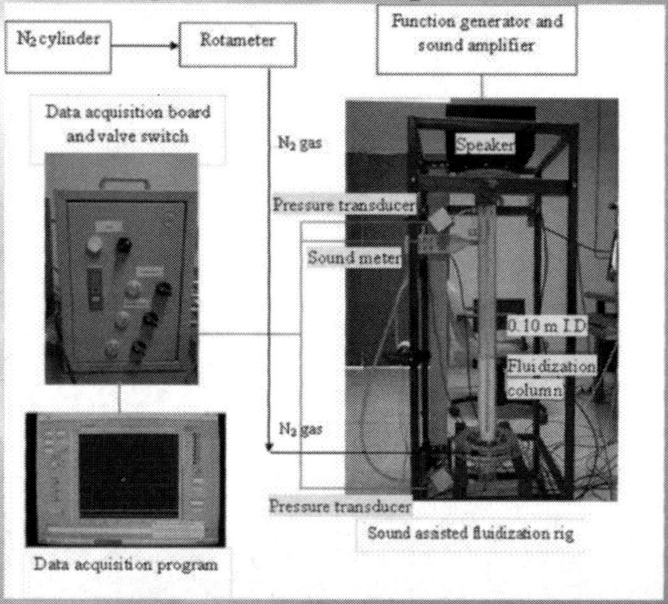

Figure 2 Sound assisted fluidization setup for 0.10 m ID fluidization column

Experiments: Materials

Table 1 Particle density and average particle size

Material	Apparent density (kg/m^3)	Average particle diameter (mm)
Glass ballotini	2,432	32.61

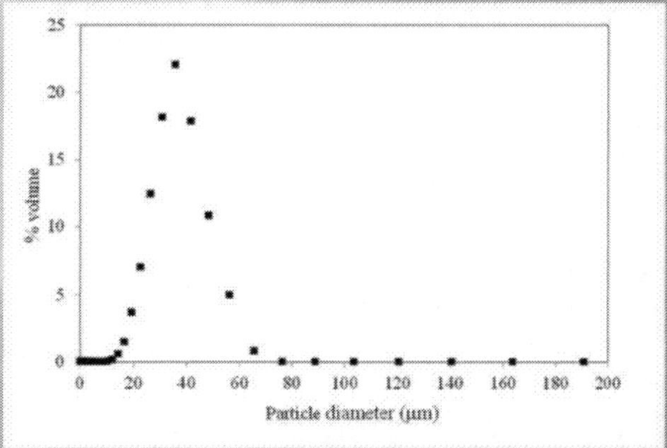

Figure 3 Particle size distribution of glass ballotini

Experiments: Methodology

- Fluidization experiment was performed by increasing inlet superficial gas velocity and measuring bed pressure drop at each gas velocity.
- Defluidization experiment was, on the other hand, by decreasing inlet superficial gas velocity.

Experiments: Experimental conditions

Table 2. Summary of experimental conditions

Experimental conditions	Inlet superficial velocity (mm/s)	Sound pressure level (dB)	Sound frequency (hz)
	0-15	0	0
	0-15	80	50,80,100,300,500

Results and discussion: Total bed pressure drop

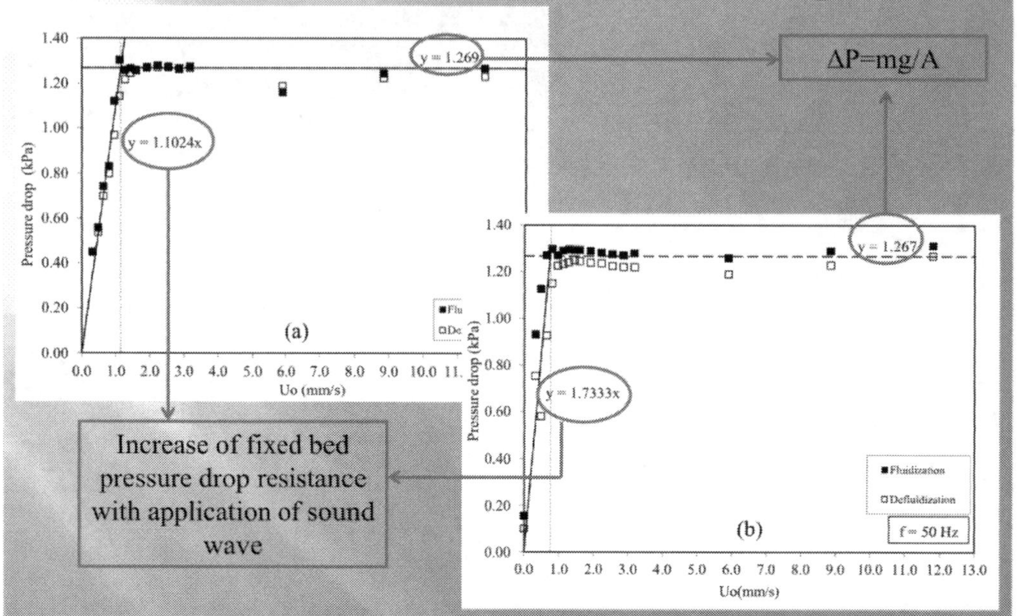

Figure 4 Total bed pressure drop versus inlet superficial velocity for (a) conventional fluidization and (b) sound assisted fluidization at sound pressure level = 80 dB and sound frequency = 50 Hz

Results and discussion: Fixed bed pressure drop

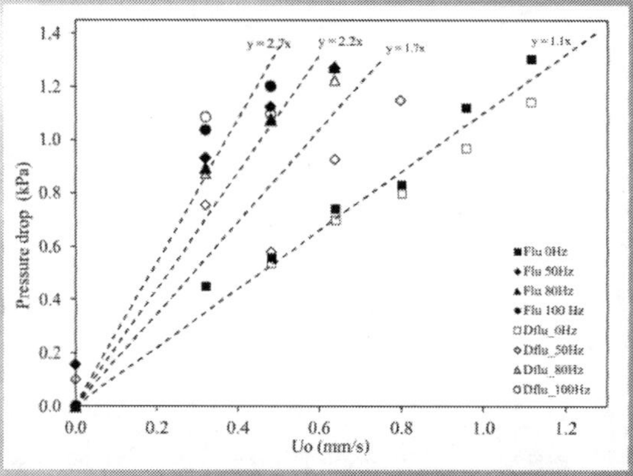

Figure 5 Fixed bed pressure drops versus inlet superficial velocity of conventional and sound assisted fluidized bed from fluidization and defluidization experiments

Increase of fixed bed pressure drop resistance with increasing sound wave frequency

Results and discussion: Fixed bed pressure drop

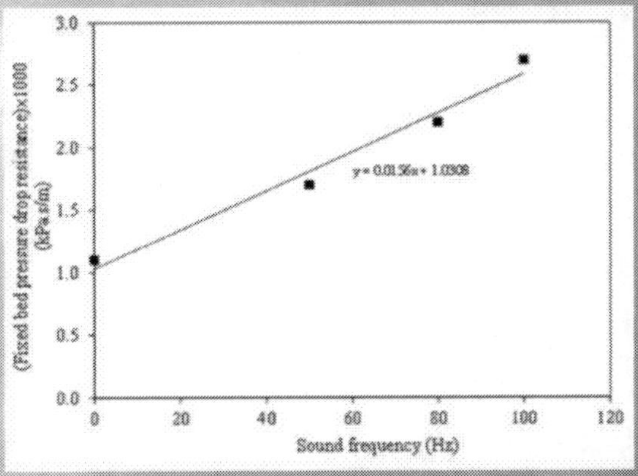

Figure 6 Relation between sound frequency and fixed bed pressure drop resistance

Linear increase of fixed bed pressure drop resistance with sound wave frequency

Results and discussion: Incipient and complete fluidization

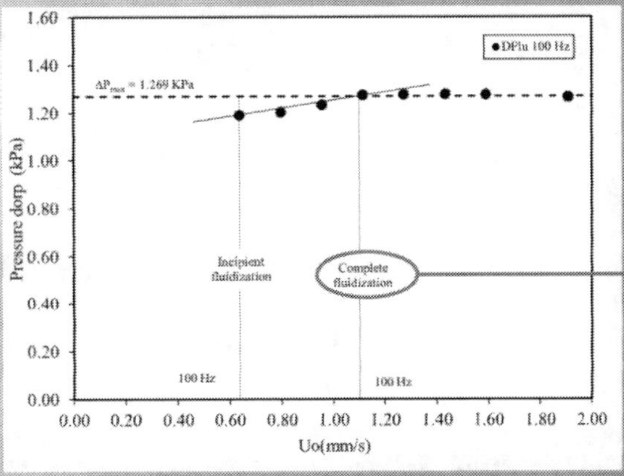

Figure 7 Incipient fluidization point and complete fluidization point

Minimum fluidization velocity

Range from incipient and complete fluidization points for transition from fixed bed to fluidized bed

Results and discussion: Minimum fluidization point

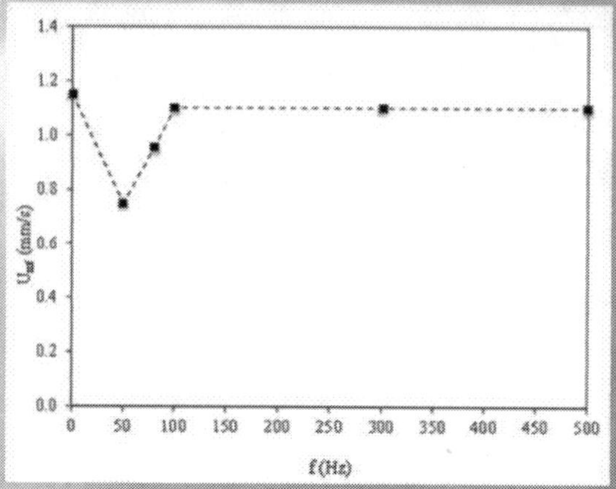

Figure 8 Relation between sound frequency and fixed bed pressure drop resistance

- With application of sound, particle bed fluidized at lower inlet superficial velocity.
- Minimum value of minimum fluidization velocity found at sound frequency of 50 Hz.

Results and discussion: Standing wave characteristics

- x= h or at bed surface, SPL =80 dB.

- With increasing distance down to distributor, SPL increased and maximum SPL was at x=0. .

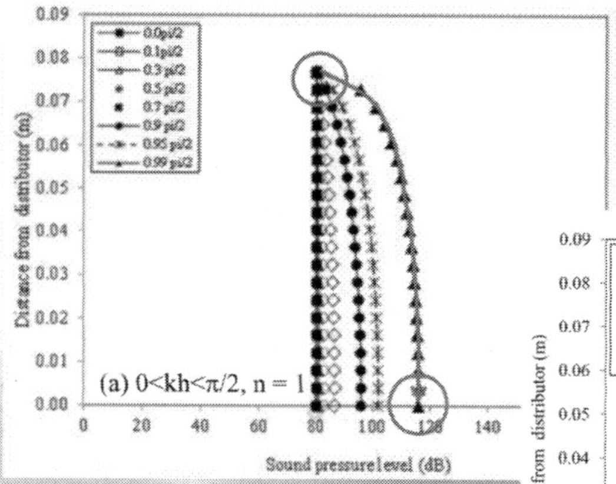

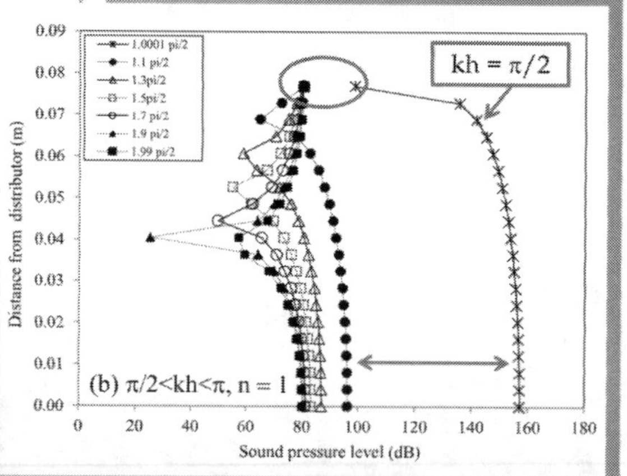

- Rapid increase in SPL when kh approached π/2.
- Maximum SPL found at kh = π/2.

Figure 9 Variation of sound pressure level along bed height (0<x<h) for n = 1

Results and discussion: Standing wave characteristics

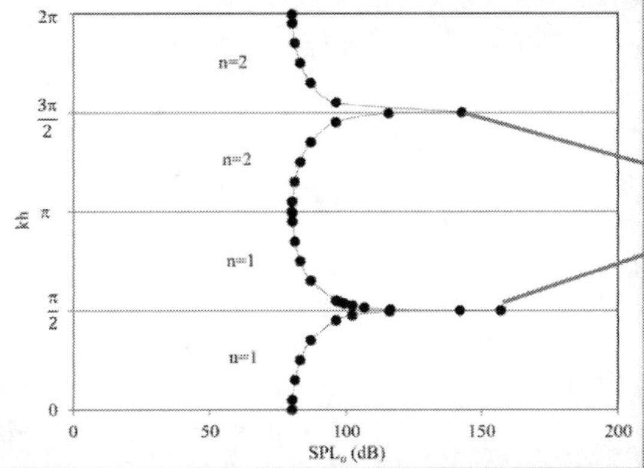

Critical kh value for each harmonic number giving high magnitude of SPL_o

Figure 10 Variation of SPL_o with kh for 0<kh<π, n = 1 and for π<kh<π/2, n = 2

- Exponential increase and decrease of SPL_o with kh for kh of less than critical value and that of higher than critical value, respectively.
- Explanation of local minimum of minimum fluidization velocity at certain sound frequency.
- Possibility of several local minima of minimum fluidization velocity, at critical kh values.

Results and discussion: Minimum fluidization velocity in relation to standing wave characteristic

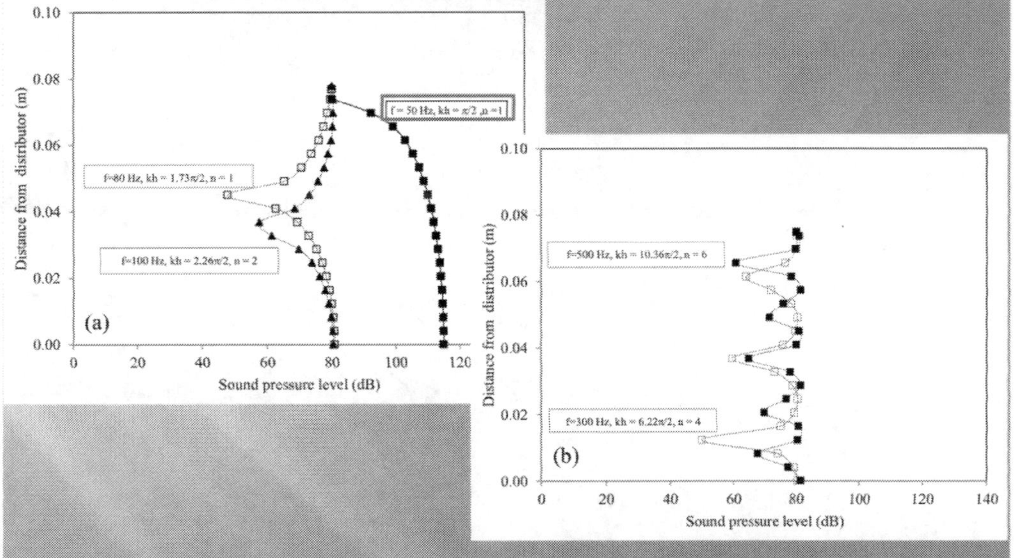

Figure 11 Sound pressure level along distance for fixed bed from distributor at different sound frequency (a) f = 50, 80, and 100 Hz and (b) f =300 and 500 Hz

Results and discussion: Minimum fluidization velocity in relation to standing wave characteristic

Table 3 Summary of standing wave properties and corresponding minimum fluidization velocity

Sound frequency (hz)	kh	Harmonic number (n)	Average SPL (dB)	U_{mf} (mm/s)
50	π/2	1	108.15	0.750
80	1.73p/2	1	74.52	0.956
100	2.26p/2	2	76.02	1.100
300	6.22p/2	4	75.13	1.100
500	10.36p/2	6	76.59	1.100

- At sound frequency of 50 Hz;
  - Highest SPL magnitude along distance x of standing wave, as well as theirs average values, causing minimum fluidization velocity to be lowest.
- At other sound frequency:
  - Approximately the same average value of SPL and, hence, minimum fluidization velocity.

Conclusions:

Standing wave characteristics was implemented to explain hydrodynamics behaviors of a sound-assisted fluidization of a rigid-microsized powder under different inlet superficial velocity and sound frequency ranging from 0-500 Hz at a fixed sound pressure level of 80 dB.

- Due to exponential and periodic variation of vibrational strength of standing wave with sound properties, minimum fluidization velocity was, then, varied in the same manner.

- Local minima of minimum fluidization velocities were, thus, multiple values rather than a single value.

Acknowledgment:

Thailand Research Fund and Chiang Mai University.

Reference:

[1] Herrera C.A. and Levy E.K., Characteristics of acoustic standing waves in fluidization beds. American Institute of Chemical Engineers. 2002; 48: 503-513.

BEHAVIOR OF FINE PARTICLES IN A COLD PLASMA-ENHANCED SPOUTED BED

Bai-qiang Zhang, Nobusuke Kobayashi and Yoshinori Itaya*
Gifu University, Gifu, Japan

Abstract

Non-thermal plasma has been used widely in the field of chemical industry processes, such as mixing, coating, granulation, and pollutant removal processes. This plasma technique applied to even fine particle, and plasma process with using fluidized bed are also proposed and demonstrated in recent years. However, behavior of particles in a fluidized bed with irradiation of plasma in detail is still unknown. In this paper, fluidization experiment of fine particles under the irradiation of non-thermal plasma was carried out in a spouted bed, and the effect of the plasma irradiation on the particle behavior was evaluated. In the experiments, initial conditions, such as initial static bed height (H), gas flow rate (u), plasma voltage, and plasma working gas, were changed and the pressure drop, bed void fraction of solid and bed temperature with plasma were measured in order to understand the fluidizing behavior of fine particles. As the results, minimum spouted velocity (u_{ms}) decreased as the plasma voltage increased regardless of initial static bed height, while the void fraction of solid bed increased with increasing plasma voltage. Fluidizing behavior varied with the working gas, and u_{ms} with N_2 was much higher than that of Ar. Temperature of the bed and gas volume did not show obvious changes when the plasma was applied.

Introduction

Non-thermal plasma process is representational plasma process. The main advantages are that electron temperature can be above 10^4 K, however, neutral gas molecules remain nearly cold. Hence, it can initiate the formation of other various radicals with the temperature increased by a few degrees. Decades, non-thermal plasma technology has aroused considerable attention in the field of chemical industry processes, such as mixing, coating, granulation, and pollutant removal processes.

Recently, lots of researchers applied this plasma process even with fine particles or with using in the fluidized bed. Karches et.al[1] have studied the coating process in a microwave plasma-enhanced circulating fluidized bed. Results shown that temperature of process can be easily controlled to vary low values with similar electromagnetic properties. Also, Pajkic et.al[2] developed a fluidized bed chemical vapor deposition (CVD) process depositing thin AlN coatings on particulate substrates. They also figured out this process could overcome limitations posed by vacuum technology, such as high capital and running costs of the equipment. Tsukada et.al[3] investigated the granulation of spherical alloy grains from metal powder mixtures in a DC plasma-spouted/fluidized bed and grained spherical alloy with 1-5 mm in diameter within several minutes. Besides, fine catalysts particles with plasma irradiation play on the significant role in the application of environment filed, especially in the decomposition processes of NOx, SOx, and VOCs Zhu et.al[4] figured out the removal efficiency enhanced by the combination of plasma with $MOx/\gamma\text{-}Al_2O_3$ catalysts, while the formation of organic by-products reduced.

Spouted bed, one of the typical fluidized beds, is predominantly applied for solid particle systems not only very small and light (fine particles) but also very big particles which are difficult to fluidize with a stable manner. The spouted bed is also well used in chemical and environmental processes with the characteristics of the effective mixing of the solid phase. However, few studies of the behavior for fine particles fluidized in the plasma-enhanced spouted bed have been reported, which is the dominate factor effected on the future study.

In this paper, the spouted bed reactor combined with plasma generator in the low frequency (20 kHz) AC electric source was used for the experiment and the effect of plasma condition on the fluidized behavior was evaluated. White fused alumina particle (WA particle), was used as the fluidized agent. Effects of particle volume, plasma voltage and working gas on behavior of fine particles were investigated in the experiments. Pressure drop, void fraction of solid bed and bed temperature with plasma were measured in order to understand the fluidizing behavior of fine particles.

Experimental

A lab-scale spouted bed with plasma enhanced was carried out to study the behavior of fine particles fluidization process (as Fig.1 shows). There are two parts of the system in this process that are spouted bed system and plasma power supply system. The spouted system is made of a 3-dimensional spouted bed (250 mm height and 36 mm internal diameter) with an inlet angle of 60° (12.6 mm internal diameter). The plasma system contains plasma jet and plasma power supplement. Plasma jet is a quartz tube (6 mm internal diameter and 100mm height), which locates at the bottom of spouted bed. There is a distributor with evenly holes distributed (0.5 mm diameter) in the top of the plasma jet tube for keeping the fine particles stable. Additionally, high frequency power is provided by plasma power supplement with the regulation of the voltage and intensity in the range of 0–10 kV. The anode is used by copper rod (1.5 mm diameter) inserted inner quartz tube, and cathode used with a sheet of copper (20 mm length) covered upon the outer quartz tube. The average diameter of WA particles is 58.6 μm and the density is 1692 kg/m^3. Prior to do experiments, a little amount of plasma working gas (Ar or N_2) is introduced from the jet into the spouted bed at atmospheric pressure, for supporting the WA particles deposited on the distributor of the spouted bed. A differential pressure gauge measures gas phase pressure fluctuation. Temperature of the spouted bed is measured by Infrared Camera (FLIR T440). The volume of gas is collected by gas collecting bags and detected by Dry Text Gas Meter. Besides, the laser sensor (IL-Intelligent) is used for measuring the surface movement for fine particles.

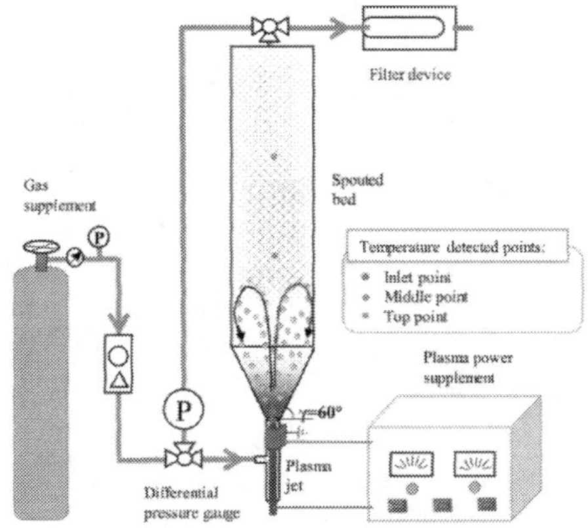

Fig. 1 Schematic of plasma enhanced spouted-bed reactor

Results and discussion

Behavior of fine particles in spouted bed

In Fig. 2, the dependencies of the bed pressure drop (ΔP) as a function of the gas flow rate (u) versus various bed height (H) are depicted. A similar general trend occurred in these curves. There were four phases for each static bed height. From point A to point B, ΔP was increased with the flow rate enhancement at the flow rate lower than minimum fluidization velocity (u_{mf}). Solid particles did not move in this period, still be as the fixed bed situation. The hollow space and fountain had appeared inside of the solid bed by increasing the u continually. ΔP kept increased until the gas broken out the solid bed. From point B to point C, when the flow rate exceeded u_{mf}, the inner fountain broken through the particles bed, which led to ΔP was declined sharply. C point referred to as initial point of fluidization, u_{mf}, and solid bed was breakthrough after C point. When the u increased to D point, referred to as initial spouting point[5], u_{ms}, WA particles spouted strongly above the fountain, which resulted in a low ΔP. After point D, the fountain height increased without changes of ΔP. u_{ms} is an important evaluation index to value the behavior of fine particles, which is related to the column diameter, bed depth and physical properties of fine particles and gas, investigated by Mathur and Gishler[6].

Fig. 2 illustrated that u_{ms} was increased with the enlargement of static bed height. Higher static bed height caused higher pressure to impede the gas break through the solid phase. Besides, there was little difference between static bed height upon and below 0.03 m static bed height, which is the height of the inlet slope part. The stage from C to D is shorter in height of 0.04 m and 0.05 m than below 0.03 m. It means that the translation time from fluidization to spouted phase is shorter than the low static bed height situation. This difference may be related to the initial ΔP of the solid bed. At the low bed height, due to the low initial ΔP, it is easy to form a tunnel in the central of the solid bed, which means the time for the stage from A to C is short. However, because of the big cohesive force for fine particles, it is difficult for fine particles to move around the boundary of the tunnel. Hence, it spends a long time from fluidization point C to spouted point D. On the contrary, at the high bed height, it spends a long time for the hollow space and fountain formation inside of the bed. The inside fountain formation results in the movement of the particles inside of the solid bed. The interaction force is becoming small. Therefore, it is easy to be spouted after breakout point.

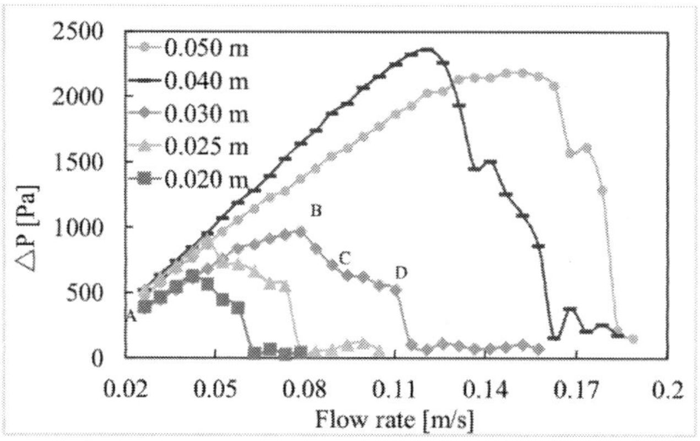

Fig. 2 WA particles pressure drop vs. flow rate in spouted bed in Ar gas

Effect of plasma on the behavior of fine particles in spouted bed

In order to study the effect of plasma irradiation on the behavior of fine particles in spouted bed, we changed various plasma voltage from 0 kV to 7 kV with the constant frequency of 20 kHz. Fig.3

compares the pressure drop versus flow rate with different plasma voltages with the bed height of 0.02 m. There were no significant changes between absence of plasma and low plasma voltage. When the plasma voltage moved to 7 kV, the u_{ms} decreased a lot. Besides, it had been found that the stage from point C to point D was shorter with plasma enhanced than without plasma enhancement. Under low plasma voltage, the energy is not enough for gas ionizing, so that there is not obviously changes between without plasma and low plasma voltage supplied. Based on the analysis of the last section, the reason for the decline of the u_{ms} may be related to the interaction between particles and particle.

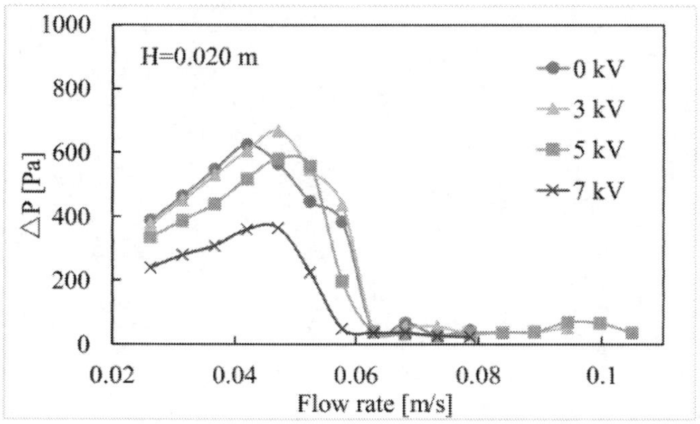

Fig. 3 WA particles pressure drop vs. flow rate with various plasma voltage in Ar gas

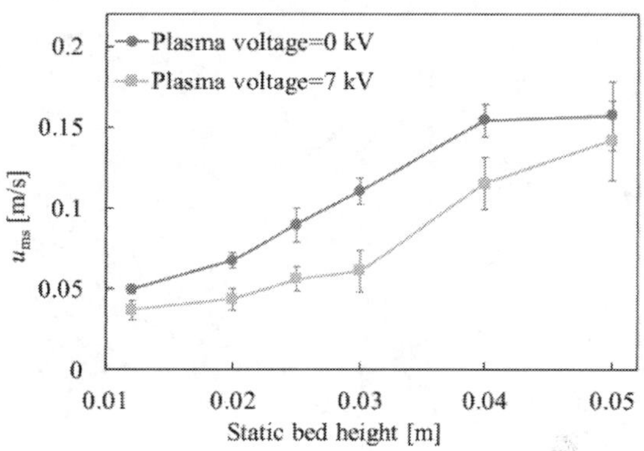

Fig.4 Minimum spouted velocity of WA particles various static bed height with the absence and presence of plasma

Fig.4 reveals that the comparison results of u_{ms} changed with static bed height with the absence and presence of plasma conditions. As the presence of plasma, the u_{ms} increased with increasing the static bed height. It was consistent with the trend of absent case. The u_{ms} decreased with plasma enhancement regardless of initial particles height. There are few reports about the u_{ms} changes in spouted bed, but lots of reports for fluidized bed. In Chevrier's[7] report, there were no apparent changes between the microwave plasma absence and presence in fluidized bed. In this study, it is obviously found that the u_{ms} decreased by the plasma irradiation of gas which is supported by Sathiyamoorthy's research. Sathiyamoorthy[8] figured out u_{mf} would be less in plasma-enhanced fluidized bed compared with without plasma added. Based on the equation of Mathur and Gishler[6], without plasma enhancement, ums related on the gas density in a same reactor. It means that the density of gas maybe changed during the plasma irradiation process.

Images of behavior of fine particles in plasma enhanced spouted bed

Plasma light emission is a well-known phenomenon in the process of plasma system. Table 1 displays several photos under various plasma voltages. The first row is the top view of spouted bed, and the second row is the left view of spouted bed. When plasma applied, two types of photos were taken with lamp and without lamp. These photos were taken under same n value, which is equal to flow rate divided minimum spouted velocity, under various plasma voltage. There is a difference especially compared with the photos without lamp. With the increment of plasma voltage, the color of the plasma gas is becoming bright. It was due to high plasma voltage provided high energy for gas, which enhanced the plasma ionization degree.

Table. 1 Images of behavior of fine particles in plasma enhanced spouted bed (n=4)

	With lamp	With lamp	Without lamp	With lamp	Without lamp	With lamp	Without lamp
Top view	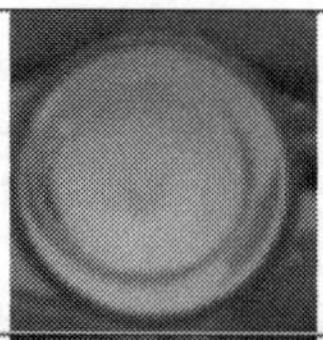	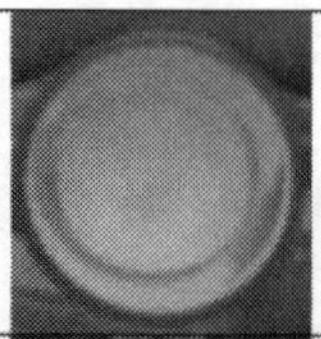		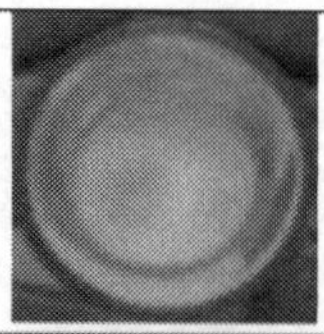	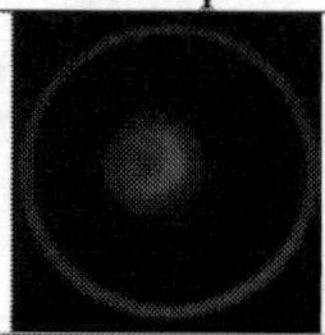	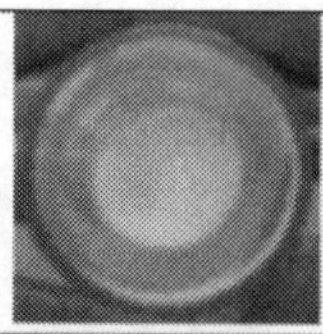	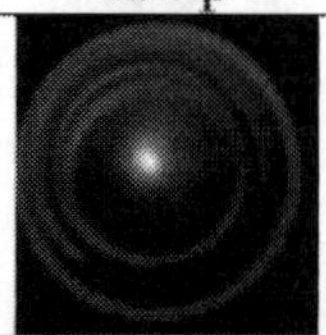
Left view	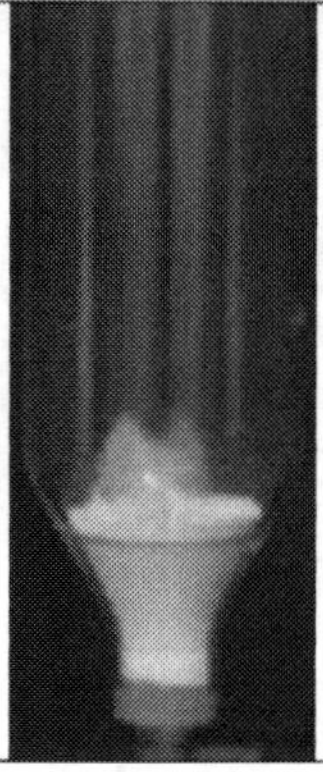	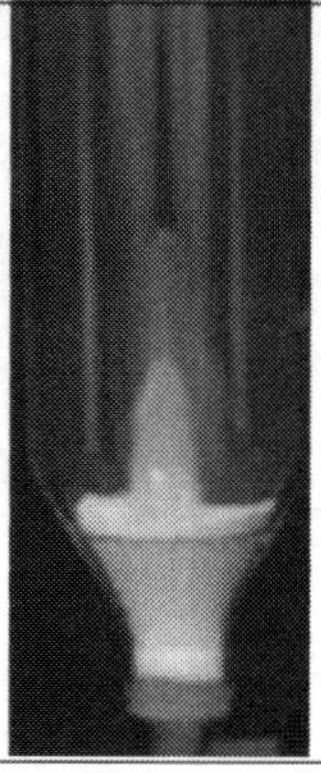		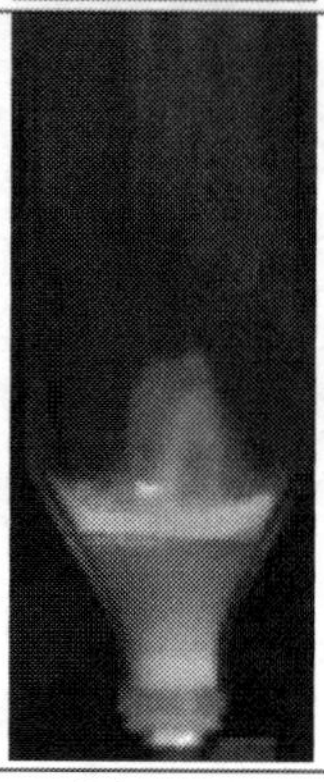		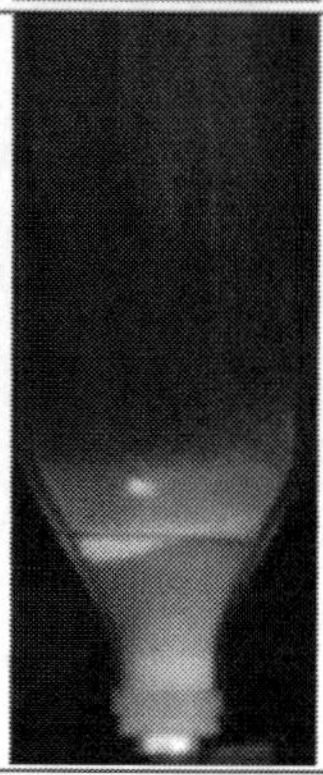	
	Without Plasma	Plasma voltage=3kV		Plasma voltage=5kV		Plasma voltage=7kV	

Effect of plasma working gas on the behavior of fine particles in spouted bed

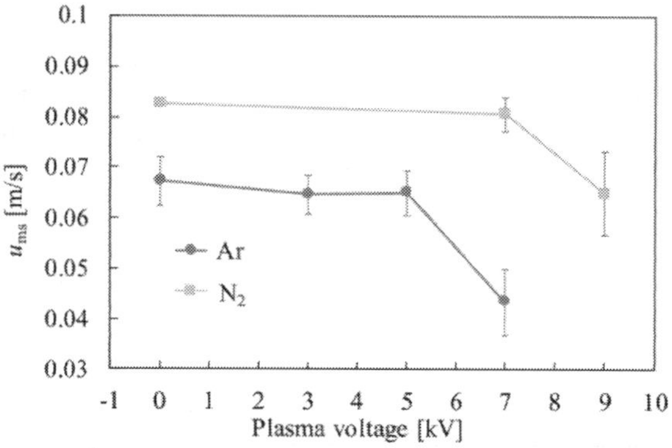

Fig. 6 Pressure drop vs. flow rate with various plasma voltage with Ar and N_2 gas

Two plasma working gases, N_2 and Ar, were investigated in the process of fine particles behavior in spouted bed. The relationships between u_{ms} and plasma voltage with Ar and N_2 gas were illustrated in Fig. 6. A certain value of plasma voltage was supplied before the experiment with a small amount of gas. Gas flow rate increased slowly to figure out the ΔP sharply down point. Due to the difference of density, the u_{ms} with N_2 was higher than that of Ar in the absence of plasma. Chevrier's[7] adapted two kinds of particles (alumina and silicon) fluidized in Ar and H_2 plasma system. Results shown that the u_{mf} was decreased for alumina particles with Ar plasmatic condition which was in accord with our results. But beyond that, u_{mf} were identical for silicon particles in Ar and H_2 plasma system, especially in the H_2 atmosphere. u_{ms} declined associated with increment of plasma voltage. Because of the difference of gas breakdown voltage, at 7kV plasma voltage, the decreasing amplitude of Ar is much larger than that of N_2. By visual observations, it can be indicated that stable N_2 plasma was more difficult to generate in the spouted bed than Ar in our study. Judging from Chevrier's report, this may be resulted by the difference of (e_r) value for N_2 (470.6×10^{-6}) and Ar (517.2×10^{-6}).

Effect of plasma voltage on plasma voltage on pressure drop

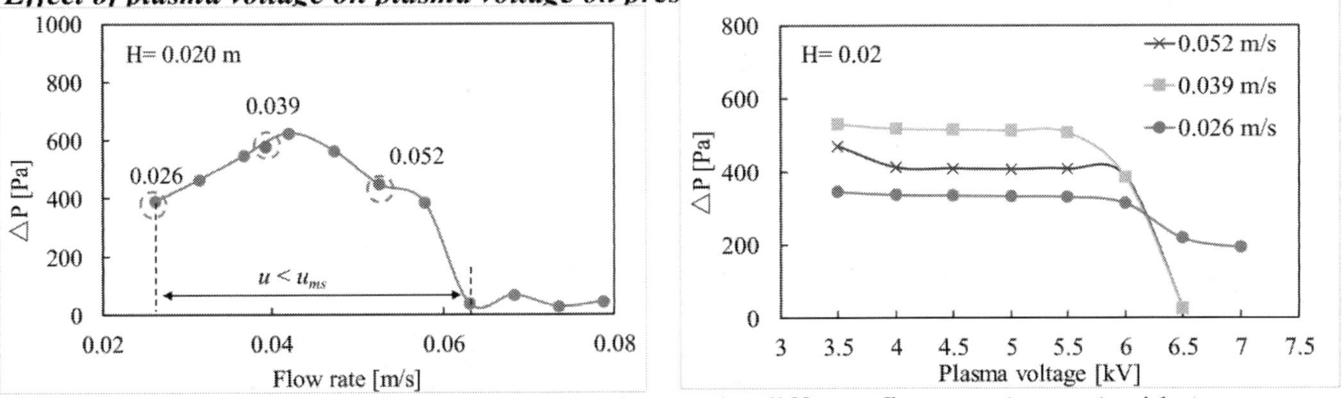

Fig.7 Effect of plasma voltage on pressure drop under different flow rate ($u < u_{ms}$) with Ar gas

In order to take account of effect of plasma voltage on pressure drop changes, initial flow rate was set up to be constant ($u < u_{ms}$) with the same static height. In this period, the particles inside of the bed do not move significantly, the pressure kept the certain value in the constant flow rate. As shown in Fg.7(a), three points were picked up as the constant value, then, plasma voltage increased slowly stepped by 0.5 kV and kept each value kept stable state. Fg.7(b) reveals that ΔP remained a stable value under low plasma voltage supply. When the plasma voltage increased to a certain value, resulted in ΔP of particles declined sharply, and the solid bed became to be the fountain situation. ΔP descent point corresponding the plasma voltage is related to the initial pressure drop not the flow rate. In comparison with the two figures, the initial ΔP in descending order: 0.039 m/s, 0.052 m/s and 0.026 m/s, which is corresponding with the decrease point with plasma voltage incensement. It means that higher pressure provided, lower plasma needed. The reason for this phenomenon may due to the situation of gas or the particles maybe changed by plasma voltage treatment.

Effect of plasma voltage on surface temperature of spouted bed

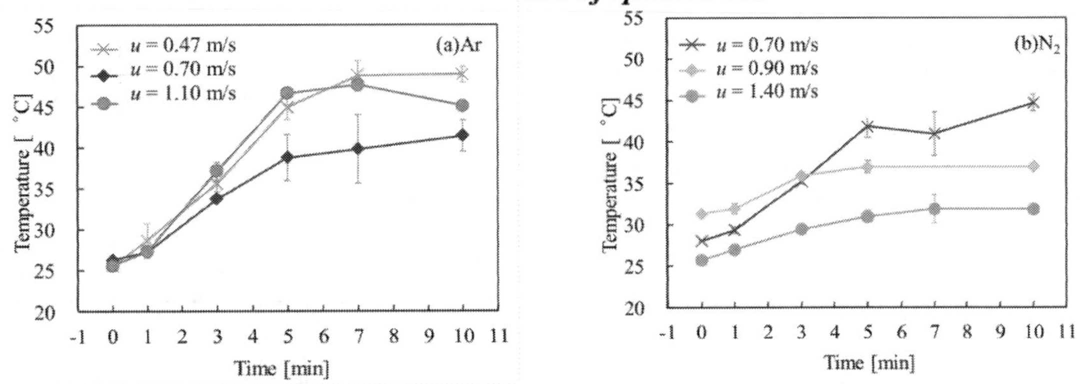

Fig.8 Surface temperature of inlet points of spouted bed (Plasma voltage =7kV)

To achieve a better understanding of the decrease reason for u_{ms} with the plasma enhancement, we measured the temperature changes during the spouted bed. It is difficult to detect the bed temperature due to the thermocouples effected by the ionization of gas. So that three points on surface (shown in Fig.1) of the spouted bed were chosen to measure the temperature by Infrared Camera (FLIR T440). It had provided the inlet point temperature is equal to the inside temperature after a long time by thermocouple. Fig.8 illustrates the inlet temperature changes versus various flow rate conditions by Ar and N_2. There was no significant difference for inlet points after 5 min regardless of Ar or N_2. Based on the temperature changes, the volume of gas can be increased just about 7% in theory, which is not enough to decrease u_{ms} so much.

Effect of plasma voltage on gas volume

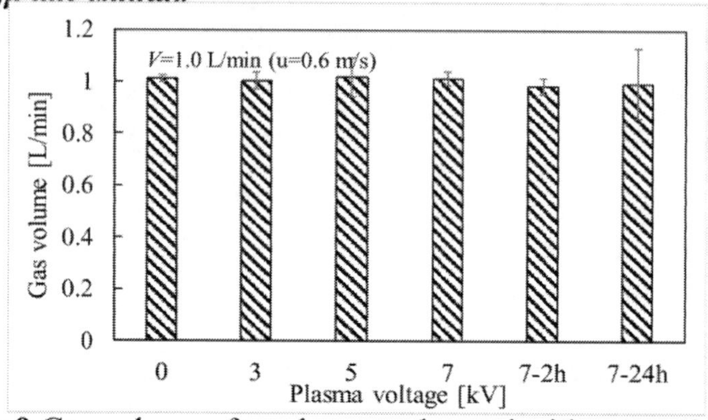

Fig.9 Gas volume after plasma enhanced without particles

Due to the gas was changed in to the plasma state by the high voltage supplied, the gas volume may be changed during this process. We measured not only the temperature of spouted bed, but also the volume of gas. The gas volumes after plasma treatment from 0 kV to 7 kV and kept the gas bags staying in air atmosphere for 2 h and 24 h were measured by using gas collecting bags. It is obviously found that the difference for several conditions even after staying for a long time (shown in Fig.9). So that, the gas temperature and gas volume do not change obviously, which are not the main reasons for the u_{ms} decline.

Effect of plasma on bed height

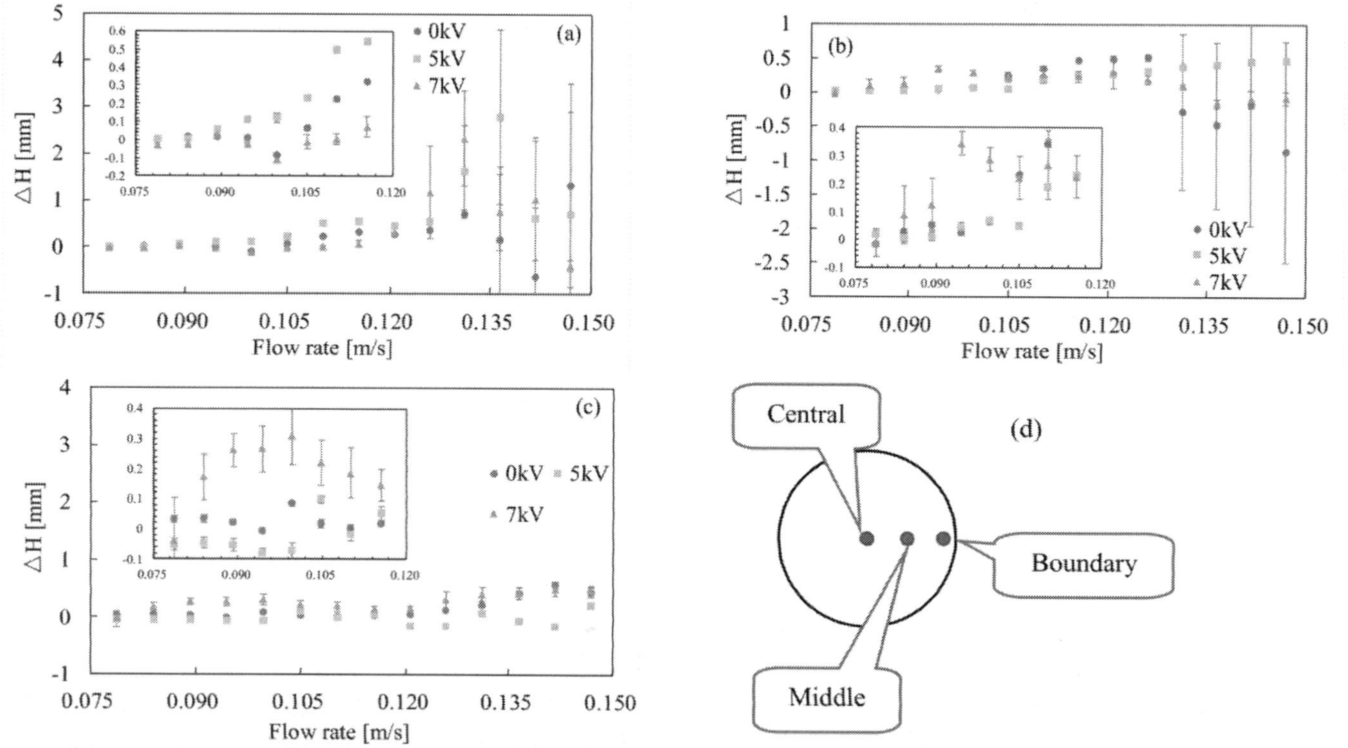

Fig.10 Bed height changes versus flow rate (H_0=0.04 m)
((a)central point, (b)middle point and (c)boundary point)

In order to trace the reason for the u_{ms} changes, we measured the surface movement by the Laser sensor for detecting the surface distance changes in the course of the flow rate changes with and without plasma enhancement. Fig.10 (a) to (c) show that the bed height changes (ΔH) changed with the flow rate respected the central point, middle point and boundary point as shown in Fig.10 (d). The initial bed height was 0.04 m and kept as stable value before flow rate of 0.08 m/s regardless of the plasma voltages. With the increment of flow rate, the ΔH increased a little bit regardless of plasma voltage. It must also be mentioned that in the comparison of detecting point. In the Fig.10 (a), central point location, ΔH increased larger as 5kV plasma voltage supplied than that of without plasma with stable values before spouted (flow rate of 0.128 m/s). But at 7kV plasma voltage, the ΔH increased not obviously with a big error bar before spouted as the detail figure in the Fig.10 (a). It means before the solid bed spouted, the solid bed in an instability state, the void fraction is not the constant. In comparison with the middle points and boundary points, it can be found ΔH increased by the plasma enhancement, especially in 7 kV with big shaking. In the whole view, the solid bed shaken strongly in whole bed with plasma enhancement, but for without plasma situation, only the central part was shaken. It means that the void fraction in high plasma voltage was not stable, which led the interaction between gas and solid particles increased and the u_{ms} declined.

Conclusion

AC plasma jet enhanced spouted bed has been developed to figure out the behavior of fine particles in this paper. The u_{ms} of fine particles was decreased by the enhancement of plasma regardless of static bed height and plasma working gas. There were no obvious changes of temperature of spouted bed surface and gas volume. The interaction between gas and particles, , which result in the enhancement of void fraction, interaction between particles and particles is the main reason for the behavior changed.

References

1. Karches M, Bayer C, and von Rohr P R (1999), "A circulating fluidised bed for plasma-enhanced chemical vapor deposition on powders at low temperatures," *Surface and Coatings Technology*, 116, pp: 879-885.
2. Pajkic Z, Willert-Porada M (2009), "Atmospheric pressure microwave plasma fluidized bed CVD of AlN coatings," *Surface and Coatings Technology*, 203(20), pp: 3168-3172.
3. Tsukada M, Goto K, Yamamoto R H, et al (1995), "Metal powder granulation in a plasma-spouted/fluidized bed," *Powder technology*, 82(3), pp:347-353.
4. Zhu, Xinbo, et al (2015), "Catalyst screening for acetone removal in a single-stage plasma-catalysis system," *Catalysis Today*, 256, pp: 108-114.
5. Gryczka, O., et al (2008), "Characterization of the pneumatic behavior of a novel spouted bed apparatus with two adjustable gas inlets," *Chemical Engineering Science,*63(3), pp: 791-814.
6. Mathur, K. B, and P. E. Gishler (1955), "A technique for contacting gases with coarse solid particles," *AIChE Journal*, 1(2), pp:157-164.
7. Chevrier G, Hanai T, Tran K C, et al (1982), "Development of a microwave plasma detector for gas-chromatogtaphy," *Canadian Journal of Chemistry-revue Canadienne De Chime*, 60(7), pp: 898-903.
8. Sathiyamoorthy D (2010), "Plasma spouted/fluidized bed for materials processing," Journal of Physics: Conference Series. *IOP Publishing*, 208(1), pp: 012-120.

490260 Magnetic Resonance Imaging of Wet Fluidization

Abstract:

In the energy, pharmaceuticals and polymer production industries, it is common for small amounts of liquid to be injected into gas-solid fluidized beds. In some cases, the liquid is a reactant, while in others it facilitates agglomeration or heat transfer. The liquid leaves to the formation of cohesive liquid bridges between particles as well as between particles and walls, often having significant effects on hydrodynamics. These effects include formation of agglomerates and in certain cases can lead to local or device-scale defluidization.

Despite the industrial importance of liquid bridging on gas-solid fluidization, the effect of liquid on fluidization hydrodynamics is still not well understood. A small number of previous experimental studies have demonstrated effects of liquid on bed on bed fluidity[1,2], particle velocities[3â€“6], bed height[1], bubble size[6,7] and minimum fluidization velocity[6]. Additionally, various computational models have been developed and utilized to understand the effects of liquid bridging on fluidization hydrodynamics. Some studies have simply lowered the coefficient of restitution[8,9] to account for the effects of liquid bridging, while others have modeled the capillary forces in the normal direction[10,11], the viscous forces in the normal and tangential directions[12], as well as the rate of liquid transfer between the surfaces of particles and liquid bridges[13]. Due to a lack of detailed experimental studies of a variety of hydrodynamic features under a variety of liquid conditions, the current literature does not fully elucidate the relative importance of liquid loading, surface tension and viscosity on fluidization behavior. This absence in the experimental literature leads to issues in understanding the validity and areas for improvement in existing computational models.

Previously, MRI has been used to study hydrodynamics in detail in gas-solid fluidized beds. Here, we use MRI to image local particle concentration and velocity with high temporal resolution in a 3D freely bubbling fluidized bed. We use these measurements to characterize both average values and variations in bubble size, bubble number density and particle speed under varying gas flow rates, liquid loading conditions, and values of surface tension and viscosity. We use these results to evaluate the importance and validity of different computational sub-models for liquid bridging.

References:

1. McDougall SL, Saberian M, Briens C, Berruti F, Chan EW. Characterization of Fluidization Quality in Fluidized Beds of Wet Particles. *Int J Chem React Eng*. 2004;2(1).

2. McDougall S, Saberian M, Briens C, Berruti F, Chan E. Using dynamic pressure signals to assess the effects of injected liquid on fluidized bed properties. *Chem Eng Process Process Intensif.* 2005;44(7):701-708.

3. Sutkar VS, Deen NG, Patil AV, et al. Experimental study of hydrodynamics and thermal behavior of a pseudo-2D spout-fluidized bed with liquid injection. *AIChE J*. 2015;61(4):1146-1159.

4. Zhang Q, Zhou Y, Wang J, et al. Particle Motion in Two- and Three-Phase Fluidized-Bed Reactors Determined by Pulsed Field Gradient Nuclear Magnetic Resonance. *Chem Eng Technol*. June 2015:n/a-n/a.

5. Passos ML, Mujumdar AS. Effect of cohesive forces on fluidized and spouted beds of wet particles. *Powder Technol*. 2000;110(3):222-238.

6. Zhou Y, Ren C, Wang J, Yang Y. Characterization on hydrodynamic behavior in liquid-containing gas-solid fluidized bed reactor. *AIChE J*. 2013;59(4):1056-1065.

7. Ma J, Liu D, Chen X. Bubble Behaviors of Large Cohesive Particles in a 2D Fluidized Bed. *Ind Eng Chem Res*. 2016;55(3):624-634.

8. Darabi P, Pougatch K, Salcudean M, Grecov D. DEM investigations of fluidized beds in the presence of liquid coating. *Powder Technol*. 2011;214(3):365-374.

9. Sutkar VS, Deen NG, Padding JT, et al. A novel approach to determine wet restitution coefficients through a unified correlation and energy analysis. *AIChE J*. 2015;61(3):769-779.

10. Mikami T, Kamiya H, Horio M. Numerical simulation of cohesive powder behavior in a fluidized bed. *Chem Eng Sci*. 1998;53(10):1927-1940.

11. Girardi M, Radl S, Sundaresan S. Simulating wet gasâ€“solid fluidized beds using coarse-grid CFD-DEM. *Chem Eng Sci*. 2016;144:224-238.

12. Kuwagi K, Mikami T, Horio M. Numerical simulation of metallic solid bridging particles in a fluidized bed at high temperature. *Powder Technol*. 2000;109(1â€“3):27-40.

13. Wu M, Radl S, Khinast JG. A model to predict liquid bridge formation between wet particles based on direct numerical simulations. *AIChE J*. 2016;62(6):1877-1897.

490409 CFD-DEM Modeling the Effect of Column Size and Bed Height on Minimum Fluidization Velocity in Micro Fluidized Beds with Geldart B Particles

Abstract:

CFD-DEM modeling the effect of column size and bed height on minimum fluidization velocity in micro fluidized beds with Geldart B particles

Yupeng Xu[a], Tingwen Li[a,b], Jordan Musser[a], Xiaoxing Liu[c], Guangwen Xu[c], William A. Rogers[a]

a. National Energy Technology Laboratory, Morgantown, WV 26505, USA

b. AECOM, Morgantown, WV 26505, USA

c. State Key Laboratory of Multiphase Complex Systems, Institute of Process Engineering, Chinese Academy of Sciences, Beijing 100190, P. R. China

Micro fluidized beds (MFBs) were first put forward by Potic et al. to refer to fluidized beds with inner diameters of a few millimeters. These are ideal systems to study because they have a large wall area per unit reactor volume which promotes excellent mass- and heat-transfer. Additionally, smaller systems allow for measurement of small quantities of solid reactants. Applications of micro-fluidized beds include the Micro Fluidized Bed Reaction Analyzer (MFBRA) and Micro Membrane Fluidized Bed Reactor (MMFBR).

The fluidization characteristics of a given kind of particles in MFBs are different from those in ordinary-size fluidized beds due to much stronger wall effects. From the viewpoint of operating and controlling a MFB kinetic analyzer, knowing minimum fluidization and minimum bubbling velocities (Umf and Umb) enables researchers to determine suitable gas velocities so that fluidization in the MFB reactor can be maintained. Limited research has been conducted to understand the fundamental hydrodynamics of micro-fluidized beds.

The fluidization behavior of Geldart B particles in micro fluidized beds is investigated numerically using Computational Fluid Dynamics coupled with Discrete Element Method (CFD-DEM) available in the open-source Multiphase Flow with Interphase eXchanges (MFIX) code. The effects of different bed diameters (D) of 8 mm, 12 mm, 16 mm and various initial static bed heights (H) were examined. It is found that both decreasing the column diameter and increasing the bed height in a micro fluidized bed increases minimum fluidization velocity (Umf). The observed overshoot in pressure drop that occurs before the onset of fluidization decreases in magnitude with increasing column diameter, however there is less sensitivity to bed height. Overall, the numerical results agree qualitatively with existing theoretical correlations and experimental studies. The simulations show that both column diameter and particle-wall friction contribute to variations in minimum fluidization velocity. These two factors are coupled and hard to separate. The detailed influences of wall friction on minimum fluidization velocity are investigated for a prescribed column diameter of 8 mm by varying the wall friction from 0 to 0.4.

Fig.1. Influence of the column diameters on the minimum fluidization velocity (static bed height of 6.88 cm)

494805 DEM and Experimental Study of the Flow of Wet Granular Assemblies Under Dynamic Conditions

Abstract:

When added to flowing particles, liquid has a significant effect on the flow of granular assemblies. Using a combination of experimental study and discrete particle method simulations (DEM), we investigate the effects of capillary force, liquid viscosity and particle size on wet granular flows, and we establish a methodology that ensures the control of the bed flow motion in a rotating drum.

We show that the strength of capillary forces between two adjacent particles can be altered through surface modification of the glass beads. Thus, under the right conditions; we demonstrate that the bed flow motion in a rotating drum can be controlled and scaled. Liquid viscosity effect on the bed flow motion is also investigated under low capillary forces. The capillary force between the particles is significantly reduced by making the glass beads hydrophobic via chemical silanization. We find that liquid-induced cohesion decreases both, the width of the flowing region and the velocity of the particles in the free surface, but increases the width of the creeping region as well as the dynamic angle of repose. We show that the local granular temperature in the flowing region is affected by the capillary force. We then propose a scaling methodology that ensures the conservation of the bed flow using the Weber number. The scaling methodology in the flow regimes considered (rolling and cascading) yields similar bed flow for different particle sizes.

494254 A Numerical Algorithm for Simulating Dense Polydisperse Gas--Particle Flows Using a Mass--Velocity Quadrature-Based Moment Method

Abstract:

Polydisperse gas-particle flows are common in many fields of engineering, such as in fluidized beds and risers, which are widely used in a variety of chemical processes. Due to the various physical and chemical interactions between particles and gas, such as particle collisions and breakage, gasification and chemical reactions, the individual particle size and the overall size distribution of particles changes as the flow develops. In this work, quadrature-based moment methods (QBMM) are applied to solve numerically the kinetic equation of the joint NDF with the hydrodynamic approximation for particle velocity, which is valid when particles are in the moderately dense and dense regime, such as in fluidized bed. By reconstructing the size-conditioned velocity distribution function, the spatial fluxes in the moment equations are treated using a kinetic-based finite-volume method. This approach was implemented in an open-source CFD package and then tested in a fluidized bed case. The results demonstrate that the ability of the code to capture the size segregation and size change due to aggregation and breakup, and that our approach is an effective way to model complicated polydisperse gas-particle flows.

495715 Effects of Wall Boundary Conditions on 3D Simulation of Pseudo Two Dimensional Fluidized Beds Using Dense Discrete Phase Model (DDPM)

Abstract:

Dense Discrete Phase Model (DDPM) is a new computational fluid dynamic (CFD) approach to simulate industrial scale fluidized bed reactors in a reasonable computational time. It is a hybrid Lagrangian-Eulerian approach tracking the particles in a Lagrangian framework according to Newtonâ€™s laws of motion while interactions between particles are estimated by kinetic theory of granular flows (KTGF). However, there are still few studies on the DDPM model and there is a need to study the effect of wall boundary conditions on the CFD simulation of bubbling fluidized beds and identify the controlling factors on simulation of wall effects.

In this work, a pseudo two dimensional fluidized bed (0.3 m Ã— 0.7 m Ã— 0.015 m) was simulated using DDPM and compared with experiments. A three-dimensional computational column was generated to study the movement of particles in third dimension and their collisions with back and front. A mesh dependency study was conducted, and minimum number of cells in depth direction was determined to observe the back and front wall frictions. Specularity coefficient which determines the elasticity of particle wall interactions was changed between 0.0001 to 0.5. Optimum coefficient was calculated in a way that velocity of particles and their circulation patterns had the best consistency with experiments. Moreover, effect of wall friction on size of bubbles, bubble rising velocity, pressure drop was observed. The results showed increase in size of bubbles and void fraction of the system by decreasing the wall friction. Besides, as particles rebounds off the wall in Lagrangian frame, the change in particles momentum can be defined by the normal and tangential coefficient of restitutions. The results showed that restitution coefficient didnâ€™t have that much effect on velocity of particles, and dynamic of bubbles. It was observed that Specularity coefficient is different at higher velocities and it should be calculated for different superficial velocities.

499371 RANS Modeling of Cluster-Induced Turbulence in Particle-Laden Channel Flow

Abstract:

A phenomenon often observed in gas-solid flows is the formation of mesoscale clusters of particles due to the relative motion between the solid and fluid phases that is sustained through the dampening of collisional particle motion from interphase momentum coupling inside these clusters. The formation of such sustained clusters, leading to cluster-induced turbulence (CIT), can have a significant impact in industrial processes, particularly in regards to mixing, reaction progress, and heat transfer. One of the challenges in successfully resolving these clusters through Eulerian-Lagrangian simulations in previous work has been the requirement for an overwhelming degree of refinement of the modeled geometry. A less computationally expensive model is needed to model and study CIT in practical applications. This work implements a two-phase Reynolds-Averaged Navier-Stokes (RANS) model to capture the wall-normal flow characteristics in fully developed channel flow with CIT. Dilute and moderately dense concentrations of particles are considered and compared with both high-resolution Eulerian-Lagrangian and Eulerâ€"Euler anisotropic Gaussian simulation results. The mean gas velocity is varied to study the relative importance of shear vs. cluster-induced turbulence for fixed mass loading of particles. The inclination of the channel geometry is also varied to examine the effect of the direction of gravity relative to the direction of flow on cluster formation and turbulence statistics.

503670 Digital Alchemy for Assembly Engineeringc

Abstract:

Back in the day, Stuart Churchill was a pioneer in the use of computers for modeling chemical and thermal processes. Since that time, the role of computing in chemical engineering has become indispensable. In this talk, we present new computational methods we call "digital alchemy" that allow for predictive modeling of thermodynamic assembly, and permit the design of building blocks optimized for assembly into target structures with desired properties. In this approach, standard thermodynamic ensembles are generalized to include alchemical variables and their thermodynamic conjugates, allowing for rapid, real-time computational investigation of large, complex design spaces.

503671 Improved Algebraic, Numerical, and Graphical Representations in Fluid Mechanics

Abstract:

The concept of equivalents is applied to fully developed flows through straight channels, and two applications are introduced: the equivalents that produce equal rates of flow through channels of different cross-section, and the diminishment that occurs between the regimes of laminar and turbulent flow. The validity of the eddy diffusivity and the invalidity of the mixing-length as measures of the shear stress due to turbulence have perhaps never before been demonstrated so clearly. The possible choices of primary variables and the consequences are also examined. The objective of these explorations is an improved understanding of the basic fluid mechanics by teachers, students, and industrial practitioners.

503672 The Scaling of Turbulence Near the Wall and the Churchill Turbulent Flux Correlation: Insights with Lagrangian Simulations

Abstract:

The evaluation of mean velocity and mean temperature in wall-bounded turbulent flows has been based on the "classical" theory of turbulence and the law of the wall. The viscous wall parameters, namely the friction velocity, the friction temperature, and the kinematic viscosity of the fluid have been used to scale not only the mean values of velocity, but also other mean turbulent quantities close to the wall, including the Reynolds stresses and the turbulent velocity and temperature fluctuations. Stuart Churchill, as he questioned the validity of the Reynolds analogy between momentum and heat or mass transfer in his works [1], he realized that extending the scaling of momentum to the scaling of scalar quantities in turbulence was not a sound approach. He proposed a scaling of turbulence that depends on the turbulent momentum flux and the turbulent heat flux. In his AIChE Institute Lecture (in Miami, 1998), Churchill presented a summary of his work on this topic, and closed form solutions for the mean velocity and temperature [2]. The turbulent Prandtl number, the most significant parameter for the modeling of turbulent transport not only in theory but also in modern CFD commercial software packages, was also interpreted physically based on these ideas [3]. We have used direct numerical simulation results and results from Lagrangian simulations [4] for the transport of passive scalars in turbulent channel flows to explore Churchill's ideas on turbulent flow scaling and to develop an updated theory for the scaling of heat transport in turbulent flows [5-7]. His passion to remove empiricisms and approximations from the development of predictive correlations in turbulent transport has been an inspiration. In this talk, we will revisit the Churchill scaling in view of our simulation results, and will discuss the counterintuitive finding that wall turbulence can be used to separate rather than mix substances under special circumstances [8].

1. Churchill, S. W., 1996, "A Critique of Predictive and Correlative Models for Turbulent Flow and Convection," Ind. Eng. Chem. Res., 35, pp. 3122–3140.
2. Churchill, S.W., 2000, "Progress in the Thermal Sciences: AICHE Institute Lecture," AIChE Journal, 46(9), pp. 1704-1722.
3. Churchill, S. W., 2002, "A reinterpretation of the Turbulent Prandtl Number," Ind. Eng. Chem. Res., 41, pp. 6393–6401.
4. Papavassiliou, D.V, 2002, "Turbulent transport from continuous sources at the wall of a channel," Int. J. Heat Mass Transfer, 45(17), pp. 3571-3583.
5. Le, P.M., and Papavassiliou, D.V., 2008, "On the scaling of heat transfer using thermal flux gradients for fully developed turbulent channel and Couette flows," Int. Commun. Heat Mass Transfer, 35(4), pp. 404-412.
6. Srinivasan, C., and Papavassiliou, D. V., 2011, "Prediction of Turbulent Prandtl Number in Wall Flows With Lagrangian Simulations," Ind. Eng. Chem. Res., 50(15), pp. 8881–8891.
7. Srinivasan, C., and Papavassiliou, D.V., 2013, "Heat transfer scaling for wall bounded turbulent flows," Applied Mechanics Review 65(3), Art. 031002.

8. Nguyen, Q., Srinivasan, C., and Papavassiliou, D.V., 2015, "Flow induced separation in wall turbulence", Phys Rev E, 91, Art. 033019.

503674 Flow Boiling Using a Piranha Pin Fin Heat Sink

Abstract:

Flow Boiling using a Piranha Pin Fin Heat Sink

Corey Woodcock, Xiangfei Yu, Yoav Peles[1] and Joel Plawsky

Rensselaer Polytechnic Institute; [1]University of Central Florida

The *Piranha Pin Fin* (*PPF*) microdevice is a heat exchanger engineered to dissipate extreme thermal loads using either dielectric fluids such as HFE7000 or water. Fabrication of the device is CMOS compatible and the first generations of the device have been fashioned from silicon substrates. The PPF operates on the principle that heat transfer would be more efficient if heated fluids could be removed nearly as fast as they are created. This process would keep the driving force as high as possible. Thus the heat exchanger is built as two layers; a primary substrate that dissipates the bulk of the heat and shunts the heated fluids to a booster section that continues to heat the fluid and extends the total surface area of the device. Three generations of devices were fabricated. The first two generations were single-layer systems designed to help understand the operation of the primary layer and optimize the shape and location of the pin fins. The third generation was the first, two-layer system to be fabricated and tested.

Figure 1 Representative PPF array

Correlations for the heat transfer coefficient, h_{tp}, based on the Boiling Number and the Weber Number in the PPF design was developed and validated with Gen II. The discrepancy of heat transfer coefficient between predicted and measured values was within 20%. The ratio of two-phase to single-phase pressure drop was correlated with the Boiling Number (Bo) and the Jakob Number (Ja). The discrepancy of the two-phase multiplier between prediction and experimental measurements were within 12%.

The third generation *PPF* heat sinks were the first series to provide both a primary substrate populated with *PPF*s, as well as a booster substrate populated with solid micro pin fins. The substrates are micro-aligned and fusion-bonded to provide a full three dimensional (3D) architecture with bulk silicon properties. Experimental results from the Gen III *PPF* microdevices provide a proof of concept of the true *PPF* heat sink concept with fluid venting into multiple substrates. At low flowrates (G = 670 [kg/s/m^2]) the Gen III *PPF* heat sink has been shown to dissipate heat loads approaching 1.1 [kW/cm^2] at an exit quality of 95%, while at higher flowrates (G = 3,300 [kg/s/m^2]) the thermal limit of the heat sink remains unknown due to exceeding the available power supply at >1.4 [kW/cm^2].

503673 Transport Problems in the Spirit of Stuart Churchill for Teaching and Research at the University of Michigan

Abstract:

Stuart Churchillâ€™s solutions to important heat transport problems have greatly influence both pedagogy and research. The correlation of Churchill and Chu for the natural convection heat transfer to or from a vertical plate, over an astonishing range of Rayleigh numbers is based partly on a similarity solution of a type common in heat, mass, and momentum transfer for problems in chemical engineering. One such problem is asphaltene or wax transport and deposition onto the wall of an oil pipeline, which can be solved using similar methods, as will be discussed. In addition, such problems, and the wealth of their applications make for excellent motivations and examples in the teaching of graduate or undergraduate transport, as will be discussed in the talk.

509054 Solve this! Fundamental Approach to Problem Solving in Industrial Processes I (Invited Talks)

Abstract:

Solving complex plant problems is an everyday exercise that affects all projects and budgets, and it is very rarely introduced to students at any level. In fact, by the time young engineers are few years into working in industry, most have lost the engineering skills that they developed in school. We hope to encourage future engineers to retain those skills and learn the importance of plant problem-solving.

489690 Investigating Cylindrical Particle Fluidization Using X-Ray Particle Tracking Velocimetry (XPTV)

Abstract:

X-ray particle tracking velocimetry (XPTV) is used to determine the three dimensional particle motion and orientation characteristics of a cylindrical particle in a bubbling fluidized bed filled with a binary mixture of cylindrical particles and fluidizing material. Experiments are carried out in a 10.2 cm ID fluidized bed. Two types of cylindrical particles, Cylinder A (diameter d_i = 4.5 mm, length l = 7.1 mm) and Cylinder B (diameter d_i = 3.0 mm, length l = 16 mm), are used to evaluate the effect of particle shape. Glass beads with d_i = 300-500 Î¼m are used at the fluidizing bed material. The cylindrical particles and bed material are carefully premixed using different mass ratios but a fixed static bed height of H = 10.2 cm. A single cylindrical particle is removed and tagged with a small metal rod as a tracer particle so that its motion and orientation can be simultaneously captured by two perpendicular X-ray imaging systems. XPTV is used to provide a detailed three dimensional trajectory map of the tracer particle. The effect of various operating parameters on the cylindrical particle position, velocity, and orientation will be discussed.

490320 Assessing Agglomerate Characteristics in a Nanopowder Fluidized Bed

Abstract:

Nanoparticles in a gaseous environment agglomerate in a hierarchical fashion; this leads to peculiar hydrodynamics when fluidizing nanopowders. From production until powder processing in the gas phase, nanoparticles typically go from individual particles to aggregates, simple agglomerates, and complex agglomerates. Even though the structures at each level have unique properties, they are commonly assessed as a whole. Additionally, the effect of external factors on the morphology of these structures during gas processing is not well understood and challenging to study due to the limited techniques for in situ analysis of the dynamic phenomena. Here, we study different ceramic oxide nanoparticles. We describe the structural characteristics of each hierarchical level of complex agglomerate formation obtained from two in situ techniques: small-angle neutron scattering for the smaller scales, and a settling tube coupled with a boroscope for the larger scales. The first scale, namely aggregates, are open structures with a fractal dimension of about 1.5, which then form simple agglomerates with a fractal dimension close to 3, that later cluster into complex agglomerates that present a fractal dimension of about 2. In addition, we characterize the contact mechanics of the nanopowders agglomerates using atomic force microscopy. Finally, we will discuss the entrainment of nanosized clusters from the fluidized bed.

494432 Conditions for Pattern Formation in Pulsed Fluidized Beds

Abstract:

Bubbling gas-solid fluidized bed reactors are employed in numerous applications in the chemical, energy, environmental, and pharmaceutical industries, where excellent interphase transport and homogenous mixing are of paramount importance. The performance of fluidized beds is highly dependent on their bubble characteristics, which govern system hydrodynamics and efficiency of the transport processes. Nevertheless, the inherently complex dynamics of fluidized beds give rise to difficult to predict temporal and spatial distributions of bubbles, due to coalescence and break-up, and possible maldistribution issues, which significantly complicates engineering design, process control and system scale-up [1].

Bringing in additional degrees of freedom provides the flexibility to manipulate the system hydrodynamics. In particular, pulsating gas-solid fluidized beds can lead to a unique fluidization state, where gas bubbles rise in a staggered way, forming regular patterns with a characteristic wavelength [2]. Within a certain range of pulsed gas velocity, the bubble distribution and the physical properties can be effectively tailored by tuning the conditions of the inlet pulsed flow, such as the offset, pulsating frequency and amplitude. This phenomenon shows great potential to be applied as a non-intrusive method to structure the bed [3].

Despite its advantages, fundamental aspects underpinning pattern formation, such as energy dissipation and momentum transfer, have remained largely unexplored. Over the past years, a few modeling attempts have been conducted to investigate its underlying dynamics [4, 5], but with limited success. Only recently, we reported the first successful simulations capable of reproducing the pattern dynamics using a CFD-DEM (computational fluid dynamics – discrete element method) approach [6]. These numerical results provide valuable physical insights into the fundamentals of fluidized bed dynamics.

In this presentation, we will discuss experiments on the formation and stability of regular patterns in 2D and 3D pulsed fluidized beds, subjected to various conditions and compare results with CFD-DEM numerical simulations. We present the impact of different pulsed flow conditions, particle physical properties, and bed depth on the pattern characteristics and regularity. A phase diagram is constructed to locate the pattern region and pointing out the favorable conditions for attaining patterns. We also discuss the propagation of the pattern structure in beds of different depth. By combining the experimental data with computational results, we interpret how the patterns propagate and are gradually disturbed when bubbles travel toward the bed surface, through analyzing gas and solid phase dynamics. These results are useful from a fundamental point of view, as well as for scale-up for applications.

[1] S. Pannala, M. Syamlal, T. J. O'Brien, Computational Gas-Solids Flows and Reacting Systems: Theory, Methods and Practice, IGI Global, Hershey, 2011.

[2] M.-O. Coppens, M.A. Regelink, C.M. van den Bleek, Pulsation induced transition from chaos to periodically ordered patterns in fluidised beds, Proceedings of the 4th World Congress on Particle Technology, Sydney, 2002.

[3] M.-O. Coppens, J.R. van Ommen, Structuring chaotic fluidized beds, Chem. Eng. J. 96 (2003) 117-124.

[4] K. Wu, L. de Martín, L. Mazzei, M.-O. Coppens, Pattern formation in fluidized beds as a tool for model validation: A two-fluid model based study, Powder Technol. 295 (2016) 35-42.

[5] X. Wang, M. Rhodes, Pulsed fluidization—a DEM study of a fascinating phenomenon, Powder Technol. 159 (2005) 142-149.

[6] K. Wu, L. de Martín, M.-O. Coppens, Pattern formation in pulsed gas-solid fluidized beds – the role of granular solid mechanics, Chem. Eng. J. (2017) (under revision).

Evaluation of Spouted Bed Flow Instabilities via High Speed Video and Pressure Fluctuations.

Steven L. Rowan, Jingsi J. Yang, Ronald W. Breault, Justin M. Weber

NATIONAL ENERGY TECHNOLOGY LABORATORY

October 30, 2017

Introduction

NATIONAL ENERGY TECHNOLOGY LABORATORY

- **Slot-rectangular spouted beds are often used in drying and coating of Geldart D materials.**
  - easy to scale up, can achieve different flow regimes via changing inlet nozzle.
- **The majority of previous spouted bed literature has focused on cylindrical spouted beds, not slot-rectangular.**
- **The limited amount of slot-rectangular literature has noted the existence of flow/spout instabilities.**
  - Asymmetric gas core, pulsations, oscillation between spouting and slugging.
- **Dogan et al. determined that these were actually separate spouting regimes with different flow characteristics.**

The objective of the current study was to examine these spouting instabilities in a laboratory-scale slot-rectangular spouted bed via a combination of high speed video and analysis of pressure signals for a variety of initial static bed heights and inlet nozzle sizes.

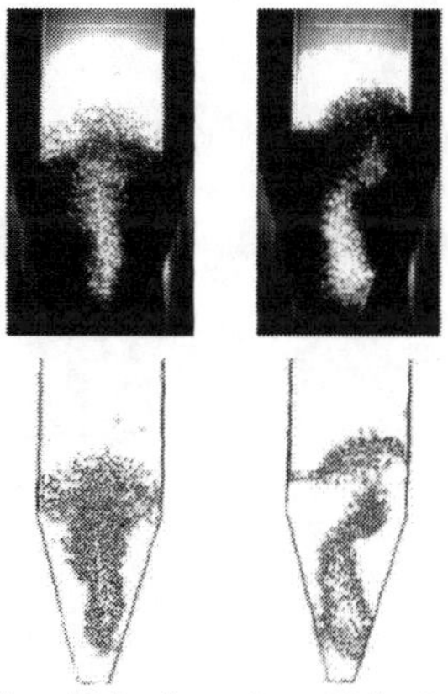

Symmetric Spouting Asymmetric Spouting

Experimental Setup

NATIONAL ENERGY TECHNOLOGY LABORATORY

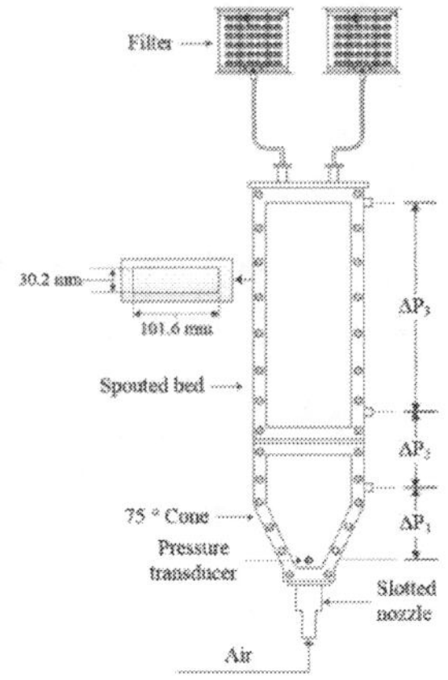

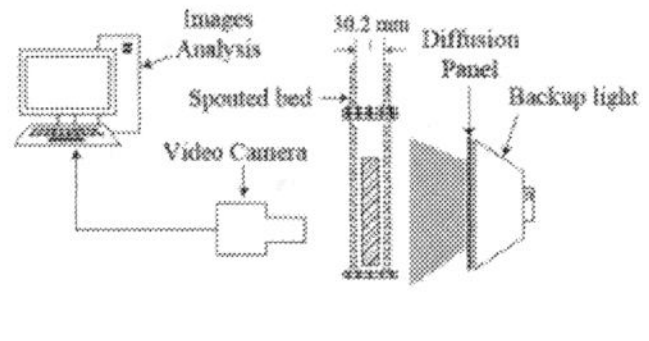

- **Slot-Rectangular Spouted Bed:**
  - Width: 101.6mm (4 inches)
  - Depth*: 30.2mm (1 3/16 inches)
  - Height: 864mm (34 inches)
  - Aluminum Frame with clear acrylic panels.
- **3D Printed Nozzle Assemblies**
  - 3/8-inch equivalent (3.8mm x 19.1mm)
  - 1/2-inch equivalent (6.6mm x 19.1mm)
- **Sony RX10-II Camera**
  - 500 fps
  - 1920 x 1080 pixels
  - 0.001s shutter time
- **Dracast LED500 light panel.**
- **3.2mm Nylon Beads**
  - SG: 1.13
  - U_{mf} = 1.08 m/s

U.S. DEPARTMENT OF ENERGY

3

Video Analysis: Determination of Bed Height

NATIONAL ENERGY TECHNOLOGY LABORATORY

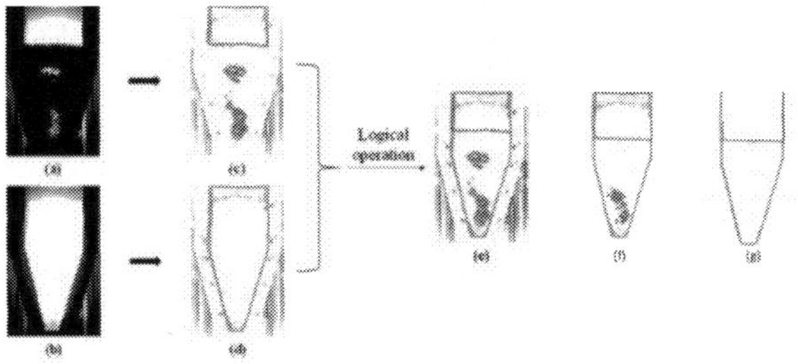

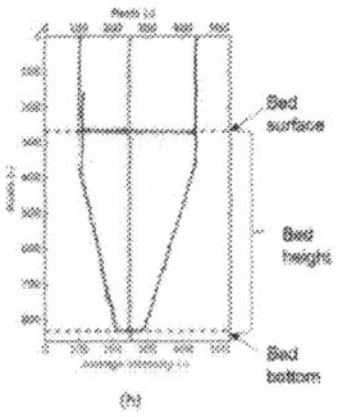

Analysis procedure:

- Acquire "blank" video (b)
- Acquire video of test condition (a)
- Convert (a) and (b) to greyscale (c) and (d)
- Merge (c) and (d) and convert to binary (e).
- Crop (e) to isolate the bed image, generating (f).
- Create "extracted binary" (g) by deleting all information except bed surface and bottom from (f).
- Calculate and plot average pixel intensity (h)
  - Bed bottom and surface correspond to peak pixel intensities.

U.S. DEPARTMENT OF ENERGY

4

Video Analysis: Sample Results

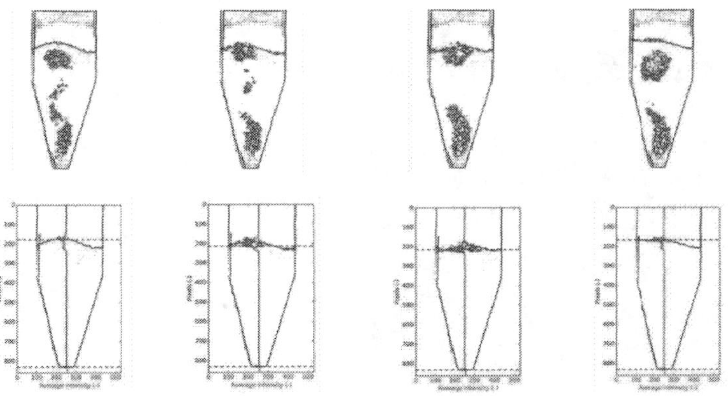

Observed Spouting Regimes/Behaviors

- **Internal Spouting Zone ($U_{ms} < U_g < U_{ex}$)**
  - **Internal Jet:** Initial formation of a gas void (or cavity) at the bottom of bed, which then expands upwards through the bed.
  - **Jet-Bubbling:** Exhibits a pattern where the jet periodically expands upwards until a portion of the jet detaches and forms a bubble that continues to rise upwards through the bed, while the remainder of the jet collapses to a lower height prior to renewed expansion leading to the formation of the next bubble.
  - **Jet Pulsating:** similar to Jet-Bubbling, but the jet penetrates much of the bed height, and bubbles form close to bed surface, eject significant amounts of solids into the freeboard region. Unlike in the case of jet-bubbling, the jet itself does not undergo significant expansion and collapse.
- **External Spouting ($U_g >= U_{ex}$)**
  - **Symmetric Spouting:** Jet completely penetrates the bed, is axially aligned, exhibits "fountain" structure. Stable with no pulsations.
  - **Asymmetric Spouting:** Jet is not axially-symmetric. Can be skewed to one side of the bed, or curved (snake-like) with triangular-shaped stagnant zones.
  - **Dilute Spouting:** Annulus appears to have the same particle concentration as the spout.
  - **Incoherent Spouting:** Exhibits extreme pulsations, with periodic solids entrainment, recirculation and fountain height fluctuations.

Visual Analysis of Spout Stability

NATIONAL ENERGY TECHNOLOGY LABORATORY

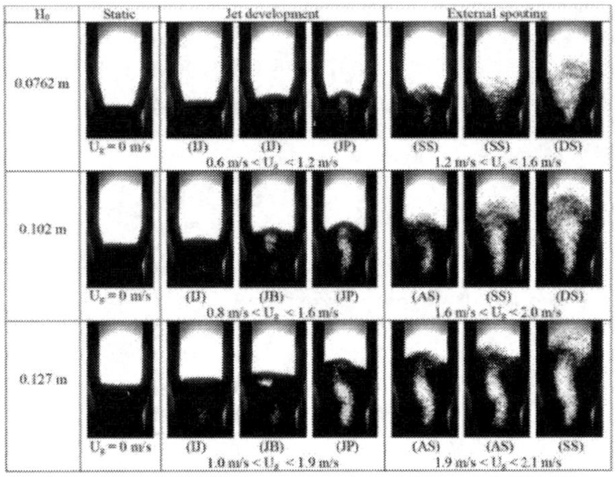

(source: Yang et al., Determination of flow patterns by image analysis in a rectangular spouted bed, submitted to Powder Tech)

- Images show progression from static bed through external spouting regimes for 3/8-in equivalent rectangular nozzle.
- Jet development (internal spouting) regions dominated by:
  - 3-in and 4-in bed: Jet-bubbling
  - 5-in bed: Jet-pulsating
- External spouting regions dominated by:
  - 3-in bed: symmetric spouting transitioning to dilute spouting.
  - 4-in bed: recoverable asymmetric – symmetric-dilute spouting
  - 5-in bed: recoverable asymmetric – symmetric spouting

Visual Analysis of Spout Stability

NATIONAL ENERGY TECHNOLOGY LABORATORY

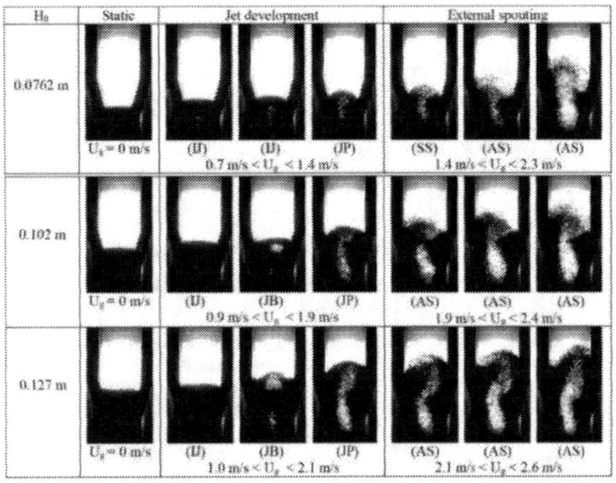

(source: Yang et al., Determination of flow patterns by image analysis in a rectangular spouted bed, submitted to Powder Tech)

- Images show progression from static bed through external spouting regimes for 1/2-in equivalent rectangular nozzle.
- Jet development (internal spouting) regions dominated by:
  - 3- , 4-, and 5-in beds: Jet-bubbling
- External spouting regions dominated by:
  - 3- , 4-, and 5-in beds: asymmetric spouting

Jet-Bubbling vs Jet Pulsating

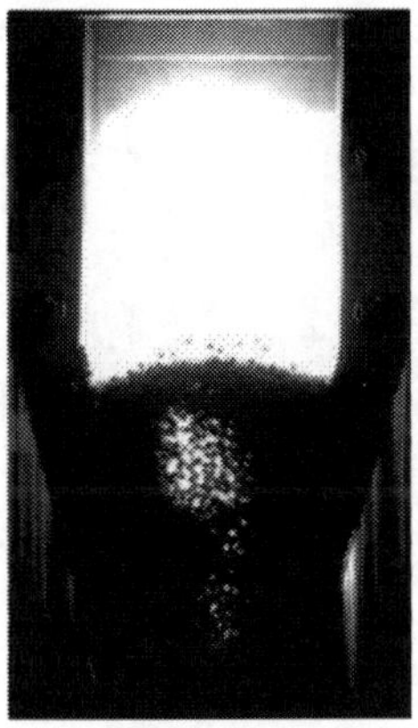
Jet-bubbling regime

Jet-pulsating regime

- **Jet-bubbling:**
  - Jet expands upwards towards the surface of the bed.
  - As it approaches the bed surface, the upper portion of the jet detaches and forms a bubble.
  - These extreme variations in jet height result in larger pressure fluctuations.
- **Jet-pulsating:**
  - Jet is mostly uniform in height.
  - Bubbles that are formed are smaller and detach from the very top of the jet.
  - Leads to smaller fluctuations in pressure.

Regime Mapping

3/8-inch-Equivalent Rectangular Nozzle

1/2-inch-Equivalent Rectangular Nozzle

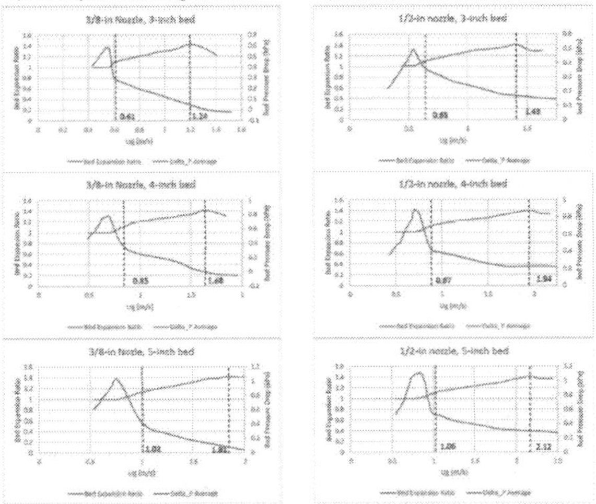

- **Minimum (internal) spouting velocity obtained from average pressure drop profiles.**
- **External spouting velocity determined from bed expansion ratio (HSV analysis)**
- **In each case, $U_{ex} \approx 1.4$**
- **In most cases, $U_{ex} \approx 2*U_{ms}$**
- For the two 5" static bed height cases:
  - **$Ums \approx U_{mf}$** (U_{mf} = 1.08 m/s)

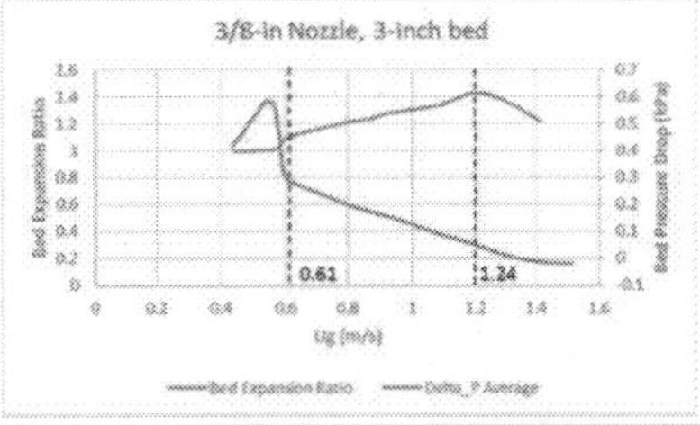

Standard Deviation of Bed Pressure Drop

NATIONAL ENERGY TECHNOLOGY LABORATORY

3/8-inch-Equivalent Rectangular Nozzle | 1/2-inch-Equivalent Rectangular Nozzle

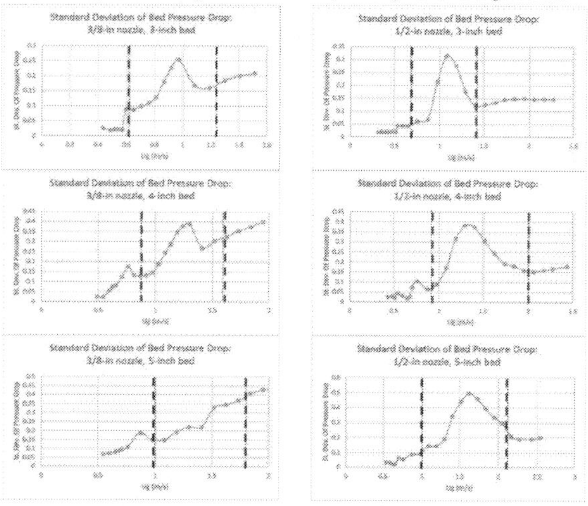

- **In all but the 3/8-in nozzle, 5-in bed case, the area bounded by the dotted lines contains a "hump"**
  - Coincides with the jet-bubbling regime
  - Caused by fluctuations of the jet height due to detaching of bubbles.
- **The 3/8-in nozzle, 5-in bed case lacks the "hump"**
  - Coincides with jet-pulsating regime.
- **Larger "humps" associated with asymmetric spouting suggest that more vigorous bubbling leads to flow asymmetries.**

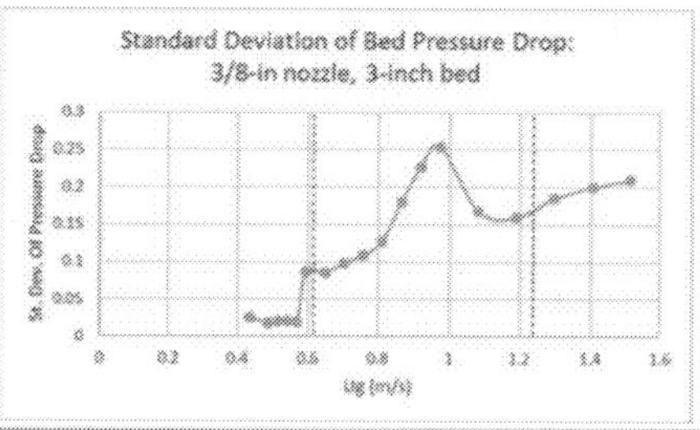

U.S. DEPARTMENT OF ENERGY

Autocorrelation Data

NATIONAL ENERGY TECHNOLOGY LABORATORY

- **Autocorrelation function of a signal:**
  - Correlation between two points within a signal separated by a time lag.
  - If the signal is period, the autocorrelation will be periodic.
  - Function is additive. (if signal has multiple periodicities, so will the autocorrelation)
  - Given by:

$$c_{xx}(k) = \sum_{n=0}^{N-|k|-1} \left(x(n)-\bar{x}\right)\left(x(n-k)-\bar{x}\right)$$

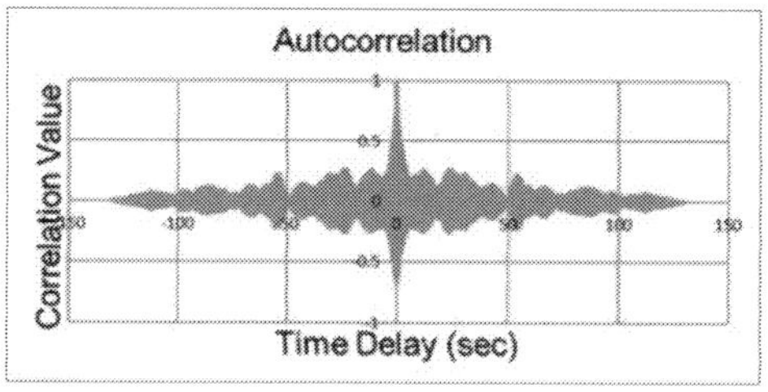

U.S. DEPARTMENT OF ENERGY

Autocorrelation Data

NATIONAL ENERGY TECHNOLOGY LABORATORY

3/8"-equivalent nozzle, 3 inch (0.0762 m) Static Bed

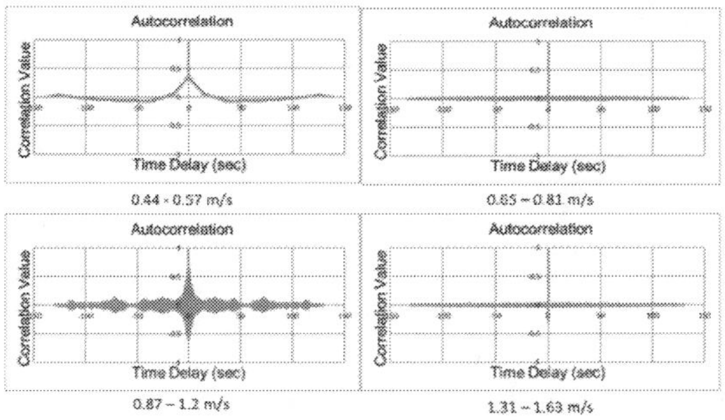

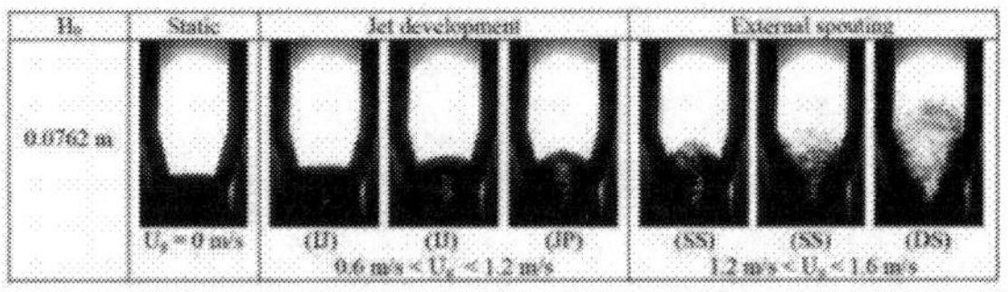

- **The region bounded by U_{ms} and U_{bx} exhibits two separate waveforms in the autocorrelation data.**
  - The first is assumed to be associated with formation of the initial gas jet.
  - The second is believed to be associated with formation of bubbles within the dense bed (i.e. the jet-bubbling regime).

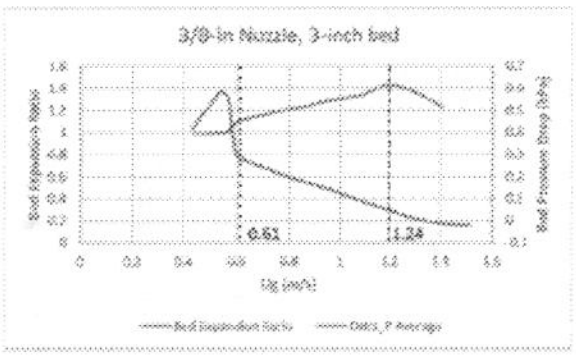

Autocorrelation in Jet Development Region

NATIONAL ENERGY TECHNOLOGY LABORATORY

3/8"-equivalent nozzle, 3 inch (0.0762 m) Static Bed

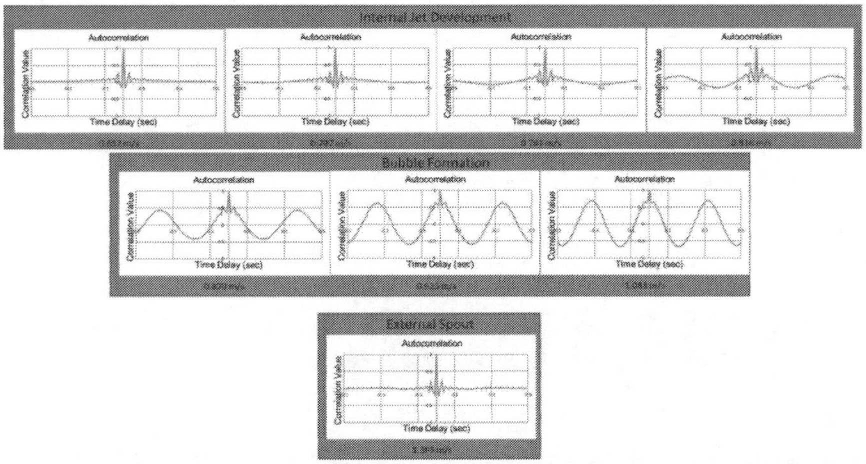

Autocorrelation Data

3/8"-equivalent nozzle, 4 inch (0.102 m) Static Bed

NATIONAL ENERGY TECHNOLOGY LABORATORY

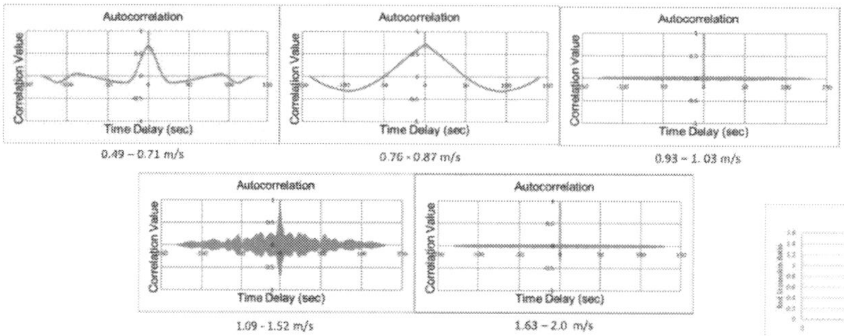

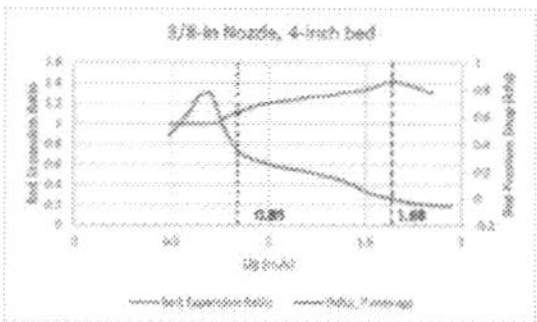

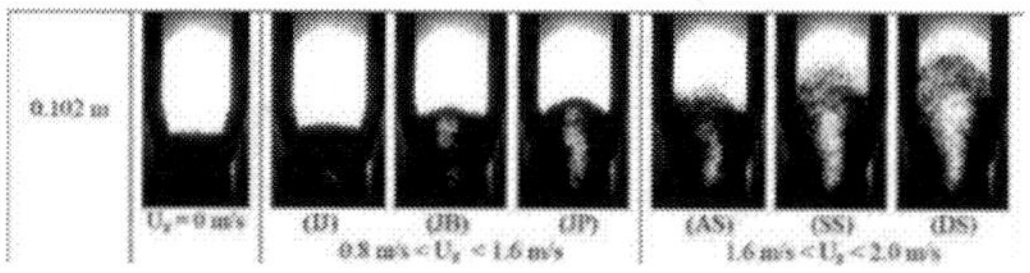

Autocorrelation Data

3/8"-equivalent nozzle, 5 inch (0.127 m) Static Bed

NATIONAL ENERGY TECHNOLOGY LABORATORY

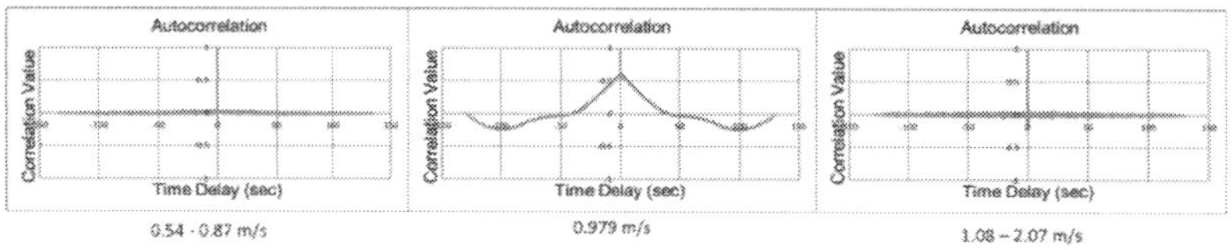

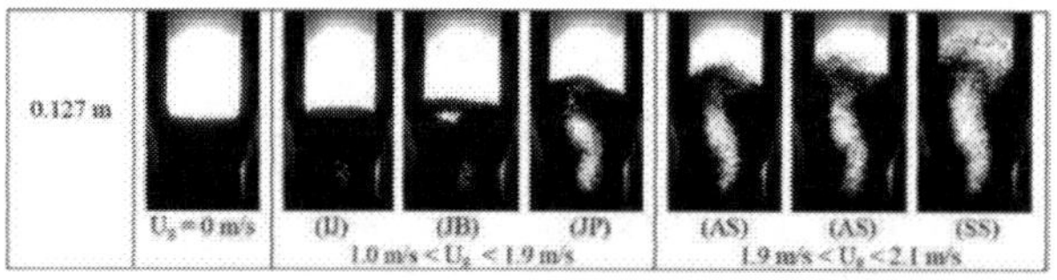

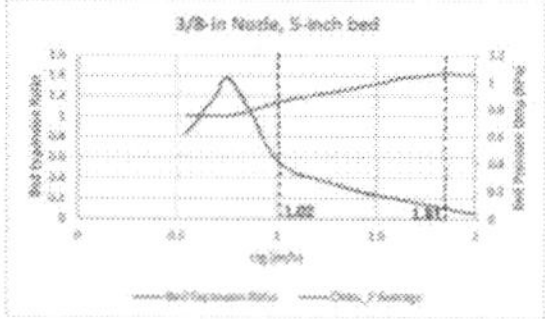

Autocorrelation in Jet Development Region

3/8"-equivalent nozzle, 5 inch (0.127 m) Static Bed

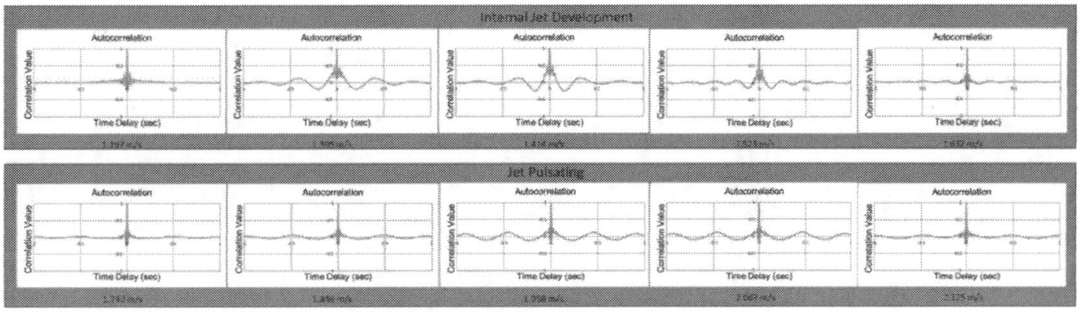

This bed height exhibits a transition from jet formation to jet-pulsating flow as opposed to jet-bubbling, as evidenced by the lack of a strongly correlated, period pressure signal.

Autocorrelation Data

1/2"-equivalent nozzle, 3 inch (0.0762 m) Static Bed

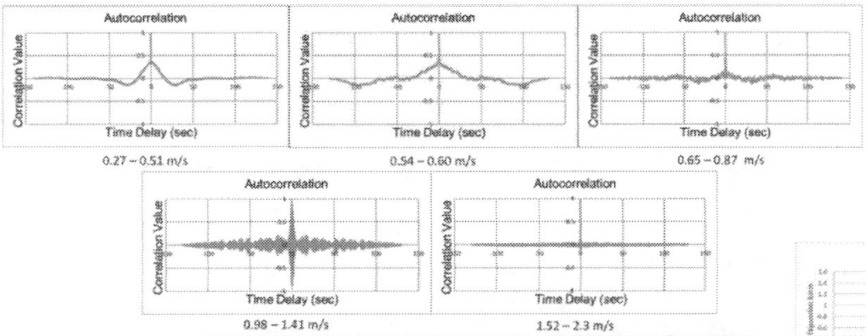

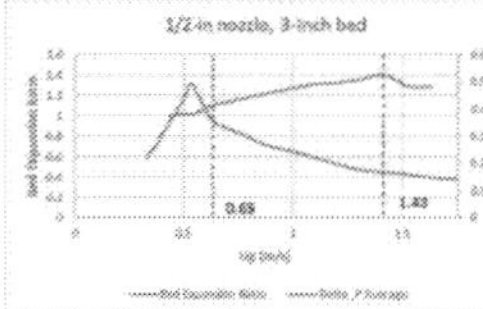

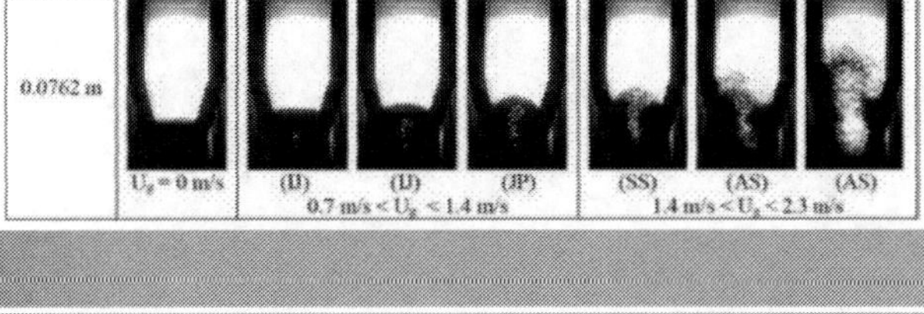

Autocorrelation Data

NATIONAL ENERGY TECHNOLOGY LABORATORY

1/2"-equivalent nozzle, 4 inch (0.102 m) Static Bed

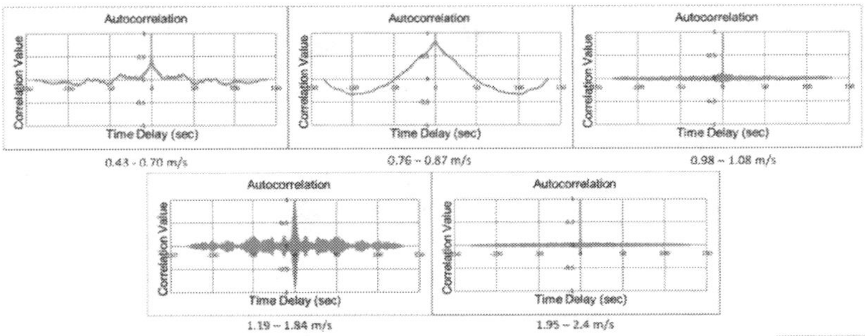

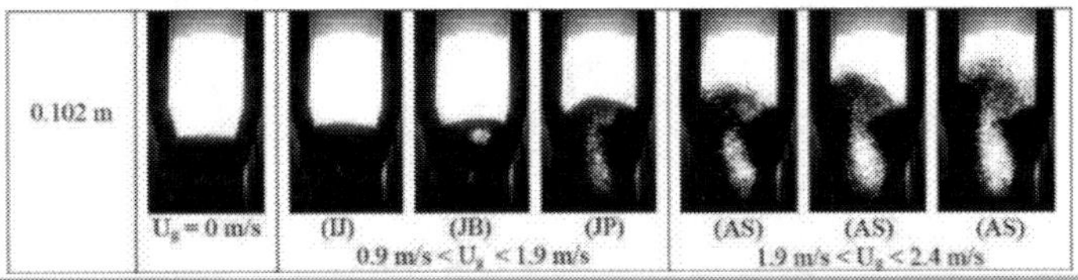

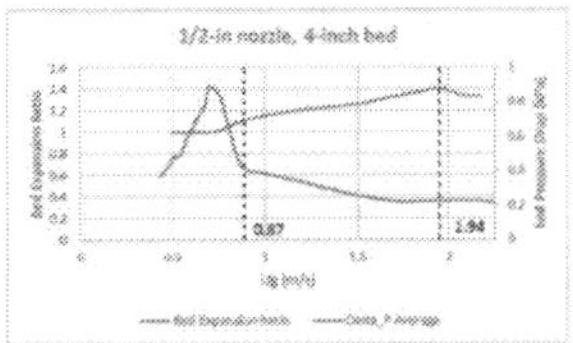

U.S. DEPARTMENT OF ENERGY

19

Autocorrelation Data

NATIONAL ENERGY TECHNOLOGY LABORATORY

1/2"-equivalent nozzle, 5 inch (0.127 m) Static Bed

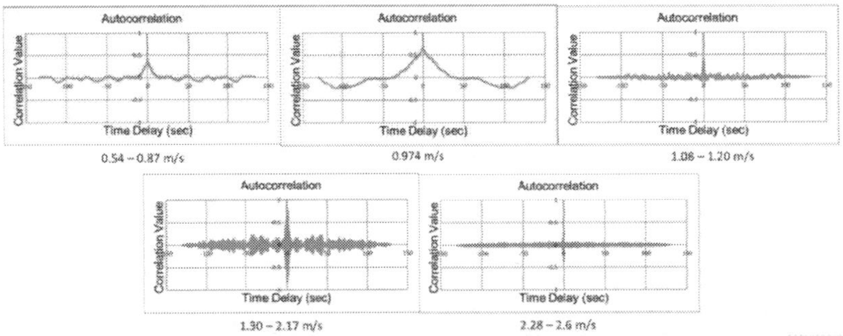

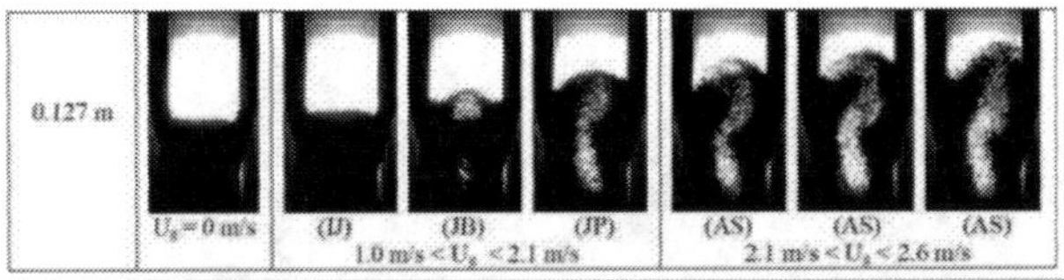

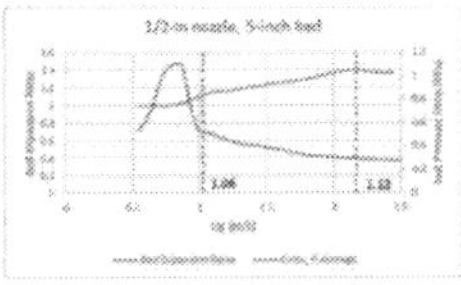

U.S. DEPARTMENT OF ENERGY

20

Power Spectral Density (PSD) Data

NATIONAL ENERGY TECHNOLOGY LABORATORY

- **Power Spectral Density:**
  - Related to the autocorrelation via:

$$S(f)=\sum c_{xx}(\tau)e^{-2\pi i f\tau}$$

  - Provides a measurement of the power associated with a given frequency.
  - Useful for determining dominant frequencies in time series data (i.e. pressure signals).

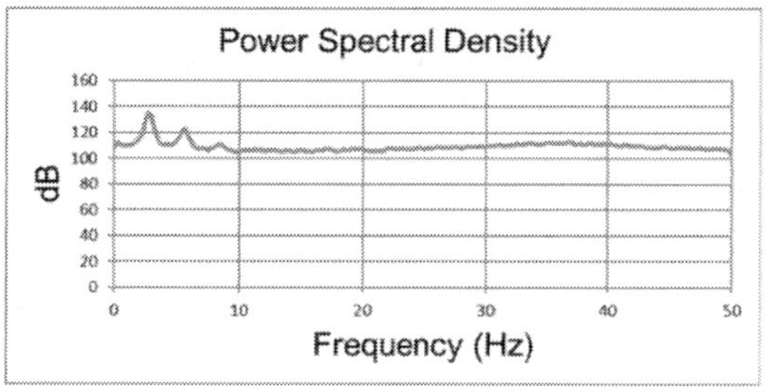

Power Spectral Density (PSD) Data

3/8"-equivalent nozzle, 3 inch (0.0762 m) Static Bed

NATIONAL ENERGY TECHNOLOGY LABORATORY

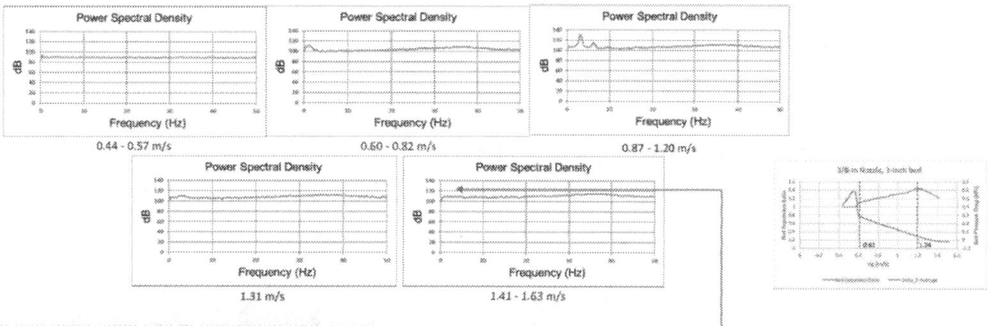

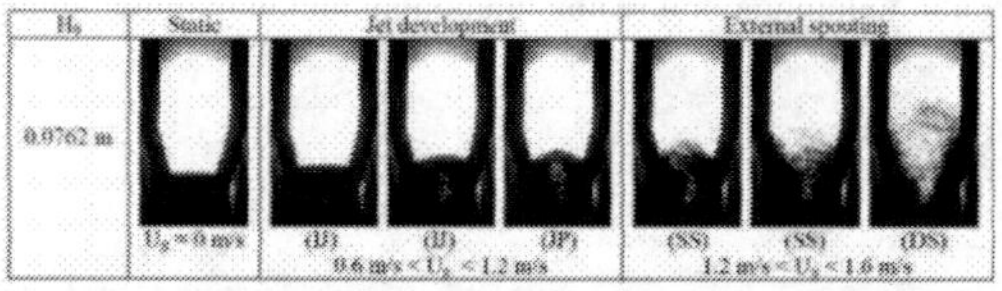

Note that the oeak frequency peak goes away.

Power Spectral Density (PSD) Data

3/8"-equivalent nozzle, 4 inch (0.102 m) Static Bed

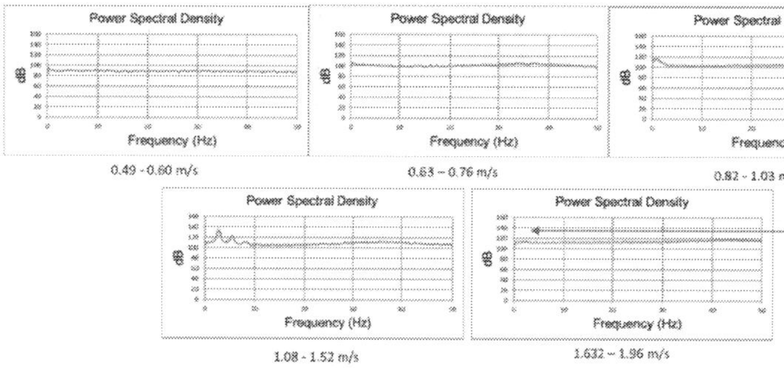

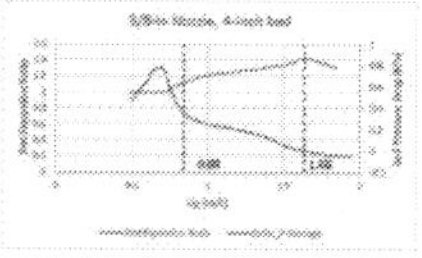

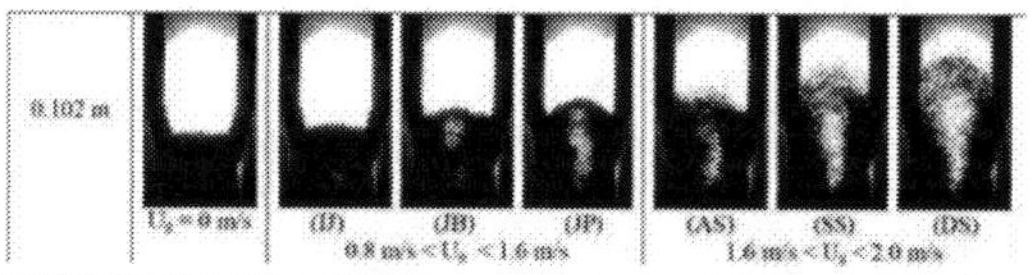

Note that the peak frequency peak goes away.

Power Spectral Density (PSD) Data

3/8"-equivalent nozzle, 5 inch (0.127 m) Static Bed

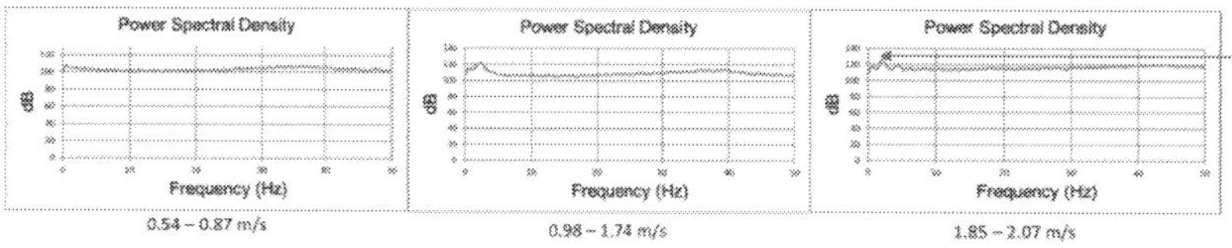

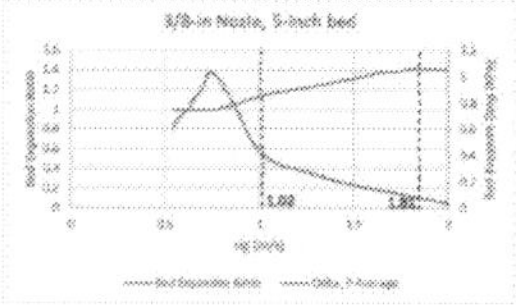

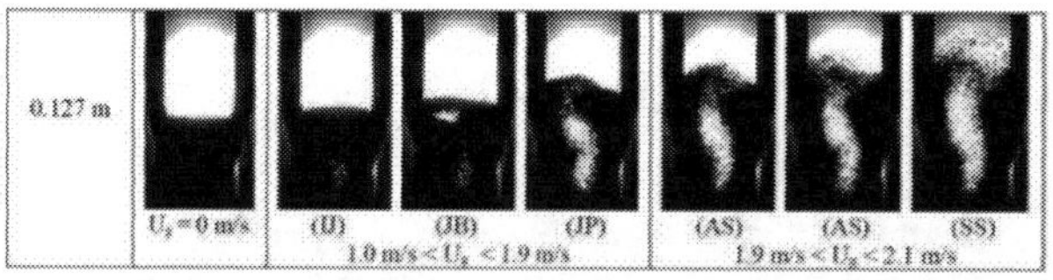

Note: this peak frequency does not go away.

Power Spectral Density (PSD) Data

1/2"-equivalent nozzle, 3 inch (0.0762 m) Static Bed

NATIONAL ENERGY TECHNOLOGY LABORATORY

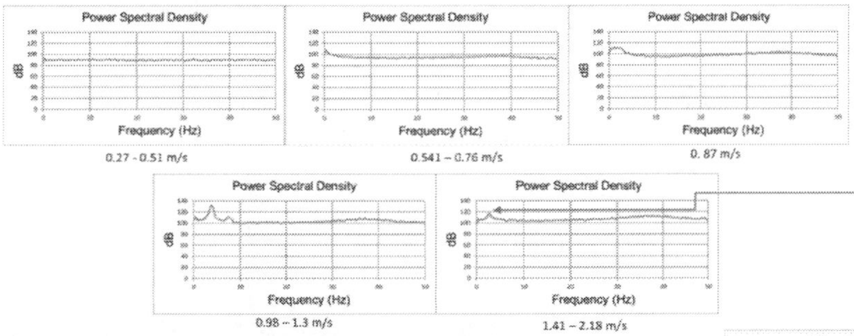

Note: this peak frequency does not go away.

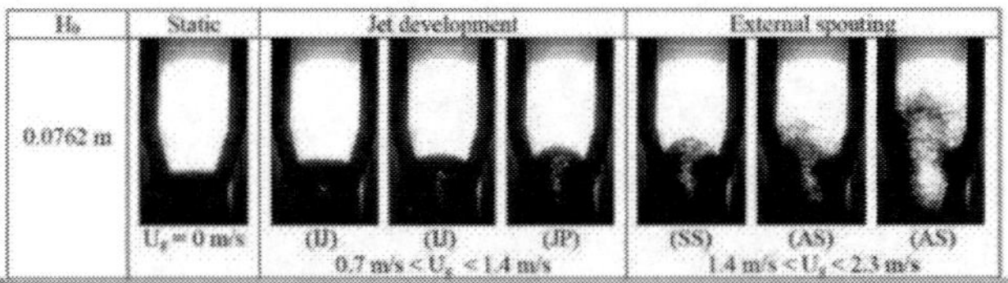

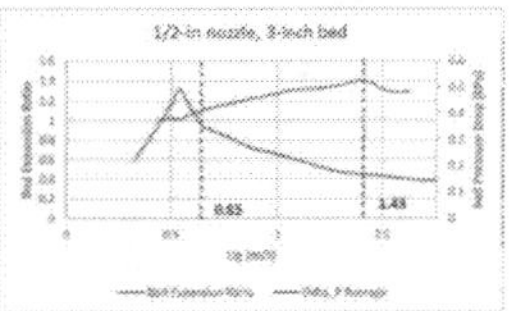

Power Spectral Density (PSD) Data

1/2"-equivalent nozzle, 4 inch (0.102 m) Static Bed

NATIONAL ENERGY TECHNOLOGY LABORATORY

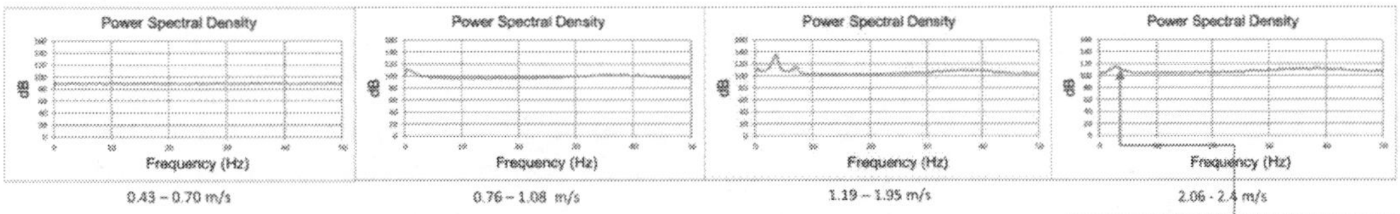

Note: this peak frequency does not go away.

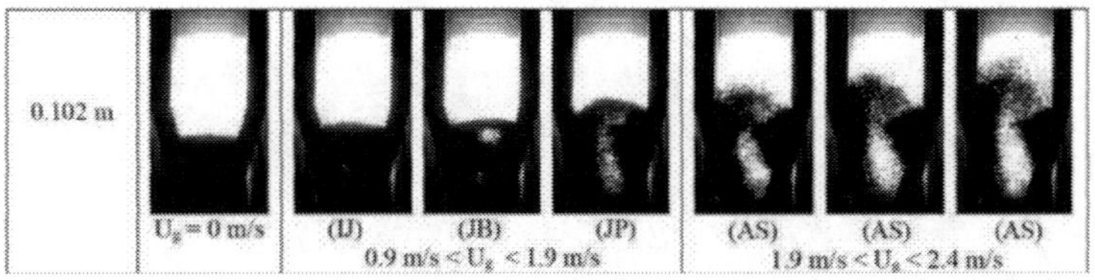

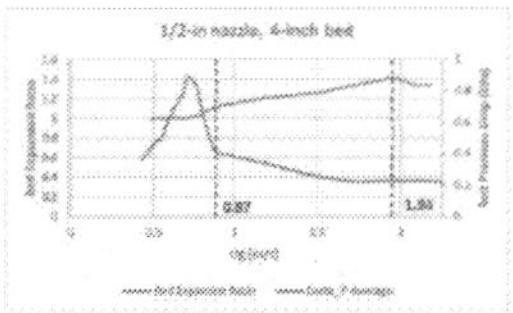

Power Spectral Density (PSD) Data

1/2"-equivalent nozzle, 5 inch (0.127 m) Static Bed

NATIONAL ENERGY TECHNOLOGY LABORATORY

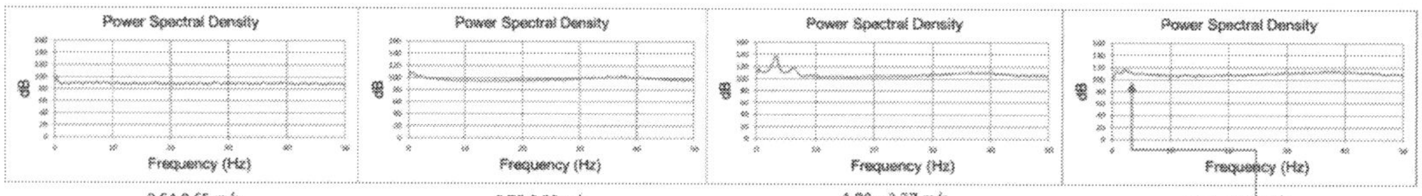

Note: this peak frequency does not go away.

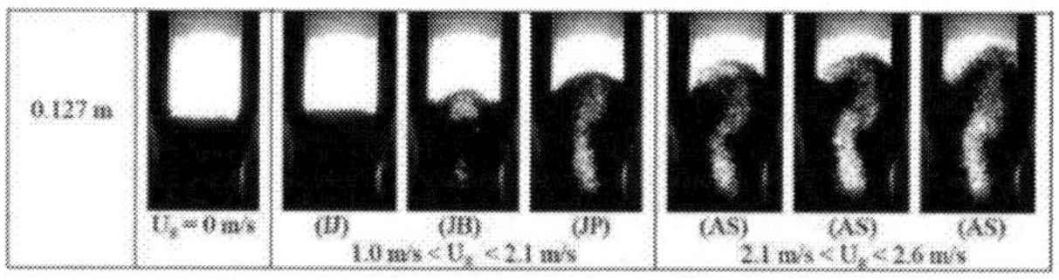

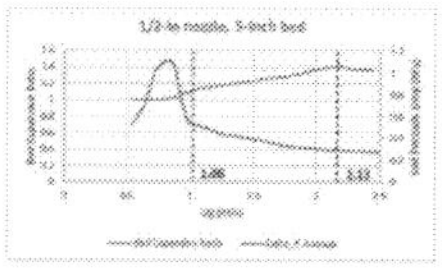

U.S. DEPARTMENT OF ENERGY

27

Making Sense of the Data

NATIONAL ENERGY TECHNOLOGY LABORATORY

Questions to answer:

- **Why does the highest bed height/smallest nozzle case bypass the jet-bubbling regime?**
- **What is the cause of the ~3hz dominant frequency in the asymmetric spouting cases?**

U.S. DEPARTMENT OF ENERGY

28

A Closer Look at Symmetric Spouting

NATIONAL ENERGY TECHNOLOGY LABORATORY

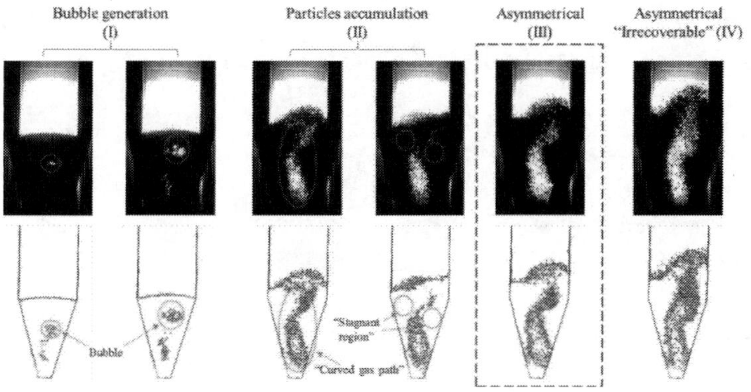

- Asymmetries begin to appear in the internal spouting regime and grow progressively worse.
- The jet "curves" around zones of stagnant particles.
- Particles tend to be ejected into the freeboard in a non axially-symmetric manner.
- This tends to constantly replenish and solids entrained out of the stagnant zones, thus attenuating the asymmetries.
- The authors believe that the interactions between the gas jet and these stagnant zones may lead to the presence of the ~3 Hz dominant frequency present during asymmetric spouting.

U.S. DEPARTMENT OF ENERGY 29

A Closer Look at Asymmetric Spouting

NATIONAL ENERGY TECHNOLOGY LABORATORY

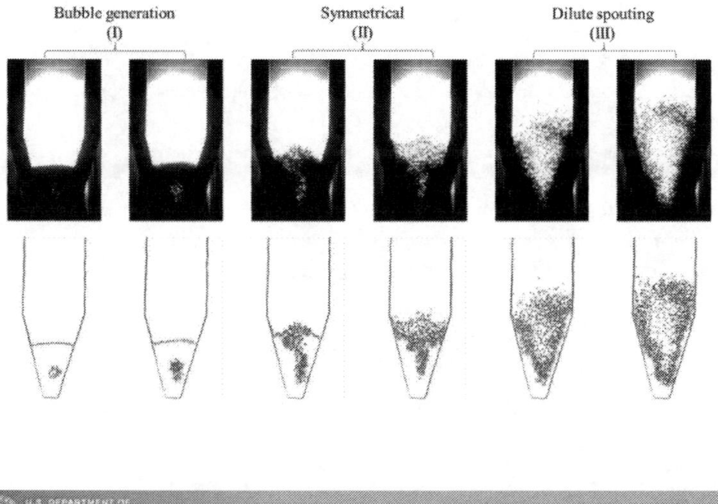

- In comparison to asymmetric spouting, symmetric spouting does not exhibit the formation of stagnant zones of solids.
- At higher gas velocities, the symmetric spout tends to transition to dilute spouting.
- It is assumed that the lack of interactions between the gas jet and stagnant solids zones, as well as the more dispersed nature of the the spouting at higher gas velocities, is what allows the ~3 Hz pressure fluctuations to die out during symmetric external spouting.

U.S. DEPARTMENT OF ENERGY 30

Conclusions:

NATIONAL ENERGY TECHNOLOGY LABORATORY

- **In this on-going work, the spouting behavior of 3.2mm Nylon beads was examined for a range of nozzle sizes and static bed heights.**
- **It was observed that:**
  - The smaller nozzle size generally led to more stable (i.e. symmetric) spouting.
  - The larger nozzle size lead to asymmetric spouting conditions, and the formation of stagnant zones of solids within the annulus region.
  - Asymmetric spouting is preceded by a more vigorous jet-bubbling phase.
  - The presence of a low frequency peak in the spectral density of the pressure data suggests an interaction between the gas jet and the stagnant solids zones under conditions of asymmetric spouting.

U.S. DEPARTMENT OF ENERGY 31

References

NATIONAL ENERGY TECHNOLOGY LABORATORY

- Mathur, K.B., Gishler, P.E., A technique for contacting gases with coarse solid particles, AICHE J. 1 (1) (1955) 157-164.
- Dogan, O.M., L., Frietas, A.P., Lim, C.J., Grace, J.R., Luo, B., Hydrodynamics and stability of slot-rectangular spouted beds, Part I: Thin bed, Chem. Eng. Comm 181, (2000) 225-242
- Mujumdar, A.S., Spouted bed technology – a brief review, in: A. S. Mujumdar (Ed.), Drying '84, Hemisphere McGraw-Hill, New York, (1984) 151-157.
- Epstein, N., and Grace, J.R. (Ed's), Spouted and Spout-Fluid Beds: Fundamentals and Applications, Cambridge University Press, New York, 2011.
- Yang, J., Breault, R.W., Weber, J.M., Rowan, S.L., Determination of flow patterns by image analysis in a rectangular spouted bed, (submitted to Powder Technology)

U.S. DEPARTMENT OF ENERGY 32

NATIONAL ENERGY TECHNOLOGY LABORATORY

ACKNOWLEDGEMENTS

This research was supported in part by an appointment to the National Energy Technology Laboratory Research Participation Program, sponsored by the U.S. Department of Energy and administered by the Oak Ridge Institute for Science and Education.

U.S. DEPARTMENT OF ENERGY 33

Questions?

NATIONAL ENERGY TECHNOLOGY LABORATORY

499538 Bed Density in a Grating Internals Fluidized Bed Stripper

Abstract:

Fluidized bed strippers are essentially flowing fluidized beds where an up-flowing gas removes (strips out) product entrained in a down-flowing stream of the emulsion phase (denser solid phase). The main industrial applications of fluidized bed strippers are in fluid catalytic cracking (FCC) and fluid coking processes where residual hydrocarbons are stripped from solids particles. The stripper facilitates the transfer of the gas in the interstices of the emulsion phase and that adsorbed on the particles into the bubbles of the stripping gas. State of the art knowledge suggests that the counter-current contacting between the down-flowing emulsion phase stream and the up-flowing bubbles stream is enhanced using an assembly of horizontal or inclined baffles. Also, previous studies have shown that in unbaffled fluidized beds the bubbles movement is unrestricted so, as they rise, interact, coalesce and grow they tend to move laterally toward the center. Those conditions result in too low of mass transfer to be an effective stripper.

As is typical for two-phase countercurrent flow unit operations, flooding can be an issue with strippers. Flooding can occur at a given gas velocity if the solids flow is too high or at a given solids flow rate if the gas velocity is too high. Installing internals reduces the risk of flooding because they promote cross-current contacting between gas bubbles and the catalyst. Therefore, as has been previously observed, usage of internals increases not only the stripping efficiency but also the flow through capacity of the stripper.

Test were conducted in a 0.9-m-diameter, 7.5 m tall column with a 3.7 m height baffled with horizontal grating trays, spaced 0.6 m apart. Pressure drops were measured across the baffled height to estimate the stripper bed density. Tests were conducted using equilibrium fluid catalytic cracking (FCC) catalyst at a superficial gas velocity of 0.3 m/s at solids fluxes of up to 150 kg/s-m^2. In this presentation, we will report stripper density variation with solids flux.

500957 Electrical Capacitance Volume Tomography Studies of Characteristics of Gas-Solid Slugging Fluidization with Geldart Group D Particles Under Elevated Temperatures

Abstract:

Slugging which occurs in a fluidized bed with a large height-to-diameter ratio is a major gas-solid fluidization regime. The slugging regime is commonly examined under the ambient condition. With the development of the electrical capacitance volume tomography (ECVT) sensors and measurement technique, the slugging phenomena can now be examined under elevated temperature conditions in a greater detail. The 3-dimensional images from the ECVT sensor can provide the slugging properties such as shape, size, volume fraction and velocity, thereby providing overall understanding of the slugging behavior in a fluidized bed at elevated temperatures. In this study, the ECVT sensor is applied to a slugging fluidized bed with a 0.075 m diameter with solids particles of 1.5 mm diameter and 2500 kg/m3 density. Square-nosed slugs are observed in the bed. It is found that the time intervals between two slugs decreases with the increase in the gas velocity and reaches a steady value at about 1s per slug when the gas velocity is about 1.7 m/s above the minimum fluidization velocity. Under the ambient temperature, when the gas velocity is lower, the bed fluctuates with a high frequency mainly due to bubbles. However, under elevated temperatures, this phenomenon is not observed. The slug rising velocity increases linearly with the gas velocity. However, its value is lower than that of the round-nosed slugs as calculated from previously proposed equations based on the two-phase theory.

502421 Local Agglomeration Measurements of Mildly-Cohesive Particles in a Dilute Riser

Abstract:

Circulating fluidized beds (CFB) are commonly used in industrial operation, and typically involve mildly-cohesive (Group A) particles (e.g. FCC particles). Here, mildly cohesive (Group A) spherical particles were carefully characterized and used in an investigation into the effects of superficial gas velocity, particle size and relative humidity on agglomerate and particle behavior. With the use of a state-of-the-art camera that can resolve particles down to 10 microns in diameter, novel local measurements of agglomerate size and fraction of agglomerate size in a dilute, experimental riser are reported, along with local measurements of solids volume fraction, granular temperature, and mean velocity of primary particles and agglomerates. The results presented here are useful to understanding the behavior of cohesive particles in risers and for model validation of cohesive-particle theories.

499834 Population Balance Modeling of Breakage of High Aspect Ratio Crystals in Stirred Slurries

Abstract:

Modeling of crystal breakage predominantly focuses on low aspect ratio crystals. When high aspect ratio crystal breakage is studied, it is often through single particle impact experiments or with multiple crystals in mills or agitated dryers. However, breakage also occurs in slurries in stirred vessels such as crystallizers. This work investigates breakage of high aspect ratio crystals, such as paracetamol, in agitated slurries. The aim is to develop realistic functions for specific rates of breakage and breakage distribution functions for high aspect ratio crystals. This work presents the development of specific rates of breakage and the breakage distribution functions from experimental data. Specifically, these functions are based on both the aspect ratio and the major axis of each crystal. A population balance model that includes the newly developed functions is used to generate crystal size distributions as a function of time. Comparisons of the model results with experimental data indicates how well these functions account for real behavior. Although this development is for a single system, the approach is of general applicability and can be used for other high aspect ratio crystals.

488045 Population Balance Equation for Calculation of the Inlet Distribution for Oil and Water Droplets in Continuous Gravity Separators

Abstract:

In this work, we use a population balance equation (PBE) to describe the distribution of oil and water droplets in the respective water- and oil-continuous phases in the turbulent inlet section of a continuous gravity separation unit. The volumes of the oil- and water-continuous layers in the inlet section are considered as ideal stirred tanks, thus the PBE for each layer can be expressed as $df(r)/dt = C_+(r) - C_-(r) + B_+(r) - B_-(r)$, where the right-hand side expressions describe the birth (index +) and death (index -) integrals for both coalescence (C) and breakage (B). We use the particle radius as inner coordinate and formulate the PBE as a dimensionless equation (Grimes, 2012). We do not regard existence and uniqueness of solutions for the above PBE in this work, however, our studies indicate that the PBE converges rapidly to a unique solution for a distinct set of parameters, but independent of the initial condition.

Thus, the equilibrium solution to the dynamic PBE could represent the inlet (boundary) condition for a simple dynamic gravity separator model as introduced in (Backi & Skogestad, 2017). The model consists of three dynamic states, namely the levels of water- and overall liquid (oil + water) as well as the gas pressure. These dynamic states are controlled to nominal values by PI controllers via the respective outflows. In addition, simple steady-state droplet balance calculations solely based in Stokes' Law are applied to the initial distribution of oil and water droplets. Hence, birth and death due to coalescence and breakage are neglected as the droplets move along the length of the separator towards the outlets; this will be subject to future work. The droplet size classes are static inside the separator and thus only sedimentation and creaming will be regarded.

Further assumptions include

- There are no liquid droplets in the gas phase and no gas droplets in the liquid phases
- No emulsion layer (dense-packed layer) is considered between the continuous water and oil phases; the droplets are regarded to directly leave into their respective bulk phases at the interface. This means that the rate of interfacial coalescence is smaller than the flux of droplets to the interface (reasonable if an effective demulsifier is properly dosed to the separator inlet stream)
- The horizontal velocity of the droplets inside the gravity separator is equal to the velocity of the respective phase (volumetric inflow divided by the respective cross sectional area); hence, no slip between dispersed droplets and continuous phase is assumed
- The velocity of the droplets in vertical direction is determined by Stokes' Law without regarding a correction factor as e.g. described in (Richardson & Zaki, 1954)

By comparing horizontal and vertical velocities of each (constant) particle class, we can calculate the total volumes of particles going into their respective bulk phases or leaving the separator through the dispersed phase's outlet.

Backi, C. J., & Skogestad, S. (2017). A Simple Dynamic Gravity Separator Model for Separation Efficiency Evaluation Incorporating Level and Pressure Control. *Proceedings of the 2017 American Control Conference.*Seattle: IEEE Control Systems Society.

Grimes, B. A. (2012). Population Balance Model for Batch Gravity Separation of Crude Oil and Water Emulsions. Part I: Model Formulation. *Journal of Dispersion Science and Technology, 33*, 578-590.

Richardson, J. F., & Zaki, W. N. (1954). The sedimentation of a suspension of uniform spheres under conditions of viscous flow. *Chemical Engineering Science, 8*, 65-73.

496712 Data Driven Model for the Prediction of Particle Size Distribution from Measured Chord-Length Distribution: Model Extensions and Application to Population Balance Model Identification

Abstract:

A new data-driven model has been developed to determine 1D/2D particle size distribution (PSD) from measured FBRM chord-length distribution (CLD) data. The structure of the model consists of three steps: first, the measured CLDs are compressed down to a small set of parameters; second, these parameters are correlated with low order moments or a small number of percentiles of the PSD using regression models; third, the PSD low order moments are used as input variables, and a two layer-network sub-model is built to predict the PSD in a form of a histogram. In this work the model is further extended using geometric models and principal component analysis (PCA) as slave models for the first step. This modeling paradigm is utilized to convert in-line measured CLD to predicted PSD by laser diffraction of particles with an elongated morphology. The de-convoluted data is then used with population balance equations to identify growth and aggregation mechanisms. There is no special requirement for the preparation of the training set, and the model is robust to potential second order artifacts in the size distribution. Thus the modeling paradigm is very robust and economic in terms of data requirements. The model steps are physically intuitive allowing using this paradigm to model multiple types of systems.

491484 Population Balance: The Auto-Catalytic Nature of Secondary Nucleation in Seeded Batch Crystallizations

Abstract:

ISIC 20 Abstract Submission

Population Balance: The Auto-Catalytic Nature of Secondary Nucleation in Seeded Batch Crystallizations

Rory Tyrrell, Brian De Souza, and Patrick J. Frawley

Synthesis and Solid State Pharmaceutical Centre

Bernal Institute

University of Limerick

Ireland

Existing models for secondary nucleation kinetics predominantly attribute the generation of nuclei to varying forms of particle breakage or attrition during the crystallization process. However, recent literature has provided new insight into a secondary nucleation mechanism in seeded batch reactions whereby the seed itself serves as a breeding ground for new secondary nuclei to form on its surface (Anwar, et al., 2016; De Souza, et al., 2016). Furthermore, it has been reported that crystal breakage can be neglected almost entirely in typical crystallization processes as a boundary layer effect between crystals and reactor features, such as impeller blades, effectively protects against any potential breakage events (Tyrrell, et al., 2017). Through this deeper understanding of the crystallization process, process modelling tools such as the Population Balance (PB) method have become gradually become more accurate as these tools are only as strong as their underlying kinetic expressions.

Presented here is a Finite Volume Method (FVM) for solving the Population Balance equations for growth, secondary nucleation, and associated mass balances (i.e. de-supersaturation). As the kinetic expressions for growth are known to be relatively well defined, focus is given mostly to the secondary nucleation expressions governing the generation of new nuclei in the early time-scales of the crystallization process. It is assumed that crystal breakage is a non-issue, and that the process is non-agglomerating. Therefore, the modelling methodology is focused around developing a kinetic expression describing the surface-nuclei, auto-catalytic mechanism for secondary nucleation proposed in the literature above.

In addition to a new kinetic expression for secondary nucleation, a full description of the formulation for the FVM-PB model is described. The necessity for high-resolution models is discussed, in combination with flux-limiting methods for managing the associated numerical oscillations typical of higher-order systems. Furthermore, a critical factor of FVM-PB models, as well as modelling in general, is the requirement for governing conservation laws to be maintained; violation of mass-balances in finite volume schemes with numerical source terms for nucleation can prove to be a challenging problem to surmount and warrants discussion.

References

Anwar, J., Khan, S. & Lindfors, L., 2016. Secondary Crystal Nucleation: Nuclei Breeding Factory Uncovered. *Angewandte Chemie,* 54(49), pp. 14681 - 14684.

De Souza, B., Cogoni, G., Tyrrell, R. & Frawley, P. J., 2016. Evidence of Crystal Nuclei Breeding in Laboratory Scale Seeded Isothermal Crystallization Experiments. *Cryst. Growth Des.,* 16(6), pp. 3443 - 3453.

Tyrrell, R., De Souza, B. & Frawley, P. J., 2017. Particle Breakage: Limiting Conditions for Crystal-Crystallizer Collisions. *Crys. Growth Des. (Under Review).*

502313 DEM-Based Prediction of Critical Impact Velocities of Aggregation and Breakage and Daughter Distributions of Cohesive Powders

Abstract:

Flowing, cohesive particles are subject to the formation and breakage of agglomerates. One approach to the continuum modelling of such flows is the population balance coupled with kinetic-theory-based conservation equations. The population balance includes sink and source terms representing aggregation and breakage. These sink and source terms depend on the success factors of aggregation and breakage, which quantify the fraction of collisions that result in aggregation and breakage. Further, the source term due to breakage includes a daughter distribution, which gives the sizes of agglomerates formed when a large agglomerate undergoes breakage. Previous works have shown that the success factors of aggregation and breakage, as well as the daughter distribution, depend on impact velocity. In this work, a fundamental approach is taken to relating the success factors and daughter distributions to the granular temperature (continuum variable in kinetic-theory-based balances), which is a measure of the impact velocity. Next, many-particle discrete element method (DEM) simulations are utilized to determine the critical impact velocities of agglomeration and breakage and the daughter distribution as a function of impact velocity and over a range of agglomerate sizes. Collectively, this work demonstrates a new DEM-based methodology for extracting the closures needed for the population balance.

496115 Towards a General Model for Twin-Screw Wet Granulation: Development and Calibration of a Novel Three Compartmental PBM Model

Abstract:

Traditional pharmaceutical solid oral dosage processes comprise of a series of batch unit operations. More recently, a transition is developing from batch unit processes to continuous manufacturing application, to cope with the inefficiencies and high cost involved in process development & scale up. Twin-screw wet granulation is a developing pharmaceutical continuous process that is being assessed for its performance in solid dosage manufacturing. In this research, the twin-screw wet granulation unit is a unit operation of the ConsiGmaTM-25 continuous powder-to-tablet process line from GEA Pharma Systems. However, since these continuous processes are still under development focus in the pharmaceutical industry, detailed process knowledge and understanding is still evolving. Application of mechanistic models can help in filling knowledge gaps by thoroughly investigating gathered detailed experimental data and unravelling the underlying mechanisms.

For this development, a unique collection of experimental datasets at different locations inside the granulator was created (after the wetting zone, after the first and second kneading zone). This gives significant insight into what mechanisms take place at what location. When granulating the preblend, the main differences are observed in the granule size distribution (GSD) when the liquid to solid (L/S) ratio changes. At low L/S ratio, the GSD exhibits bimodality, whereas at high L/S ratio the GSD shows unimodal behaviour.

Based on these observations, a novel Population Balance Model (PBM) was developed for predicting the GSD at the end of the outlet of the granulator starting from the pre-blend. In PBM, physical mechanisms, such as aggregation and breakage of granules, are represented by kernels which are often semi-empirical in nature. This work pointed out that none of the traditional kernels from literature could reproduce our experimental observations (i.e. calibration failed), mainly because their shape was contradicting the observed trends and no single set of kernels could predict both unimodal and bimodal behaviour. Therefore, novel kernels had to be developed.

In the model we propose, prediction of both the unimodal and the bimodal behaviour with one single aggregation kernel structure is possible. The developed kernel consists of a convolution of a continuously monotonic increasing function (to drive the GSD to higher granule sizes) and a smooth stepping function to drive the GSD to bimodal behaviour when needed. This stepping function can be reduced in weight or completely switched off in case of unimodal behaviour. The breakage kernel (only used in the kneading zones) is a combination of attrition and uniform breakage.

The hypothesis for this aggregation kernel is that aggregation is the source of bimodality, whereas in literature it is often suggested to be caused by breakage. However, from the performed measurements of the GSD in the wetting zone, it is evident that bimodality is induced in this zone, where breakage is highly unlikely. With changing L/S ratio, the amount of available liquid is quite different. When low amounts are available in the system (low L/S ratio), particles cannot grow larger than a certain size because there simply is not sufficient liquid available to promote further granulation to larger sizes. In this way, hypothesised mechanistic knowledge is built into the mathematical model. If breakage would be the source of bimodality, there is no reason why there are differences at different L/S ratios, as the observed friability is expected to be similar in both cases.

The full model is in fact a combination of three PBM models in series: a PBM compartmental model. For each zone for which data is gathered, a compartment is made: the wetting zone (preblend to after the addition of the liquid binder), kneading zone 1 (from after the wetting zone to after the first kneading elements), kneading zone 2 (just before the second kneading elements up until the end of the granulator barrel).

The model structure proposed for the aggregation kernel yields good results for a range of L/S ratios. The use of one single kernel to achieve this is a major step forward as it points out that similar mechanisms are occurring but at a different rate for different L/S ratios.

496127 Towards a General Model for Twin-Screw Wet Granulation: Validation of a PBM Model through Linking Calibrated Model Parameters to Process Conditions

Abstract:

In a newly developed three compartmental PBM model with a novel kernel for aggregation, prediction of both unimodal and bimodal behaviour for different L/S ratios with one single aggregation kernel structure is achieved. The use of one single kernel to achieve this is a major step forward as it points out that similar mechanisms are occurring but at a different rate for different L/S ratios. The model structure and calibration is presented in another abstract.

In the work presented here, the link between operational conditions (screw speed, mass flow rate, and L/S ratio) and the calibrated model parameters of the novel PBM is investigated. Since the experimental data is gathered using a three level full-factorial design of experiments (DOE), the effect of the experimental conditions screw speed, mass flow rate, and L/S ratio on the calibrated model parameters can be investigated. Separate relationships were retrieved and built into the PBM. The latter can now predict granule size distributions taking the experimental conditions as input.

For the validation, two new independent experiments were performed that were not used during model calibration. The L/S ratio of the centre point was perturbed in both directions leading to different levels of bimodality.

The predictions of the validation experiments were found in good agreement with the experimental data. This is a proof that the model is applicable in a whole DOE space, and that the PBM for TSWG of this hydrophobic compound can be considered as validated. Further, this means that the addition of mechanistic knowledge into the aggregation kernel is a vital part of modelling the correct behaviour inside the granulator. It significantly increased the predictive power of the model.

496135 Towards a General Model for Twin-Screw Wet Granulation: Application of a Novel Validated PBM Model to a Hydrophilic Compound and Linking Model Parameters to Blend Properties

Abstract:

In a newly developed three compartmental PBM model with a novel kernel for aggregation, prediction of both unimodal and bimodal behaviour for different L/S ratios with one single aggregation kernel structure is achieved. The use of one single kernel to achieve this is a major step forward as it points out that similar mechanisms are occurring but at a different rate for different L/S ratios. The model structure and calibration is presented in another abstract. Moreover, the model was extended to include dependencies on operational conditions (screw speed, mass flow rate and L/S ratio) and validated using independent experimental data. This work is presented in a second abstract.

In this third abstract, the same novel PBM model is applied to a hydrophilic compound based on the assumption that the granulation behaviour is similar, i.e. the shape of the GSD is mainly dominated by L/S ratio.

From the calibration effort, it is observed that the GSD can be well predicted by the novel PBM structure, similar to the hydrophobic compound. This leads to the conclusion that the granulation mechanisms in TSWG are similar for different blends with different properties, but the process rates depend on the formulation.

Ultimately, the novel PBM forms the basis for a model that not only takes into account the operational settings screw speed, mass flow rate, and L/S ratio, but also the properties of the preblend. When linking the three parts (model parameters, blend properties, and experimental conditions), new blends can be tested in a more efficient manner by using the model to predict the expected behaviour. The model can help in predicting the GSD, and the operational conditions that lead to the optimal GSD. Moreover, the model can be used to indicate the locations in the operational settings space for which experiments would yield a high information content different mechanisms of the granulation. This is called optimal experimental design (OED) which can, opposed to traditional Design of Experiments (DOE), significantly reduce the amount of experiments required for new formulations.

492973 Population Balance Modeling Applied to the Milling of Hot-Melt Extrudates

Abstract:

Hot-melt extrusion has become a common process in the pharmaceutical industry to improve the solubility and bioavailability of low solubility drugs. Milling following the extrusion process is a necessary unit operation so that the extrudate can be blended with excipients for production of capsules and tablets. Despite the importance of milling processes, as in the production of drug products utilizing extrudate, they are still poorly understood and difficult to control. Population balance modeling (PBM) has traditionally been applied to milling processes for control, optimization, simulation, and scale-up. The PBM is composed of the selection function, which describes the propensity for a material to undergo breakage, and the distribution function, which describes the progeny of particles created once a particle is broken. The parameters of these functions depend on the milling conditions (e.g. impeller speed, screen size, feed rate, etc.) and the properties of the material to be milled (e.g. hardness, facture toughness, etc.). Once the parameters of these functions are determined, the PBM can readily predict the particle size distribution (PSD) for any material and any milling condition.

In this work, the so-called time-discrete event-based breakage PBM is applied to model the size reduction of copovidone extrudate using an impact-type mill (Fitz mill). The impeller speed and the screen size of the Fitz mill were varied to determine the dependence of the PBM model parameters to milling conditions. Material properties and process parameters are successfully separated in the PBM which allows for accurate prediction of PSD. The PBM was extended to predict the PSD using a using a pilot-scale mill and found to be useful for scale-up Accordingly, the application of PBM should greatly reduce the effort related to milling development.

All authors are employees of AbbVie. The design, study conduct, and financial support for this research was provided by AbbVie. AbbVie participated in the interpretation of data, review, and approval of the publication.

499348 Discrete and Continuum Plane-Hopper Flow Simulations with Experimental Validation Using x-Ray Imaging, Arch Profilometry, and Wall Pressure Measurements

Abstract:

The flow of spherical glass beads, approximately cubic wood crumbles®, and wood microchips are investigated using a combination of discrete element modeling (DEM), finite element analysis (FEA), and physical experiments using a custom plane-flow hopper in which the flow is characterized using x-ray imaging, profilometry to measure arching geometry and pressure measurements at the wall. The purpose of this study is to investigate fundamental flow behavior, and for this purpose the simulations and physical tests are specially designed. During the physical tests and simulations, the hopper geometry begins as a V shape with the two opposing side walls touching at the bottom of the V. Basic tests are performed by first filling the hopper with material, and then rotating the opposing side walls inward about their common intersection axis to cause an approximately uniform stress condition throughout most of the hopper. The side walls then slide upward to effectively increase the size of the opening at the bottom of the hopper until the arch of material in the hopper breaks, and all of the material falls out of the hopper. More advanced tests are also performed in which the initial rotation of the walls is slightly reversed to decrease the pressure in the hopper, corresponding to "preshear" and "shear" conditions in standard shear tests. Shear tests of the materials are also performed using a Schulze automated ring shear tester with a standard size M shear cell, and the shear tests of the glass beads and wood crumbles® are also simulated using DEM to calibrate the DEM with the FEA. Reasonable agreement between predicted and measured flow behavior is obtained indicating that this type of physical tests and modeling shows promise for characterizing and predicting the flow behavior of compressible and anisotropic materials.

RUTGERS
THE STATE UNIVERSITY OF NEW JERSEY

Predicting Feeder Performance during Hopper Refill Based on Material Flow Properties

Tianyi Li, Yifan Wang, Benjamin Glasser, Fernando Muzzio
Oct. 30, 2017

RUTGERS

Outline

- Motivation and Goals
- Materials and Methods
  1. *Materials*
  2. *Material flow property measurements*
  3. *Feeder refill characterization*
  4. *Multivariate analysis*
- *Results and Discussion*
- Conclusion and Future work

Image source: www.basf.com

2

RUTGERS

Motivation

- Feeder has been recognized as an critical unit operation to many powder-based processes across industry.
- The refilling of feeder may cause deviation in feeding consistency and affects the downstream process.
- *Variability of flow properties has major impact on manufacturing process and product quality.*

Goals

- To build up a multivariate model to predict feeder's performance based on materials flow properties.
  i. PCA-SS (Principal component analysis followed by similarity scores)
  ii. PLSR (Partial least squares (PLS) regression)
- To predict deviations during feeder refill based on material's flow properties.

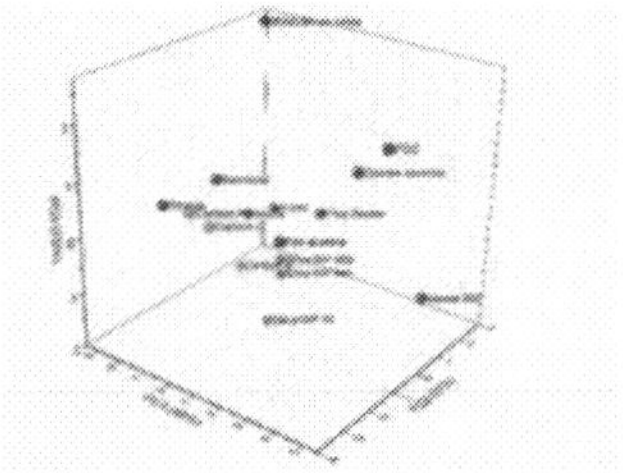

3

RUTGERS

Material Library

- Lactose Monohydrate (ERC-SOPS)
- Fine Alumina (Albemarle)
- Molybdenum Dioxide (Albemarle)
- Zeolite (BASF) – Zeolyst Y-type CBV100
- Zeolite Y-655, calcined (Albemarle)
- Fine Zeolite (Albemarle) – hydrophobic, small spherical particles
- Satintone (BASF) – calcined kaolin BASF Grade Satintone SP-33

4

RUTGERS

Characterization Techniques for Material Library

- Particle size distribution
  - *d10, d50, d90*
- Shear cell test
  - *Cohesion, Unconfined Yield Strength, Major Principal Stress, Flow Function Coefficient, and Angle of Internal Friction at initial consolidation stresses of 3kPa, 6kPa, 9kPa, and 15kPa*
- Compressibility test
  - *Conditioned bulk density, Compressibility % at consolidation of 15kPa*
- Permeability test
  - *Pressure Drop at consolidation of 15kPa*
- Stability/ Variable Flow Rate test
  - *Basic Flow Energy, Stability Index, Specific Energy, Flow Rate Index*

5

RUTGERS

Loss-in-weight feeder experimental set-up

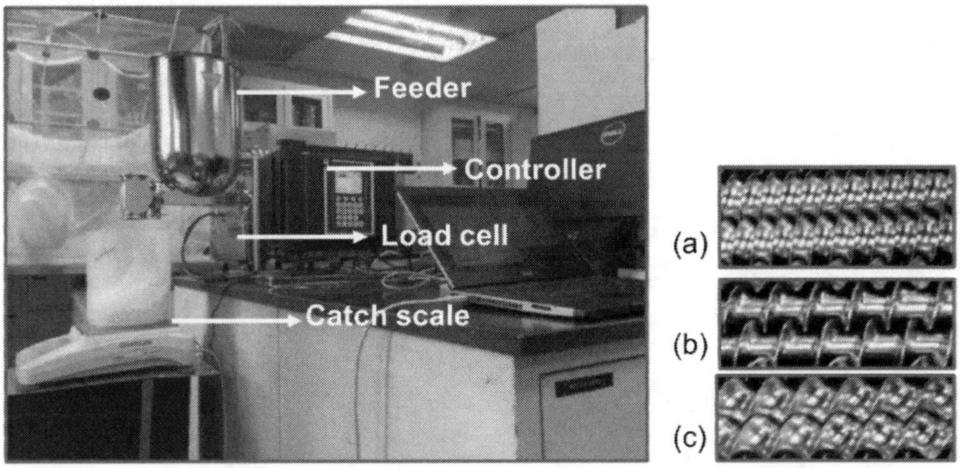

Experimental set-up of characterizing feeding performance for each material. Available screws for feeder experiment: (a) screw type 1, fine concave screw, (b) screw type 2, fine auger screw, and (c)screw type 3, coarse concave screw.

RUTGERS **Previous work**

Predicting feeding performance based on material flow properties

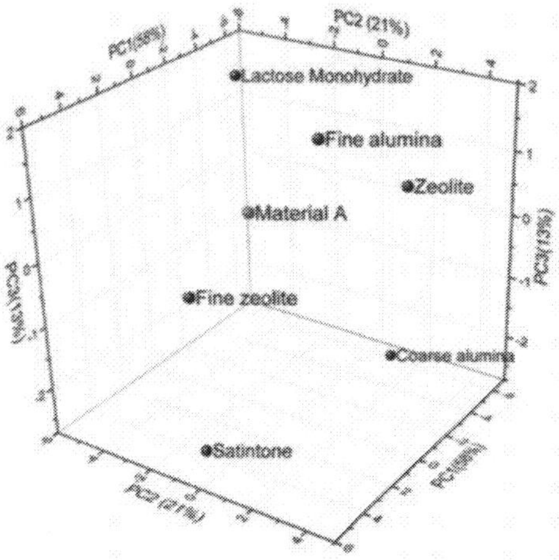

Fig. PCA model based on previous material flow properties' library

Reference: Y. Wang, et al., Predicting feeder performance based on materials flow properties, Powder Technology 308 (2017) 135–148

Powder Technology

Predicting feeder performance based on material flow properties

7

RUTGERS

PREDICTING FEEDING PERFORMANCE DURING FEEDER *REFILL*

8

RUTGERS

Before we start:

1. **What would happen during hopper refill?**
2. **How to measure the feeder's performance during hopper refill?**
3. **What powder and process variables may result in different feeding performance during hopper refill?**
4. **Can we predict the feeder's performance based on different material flow properties?**

Image from: http://www.ctvnews.ca/canada/new-beer-policy-brews-controversy-in-new-brunswick-1.1943649

9

RUTGERS

How to measure the feeder's performance during hopper refill?

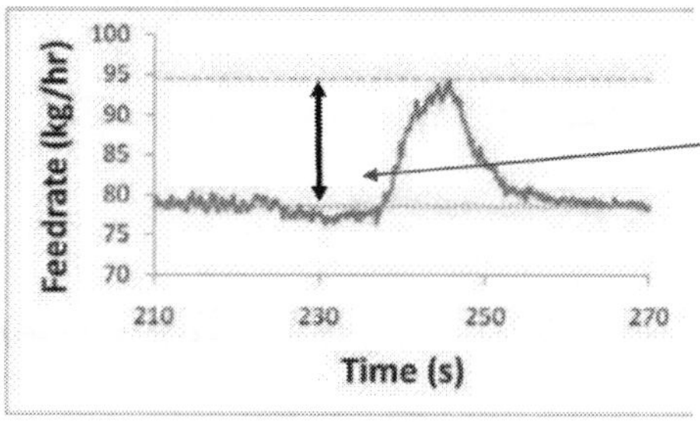

Maximum Deviation: The maximum deviation between the actual feedrate and the target setpoint.

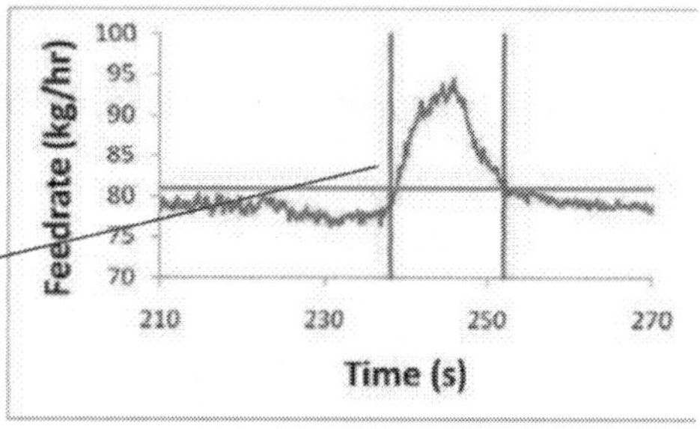

Deviation Time: The time that the feedrate is out of specification.

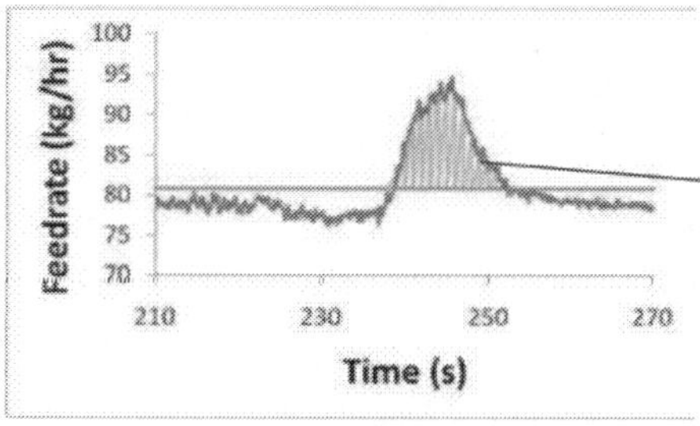

Total Deviation: the amount of powder fed in excess or less than expected. As we are comparing different materials, we are doing % total deviation in this work.

10

Figures: W. Engisch, feedrate deviation caused by hopper refill of loss-in-weight feeders, Powder Technology 283 (2015) 389–400.

RUTGERS

What powder and process variables may result in different feeding performance during hopper refill?

- **Refill Volume (Refill level):** As the figure shows, feeding performance during refill is significantly affected by refill level.
- **Refill Weight:** Density = $\frac{Weight}{Volume}$
- **Feedrate:** For our work, feedrate is constant for all the materials at 8kg/hr.

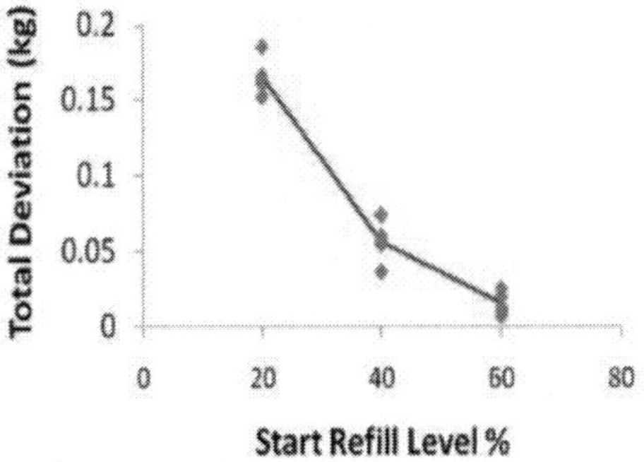

Figures: W. Engisch, feedrate deviation caused by hopper refill of loss-in-weight feeders, Powder Technology 283 (2015) 389–400.

- **Agitator RPM:** Agitator RPM inside the hopper would be at a certain percentage of the feeder's screw speed/drive command, which will differ for each material.
- **Material Properties: ?**

11

RUTGERS

Prediction using PLS regression

A PLS regression model relates material flow properties directly to feeder performance, quantified by Maximum Deviation, Deviation Time and %Total Deviation.

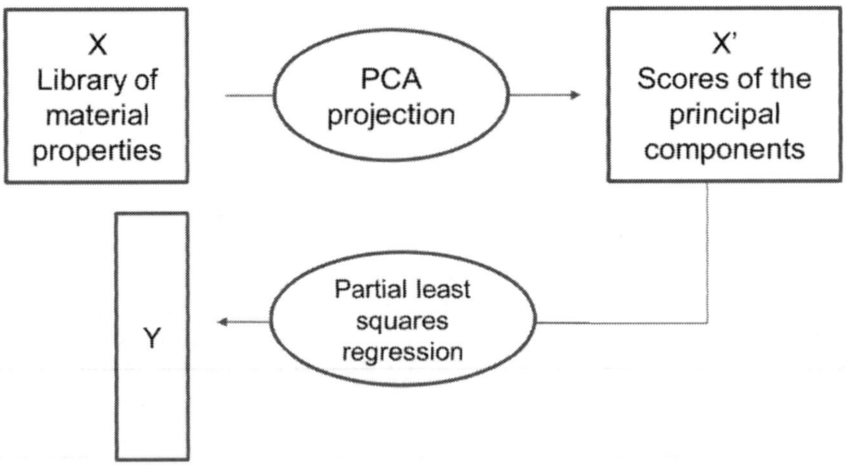

12

RUTGERS

Predicting using PLSR model

Set each material as Material A (to be predicted), so all the materials in the material library have a measured value and a predicted value of Maximum Deviation, Deviation Time and % Total Deviation.

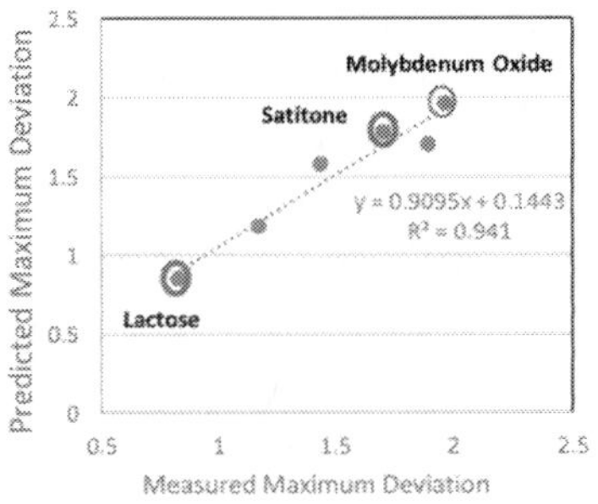

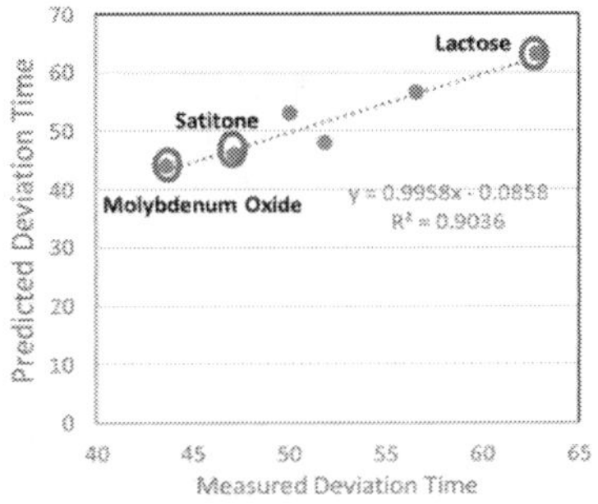

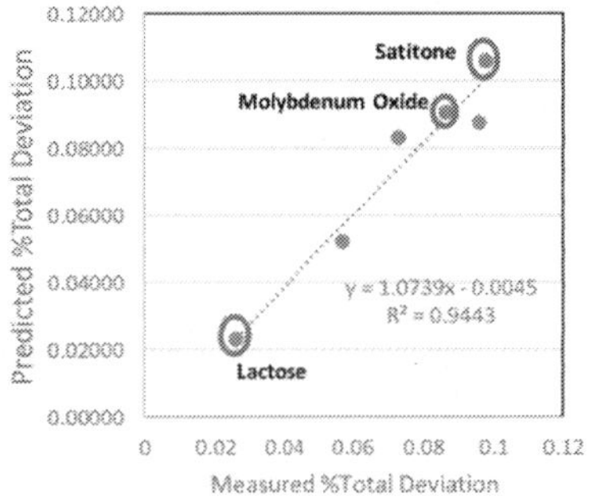

Maximum Deviation | Deviation Time | %Total Deviation

13

RUTGERS

Methodology for predicting feedrate deviation caused by HOPPER REFILL

Receive new material → Characterize new material properties → Project data to PLSR models to predict feeder performance → Predict the initial feedfactor → Predict optimal screw based on model prediction → Predict feeder's performance during hopper refill

14

RUTGERS

Conclusion

- *The correlation between material flow properties and feeding deviation caused by hopper refill has been studied.*
- *The deviation during hopper refill, quantified by Maximum Deviation, Deviation Time and %Total Deviation can be predicted using multivariate analysis tools.*

Future Work

- *Expand material library and optimize the predictive model by including more materials*
- *Identify reduced sets of measurements*
- *Correlate material flow properties to other unit operations of interest.*

15

RUTGERS

Acknowledgement

- Yi Tao
- Sara Koynov
- Engineering Research Center for Structured Organic Particulate Systems (CSOPS)
- Rutgers Catalyst Manufacturing Consortium (CMC)

ENGINEERING RESEARCH CENTER FOR
STRUCTURED ORGANIC PARTICULATE SYSTEMS
RUTGERS UNIVERSITY
PURDUE UNIVERSITY
NEW JERSEY INSTITUTE OF TECHNOLOGY
UNIVERSITY OF PUERTO RICO AT MAYAGÜEZ

16

RUTGERS

Thank you for your attention.

Please feel free to contact Tianyi at tianyi.li@rutgers.edu for further discussion.

17

RUTGERS

Why is powder flow complex?

The influence of processing variables:
Consolidation stress, aeration, moisture content, storage time, etc.

The influence of particle properties:
Size distribution, shape distribution, density, cohesion, adhesion, pore volume, hygroscopicity, electrostatics, etc.

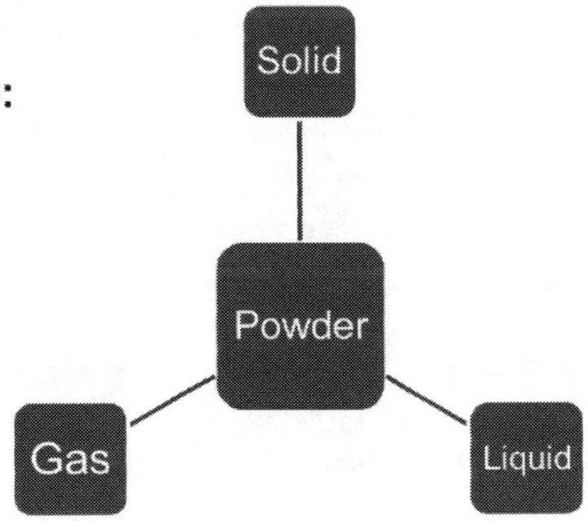

- Powder flow behavior is multi-dimensional, and depends on many characteristics.
- There is no unifying index to define flow properties.
- Multiple tests need to be performed to fully characterize flowability.

18

Introduction to PCA

- The direction of the first principal component (PC1) points in the direction of the maximum variation.
- PC1 may not be enough to describe the data. Repeating the projection process in the remaining part of the space gives the second principal component (PC2).
- The 2-D plane describes best the variation of the data. The process can be continued to find more PCs.

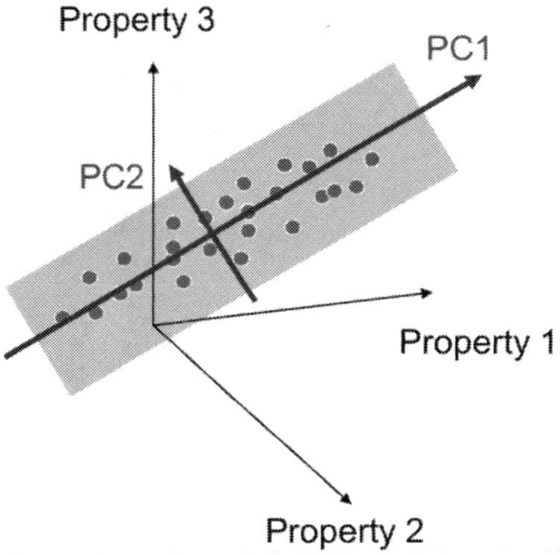

Feeder Performance Evaluation

- Material properties have a significant effect on feeder performance.
- A loss-in-weight feeder consists of three parts:
  volumetric feeder, weighing platform, and gravimetric controller.
- The target mass feed rate is set at ***m***, the actual mass feed rate is calculated by

$$\dot{m} = \frac{\Delta m_i}{\Delta t}$$

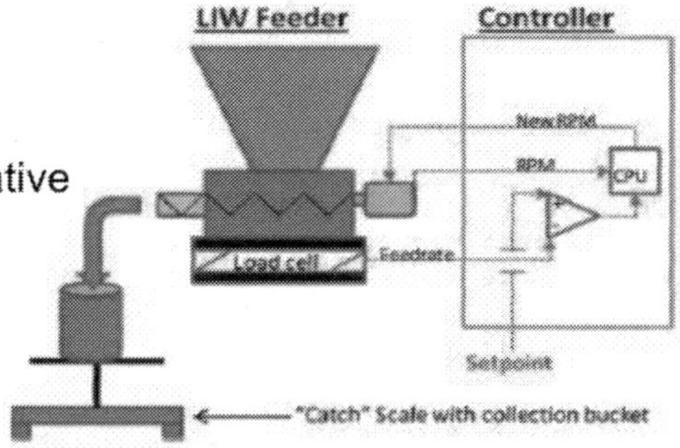

- Relative standard deviation (RSD) and relative difference from the mean (RDM) is used as criteria for feeder performance:

$$\sigma = \sqrt{\frac{\sum_{i=1}^{n}(\dot{m}_i - \bar{\dot{m}})}{n-1}}$$

$$\mathrm{RSD} = \frac{\sigma}{\bar{\dot{m}}} \qquad \mathrm{RDM} = \frac{|\dot{m} - m|}{\dot{m}}$$

σ is standard deviation, $\bar{\dot{m}}$ is the arithmetic mean of the actual mass feed rate, and n is the number time points.

Reference: Engisch, William E., and Fernando J. Muzzio. "Method for characterization of loss-in-weight feeder equipment." *Powder technology* 228 (2012): 395-403.

RUTGERS

Backup Slide

moto

Feeder Screws and Feed Rates

The feed rate indications are approximate values. Actual feeder screws will be determined by the material being fed. Speed range (rpm) and the type of gear are determined by the feed rate. Feeder screws are easily interchangeable.

Pitch	Type of drive	Twin concave screws 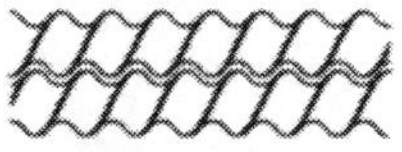		Twin auger screws 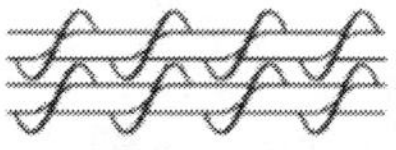		Speed range	Gearbox type
		dm^3/h	ft^3/h	dm^3/h	ft^3/h	RPM	
coarse	(=)	1.5 - 149	0.05 - 5.3	1.6 - 164	0.06 - 5.8	8 - 745	A
		0.7 - 70	0.02 - 2.5	0.8 - 79	0.03 - 2.8	4 - 357	B
		0.3 - 31	0.01 - 1.1	0.3 - 34	0.01 - 1.2	2 - 154	C
	(~)	10 - 182	0.4 - 6.4	11 - 200	0.4 - 7.1	52 - 909	A
		5 - 87	0.2 - 3.1	5.5 - 96	0.2 - 3.4	25 - 436	B
		2 - 38	0.07 - 1.3	2.3 - 41	0.08 - 1.4	11 - 188	C
		dm^3/h	ft^3/h	dm^3/h	ft^3/h	RPM	
fine	(=)	0.6 - 56	0.02 - 2	0.6 - 60	0.02 - 2.1	8 - 745	A
		0.27 - 27	0.01 - 1	0.3 - 29	0.01 - 1	4 - 357	B
		0.12 - 12.4	.004 - 0.4	0.12 - 12	.004 - 0.4	2 - 154	C
	(~)	4 - 68	0.14 - 2.4	4.2 - 73	0.15 - 2.6	52 - 909	A
		2 - 33	0.07 - 1.2	2 - 35	0.07 - 1.2	25 - 436	B
		0.8 - 14	0.03 - 0.5	0.9 - 15	0.03 - 0.53	11 - 188	C

21

RUTGERS

Measurements to characterize powder flow properties

- Particle size distribution
- Compressibility test

- measures change of powder bed volume as normal stress changes.

- Shear cell test

- Measures powder yield stress as a function of normal stress.

- Permeability test

- Measures flow of air through a powder bed.

- Dynamic flow test

- Measures powder flow in terms of energy.

- Electrostatics test

- Measures impedance of a powder bed.

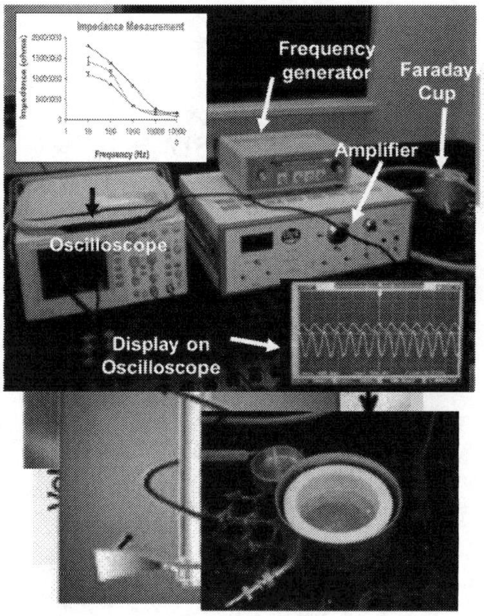

Image source: Freeman Technology Inc.
http://www.freemantech.co.uk/powders/about the FT4 powder rheometer

22

RUTGERS

PCA Model used for Feeder refill prediction

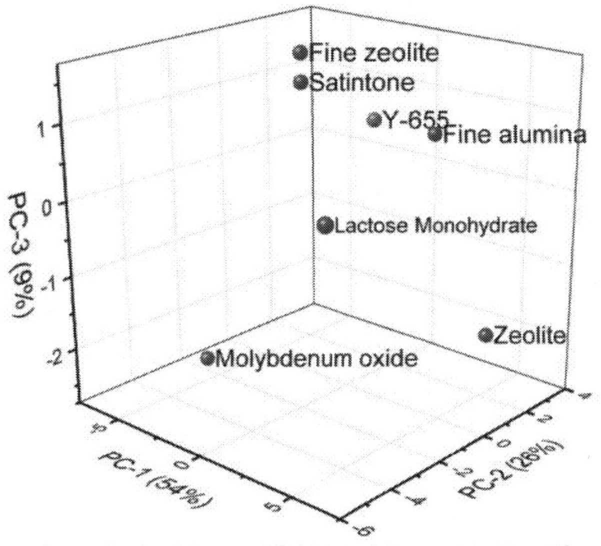

- Each material was first filled up to 80% of the hopper and then run at 8kg/hr, till the hopper level reaches the refill point which was set as 40% hopper level. The feeder automatically changed into volumetric feeding, with the screw speed and the agitator RPM staying constant until the hopper level reached 80% again which means the refill was done. The speed for screw and agitator during refill are based on the initial feed factor.
- We will predict the feeder's refill performance for Y-655.

23

RUTGERS

Prediction using PLSR model

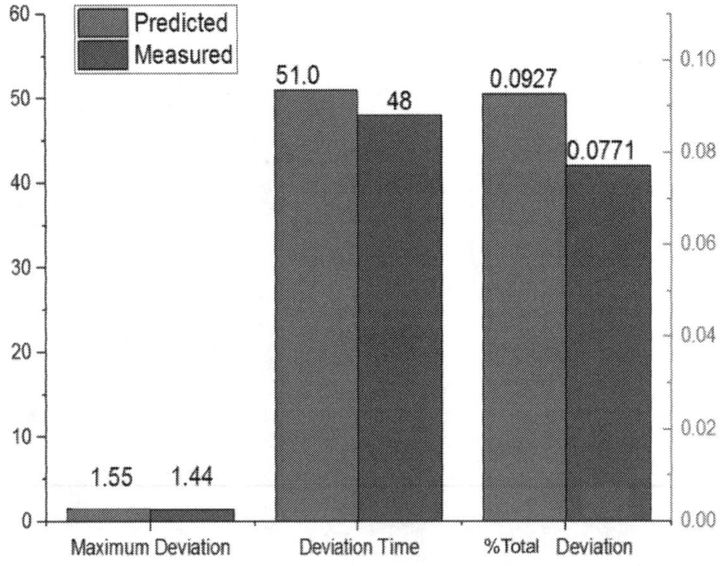

- Material A, Calcined Zeolite Y-655 feeder's refill performance was predicted.
- Feeder's drive command/screw speed during feeder refill under volumetric feeding was added to the library of factors used for prediction.
- We got good results for all the three measurements of deviation caused by feeder refill.

24

Transfer Chute Design Considerations for Dust Control Using Computational Fluid Dynamics (CFD)

Xiaoling Chen[1] , Craig Wheeler[2]

[1]*Beijing Key Laboratory of Process Fluid Filtration and Separation, College of Mechanical and Transportation Engineering, China University of Petroleum-Beijing, China*

[2] *Centre for Bulk Solids and Particulate Technologies, The University of Newcastle, Australia*

Abstract: Dust emission from belt conveyor transfer chutes is one of the major sources of fugitive dust in material handling plants. Among the transfer chute configurations investigated in authors' previous studies, stilling chambers show benefits in terms of minimizing fugitive dust emissions compared to conventional designs. This study focuses on the optimal design of a stilling chamber using the two-phase three-dimensional Euler-Euler model. Typical design parameters that will be investigated include; the volume of the stilling chamber, the length of the stilling chamber and the pattern of the baffles inside the enclosure. In addition, the application of a self-adjusting flap at the outlet of the chute, both with and without the stilling chamber fitted, is also investigated. The results show that: the stilling chamber reduces likely dust emission from the chute; increasing the height of the stilling chamber does not necessarily reduce dust generation; extending the length of the stilling chamber can reduce likely dust emission, however it is not proportional to the length of the stilling chamber; the location of the baffles within the stilling chamber have a significant impact on likely dust emission; and, the self-adjusting flap has a significant effect on controlling the dust, especially in a chute without a stilling chamber.

Keywords: Computational Fluid Dynamics(CFD); Transfer Chute; Dust Emission

1. Introduction

Materials handling is one of the main causes of dust emissions in mining operations. Typically, dust is generated when bulk materials are loaded, dumped and transferred. Ninety percent of diffuse dust emissions are produced during the storage and transport of bulk material[1]. According to the literature[2], one of the most important sources resulting in dust emissions is due to the transportation of bulk material on belt conveyors. In the case of belt conveyors, an area of particular concern for dust control occurs during the transfer of bulk material from one conveyor to another called the transfer point where usually a chute is employed. Therefore, the performance of transfer chutes has a significant impact not only on the efficiency of conveyor belt systems, but on the level of fugitive dust emissions.

Currently, the majority of dust control systems for conveyor transfers are "active", which need external energy and equipment or suppressants. The active dust control system depends on a variety of water and chemical materials to decrease the amount of the dust emissions, such as the application of chemical surfactants, including water spray systems or foam and water misting systems. These well-known systems are of broad application at present and attract many researchers' attentions [3-5] . However, a great deal of disadvantages and limitations exist with these types of systems. For example, many coal handling plants use process water to control dust emission, which not only results in a reduction in the calorific value of the coal per tonne as the moisture content of coal is increased, but significantly increases the potential for material handling issues throughout the rest of the plant. Often process constraints limit the amount of moisture that can be added to bulk materials, this combined with the importance of minimising water usage places a much greater emphasis on the need to develop "passive" dust control systems, where no external energy and equipment nor suppressants are required. Passive control is achieved through the efficient integration of the moving stream of bulk material with the induced air flow. In other words, the passive control relies on the proper design of equipment.

The design of the transfer chute is often the product of design guidelines arriving from a combination of flow test data, scaled testing, empirical relationships and rules-thumb which have a lot of limits. Understanding the behaviour of granular materials flow in transfer chute is pertinent to transfer chute design. Predicting, and therefore controlling, the flow of a multiphase is a challenge to the theoreticians as well as experimenters. With increasing computational power, process simulation has been widely used in science and industry as a cost-effective and flexible research and design tool, there is no exception to transfer chute design. Discrete Element Method (DEM) has been proven as a powerful tool to assist engineers to design transfer chutes, references [6-10] show a small outline of this development. There are many simulation programs and software based on DEM for chute design are on the market, like Chute Maven, Chute Analyst, Rocky, EDEM and so on. With the aid of DEM it is not only possible to simulate and visualize the material flow, but it also enables the analysis of the forces and moments acting on different elements of the equipment due to the effect of inter-particle contacts and the effect of particle-wall interactions. However, it cannot be used to predict the air flow which is directly related to the dust emission due to the limited of the DEM method. Computational Fluid Dynamics (CFD) also shows good potential for simulating granular flows by predicting flow shape, stream velocity, and location of flow recirculation zones due to the effect of the bulk material stream on air flow[11-16]. From the point of view of dust emission prediction, the CFD method is more suitable than DEM since the air flow plays a critical role in dust circulation and emission.

Previous studies[17, 18] undertaken by the authors indicate that two-phase CFD simulations correlate well with scale model experimental tests, indicating that this approach can be used to effectively predict likely dust emissions from transfer chutes. Among the transfer chute configurations investigated in previous studies[19], stilling chambers receive fairly good effect in terms of minimizing fugitive dust emissions compared to conventional designs because the flaps in the stilling chamber can lower the air velocity to allow dust particles to drop out of the air and onto the receiving conveyor belt. Furthermore, stilling chambers are a more suitable option to retrofit to existing chute installations than other options investigated, such as restrictor plates, etc. In addition, stilling chambers are less likely to be worn comparing with restrictor plates. Therefore, this paper focuses on the optimal design of a stilling chamber using the two-phase three-dimensional Euler-Euler model using ANSYS FLUENT 14.5. Typical design parameters that will be investigated include; the volume of the stilling chamber and the pattern of the baffles inside the enclosure. In addition, the application of a self-adjusting flap at the outlet of the chute, both with and without the stilling chamber fitted, is also investigated.

2. CFD Simulation

There are two commonly used approaches to model gas-solid flow, namely the Discrete Phase Model (DPM) and the Two-Fluid Model (TFM). These approaches are represented as the Euler-Lagrangian model and Euler-Euler model respectively in ANSYS FLUENT. The former model is suitable for dilute gas-solid flow, while the later for dense phase simulations and therefore being much more applicable to complex multiphase flows. For the TFM approach, the different phases are treated mathematically as interpenetrating continua, and the conservation equations have a similar structure for each phase. The air is assumed to be incompressible and at a constant density, while the particles are assumed to be spherical and have a uniform density. Due to the complexity of the particle volume fraction in the transfer chute problem a TFM approach was adopted.

We propose in the current work to solve the governing equations of mass and momentum by means of a multiphase Eulerian model incorporating the Kinetic Theory of Granular Flow (KTGF) available in the software ANSYS FLUENT. The energy conservation equation was ignored as the flow is isothermal. The fundamental equations of mass and momentum conservation for the two-fluid Euler-Granular model are well known and therefore will not be presented here.

2.2 Boundary conditions and numerical procedure

The air is assumed to be incompressible and at a constant density, while the particles are assumed to be spherical and have a uniform density. A particle diameter of 4 mm was selected for the simulations. The particle diameter was selected to ensure the reasonable computational time for each chute configuration, in addition with reference to the PSD results, where more than 75% of the bulk material was less than 4 mm.

At the inlet all velocities and volume fractions of both phases were specified. Since there were no experimental observations describing the gas-solid distribution at the inlet, the inlet velocity profile is assumed to be uniform with a constant volume fraction. The average solids velocity at the inlet is dependent on the discharge properties of the feed kibble and it is assumed that the inlet air has the same velocity as the solid phase. In order to predict the inlet gas-solid distribution a Discrete Element Method (DEM) simulation using commercial software Rocky was undertaken to calculate the inlet solid volume fraction. Based on the simulation results the inlet solid volume fraction was set to 0.45. Furthermore, the maximum packing limit was set at 0.48, and the maximum frictional limit specified as 0.46. At the outlet the pressure was assumed to be atmospheric while the remaining opening areas were defined as pressure inlets. Based on the previous parameters study [17], the specularity coefficient and particle-particle interaction coefficient were specified as 0.20 and 0.45, respectively. The receiving belt was defined as moving wall boundary with the constant velocity 0.62 m/s.

Based on the investigation of a number of different sub-models for viscosity, turbulence and drag, Syamlal-O'Brien granular viscosity model, SST k-ω model and Di-Felice model were selected to conduct this study. User Defined Functions (UDFs) were used to implement Di-Felice drag model in the software.

Unsteady simulations were performed with an improved mesh compared with the previous parameters study. A larger time step of 1×10^{-3} s was used, while 20 iterations per time step were employed to ensure numerical stability. A convergence criterion of 1×10^{-3} was specified for the relative error between two successive iterations. The same mesh size that was used in the previous study [17] was used to generate the new mesh, thus the mesh independence study was not repeated.

3. Results and Discussion

The multiphase CFD approach is limited in this particular application, since it is not possible to directly predict dust emissions; rather it is qualitatively predicted based on the air velocity and the volume of air exiting the transfer chute. The prediction is obtained based on the assumption that a high outlet air velocity results in a high level of dust emitted. Additionally, the air mass flow rates at different locations were also compared, since the air flow rate significantly influences chute performance in relation to dust emission. This is necessary to ensure the effect of different outlet shapes and sizes, multi-outlet and asymmetric velocity distributions are considered.

3.1 Stilling chamber Height

As recommended by Swinderman et al. [20], the large enclosure allows internal circulation of the dust-laden air, which allows dust particles to drop out of the air and back into the material stream due to the dissipation of the energy. To understand the influence of the volume of the stilling chamber, a chute with high stilling chamber named Chute E1 (as shown in Fig. 1 (b)) was used to carry out the simulation. The outlet air velocity distributions and mass flow rate were compared with the original Chute E (as shown in Fig.1 (a)) shown in Fig.2 and Fig.3 respectively.

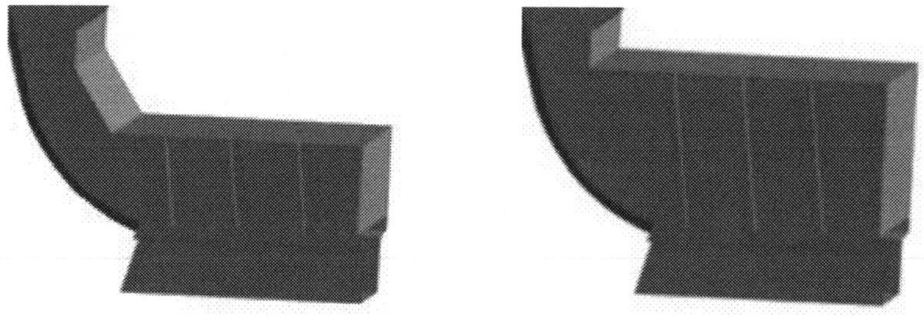

(a) (b)

Fig.1 (a) Chute E (b) Chute E1: incorporating higher stilling chamber.

Figure 2 shows the outlet air velocity distributions for Chute E and Chute E1. The outlet air velocities at the top section of Chute E are higher than that of Chute E1, while the outlet air velocities at the bottom section are slightly lower than that of Chute E1. From Fig. 3, it is clear that there is no obvious difference in outlet air mass flow rate between Chute E and Chute E1. The results indicate that the increased volume does not lead to a lower outlet air velocity and mass flow rate as expected.

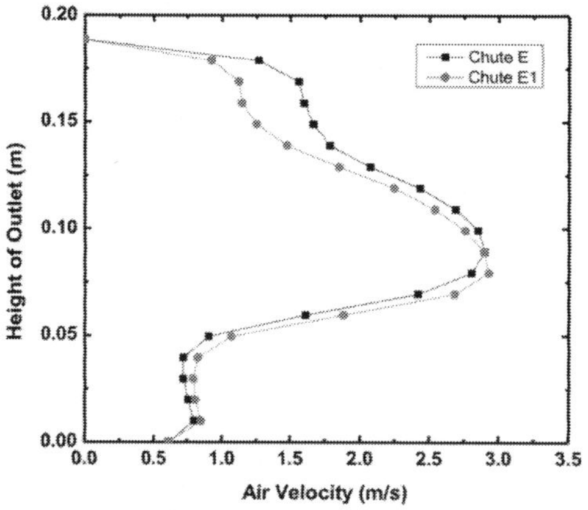

Fig.1 The outlet air velocity distributions for Chute E and E1.

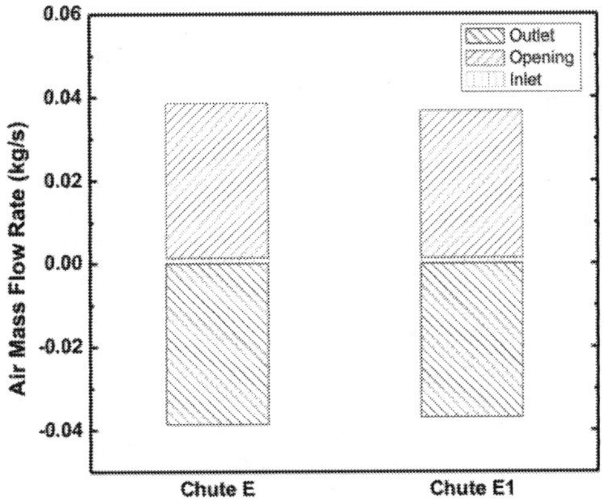

Fig. 2 The outlet air mass flow rates for Chute E and E1.

3.2 The Length of the Stilling Chamber

The purpose in investigating stilling chambers with different lengths is to understand the influence of stilling chamber length on the performance in terms of dust control. In addition to the original length used in the previous experimental study two different stilling chambers, with 1.5 times and 2 times the original length, were conducted using simulation. The two chutes are the same as Chute E, except the length of the stilling chamber, so they are referenced as Chute E2 and Chute E3 in the following content. The baffles inside the stilling chamber stick to the original design, the number of the baffles are increased in direct proportion to the length of the stilling chamber. The structures of the simulated stilling chambers are shown in Fig. 4.

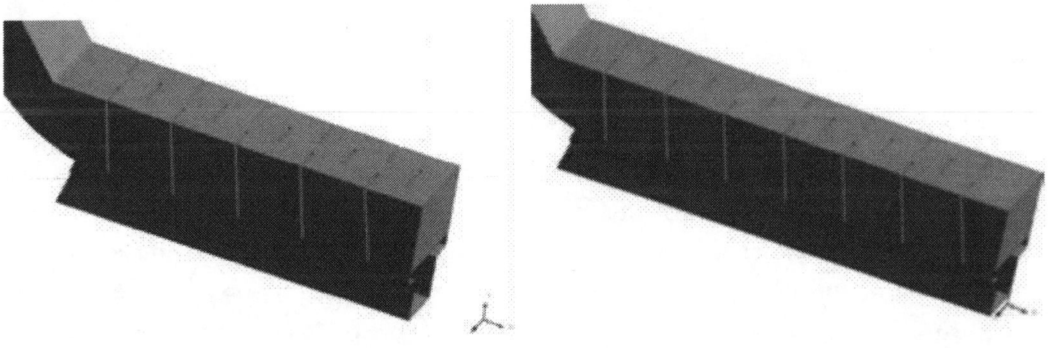

(a) (b)

Fig.4 (a) Chute E2:1.5 times stilling chamber (b) Chute E3:2 times stilling chamber.

The outlet air velocity distributions and mass flow rates for Chute E and the two chutes with longer stilling chambers are shown in Fig. 5 and Fig. 6 respectively. From Fig. 6, it is clear that Chute E has the highest outlet air velocity, while it is hard to classify which chute has the lowest outlet air velocities. It is also demonstrated by the mass flow rate because there is no notable difference between the two chutes with longer stilling chamber. Thus, the results reveal that the longer stilling chamber has a better performance in regards to dust emission, but the effectiveness has grown disproportionately. The results reveal that the performance of a transfer chute in regards to dust control can be improved by extending the length of the stilling chamber. However, after a certain distance there are no obvious benefits achieved by further extending the length of the stilling chamber.

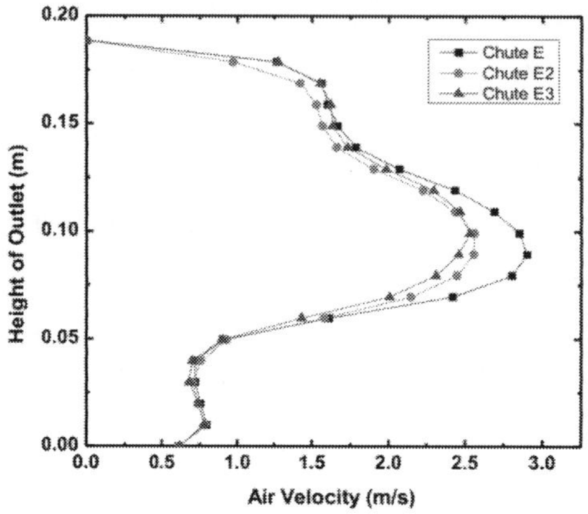

Fig. 5 The outlet air velocity distributions for Chute E, E2 and E3

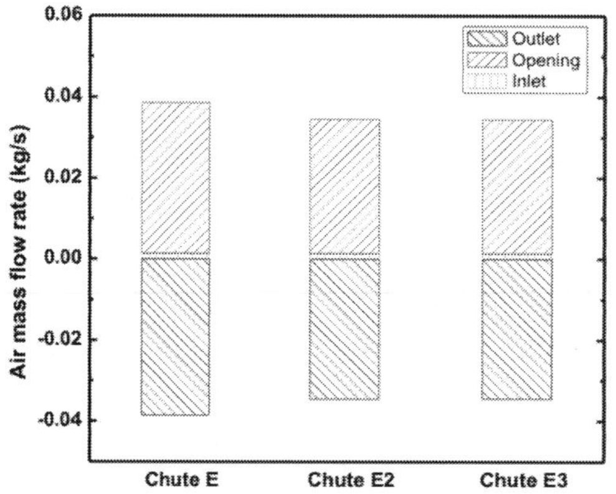

Fig. 6 The outlet air mass flow rates for Chute E, E2 and E3.

3.3 Stilling Chamber Internal Flaps

The flaps installed inside the stilling chamber were designed with the aim of lowering the air velocity at the stilling chamber exit in order to reduce the amount of fugitive material carried from the chute exit with the escaping air. In order to understand the influence of flaps on the outlet air velocity and mass flow rate, four chutes with different flaps were investigated. They are labelled E4 through to E6 and include; E4 – stilling chamber with large middle flaps (2 times the length of the original design), E5 – stilling chamber with large flaps which are about two-thirds of the enclosure width, E6 – stilling chamber with large flaps which are about five-sixths of the enclosure width in addition to small intervals. The structures of these three different stilling chambers are illustrated in Fig.7.

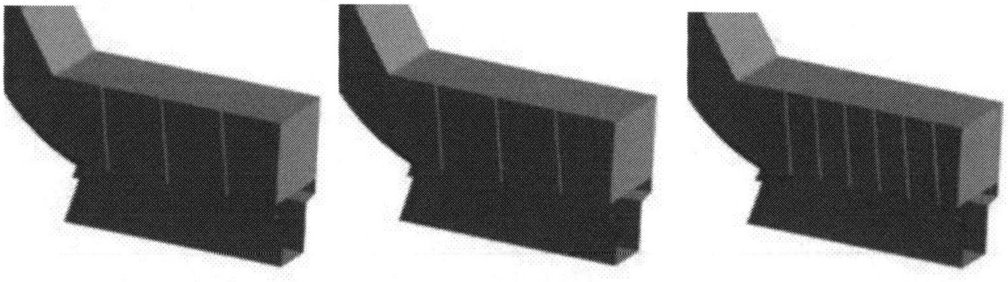

(a) Chute E4 (b) Chute E5 (c) Chute E6

Fig. 7 Illustrations of analysed chutes E4-E6.

In order to understand the effectiveness of the flaps on the control of dust emission, the outlet air velocity distributions and air mass flow rates obtained from different chutes are illustrated in Fig. 8 and Fig. 9 respectively. From Fig. 8, Chute E has the largest outlet air velocity, while Chutes E4, E5 and E6 are successively smaller. The maximum outlet air velocity decreases from 2.90 m/s at the outlet of Chute E to 1.60 m/s for Chute E6. Both the maximum and average velocity of Chute E6 are much lower than the rest of chutes, showing that the size and distribution of flaps in the stilling chamber reduces the outlet velocity significantly, indicating less potential for dust emission. It can be seen that the mass flow rates shown in Fig. 9 also supports the conclusion obtained from Fig. 8. It is noted that there is no significant difference between Chute E4 and Chute E5 in outlet air velocity distributions, while the outlet air mass flow rates do show a notable difference. The cause of the inconsistency is the asymmetric distribution of outlet air velocity. As stated before the outlet air velocity distributions shown in Fig. 8 were obtained from the centreline of the outlet, therefore, it cannot represent the whole outlet situation under the condition of an asymmetric distribution.

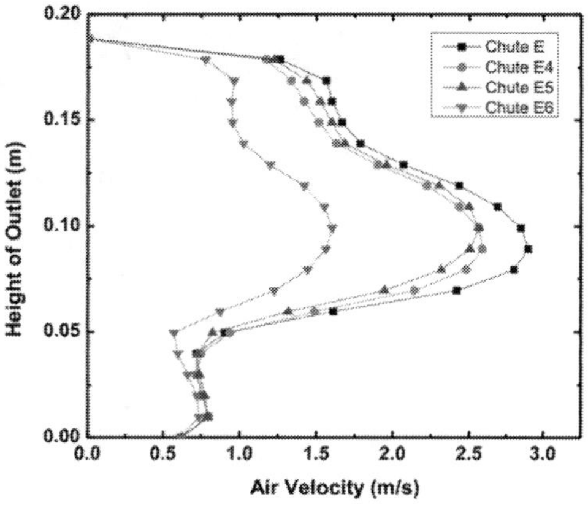

Fig. 8 Outlet air velocity distributions for Chute E, E3, E4 and E6.

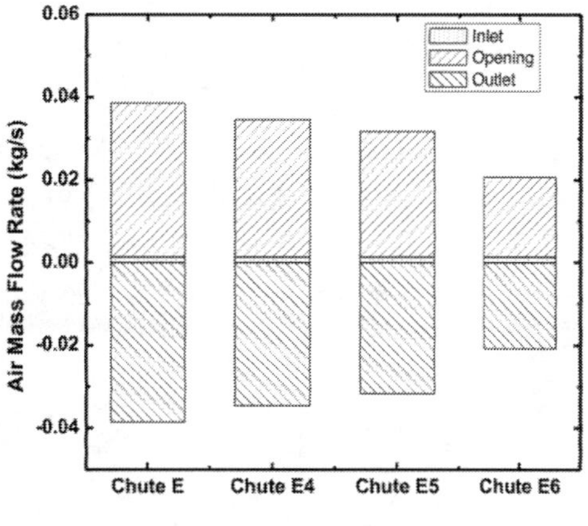

Fig. 9 Outlet air mass flow rates for Chute E, E4, E5 and E6.

As an example to show the influence of curtains on the air flow pattern, Fig. 10 shows a top view of snapshot of outlet air velocity vectors for Chute E and E6. The results indicate that the design of the flaps has a significant effect on outlet air velocity at the stilling chamber exit. The flow patterns of air are changed by the baffles which results in different pressure distributions in the different chutes. As shown by the mass flow rate results, most of the air enters the chute from the opening area. When the air velocities are dampened by the baffles, more and more air accumulates in the chute which leads to a pressure increase in the chute. Therefore, only a small amount of air is sucked into the chute from the top opening area where the inlet pipe meets the standpipe due to the small pressure difference.

(a)

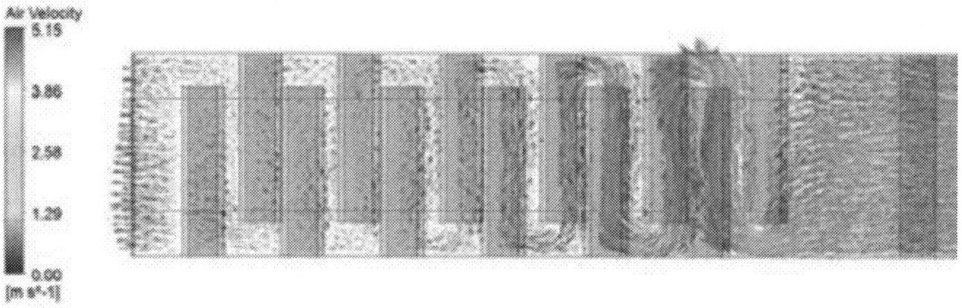

(b)

Fig. 10 Vectors of air velocity in stilling chamber (a) Chute E, (b) Chute E6.

Figure 11 shows the snapshots of the top opening area pressure contours for Chute E, E4, E5 and E6. It shows the relative pressure with atmospheric pressure. The negative means the chute pressure at the top opening area is lower than atmospheric pressure which leads to air being sucked into the chute. The positive number means the chute pressure at the opening area is higher than atmospheric pressure which results in air coming out of the chute. Zero pressure means no air comes in or out the chute from the opening area. It is clear that Chute E has the smallest pressure, while Chutes E4, E5, and E6 are successively higher. For the Chute E, the pressure for most of the opening area is smaller than -2.5 Pa. However, the pressure for almost all of the opening area is in the range of 0 to -2.5 Pa for Chute E6. This gives a good indication of why only about half the volume of air gets sucked into the chute from the opening area compared with Chute E.

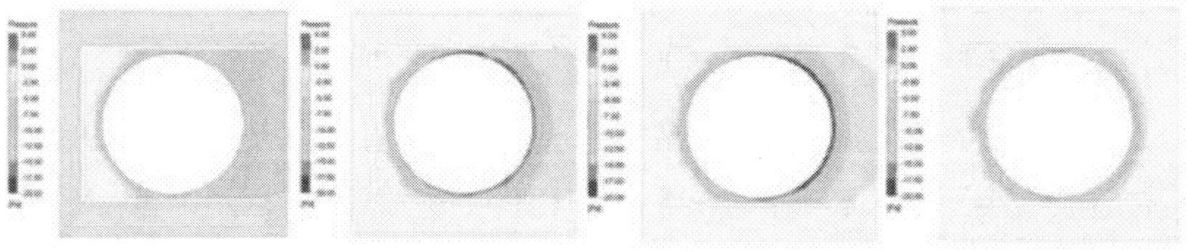

(a) Chute E (b) Chute E4 (c) Chute E5 (d) Chute E6

Fig. 11 Pressure contours of top opening area.

3.4 The application of Control Flap

A control flap was designed with the aim of lowering the air velocity at the spoon exit in order to reduce the amount of fugitive dust carried from the chute exit with the escaping air. This design was influenced by the success of Wheeler et al. [21] in using a similar flap for a grain ship loader.

The control flap is designed to pivot on its hinge such that the lower rubber flap rests slightly above the material stream and directs air flow containing entrained dust back into the material stream. From the previous experimental and simulation study, Chute C, which is composed of a baseline case with a lower removable cover and lower openings fitted, does not show an obvious improvement when comparing with the baseline Chute. In order to evaluate the effectiveness of the control flap, a control flap was introduced to a chute named Chute C1 to carry out the simulation study. The only difference between Chute C1 and Chute C is the control flap. The configurations of Chute C and C1 are illustrated in Fig. 12.

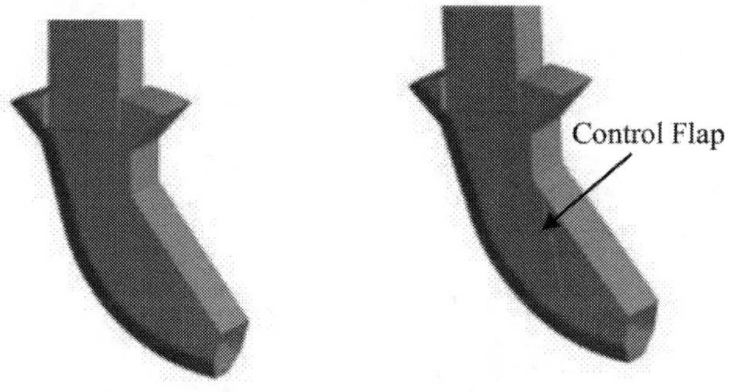

Fig. 12 Illustrations of Chute C and C1.

Figure 13 shows the outlet air velocity distributions for Chute C and Chute C1. It is obvious that both the maximum and average outlet air velocity is effectively dampened by employing the control flap in Chute C1. In view of the relative symmetrical distribution of outlet air velocity, it is easy to understand that the mass flow rate of Chute C1 with the value of 0.0449 kg/s is remarkably smaller than that of Chute C with the value of 0.0547 kg/s, as shown in Fig. 14. The result stated above reveals that the control flap has a significant effect on reducing the amount of fugitive dust in a Chute without a stilling chamber.

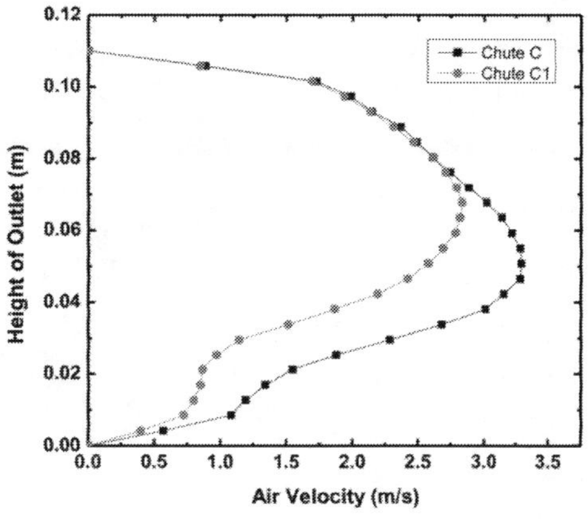

Fig. 13 The outlet air velocity distributions for Chute C and Chute C1.

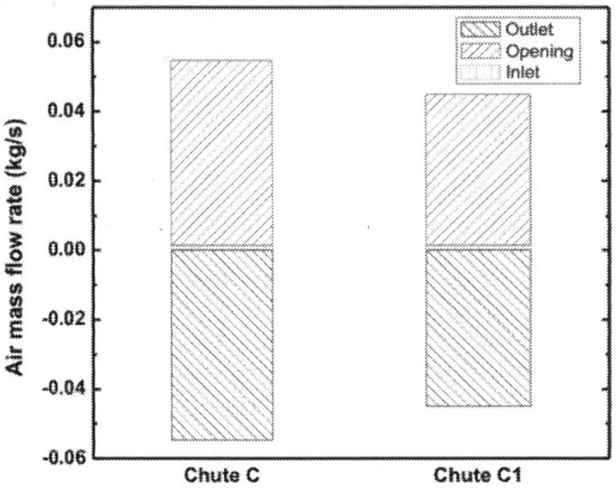

Fig. 14 Air mass flow rates for Chute C and Chute C1.

Both the control flap and curtains in the stilling chamber have been proven to be able to lower the air velocity at the chute exit. The effect of combining them is also of interest, with a simulation undertaken of a chute with a control flap and stilling chamber, named Chute E7 as shown in Fig. 15. The outlet air velocity distributions and mass flow rates for Chute E and Chute E8 are illustrated in Fig. 16 and Fig. 17, respectively.

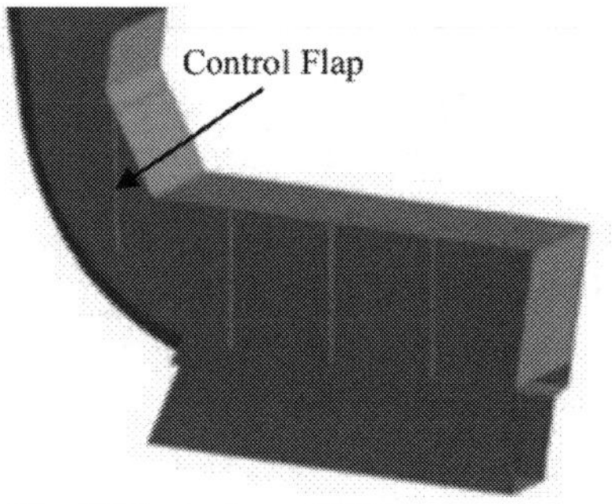

Fig. 15 Illustrations of Chute E7.

As shown in Fig. 16, the outlet air velocities obtained from Chute E7, which uses the control flap, are smaller than that of Chute E. It is also supported by the mass flow rate shown in Fig. 17. The mass flow rate of air at the outlet for Chute E7 is 0.0342 kg/s which is slightly smaller than that of Chute E with value of 0.0386 kg/s. Comparing the difference between Chute C and Chute C1, it is obvious that the effect of the control flap application in the stilling chamber is not as dramatic as it is in the chute without the stilling chamber.

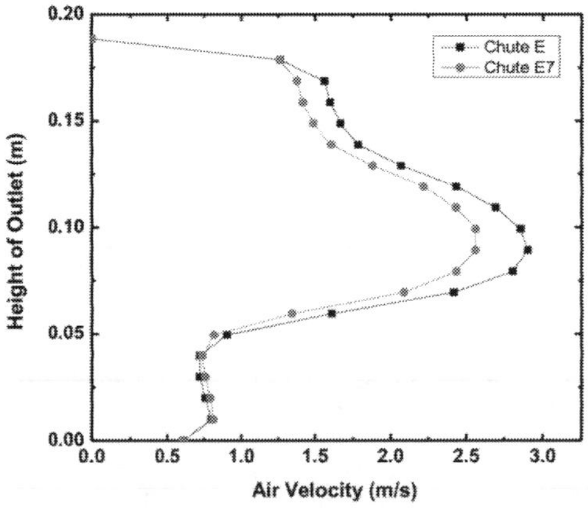

Fig.16 Outlet air velocity distributions for Chute E and Chute E7.

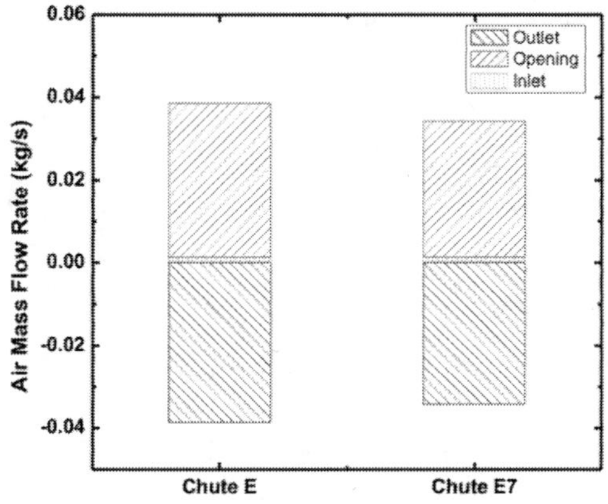

Fig.17 The mass flow rates for Chute E and Chute E7.

4. Conclusion

The feasibility of the Two Fluid Model to conduct the transfer chute simulation has been verified. Using the parameters and models determined from the previous studies, the influence of the variable geometry of stilling chambers and control flaps were investigated. The results were presented and discussed. By comparing with the original design, the following conclusions are summarised:

1) Increasing the height of the stilling chamber does not obviously reduce the out let air velocity and air mass flow rate which means it does not necessarily reduce dust generation.
2) Extending the length of the stilling chamber can improve the performance of the chute regarding dust control. However, it is not proportionate to the increase in the length of the stilling chamber, and after a certain length, no obvious improvement is achieved.
3) A control flap has a significant effect on controlling the dust, especially in a chute without a stilling chamber.
4) The Two Fluid Model approach in Computational Fluid Dynamics can be used as a suitable design tool for predicting the dust emission of transfer chutes.

Acknowledgements

The authors would like to acknowledge the financial support from National Natural Science Foundation (No.51504273) and Science Foundation of China University of Petroleum, Beijing (No. 2462014YJRC014 &No.2462015YQ0402).

Reference

[1] Junemann, R and Holzhauer, P. The Reduction of Dust Emissions in Bulk Handling Installations. Bulk Solids Handling, 1992. **12**(2): p. 217-226.

[2] Mody, V and Jakhete, R. Dust Control Handbook for MIinerals Processing. 1987, Martin Marietta Laboratories. p. 220.

[3] Tessum, M W, Raynor, P C and Keating-Klika, L. Factors Influencing the Airborne Capture of Respirable Charged Particles by Surfactants in Water Sprays. Journal of Occupational & Environmental Hygiene, 2014. **11**(9): p. 571-582.

[4] Swanson, J G and Langefeld, O. Fundamental research in water spray systems for dust control. Mining Technology, 2015. **124**(2): p. 78-82.

[5] Kou, Bao-fu, Liu, Qiu-zu, Cao, Shi-chen, Hu, Xiao-he, Li, Yan-feng, Wang, Yi-ran and Zhao, Bo-hui. Experimental

investigation on atomization and collecting efficiency of wind-spray dust controller and its parameters optimization. Journal of Central South University, 2015. **22**(11): p. 4213-4218.

[6] Cleary, P. Large scale industrial DEM modelling. Engineering Computations, 2004. **21**(2/3/4): p. 169 - 204.

[7] Cleary, P, Sinnott, M and McBride, W. Prediction of particle flows and blockage problems in realistic 3D transfer chutes. 2005. Institution of Chemical Engineers (IChemE).

[8] Kessler, F and Prenner, M. DEM - Simulation of Conveyor Transfer Chutes. FME Transactions, 2009. **37**: p. 185-192.

[9] Qiu, X and Kruses, D. Design of Conveyor Chute Using Discrete Element Method, in Proceedings of the Fourth U.S. National Congress on Computational Mechanics. 1997: San Francisco, USA.

[10] Gröger, T and Katterfeld, A. Application of the discrete element method in materials handling - Part 3: Transfer stations. Bulk Solids Handling, 2007. **27**(3): p. 158-166.

[11] McILVENNA, P and Mossad, D R. Two dimensional transfer chute analysis using a continuum method in 3rd International Conference on Computational Fluid Dynamics in the Minerals and Process Industries (CFD2003). 2003. Melbourne, Australia.

[12] Witt, P J, Carey, K G and Nguyen, T V. Prediction of Dust Loss from Conveyor Using CFD Modelling. in Second International Conference on CFD in the Minerals and Process Industries. 1999. CSIRO, Melbourne, Australia.

[13] Donohue, T J, Roberts, A W, Wheeler, C A and McBride, W. Analysis of transfer chute performance focusing on reduced dust emissions. in 8th World Congress of Chemical Engineering 2009. 2009. Montreal, Quebec.

[14] Katterfeld, A, Donohue, T J and Wheeler, C A. Simulation Based Dust Prediction of Transfer Chutes. Bulk Solids Handling, 2010. **30**(5): p. 240-245.

[15] Kessler, F P, M. DEM: Simulation of conveyor transfer chutes. FME Transactions, 2009. **37**: p. 185-192.

[16] McBride, W and Ilic, D. Design of Self Compensating Soft Loading Hood. Particle & Particle Systems Characterization, 2007. **24**(4-5): p. 370-374.

[17] Chen, X and Wheeler, C. Computational Fluid Dynamics (CFD) modelling of transfer chutes: A study of the influence of model parameters. Chemical Engineering Science, 2013. **95**(0): p. 194-202.

[18] Chen, X and Wheeler, C. Computational Fluid Dynamics (CFD) modelling of transfer chutes: Assessment of viscosity, drag and turbulence models International Journal of Multiphase Flow, 2015. **69**: p. 42-53.

[19] Chen, X, Wheeler, C, Donohue, T, McLean, R and Roberts, A. Evaluation of dust emissions from conveyor transfer chutes using experimental and CFD simulation. International Journal of Mineral Processing, 2012. **110–111**(0): p. 101-108.

[20] Swinderman, T, Goldbeck, L.,and Marti, A. . Foundations 3 - The Practical Resource for Total Dust and Material Control. Illinois: Martin Engineering. 2002.

[21] Wheeler, C, Krull, T, Roberts, A and Wiche, S. Design of ship loading chutes to reduce dust emissions. Process Safety Progress, 2007. **26**(3): p. 229-234.

501396 Effect of Baffles on Heat Transfer and Temperature Distribution in Granular Materials in Rotating Drums

Abstract:

Rotating drums are one of the most common devices used for industrial processes that involve thermal treatment of granular materials. One of the methods used to improve efficiency of these processes is to fit baffles to the drums. Prior experimental and numerical studies have shown that baffles improve particle mixing and facilitate faster particle treatment. In the current study, we use discrete element method (DEM) simulations, in order to investigate the effect of baffles on heat transfer of granular materials in rotating drums. Particularly, we investigate the effect of baffles on the scale up of heat transfer based on the recent model developed by Yohannes et al (2016)* for granular flows in rotating drum with no baffles. The simulations were run at low temperatures, where conduction through particle-particle and particle-wall contact area is the dominant heat transfer mechanism. We study the effect of size, number, and shape of baffles on the scale up model and important timescales. A wide range of several particle properties and operating conditions were considered in the simulations. In this talk, we will present results from our simulations and comparison of the results to prior models and experimental results. *B. Yohannes, H. Emady, K. Anderson, I. Paredes, M. Javed, W. Borghard, F. J. Muzzio, B. J. Glasser, and A. M. Cuitiño (2016), Physical Review E 94(4) 042902.

498932 Modeling of the Flow Dynamics of High Porosity Filter Media in Depth Filtration

Abstract:

Depth filtration is a separation process that selectively removes either solids or fine liquid droplets from another liquid phase. The analysis of this process is well developed, being widely applied in groundwater flow, oil transport in porous rock and other engineering applications. Specifically, the Kozeny-Carman (KC) model is commonly used to predict the flow dynamics during such processes. In recent work, however, we have found that the KC model cannot be directly applied to the flow through certain poly-disperse particle beds or, more generically, porous materials that have complex pore size distributions. In these cases, the KC model requires both aphysical average particle sizes and/or tortuosities in order to fit our experimental data. In contrast, our modified model introduced two intuitive fitting parameters that describe the cake/pore structure â€" the fraction of expanded voids (kappa) and the ratio of void sizes (beta) â€" and was found to dramatically improve predictions of flow through packed spheres. Here we adapt this model to beds comprised of high void fraction materials (diatomaceous earth). By formally accounting for the complex pore size distribution, we predict flow dynamics that are much closer to our experimental results than the predictions of the KC model and show that this approach is viable for both statically formed and evolving (dynamic) beds. Interestingly, for a best fit, we find that the model predictions require that we ignore the pores that are intrinsic to the individual filter media particles and instead only account for pores formed between particles. Finally, in an effort to understand the relationship between flow dynamics and pore size distribution more fully, we built a dynamic filter cake model that continuously modifies the pore size distribution as contaminants are deposited. The predicted flow dynamics of this new model match the dynamic experimental results remarkably well, setting the stage for a priori prediction of filtration times, flow-rates, and pressure requirements from simple measurements of the size distribution of both the filter media pores as well as the contaminant particles/droplets.

496249 Drying Behaviour of Complex Amorphous Solids

Abstract:

Drying Behaviour of Complex Amorphous Solids

Daryl R. Williams

Surfaces and Particle Engineering Laboratory, Department of Chemical Engineering, Imperial College London, South Kensington Campus, London SW7 2AZ, UK

(d.r.williams@imperial.ac.uk+44 (0) 0207 594 5611)

In fully understanding drying behavior of solids, it can be of enormous fundamental importance to understand the process by which water is initially sorbed. The uptake of water by amorphous or partially amorphous organic powders is a remarkably ubiquitous phenomena, yet one which is surprisingly poorly understood. Although this phenomena has been studied by notable scientists for 100 years, the literature, including many current publications, is awash with papers which misunderstand fundamentally the water sorption behaviour of solids which have significant amorphous content. This includes freeze dried and spray dried solids, as well as many natural materials such as cotton and other cellulosic materials, keratin and other cellular solids a well as many polymeric materials Its importance in solids engineering is that moisture sorption is often the precursor event to powder caking, poor powder flow, batch to batch variabilityâ€™s, phase changes, changes in T_g, re-crystallisation and a myriad of other often undesirable property changes in solid state materials systems.

The paper will commence by reporting on a detailed experimental study of water vapor sorption in freeze dried solids as well as a range of natural solid materials.Â The kinetics and equilibrium uptake of moisture by these materials will then be analysed using a range of polymer-solution based thermodynamics models, starting with the classic Flory-Huggins model and then leading onto the more recent lattice fluid and non-equilibrium lattice fluid theories. These models are far superior and fundamentally sounder than the surface adsorption models such as GAB and BET which are commonly (and erroneously) used to model sorption isotherms in amorphous powders. These thermodynamic models provide a highly relevant theoretical framework which can allow the complexities of water sorption within amorphous solids to be understood on a fundamentally sounder basis. Finally this paper highlight how this new understanding of the importance of water-amorphous solid state thermodynamics will help us gain a sounder understanding of how water sorbs and dissolves in amorphous organic powders.

497778 Effect of Particle Surface Roughness on Wall to Particle Heat Transfer

Abstract:

Solid drying commonly involves heat transfer between a hot wall and particles. Hence, accurate prediction of overall wall to particle heat transfer coefficient is crucial to the process. For granular materials, microscopic particle properties (size, shape, surface roughness, etc.) play a significant role in determining the bulk (macroscopic) behavior. Therefore, careful consideration of each of the particle-scale properties is critical in particulate heat transfer. The effects of particle size on particle-wall heat transfer have been investigated extensively in past works; however, the effects of particle surface roughness and shape have not been well established. This work is mainly focused on linking particle surface roughness to macroscopic particle-wall heat transfer, which impacts the contact area between particles and their separation distance. Experiments were designed to operate in wall- conduction dominant regime. For particle characterization, the roughness of spherical glass particles was modified and asperity heights were measured. Asperity heights of the particles were determined from atomic force microscopy topographic maps of the surfaces. Wall-to-particle heat transfer coefficients for different particle roughness levels were compared, allowing for the sensitivity of heat transfer to surface roughness to be established. The experimental results were then validated using the discrete element simulations, which take the measured roughness values as inputs.

500673 Chiral Control of Enantiomers through Crystallization: Moving from Ternary Phase Diagrams to Design Spaces

Abstract:

Although chiral purity is a critical quality attribute for the majority of synthetic drugs being developed, the understanding of process design for chiral purity in terms of conventional batch crystallization processes is limited. The industrial implementation within this area has relied heavily on the construction of ternary phase diagrams, system classification, and the use of subject matter experts to determine thermodynamic operating policies. These operating policies have been developed to yield resulting solids with 100% ee; regardless of the product specification/target and the potential upgrades to ee remaining between the molecule being isolated and the drug substance. In the present work (which for simplicity is limited to the case of a single chiral center in the absence of salts, polymorphs, and solvates/hydrates), a new policy is proposed for obtaining resulting solids with ee < 100% but passing specifications for arbitrary systems. This operating policy can provide significant isolation yield improvements when compared to the conventional policy. Additionally, this work demonstrates an approach for the creation of design spaces for arbitrary ternary systems based on the collection of limited small (<10 g total) scale solubility data. This approach has been extended to processes with sequential re-crystallizations and can be used to determine â€œglobalâ€ optimal policies for an entire synthetic route.

499039 Leveraging Mechanistic Understanding of Impurity Rejection during Crystallization for Rational Process Design

Abstract:

The control of impurities in the chemical synthesis of a new drug substance is a requirement in registering pharmaceuticals with worldwide regulatory agencies. Crystallization is an essential tool in controlling the purity profile of a drug substance and its intermediates. An overview of the various mechanisms of impurity incorporation during crystallization will be discussed along with remediation strategies. A case study will be provided on the replacement of chromatography with crystallization during the development of a second generation commercial manufacturing route for Ceftolozane, the cephalosporin antibiotic found in Zerbaxa. The implementation of a crystallization based purification process for Ceftolozane, replacing a chromatography/nanofiltration/diafiltration sequence, resulted in a more productive, greener, and more portable manufacturing process. Another case study will be presented on limiting substrate concentration on product crystals during a reactive crystallization via engineering design.

488309 Development of Combined Continuous Reaction to Crystallization Process Using PAT Tools

Abstract:

Introduction

Figure 1. Synthesis of (*Z*)-3-chloro-2-(phenylthio)-*N*-(p-tolyl)acrylamide

Traditionally the majority of active pharmaceutical compounds have been manufactured at large plant scales using batch methodologies. Continuous manufacturing is becoming an increasing attractive alternative due to apparent benefits such as smaller manufacturing footprints, increased process safety, lower manufacturing costs and increased supply-chain flexibility. In this work, a new approach to the design of continuous manufacturing platforms is being presented which is not restricted to using such conventional equipment but which is based around bespoke design and manufacture. As a model compound, 2-chloro-*N*-(4-methylphenyl)propanamide (**S1**), a precursor in the multi-step synthesis of (*Z*)-3-chloro-2-(phenylthio)-*N*-(p-tolyl)acrylamide (**P1**, Figure 1) is used. The aim of this work is to design a combined continuous reaction to crystallization process for **S1** and overcome the existing challenges between the two process.

Solubility Studies for S1

Figure 2. Solubility data of **S1** in toluene + hexane mixtures: *, $w_2 = 1$; +, $w_2 = 0.8$; ◊, $w_2 = 0.6$; □, $w_2 = 0.4$; -, Phase Equilibria with NRTL equation calculated. w_2 is the mass fraction of toluene in binary toluene + hexane mixtures.

The solubility behaviour of **S1** in ethyl acetate + hexane, toluene + hexane, acetone + hexane and butanone + hexane solutions were measured as the solubility characteristics have considerable influence upon the design and optimisation of any crystallisation process, affecting both the resulting crystal structure and crystal size distribution (CSD)[1]. The solubility study is done using the polythermal method and conducted in EasyMax (Mettler Toledo) reactor with in situ Focused Beam Reflectance Measurement (FBRM G400, Mettler Toledo) to characterize the saturation temperature of **S1**. The experimental data are correlated with Apelbat, λh and the Phase Equilibria with NRTL (non-random two liquid) model equations using the function *lsqcurvefit* in Matlab's optimization toolbox to enable both interpolation and extrapolation of the measured data, as shown in Figure 2.

Design of Continuous Reaction and Crystallization Process

Reaction: Using a modified procedure developed by Li *et al*[2], a batch process is employed where a-chloropropionyl chloride (12 mmol) was added dropwise to a suspension of *p*-toluidine (10 mmol) in toluene (20 mL) at 0°C and the resulting solution was heated to reflux for 1 h with vigorous stirring. After cooling, the solvent was removed under vacuum and the resulting off white solid was collected by filtration and washed thoroughly with cold cyclohexane. Spectral data of the obtained product matched that reported in the literature[3]. Therefore, it has been decided to use toluene as the reaction solvent as it can also be used as a crystallization solvent based on the solubility studies. However, this developed procedure produces a small amount of HCl as a by-product, which can be problematic for reaction product purity and its subsequent crystalisation steps. Therefore, it is decided to run the reaction in a novel 3D printed continuous biphasic reaction system in the presence of aqueous sodium hydroxide to neutralize HCl salt. The design of the biphasic reactor is based on the specific needs of the reaction which is capable of achieving extremely high dispersion rates between the two phases under the very low Reynolds numbers typically encountered in the pharmaceutical industry ($Re < 5$). Process steps are quantified in detail using methods such as computational fluid dynamics (CFD) leading to tailored equipment designs which can be manufactured with high precision using metallic 3D printing. The biphasic reactor is coupled with liquid-liquid separation such that both unit operations are designed using fundamental engineering principles combined with numerical analyses (based upon the requirements of the specific process), prior to being 3D printed and combined to create a bespoke, sealed, hard-piped manufacturing system.

Reaction to Crystallization: From the liquid-liquid separator, the organic phase is transferred in the feed/dissolution tank while aqueous phase goes to waste. In situ FTIR (React IR 15, Mettler Toledo) is used on the feed/dissolution tank to determine the concentration of **S1** prior to crystallization and the crystallization kinetic data obtained (explained below) enables the adjustment of crystallization parameters for better control and set up of the process.

Crystallization: Using the solubility curve obtained from solubility studies for toluene, a continuously operated single stage mixed suspension, mixed product removal (MSMPR) crystallizer with intermittent withdrawal via dip pipe using an automated pressure supply[4-6] is set up for the cooling crystallization of **S1**. Acquisition of reliable kinetic correlations that describe how rates of nucleation and growth vary over the design space for the crystallization process is necessary for enabling robust prediction of crystallizer performance. Therefore, simple extraction of the growth and nucleation parameters from the single stage MSMPR operating at steady state is performed using the population balance model. Comprehensive experimental data for model fitting and parameter estimation is obtained by varying crystallizer operating conditions (temperature, supersaturation and suspension density) to induce changes in the rate-affecting variables for growth and nucleation. FBRM is used for real-time monitoring of size distribution and identification of steady state and hence enabling on-demand adjustment of operating conditions to transition multiple steady-states for rapid acquisition of kinetic data while PVM (ParticleView V819, Mettler Toledo) provides microscopy quality images in real time to monitor the crystal habit. Malvern Mastersizer is used as an offline analysis to determine the crystal size distribution (CSD) at steady-state while the concentration of mother liquor at steady-state is determined gravimetrically. By estimating the six kinetic parameters for crystal growth and nucleation, accurate prediction of CSD and concentration results at steady-state can be achieved.

References:

[1] Mullin, J. W. *Crystallization*; Butterworth-Heinemann, 1997

[2] Li, X.; Wang, D.; Wu, J.; Xu, W. Novel Approach to 3-Methyl-1H-quinoxalin-2-ones. *Synthetic communications* **2005,** *35* (19), 2553.

[3] Murphy, M.; Lynch, D.; Schaeffer, M.; Kissane, M.; Chopra, J.; O'Brien, E.; Ford, A.; Ferguson, G.; Maguire, A. R. Investigation of the synthetic and mechanistic aspects of the highly stereoselective transformation of [small alpha]-thioamides to [small alpha]-thio-[small beta]-chloroacrylamides. *Organic & Biomolecular Chemistry* **2007,** *5* (8), 1228.

[4] Hou, G.; Power, G.; Barrett, M.; Glennon, B.; Morris, G.; Zhao, Y. Development and Characterization of a Single Stage Mixed-Suspension, Mixed-Product-Removal Crystallization Process with a Novel Transfer Unit. *Crystal Growth & Design* **2014,** *14* (4), 1782.

[5] Power, G.; Hou, G.; Kamaraju, V. K.; Morris, G.; Zhao, Y.; Glennon, B. Design and optimization of a multistage continuous cooling mixed suspension, mixed product removal crystallizer. *Chemical Engineering Science* **2015,** *133*, 125.

[6] Morris, G.; Power, G.; Ferguson, S.; Barrett, M.; Hou, G.; Glennon, B. Estimation of Nucleation and Growth Kinetics of Benzoic Acid by Population Balance Modeling of a Continuous Cooling Mixed Suspension, Mixed Product Removal Crystallizer. *Organic Process Research & Development* **2015,** *19* (12), 1891.

485308 Process Design to Preserve Particle Properties during Crystallization of Energetic Material

Abstract:

Obtaining desired particle size and morphology of isolated material from a quenched crystallization can be challenging during scale up. This work focuses on the scale-up of a quenched crystallization of an energetic compound while maintaining a large blocky morphology of the isolated material. The compound has low solubility in most solvents, therefore recrystallization options were limited. In this case, the key to obtaining the desired particle size and morphology was quenching the reaction mixture into a nitrate salt solution. This talk will discuss the process optimization and the effects of mixing, temperature, and salt concentration on this quenched crystallization. Optimization and development efforts were performed in the 1-L Mettler Toledo OptiMax reactor before being scaled up to the pilot plant.

491816 Multi-Scale and Multi-Purpose Modelling for *in silico* Designing and Optimization of Pharmaceutical Crystallization Process

Abstract:

Crystallization often serves as a key final and intermediate separation/purification unit operation in pharmaceutical production. More than 90% of small molecule drugs are produced in crystalline form[1]. Operation conditions of crystallization process determine process metrics such as crystal size distribution, yield, space-time yield (STY) and crystal quality attributes like crystal shape, morphology as well as solvates. The crystal product properties affect the efficiency of downstream unit operations such as filtration, drying and formulation. Moreover, in pharmaceutical products the crystal properties affect the stability and efficacy of drugs. Therefore, a comprehensive multi-scale and multi-purpose thermodynamic and kinetic modeling is the main objective of this work in order to understand crystallization mechanism and define optimum operation conditions. Optimization of crystallization operation conditions includes selection of the best solvent or mixture of solvents, cooling rate, anti-solvent flowrate, pH profile, mixing and seeding parameters as well as selection of mode of operation such as batch or continuous. Figure 1 shows a comprehensive multi-scale and multi-purpose modeling framework for *in silico* designing and optimization of pharmaceutical crystallization process.

Figure 1: A schematic diagram illustrating comprehensive multi-scale and multi-purpose modeling framework for *in silico* designing and optimization of pharmaceutical crystallization process

The framework will be implemented in Matlab-Simulink environment as a tool to support different engineering tasks from solvent selection to crystallization process design, optimization and control in life science and biopharmaceutical industries.

Acknowledgment: We would like to thank the Danish Council for Independent Research (DFF) for financing the project with grant ID: DFF-6111600077B.

[1] Narayan Variankaval, Aaron S. Cote, *AIChE* 54 (2008) 1682 -1688

492269 Sensitivity Analysis and Optimization Study of Pharmaceutical Crystallization Process

Abstract:

Crystallization is a prevalently applied technique in many industries including food, bulk and fine chemicals as well as manufacturing of active pharmaceutical ingredients (APIs). In the pharmaceutical industry, crystallization processes are employed mainly for separation,, purification as well as for formulation of APIs. Moreover, the process should guarantee the performance of the delivered APIs suitable for downstream processing, handling and drug efficacy.

The performance of an API crystal is strongly affected by the product quality attributes such as crystal size distribution, morphology, shape and purity, which are the main targets of the process output to be achieved within specific range. However, in a crystallization process, complex feedback interactions take place between the formed crystals and the growth and nucleation kinetics. These are at the same time dependent on the supersaturation and hydrodynamic conditions to which the crystals are subjected. Consequently, the quality characteristics of the crystal product have a complex dependency on multiple operational variables and input parameters.

In order to evaluate and quantify which process factors or variables have the most impact on the final quality attributes of the crystal product, a comprehensive sensitivity analysis of the system model is required. This study therefore focuses on a global sensitivity analysis of a predictive pharmaceutical crystallization model through assessing the sources of uncertainty and their influence on crystal size distribution as the measured process outcome/performance criteria in order to guide design and suggest the optimized operation conditions.

This process model of pharmaceutical batch cooling crystallization is developed based on compartmentalization technique which takes into account mixing phenomena as well. Several global sensitivity analysis methods including Morris screening, global derivative based sensitivity analysis (DGSM), variance decomposition methods such as Sobol's index and linear regression of Monte Carlo outputs are used to perform the sensitivity analysis. The developed crystallization process model is analyzed for a range of values for parameters like cooling rate, growth and nucleation kinetics etc. to rank their contributions on the measured outcome of the system "crystal size distribution". Comparison of two sensitivity analysis methods and evaluation of the results are used as a guide for process optimization and robust system design.

Acknowledgment

We would like to thank the Danish Council for Independent Research (DFF) for financing the project with grant ID: DFF-6111600077B.

495572 Segmented Continuous Crystallization Process Optimization

Abstract:

The continuous crystallization behavior of glycine in segmented tubular crystallizer has been investigated. The factors of liquid slug length and shape, Reynolds number, suspension density and arrangement were investigated respectively on the crystal growth rate, size distribution and morphology. The results found that the slug being closer to a sphere both in length and shape is the best compared to the other slug geometries, in which mixing within each stable self-circulating slug is maximized, thus the crystal growth rate is faster and the particle size distribution is more uniform. Besides, Higher Reynolds number improves the growth rate and uniform size distribution at the same residence time. The mass fraction of seed crystals at 2% was also determined to be the optimal condition. The arrangement of the tube in vertical and horizontal direction has little influence on the final yield and crystal characteristics. These results provide an insight into the segmented tubular continuous crystallization.

503244 Modified-Laguerre Polynomials for Distribution Reconstruction from Moments

Abstract:

It has been shown in an earlier paper that the 8 possible combinations of accretional growth, collisional growth and breakage are attracted to either stationary states or similarity solutions at long times given that the rate kernels are power-law and the fragment distributions are self-similar. A complete mapping of the collisional growth (aggregation) solution space has also been given in terms of modified-gamma distribution parameters as approximate solutions. In this paper, a set of modified-Laguerre orthogonal polynomials is derived expressly for the purpose of reconstructing these stationary or similarity solutions from their moments. This paper discusses the derivation and illustrates the application to similarity solutions for aggregation including an assessment of how well the modified-gamma approximation captures the system behavior.

503245 Solution Space Mapping for Power-Law Aggregation Rate Kernels

Abstract:

Using the scaling presented in an earlier paper at this meeting titled â€œMapping the General 1-D Population Balance Solution Spaceâ€, the aggregation problem with power-law rate kernels is examined from the point of view of the Pulvermacher-Ruckenstein criterion for similarity solution existence. In this form one can map the solution space into two axes, one for kernel order and one for its internal structure. In this space, maps of approximate solution parameters in the form of modified-gamma distributions or the log-normal distribution can be identified in the existence region using a moment model. Davies, et al. studied and mapped gelation behavior in the non-existence region so that their results can be superimposed on the similarity solution map to generate complete mapping of the aggregation solution space.

492251 Nucleation and Agglomeration of Hydrate Particles in Gas-Liquid Multiphase Flow Systems

Abstract:

In recent years, gas hydrate has been a research hotspot in the fields of CO_2 capture, refrigeration and oil/gas flow assurance and so on. Many researchers are focusing on the nucleation and agglomeration properties of gas hydrate, which are of great significance to the hydrate application and risk management. In this work, hydrate nucleation and agglomeration properties in gas-liquid multiphase flow systems were investigated using a high pressure flow loop. By adjusting the gas and liquid flow rates, four different flow patterns were obtained during the experiments, including stratified flow, bubble flow, slug flow and annular flow. The hydrate formation critical driving force, water conversion rate, hydrate growth rate and the hydrate agglomeration degree in different flow patterns were compared. The critical driving force and water conversion rate showed the same trends in different flow patterns, which from high to low in order was: slug flow, bubble flow, stratified flow and annular flow. The order of hydrate growth rate in different flow patterns from high to low was: bubble flow, slug flow, stratified flow and annular flow. According to the above comparison and analysis, it is harder for hydrate to nucleate in slug flow and bubble flow conditions, while the formation amount and growth rate in these two flow patterns are larger. However, the opposite case occurred in the stratified flow and annular flow. In addition, two methods were proposed based on the hydrate particle chord length distributions to estimate the hydrate agglomeration degree. The calculation results showed that the hydrate agglomeration degree from high to low in order was: slug flow, stratified flow, bubble flow and annular flow.

503246 Simplifying the Population Balance for Teaching Kinetics & Contactor Design in Particle Processing

Abstract:

In the paper presented at this meeting titled â€œMapping the General 1-D Population Balance Solution Spaceâ€, a unified scaling for the population balance was presented which allows identification of the following for all possible combinations of accretional growth, collisional growth and breakage given power-law rate kernels and self-similar fragments:

- A scaled time variable from which the characteristic time for the system can be extracted
- Reduced equations for evolution of the scaled number mean particle volume
- Scaled mean particle size trajectories that depend only on kernel order
- Expressions for the scaled moments of the size distributions for the long-time attractor of the system (either stationary state or similarity solution)

This paper presents examples of size-evolution trajectory dependence on kernel order which may be used to extract rate parameters from data (the inverse problem) or for designing ideal contactor systems given a knowledge of the rate parameters. The trajectory dependence on kernel order is shown to be analogous to a reaction trajectoryâ€™s dependence on reaction order. Thus, this approach is a step toward the Hounslow program which called for teaching Particle Technology like Reaction Engineering

503247 Mapping the General 1-D Population Balance Solution Space

Abstract:

The general 1-D population balance model in particle size contains nucleation, accretional growth, collisional growth and breakage terms. When there is nucleation, there can never be similarity solutions or stationary states. In the absence of nucleation, there are 8 possible combinations of the other 3 mechanisms. Under the presumption of power-law rate kernels and self-similar fragment distributions in breakage, the attractors for these 8 combinations at long times are either stationary states or similarity solutions. This paper presents the development of a unified scaling for the population balance which allows identification of the following for each of the cases:

- A scaled time variable from which the characteristic time for the system can be extracted
- Reduced equations for evolution of the scaled number mean particle volume
- Scaled mean particle size trajectories that depend only on kernel order
- Expressions for the scaled moments of the size distributions for the long-time attractor of the system (either stationary state or similarity solution)

494482 Noble Method for Evaluating the Distributive Mixing of Bimodal Lagrangian Particles

Abstract:

Mixing operation have been one of the most important process at chemical industry for a long time. Several equipment such as static mixer or twin screw extruders are developed to obtain uniformity of product. Also considerable researches have been conducted to increase the mixing efficiency. In the most researches about mixing, evaluating the degree of mixing was an issue. For evaluating the degree of mixing quantitatively, the mixing is classified with phases of mixing materials[1]. Among phases, particle-particle mixing (powder mixing) takes considerable part of research, because of its discretizational characteristic.

Since Danckwerts proved that the variance among samples can represent the scale of segregation, the variance among bin counts is useful and popular index to quantify the spatial distribution of Lagrangian particles[2]. With the domain which is divided by several bins (cells), the variance among particles number in each bin is easy to be calculated and understood. When the spatial distribution of particles appear, the variance tend to be smaller. Although the variance among bin counts is effective to quantify the spatial distribution of unimodal particle systems, it is hard to applicate the method at bimodal particles systems. However, most of the practical particle mixing processes are bi- or multimodal particles systems. In this research, using the variance among bimodal bin counts, mixing between bimodal particles was evaluated. Bimodal bin counts is defined by subtraction of number of two particles species in each cell. Therefore bimodal bin counts values represent quantitative difference between two particles species in each cell and variance among bimodal bin counts represent the degree of mixing between two particle species. When the system reach almost evenly mixed state, variance becomes to be zero. By adapting conventional and bimodal bin counting method simultaneously, utility of the variance among bimodal bin counts is acknowledged with 4 examples; systems with balanced and imbalanced number of particles with 2D mixer and fluidized bed. With variation of the form of variance, 4 examples were analyzed. And examples proved effectiveness of variance among bimodal bin counts as index to quantify the mixing between bimodal particles. Also, to estimate the effect of the number of bins, case studies were conducted with changing the total amount of bins. For 2D mixer, 6 kind of bin sizes were adopted, and for fluidized bed, 5 kind of bin sizes were adopted based on the size of domain and the number of total particles. As a result, variance among bimodal bin counts evaluate systems appropriately which have balanced number of particles, whereas variance among bimodal bin counts with sensitivity factor and variance among conventional bin counts only for minor particles evaluate systems appropriately which have imbalance number of particles. The effect of the number of bins is minor factor within reasonable region; smaller amount than total number of particles with small size enough to evaluate the mixing.

References

1. Harnby N, Edwards FM, Nienow WA. Mixing in the process industries (2nd edition). Butterworths-Heinemann, 1997.

2. Danckwerts PV. The definition and measurement of some characteristics of mixtures. *Appl Sci Res*. 1952;A3;279â€“296.

499936 Cyclone Dipleg Design and Operation Considerations

Abstract:

The correct cyclone design is often critical to the successful operation of many chemical processes. Cyclone system design most commonly refers to the design of the cyclone and its associated dipleg. In many (if not most) cases, poor cyclone system operation is due to poor operation of the cyclone dipleg â€“ especially the secondary cyclone dipleg.

Cyclone dipleg operation is often the primary culprit in causing excessive carryover of solids from fluidized catalytic cracking (FCC) and other catalytic processes. Poor cyclone dipleg operation can also cause solids losses from other processes (gasification and circulating fluidized bed combustion).

How the correct design of various diplegs can mitigate excessive loss of solids will be discussed. This discussion will address the role and design of trickle valves, flapper valves, loop seals and dipleg configuration on dipleg design and operation.

502988 Effect of Particle-Wall Interactions in Freeboard of Gas-Solid Fluidized Beds on Tribo-Electrification and Entrainment of Fines

Abstract:

Elutriation of fines from fluidized beds can lead to loss of valuable materials, such as catalysts, and potential release of fine particulates into the atmosphere. We have recently shown that electrostatic forces play a key role in controlling the entrainment of fine materials carrying electrostatic charges. However, it is not yet clear how electrostatic particle-wall interactions in the freeboard affect the entrainment process. This subject is investigated experimentally in this study by comparing the electrostatic charge density and entrainment flux of uncoated and coated fine glass beads in two columns made of stainless steel and acrylic.

490001 Innovative Cyclone Technology - a Common Denominator for Many Successful Fluidization Processes

Abstract:

Cyclones are century old technology that it is easy to take it for granted as a "proven technology". However, many fluidization processes demand much higher performance than the conventional cyclones were originally intended for. This paper will use FCC as an example to highlight the challenges that many fluidization processes commonly face and how the novel Shell cyclone technology can overcome those challenges. These challenges may include, but not limited to

- Improving catalyst retention and reducing catalyst cost/particulate emission
- Improving unit reliability and on-stream factor

491451 A Fluidized Bed "Thermochemical Battery" for Concentrated Solar Power Applications

Abstract:

ABSTRACT

Extensive R&D is in progress to exploit the huge amount of solar energy falling on Earth. The potential advantage of Concentrating Solar Power (CSP) over its direct photovoltaic (PV) competitor relies on easier integration with energy storage. CSP coupled with Thermal Energy Storage (TES) systems is as yet more convenient than PV with batteries. Sensible heat storage, the less complex technology among TES systems, is already available at commercial scale. TES based on molten salts is well assessed and commercially available, but suffers from fairly small density of energy storage and relatively low upper temperature limit (560 °C) which hampers the efficiency of the associated energy conversion thermodynamic cycles. The use of granular solids as radiative flux collection and TES media is gaining much interest, as it enables to overcome temperature limitations of molten salts. Thermochemical energy storage (TCES) is more ambitious than TES, striving for much higher energy density storage and improved stability over long time-scales. Reversible chemical reactions such as hydration/dehydration, carbonation/calcination, oxidation/reduction of inorganic compounds are being considered for TCES [1]. Many processes entail gas-solid reversible reactions and are performed in multiphase chemical reactors, frequently of the Fluidized Bed (FB) type [2-3].

A fluidized bed "thermochemical battery" is presented in this study, based on a novel design of a fluidized bed solar collector. The peculiar design of the reactor enables to lump in a single unit all the basic goals of a solar collector with TCES: collection of solar energy by direct radiation of the granular solid; course of the endothermic reversible heterogeneous chemical reaction U->C promoted by solar radiation; storage of both the thermochemically "charged" solid compound C, as well as of the original "uncharged" solid compound U, in the very same vessel. Energy recovery associated with the reverse reaction C->U (i.e. the thermochemical battery discharge) is accomplished in a separate reactor, where heat extraction is optimized during the course of the exothermal energy production step. The battery consists of four sections: i) storage of the uncharged compound U at the bottom of the bed; ii) storage of the charged compound C at the top of the bed; iii) a disengagement section where collection of solar energy takes place, supporting the course of the "charge" endothermal reaction U->C; iv) a draft tube where the uncharged material U is conveyed into the disengagement section and to the top of the bed. Solids segregation along the bed is promoted by density/size differences between U and C particles. Operational conditions of the fluidized bed are set so as to maximize solids segregation, while preserving the required mobility of granular solids. Operation of the central draft duct is directed to ensure the required solids recirculation rate which, in turn, is strictly related to the incoming solar flux. Altogether, operation of the thermochemical battery is a dynamic process where the relative inventories of the charged and uncharged materials vary, and the interface between U and C moves during the charge and discharge phases much like a thermo/chemo-cline. The system is complemented by another fluidized bed reactor, where the exothermal "discharge" reaction C->U takes place.

The proof-of-concept of the fluidized bed thermochemical battery is given with reference to the CaO-$CaCO_3$ cycle. A directly irradiated 0.1 m ID FB reactor exposed to a 12 kW_{el} simulated solar furnace has been used to demonstrate operation of the fluidized bed thermochemical battery. Key variables of the thermochemical battery (temperatures, inventories of CaO and $CaCO_3$, quality of fluidization and extent of segregation) have been recorded during operation to verify and support further optimization of the proposed concept.

References

[1] L. André, S. Abanades, G. Flamant. Renewable and Sustainable Energy Reviews, 64 (2016) 703–715.

[2] C. Tregambi, F. Montagnaro, P. Salatino, R. Solimene. Solar Energy, 120 (2015) 208–220.

[3] N. Gokon, T. Izawa, T. Kodama. Energy, 79 (2015) 264–272.

491025 An Innovative Hydrogen and Power Coproduction System Using Integrated Â Exergy Recuperative Biomass Gasification and SOFC

Abstract:

A novel hydrogen and power coproduction system based on the exergy recuperative biomass gasification integrated with solid oxide fuel cell was proposed. In this system micro-tubular SOFC stacks are immersed in the internal circulating fluidized bed gasifier. The exothermic heat from SOFC is recuperated and utilized for endothermic heat of biomass steam gasification. In addition, providing the recirculation system of unreacted hydrogen separated from SOFC anode exhaust, gas turbine can be eliminated. Since the exergy loss of SOFC is much smaller than that of gas turbine, the net power generation efficiency of the combination of gasification and SOFC power generation system can be considerably improved. The internal circulating fluidized bed was composed of a moving bed high-temperature pyrolyzer with the function of tar cracking, a bubbling bed char gasifier equipped with micro-tubular SOFC stacks, and a fluidized bed combustor of remained char. A 20 kW-class pilot plant was constructed and tested. In the present study, the conceptual design of a integrated exergy recuperative biomass gasification and SOFC system for hydrogen and power coproduction and the performance of exergy recuperative biomass gasification are described.

503524 Flow Structures in Circulating Fluidized Bed Risers: From Dilute to High-Density Gas-Solid Flow

Abstract:

Circulating fluidized beds play an important role in a large variety of processes involving gas-solid reactions. A major characteristic of circulating fluidized beds is intense contacting of the gaseous and the solid phase, providing excellent conditions for heat and material transfer. Moreover, the continuous interaction between the two phases keeps the suspended solids motion and thus facilitates stable solids transfer to adjacent unit operations. For these reasons, circulating fluidized beds have become an integral part of the process chain of numerous operations in the chemical and metallurgical industry as well as in the energy sector. Depending on the application, the desired reaction condition can vary significantly. Accordingly, the flow condition within the circulating fluidized bed reactor needs to be adjusted to operate at maximum efficiency. At a given gas velocity, the flow condition within the riser section of circulating fluidized beds is commonly categorized on the basis of the solids holdup and can be attributed either to the dilute pneumatic conveying regime, the fast fluidization regime or to the dense suspension upflow regime. With the aim of developing a detailed understanding of the dependence of the resulting flow condition on the applied operating parameters, a pilot plant scale circulating fluidized bed unit was developed, which allows for adjusting flow conditions in the range from dilute conveying to highly dense gas-solids flow. Within the scope of the fluid-dynamic investigations, presented in this work, the solids holdup within the riser was varied in the range from 2 to 32Â vol.%. For determination of the spatial solids distribution on the cross section of the riser in the different flow regimes, the plant is equipped with an Xâ€'ray computer tomographic system. This allows for non-invasive scanning of the riser cross-section, by which the time-averaged distribution of particles can be derived with high spatial resolution. Additionally, the motion of particle clusters is investigated using capacitance probes. Based on the gained insight into the flow structures prevailing in high-density risers conclusions are drawn on the interaction mechanism of the fluid and the solids phase in vertical gas-solid upflow.

501488 Prediction of Bulk Solid Properties of Formulations

Abstract:

Bulk solids flowability is important for pharmaceutical oral dosage formulations manufacturing. The ability to predict the performance of a formulation apriori can be a huge benefit for the industry in optimizing the formulation. This study presents some of the techniques developed to predict the bulk solid properties of the formulations based on the individual components. Several methods to evaluate the flowability of the formulations are also presented and compared in this work. Some of these include particle size distributions, bulk density, compressibility and permeability and hopper design. These techniques are used to present the comparison of flowability of various formulations with the traditional METHOCELâ„¢ CR and the recently developed METHOCELâ„¢ DC2.

GHENT UNIVERSITY

FACULTY OF PHARMACEUTICAL SCIENCES

Extension of a Mechanistic Model for Drying of Single Pharmaceutical Granules to Semi-Continuous Fluid Bed Drying

Michael Ghijs

October 31th, 2017
2017 AIChE Annual Meeting

LP PAT BIOMATH janssen

Outline

1. Fluid bed dryer system and experimental data
2. Single particle drying model, and extension
3. Additional experimental data on drying behaviour single granules
   › Model calibration
4. From single granules to batch data
5. Conclusions and future work

1

Semi-continuous fluid bed dryer in continuous wet granulation process

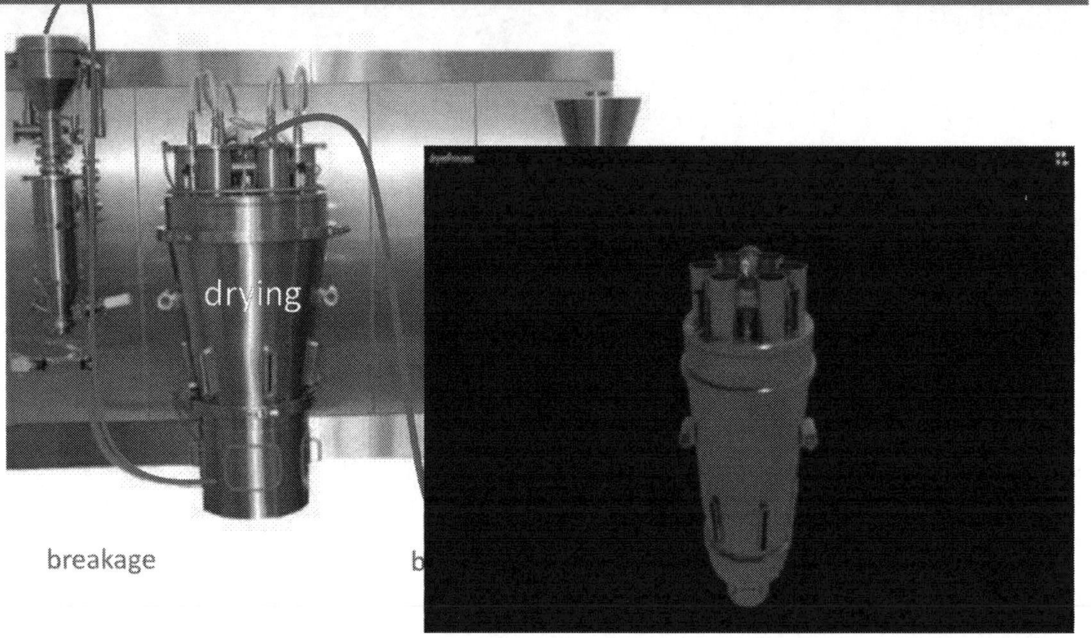

Fluid bed dryer experiments

Wet granules from twin-screw granulator
Powder feed rate: 20 kg/h
Liquid feed rate: 62.7 g/min
Screw speed: 900 rpm
Screw configuration: 2 kneading zones of 6 kneading elements at 60° angle

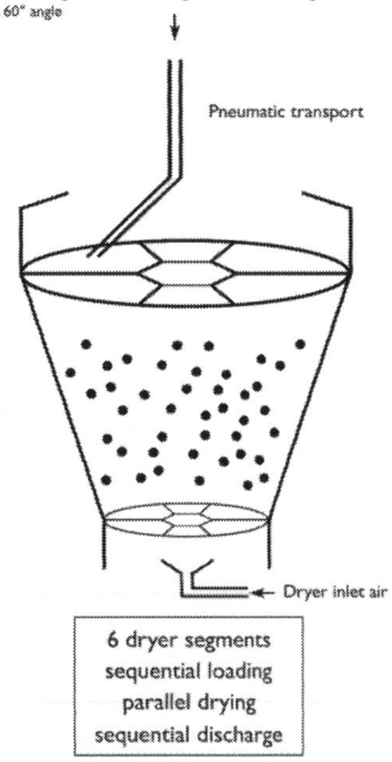

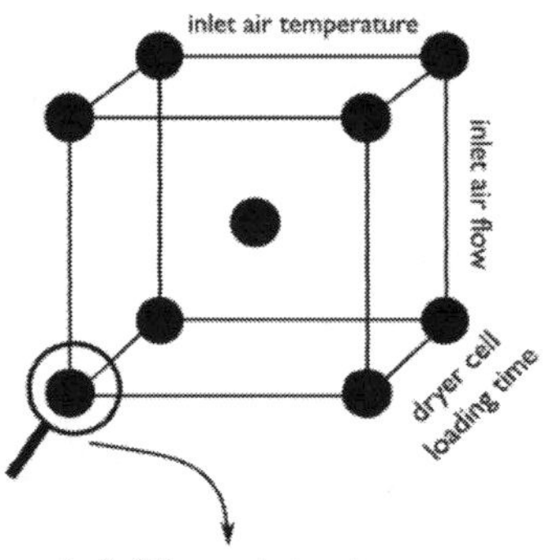

At 5 different drying times:
- PSD (QICPIC, Sympatec)
- moisture content of each of 5 sieve fractions (loss-on-drying)

sieve fractions

< 150	150-300	300-500	500-800	800-1000	1000-2000	>2000

3

Experimental moisture content data

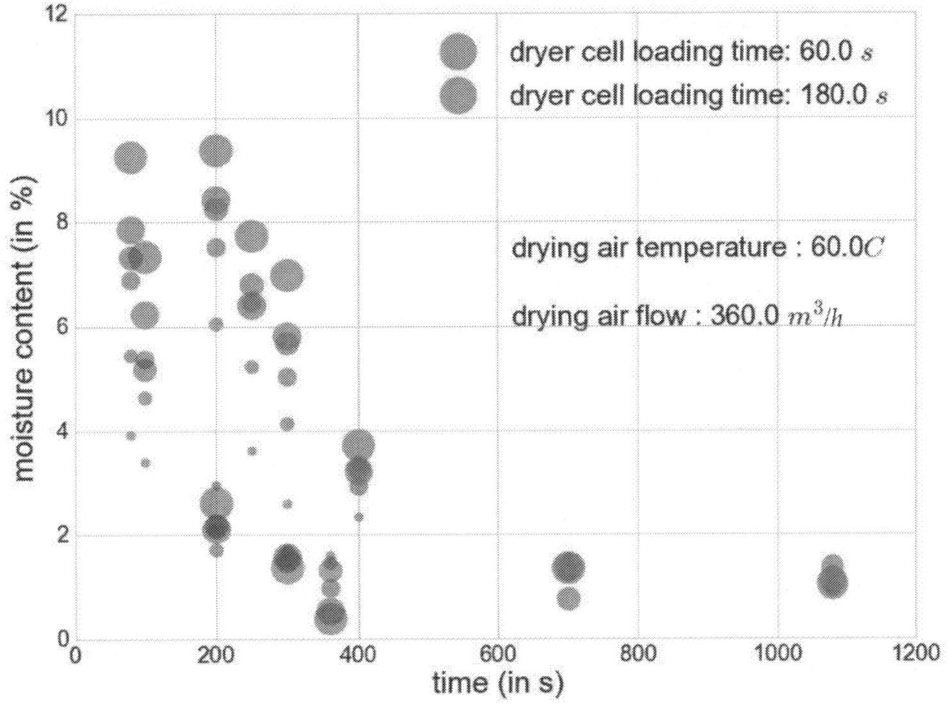

4

Single particle drying model (SPDM)

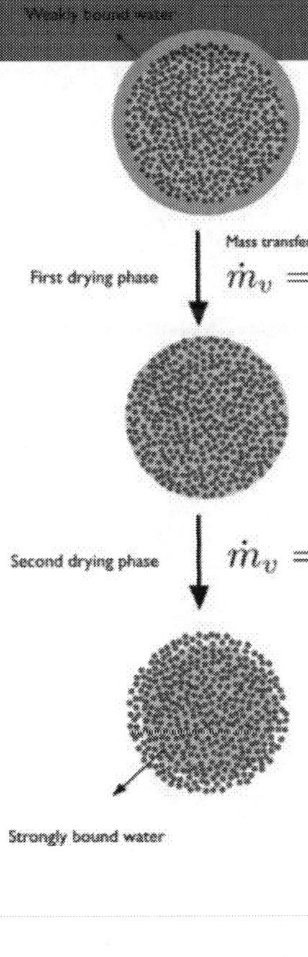

Mass transfer rate:

$$\dot{m}_v = h_D(\rho_{v,s} - \rho_{v,\infty})A_d$$

Empirical power coefficient for porosity
Calibrated to process settings

$$\dot{m}_v = -\frac{8\pi\epsilon^{\beta} D_{v,cr} M_w p_g}{\Re(T_{cr,s}+T_{wc,s})} ln[\frac{p_g - p_{v,i}}{p_g - (\frac{\Re}{4\pi M_w h_D R_p^2}\dot{m}_v + \frac{p_{v,\infty}}{T_g})T_{p,s}}]$$

Mortier, S. T. F. C., De Beer, T., Gernaey, K. V, Vercruysse, J., Fonteyne, M., Remon, J. P., Nopens, I. (2012). Mechanistic modelling of the drying behaviour of single pharmaceutical granules. *European Journal of Pharmaceutics and Biopharmaceutics : Official Journal of Arbeitsgemeinschaft Für Pharmazeutische Verfahrenstechnik e.V, 80*(3), 682–9

5

SPDM application to batch data

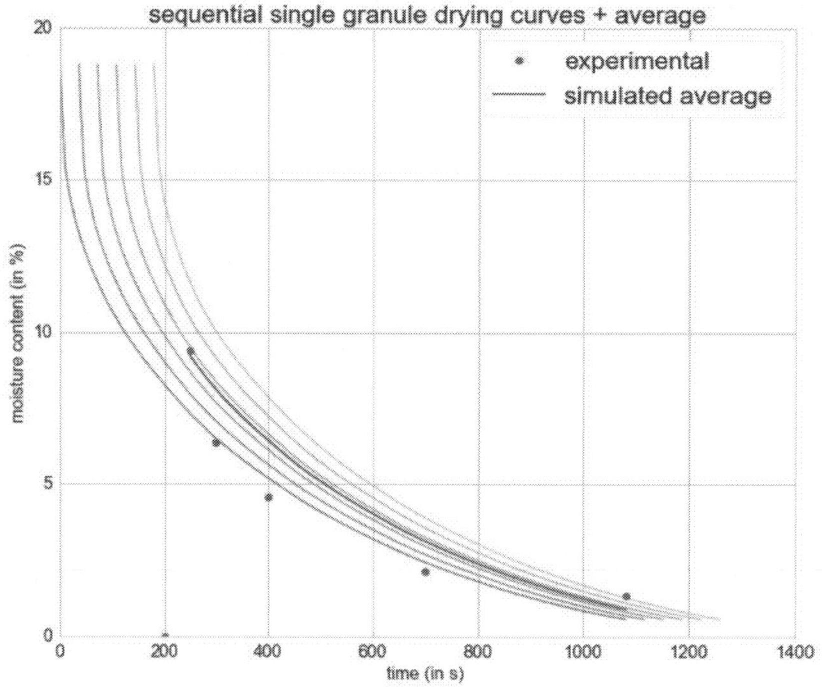

6

Missing element:
drying not slowed down at low moisture content

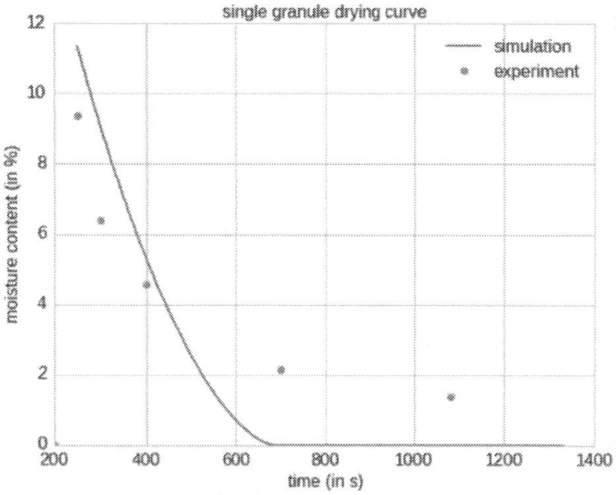

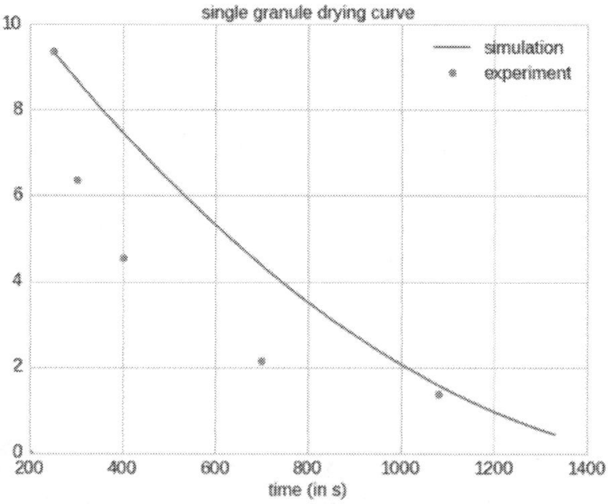

Modeled moisture content reaches zero → not experimentally observed

7

Single particle drying model (SPDM)

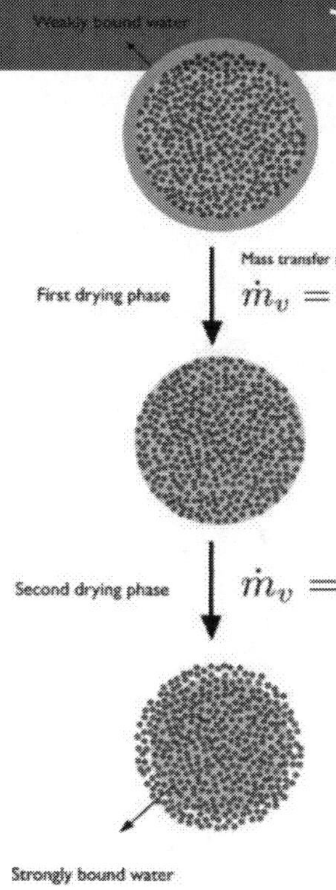

Mass transfer rate:

$$\dot{m}_v = h_D(\rho_{v,s} - \rho_{v,\infty})A_d$$

Empirical power coefficient for porosity
Calibrated to process settings

$$\dot{m}_v = -\frac{8\pi e^{\beta} D_{v,cr} M_w p_g}{\Re(T_{cr,s}+T_{wc,s})} ln[\frac{p_g - p_{v,i}}{p_g - (\frac{\Re}{4\pi M_w h_D R_p^2}\dot{m}_v + \frac{p_{v,\infty}}{T_g})T_{p,s}}]$$

extended with equilibrium moisture content

$$\frac{dX}{dt} = \frac{X - Xe}{X} * \left(\frac{dX}{dt}\right)_{SPDM}$$

8

Verifying approach: drying data of single granules

Dynamic Vapour Sorption (Surface Measurement Systems Ltd.)

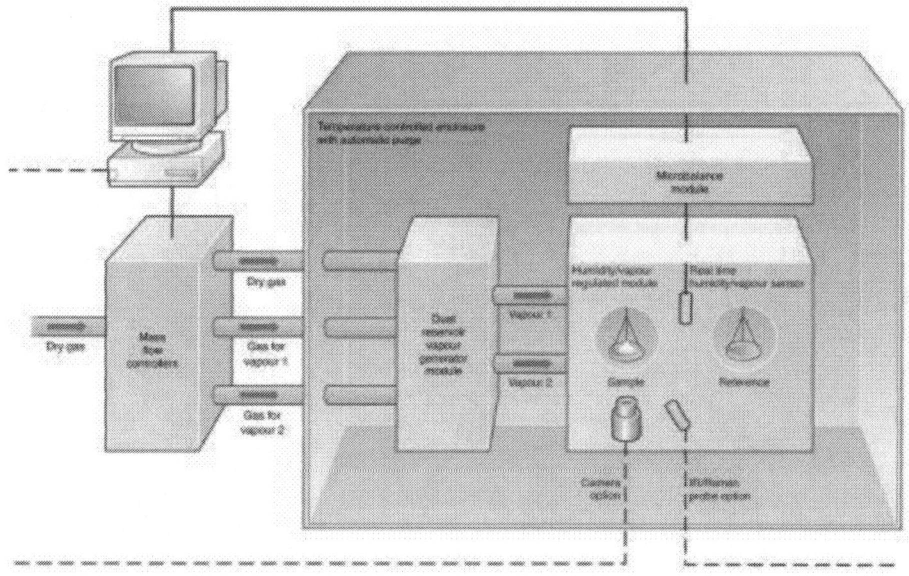

9

DVS data + calibration

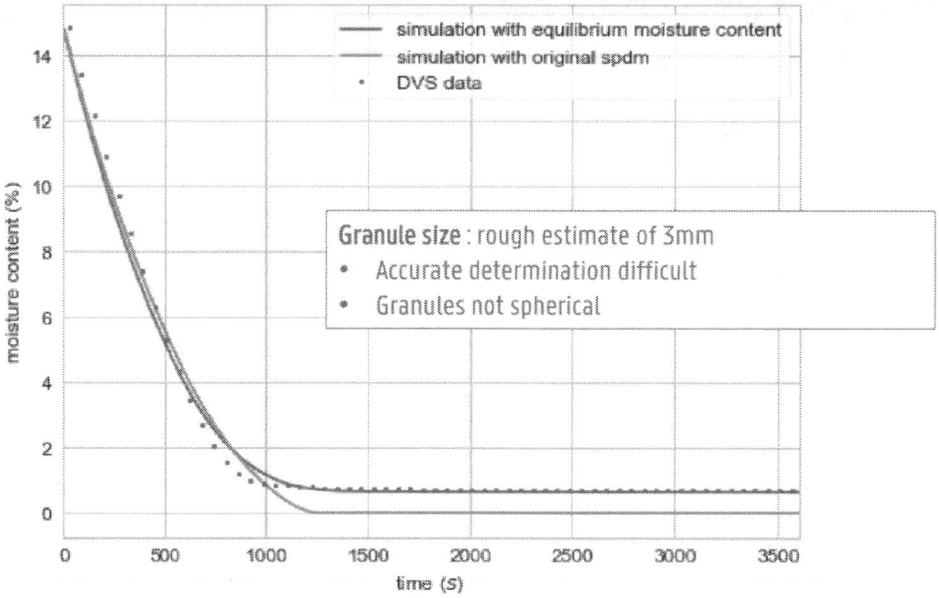

SPDM can describe drying behaviour of single granule

10

DVS data + calibration

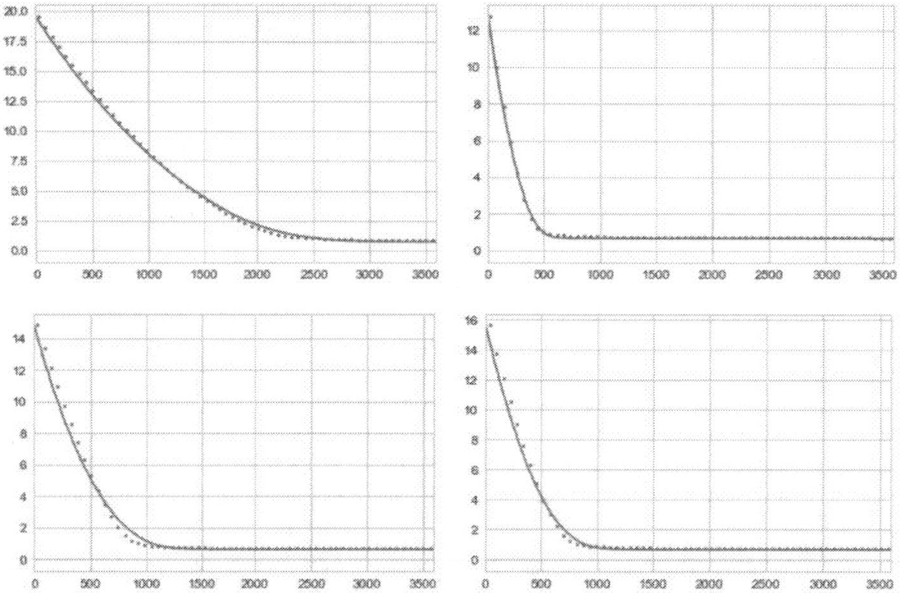

DVS experiments with different granules, at 14% rel. air humidity and 40°C

11

Calibration to batch data

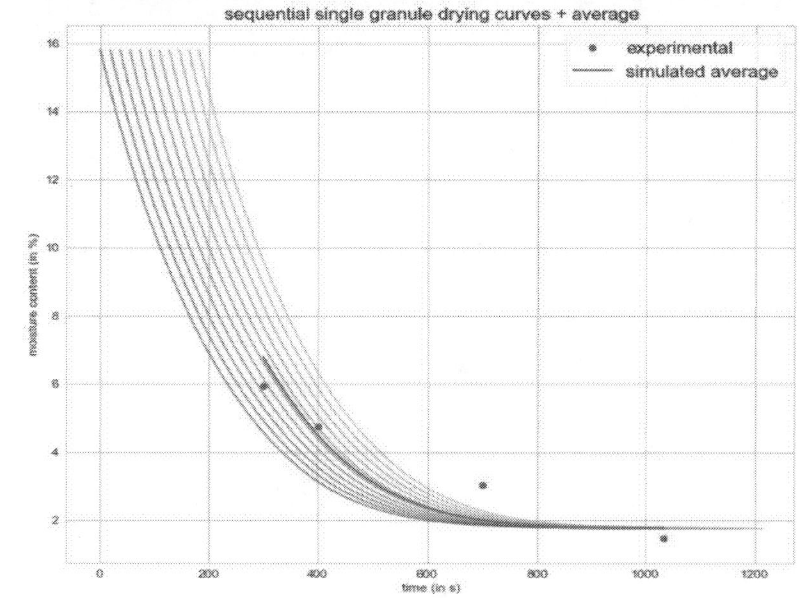

Clear effect of equilibrium moisture content

Calibration to batch data

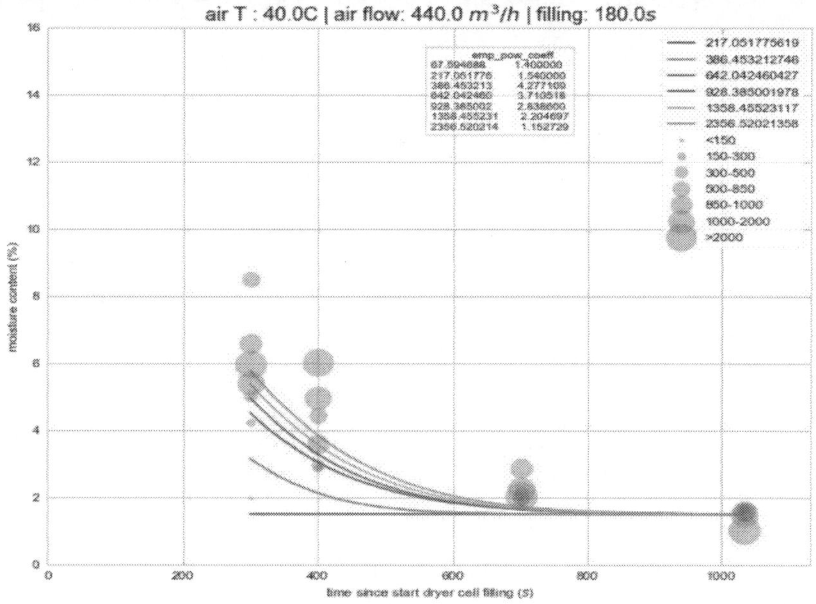

Calibration with objective function of SSE

Equilibrium moisture content

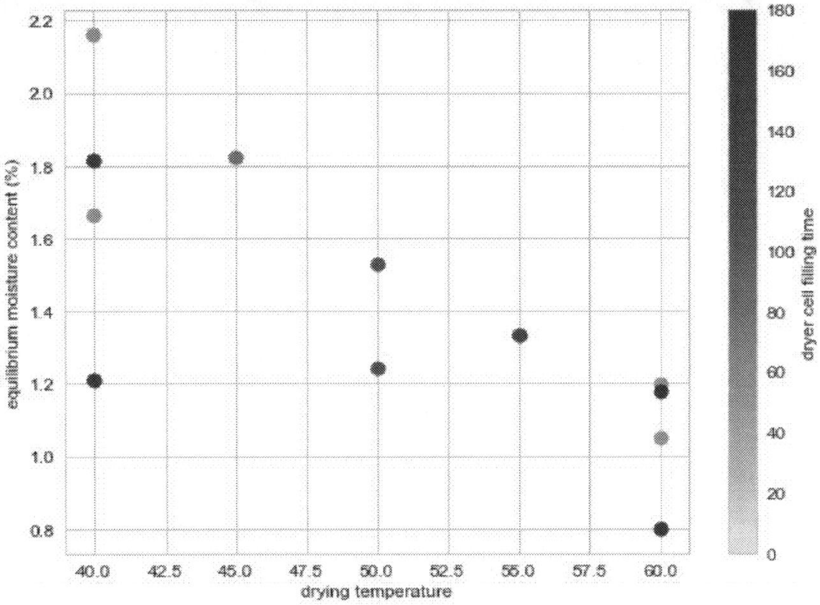

Measurement in FBD batch:

Average LOD of all size fractions at longest drying time of experiment

14

Equilibrium moisture content

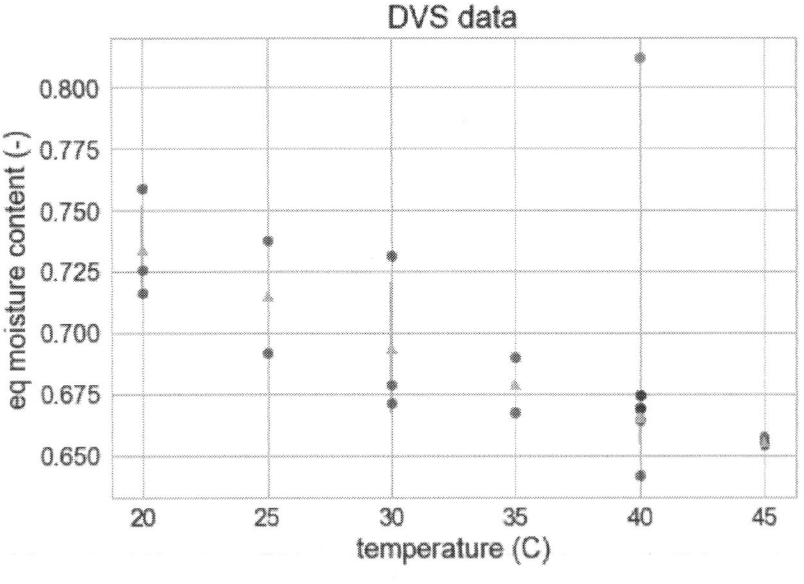

Measurement in DVS: Granule mass percentage difference between **relative humidity** of FBD drying air (14%) and 0%

15

Conclusions

1. Extended SPDM can describe drying behaviour of single pharmaceutical granules

2. Application to batch data requires different modelling approach

3. Equilibrium moisture content important model parameter

   › Dependent on ambient temperature and relative humidity

16

Future work

1. Research connection granule characteristics (size) and value of calibrated model parameter

2. Apply SPDM to batch data:
   1. different calibration conditions (objective function, optimisation algorithm)
   2. possibly look into population balance modelling as alternative

3. Model equilibrium moisture content in function of relevant factors

17

Acknowledgment

Prof. dr. Thomas De Beer
apr. Fien De Leersnyder
dr. apr. Valérie Vanhoorne

Prof. dr. ir. Ingmar Nopens
dr. ir. Séverine T.F.C. Mortier

Elisabeth Schäfer
Ashish Kumar
Philippe Cappuyns
Ivo Van Assche

LP PAT
BIOMATH
janssen

18

Thank you for your attention!

19

DVS data

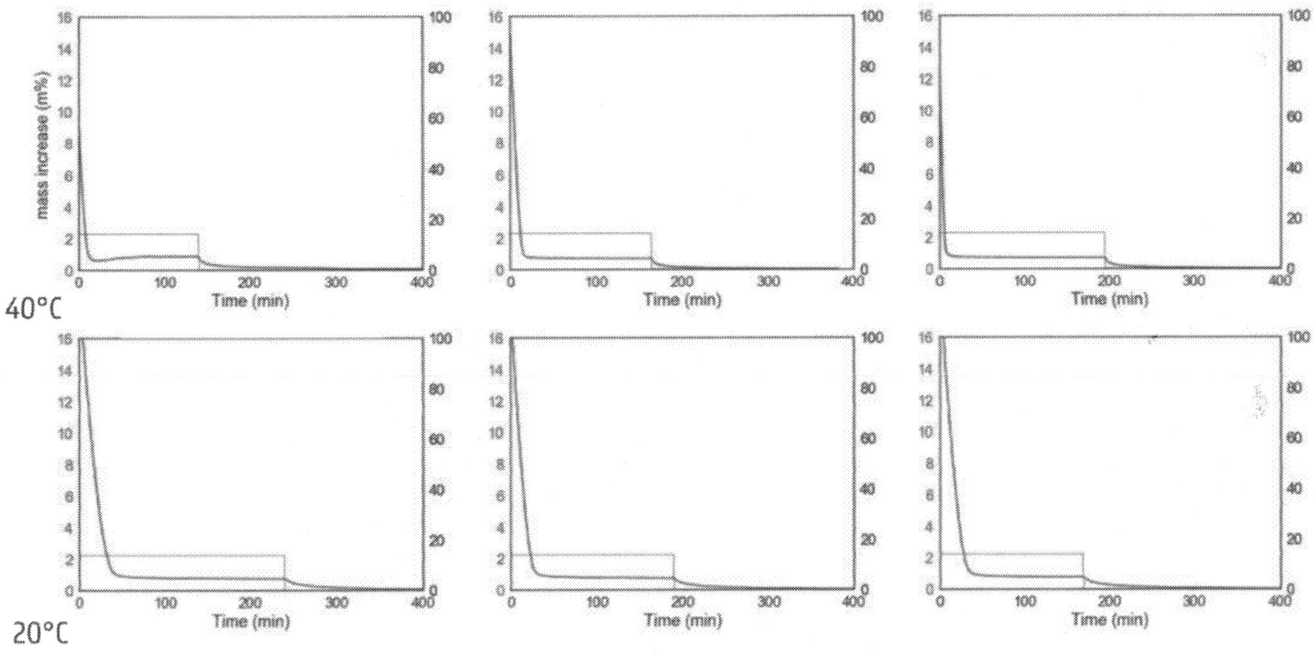

20

498693 Impregnation of Metal Solutions in Catalytic Porous Particles and Axial Dispersion Studies in Rotating Drum Using Experiments and DEM Simulations

Abstract:

Impregnation of active metals onto a porous support is an important step in the preparation of heterogeneous catalysts. In a typical impregnation process, metal solutions are sprayed over a particulate bed in a mixing vessel until the pore volume is reached. The inter-particle variability of the impregnated liquids inside the particles and metal content may significantly affect the activity and selectivity of the resulting catalyst. Current scale-up practices lead to poor fluid distribution and inhomogeneity in metal content. The aim of this work is to understand the dynamic behavior of the particles under the spray nozzle, which is essential for a desired content uniformity in the impregnation process and to optimize the process variables to achieve the best mixing and fluid uniformity in the particle bed.

Impregnation experiments were conducted in a rotating drum of D=20cm and L=30cm. The vessel was placed on a roller and the rotation speed was set at 9 rpm. Î³-Alumina spheres of d=4.9mm contained a surface area of 215 m^2/g and a pore volume of 0.6 cc/g. Each experiment was conducted with 1.75kg of alumina spheres. Nickel nitrate hexahydrate was dissolved in aqueous media and used as the metal precursor. The spray rate was set at 3.5 L/hr. After the metal solution has been sprayed, mixing of impregnated particles was continued for additional 10 minutes. During the impregnation process, higher metal content per unit mass is observed in the center of the vessel than at the edges of the vessel. This inhomogeneity levels off with additional rotations. The dispersion of material in the axial direction is the most critical factor determining the final uniformity of metal content in the particle bed. The axial dispersion coefficients are also calculated experimentally for the cases of both metal transfer and water transfer in the rotating drum.

In addition we developed a novel Discrete Element Method (DEM) model for the transfer of metal solutions from the nozzle to the particles and between particles. When a droplet of metal solution comes into contact with a catalyst support particle, the droplet penetrates into the pores. The deposition of the metal ions takes place during the equilibration of the liquid phase inside the pores and solid surface of the pores. The adsorption mechanism can be modeled by a Langmuir isotherm, which depicts a relationship between the extent of adsorption of the surface (or the fraction of the adsorption sites occupied) and the molar concentration of the solution. Solving the Langmuir model coupled with the mass balance, the equilibrated absorbed metal and the concentration of metal in the liquid phase can be calculated. Simulation results were compared with experiments with identical setup. Results obtained from our model show good agreement in the metal content in particles along the axis of rotation with experiments, and therefore it was demonstrated that a novel metal transfer algorithm incorporated into DEM could be used to accurately model catalyst impregnation.

KANSAS STATE UNIVERSITY | Bulk Solids Innovation Center

Performance Evaluation of Vacuum and Positive Pressure System for Dilute Phase Pneumatic Conveying of Barium Sulfate

Amit K. Gautam, Jason Brantley, John Lawrence*

* Research Director at BSIC, Salina, KS

KANSAS STATE UNIVERSITY | Bulk Solids Innovation Center

Outline

- Introduction
- Objectives
- Methods and Materials
- Results and Discussions
- Conclusions

KANSAS STATE UNIVERSITY | Bulk Solids Innovation Center

Introduction

- Barium Sulfate is white crystalline solid, odorless and insoluble in water
- Barium sulfate is commonly used as a filler for plastics to increase the density of the polymer in vibrational mass damping applications
- Filler (upto 70%) in polypropylene and polystyrene plastics
- It can help in increasing acid and alkali resistance and opacity
- Polytetrafluoroethylene (PTFE) filled with barium sulfate is resistant to chemicals

Source: Encyclopedia of Industrial Chemistry

KANSAS STATE UNIVERSITY | Bulk Solids Innovation Center

Problem Statement

- Inconsistency in mass flow rate was observed with probe type vacuum pneumatic conveying
- Group C material based on Geldart Classification
- Low flow function value (2.1 at the consolidation stress of 9.75 kPa), that leads to steeper hopper and wider outlet dimension for mass flow
- It needs flow aids for effective flow in hopper or silo

KANSAS STATE UNIVERSITY | Bulk Solids Innovation Center

Pneumatic Conveying System

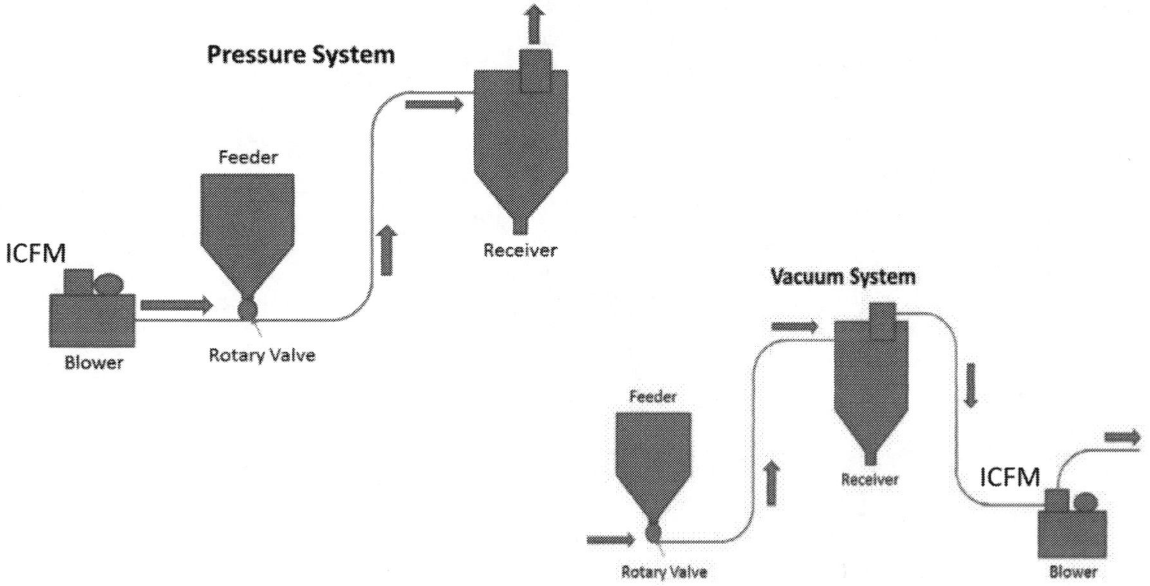

KANSAS STATE UNIVERSITY | Bulk Solids Innovation Center

Pneumatic Conveying System

Typical system parameters:

- Optimum pick-velocity to convey the material
- Pressure drop generated during operation
- Airflow (cfm or m3/s) needed at that pressure drop to create optimum pick-up velocity
- Blower selection for that location based on elevation and air density

KANSAS STATE UNIVERSITY | Bulk Solids Innovation Center

State or Phase Diagram

Operation of Pneumatic Conveying System:

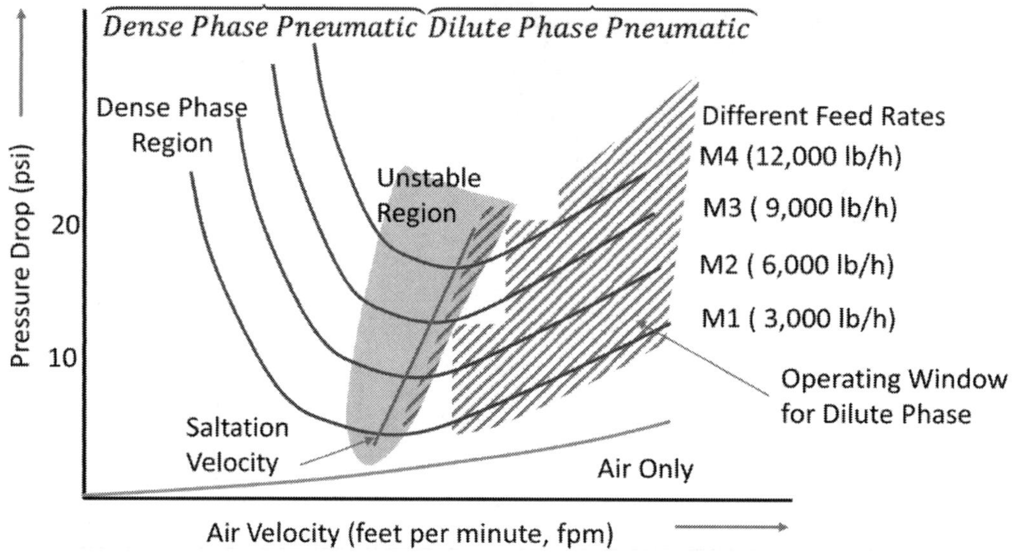

KANSAS STATE UNIVERSITY | Bulk Solids Innovation Center

Objectives

- To find out the optimum pick-velocity to convey $BaSO_4$ and maintain the target mass flow rate
- To find out which pipeline size set up positive or vacuum pneumatic conveying system in terms of saving energy

KANSAS STATE UNIVERSITY | Bulk Solids Innovation Center

Materials & Methods

Testing Set-up:

Material: Barium Sulfate

Conveying Parameters:

Target Conveying Rate : 5100 lb/hr (2313 kg/hr)

Conveying Distance: 157′ horizontal and 32′ vertical

Pipe Size: Carbon Steel 3" and 4" pipe with sch 10 thickness

Elbows & Type: Six long radius (CLR=48") elbows

Blower: PD type, Tuthill Model 4610-AAN3CV Qx-Series(3-lobe)

Feeder: Rotary valve (Model 98 Aerolock valve, 203 mm valve size and 0.23 ft3/rev (0.0065 m3/rev)

Kansas State University | Bulk Solids Innovation Center

BaSO4 Characteristics

- Loose Density = 1182.7 Kg/m^3
- Packed Bulk Density = 2332.3 Kg/m^3
- Compressibility = 49.3 %
- Particle size distribution

D50=102.3 µm; D10=40.6 µm ; D90=228 µm

Kansas State University | Bulk Solids Innovation Center

Pipeline Layout

Segment	Distance [ft]	(m)	Horizontal	Vertical	Angle
1	30	9.14	X		90
2	24	7.31		X	90
3	16	4.87	X		90
4	47	14.32	X		90
5	32	9.75	X		90
6	28	8.53			90
7	8	2.44		X	150
8	4	1.22	X		Receiver
Total	189	57.61			

KANSAS STATE UNIVERSITY | Bulk Solids Innovation Center

Pipeline Layout

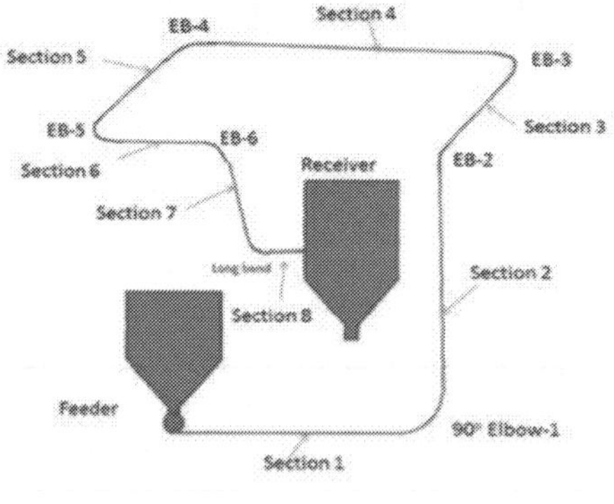

KANSAS STATE UNIVERSITY | Bulk Solids Innovation Center

Equations Used For Calculations

$$ACFM = SCFM * \frac{P_{std}}{P_{act} - (P_{sat} * \phi)} * \frac{T_{act}}{T_{std}}$$

- P_{act}-Actual absolute air pressure, psia (14.1 psia for Salina, KS)
- P_{std}- Standard absolute air pressure, psia (14.696 psia)
- P_{sat}- Saturated absolute air pressure at the actual temperature, psi (0.256318 psi)
- ϕ - Relative humidity, decimal (0.5)
- T_{act} -Actual temperature, °R (60°F or 520°R)
- T_{std} -Standard temperature, °R (530°R)

KANSAS STATE UNIVERSITY | Bulk Solids Innovation Center

Equations Used For Calculations

For Pressure

$$ICFM = ACFM * \left(\frac{P_a}{P_a + P_g}\right) * \left(\frac{T_a + 460}{T_a + 460 + 14.24 * P_g}\right)$$

For Vacuum

$$ICFM = ACFM * \left(\frac{P_a}{P_a - P_g}\right)$$

T_a -Ambient Temperature, °R (60°F or 520°R)
P_a -Atmospheric Pressure, psia (Altitude 1250ft, 14.1 psia for Salina)
P_g - System pressure drop in the system, psig

KANSAS STATE UNIVERSITY | Bulk Solids Innovation Center

Experiment Details

For target mass flow rate: 5100 lb/hr (2313 kg/hr)
Operated with fluidizer and impactor (5 pulses/min)

Cases	Pipe Size	Pressure/Vacuum	Rotary Speed (rpm)	Blower Speed (rpm)
1	4" Sch 10 4.26" (ID)	Pressure	7.3	2343, 2723,3103,3293,3483 (37-55 Hz)
2	4" Sch 10	Vacuum	5.2 5.7	2343, 2723,3103,3293,3483
3	3" Sch 10	Pressure	7.3 7.5	1520,1710,1900,2090 (24-33 Hz)
4	3" Sch 10	Vacuum	5.5	1520,1710,1900,2090

Five data points with 2 minutes interval each

KANSAS STATE UNIVERSITY | Bulk Solids Innovation Center

Results and Discussion

KANSAS STATE UNIVERSITY | Bulk Solids Innovation Center

Case 1: Pressure & 4" Line

Salina, KS condition, 4" Sch 10(4.26") **Pressure system**							
Tuthill Blower Speed, rpm	Rotary Valve Speed, rpm	Avg. Mass Flow Rate, lb/hr	Avg. Pressure Drop, psig	Avg. Airflow, S-ICFM	Avg. Airflow, A-ICFM	ACFM at pickup point	Avg Pickup Velocity (ft/min)
2343	7.3	5310	3.4	519	536	447	4513
2723	7.3	5050	3.9	605	624	516	5216
3103	7.3	4990	4.3	700	722	593	5996
3293	7.3	5080	4.6	740	764	624	6300
3483	7.3	5250	4.8	790	815	663	6703

KANSAS STATE UNIVERSITY | Bulk Solids Innovation Center

Case 1: Pressure & 4" Line

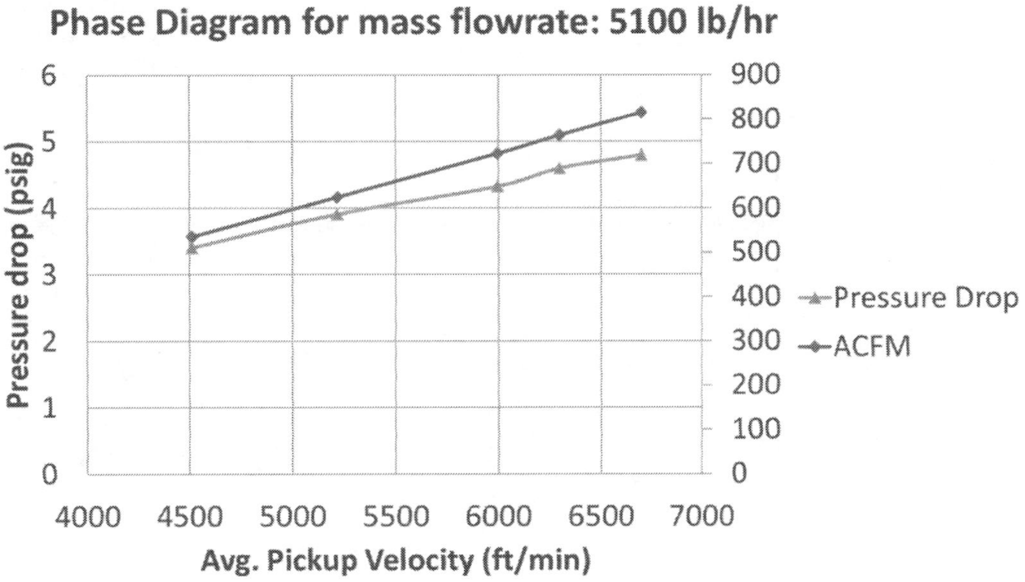

KANSAS STATE UNIVERSITY | Bulk Solids Innovation Center

Case 1: Pressure & 4" Line

	Mass Flow Rate, lb/hr				
Blower Speed, rpm	2343	2723	3103	3293	3483
Average (lb/hr)	5310	5050	4990	5080	5250
Lowest (lb/hr)	4860	4500	4920	4920	4920
Highest (lb/hr)	6840	5400	5100	5280	5640

KANSAS STATE UNIVERSITY | Bulk Solids Innovation Center

Case 2: Vacuum & 4" Line

For Rotary Valve Speed of 5.7 rpm:

Above 5100 lb/hr target

	Mass Flow Rate, lb/hr				
Blower Speed, rpm	2343	2723	3103	3293	3483
Average (lb/hr)	5330	5760	5857	5500	5570
Lowest (lb/hr)	4020	4920	5340	4440	5340
Highest (lb/hr)	6720	7260	6180	6240	5880

KANSAS STATE UNIVERSITY | Bulk Solids Innovation Center

Case 2: Vacuum & 4" Line

For Rotary Valve Speed of 5.3 rpm:

Within 5100 lb/hr range

	Mass Flow Rate, lb/hr		
Blower Speed, rpm	2343	3293	3483
Average (lb/hr)	4270	5230	5150
Lowest (lb/hr)	3540	4980	4800
Highest (lb/hr)	6120	5580	5520

Kansas State University | Bulk Solids Innovation Center

Case 2: Vacuum & 4" Line

Salina, KS condition, 4" Sch 10(4.26") **Vacuum system**							
Tuthill Blower Speed, rpm	Rotary Valve Speed, rpm	Avg. Mass Flow Rate, lb/hr	Avg. Pressure Drop, "Hg	Avg. Airflow, S-ICFM	Avg. Airflow, A-ICFM	ACFM at pickup point	Avg Pickup Velocity (ft/min)
3293	5.2	5230	8.0	735	712	514	5196
3483	5.2	5150	8.3	785	761	540	5458

Kansas State University | Bulk Solids Innovation Center

Case 3: Pressure & 3" Line

Salina, KS condition, 3" Sch 10(3.26") **Pressure system**							
Tuthill Blower Speed, rpm	Rotary Valve Speed, rpm	Avg. Mass Flow Rate, lb/hr	Avg. Pressure Drop, psig	Avg. Airflow, S-ICFM	Avg. Airflow, A-ICFM	ACFM at pickup point	Avg Pickup Velocity (ft/min)
1520	7.5	5080	5.7	296	305	227	3909
1710	7.5	5170	6.0	340	351	261	4511
1900	7.5	5170	6.3	389	402	301	5187
2090	7.5	5070	6.7	430	444	331	5715

KANSAS STATE UNIVERSITY | Bulk Solids Innovation Center

Case 3: Pressure & 3" Line

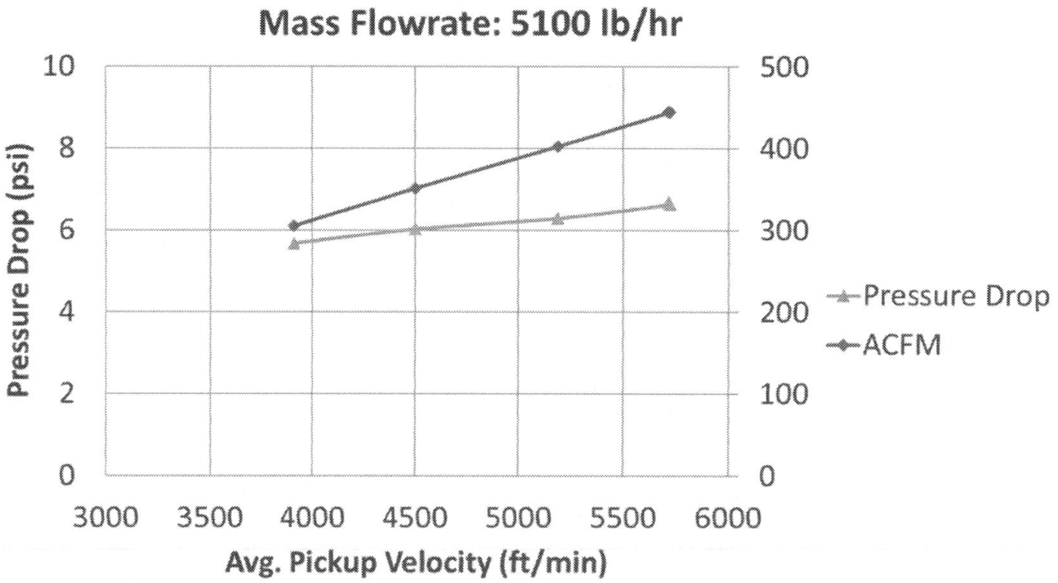

KANSAS STATE UNIVERSITY | Bulk Solids Innovation Center

Case 3: Pressure & 3" Line

	Mass Flow Rate, lb/hr							
Rotary Valve Speed	7.3 rpm				7.5 rpm			
Blower Speed, rpm	1520	1710	1900	2090	1520	1710	1900	2090
Average (lb/hr)	4900	4890	4890	5070	5080	5170	5170	5160
Lowest (lb/hr)	4560	4560	4680	4860	4560	4740	5040	4980
Highest (lb/hr)	5340	5400	5340	5340	5820	5520	5340	5520

KANSAS STATE UNIVERSITY | Bulk Solids Innovation Center

Case 4: Vacuum & 3" Line

Salina, KS condition, 3" Sch 10(3.26") **Vacuum system**							
Tuthill Blower Speed, rpm	Rotary Valve Speed, rpm	Avg. Mass Flow Rate, lb/hr	Avg. Pressure Drop, "Hg	Avg. Airflow, S-ICFM	Avg. Airflow, A-ICFM	ACFM at pickup point	Avg Pickup Velocity (ft/min)
1710	5.45	5230	11.7	321	311	185	3192
1900	5.45	5072	11.8	371	360	212	3661

KANSAS STATE UNIVERSITY | Bulk Solids Innovation Center

Case 4: Vacuum & 3" Line

For Rotary Valve of 5.5 rpm:

	Mass Flow Rate, lb/hr			
Blower Speed, rpm	1710	1900	2090	2280
Average (PPH)	5230	5072	5406	5438
Lowest (PPH)	4601	4620	5100	5192
Highest (PPH)	5520	5400	5820	5640

KANSAS STATE UNIVERSITY | Bulk Solids Innovation Center

Summary & Conclusions

- Performance evaluation test of dilute phase pneumatic conveying (positive pressure & vacuum) system for conveying barium sulfate powder was conducted
- Different blower speed (pick-up velocities) and rotary value speed were selected to find out the required mass flow rate
- For pressure system, optimum pick-up velocities to achieve target mass flow rate of 5100 lb/hr were 5200 ft/min (4" line) and 4000 ft/min (3" line), respectively
- For vacuum system, optimum pick-up velocities to achieve target mass flow rate of 5100 lb/hr were 5200 ft/min (4" line) and 3200 ft/min (3" line), respectively

KANSAS STATE UNIVERSITY | Bulk Solids Innovation Center

Summary & Conclusions

- For pressure system, rotary valve speed was found to be higher compared to vacuum system (due to back pressure resist the flow for positive pressure system)
- Pipeline 4" consumes almost 1.2 times higher airflow rate (cfm) when compared to 3" line
- In terms of energy consumption, 3" line is better than 4" line (more than 1.2 times higher energy save)

KANSAS STATE UNIVERSITY | Bulk Solids Innovation Center

Thank You

Questions?

Contact Info:
John Lawrence
Research Director
jlawren@ksu.edu
785-404-4918
607 N Front St
Salina, KS 67401

KANSAS STATE UNIVERSITY | Bulk Solids Innovation Center

492273 Swirl Number Analysis for Performance Data Prediction of Uniflow Cyclones

Abstract:

Uniflow cyclones are used for the separation of a dispersed solid phase from a continuous fluid phase e.g. in fluid catalytic cracking or combustion engines in filter applications. Therefore particles enter the separator through a vane guide pack. Due to the resulting centrifugal forces the particles move in radial direction to the particle discharge in downstream direction. To achieve high separation efficiency and low pressure drop a systematic parameter study is carried out to identify the main geometric influence factors e.g. vane angle Î± or core ratio dc/dz. Stereoscopic PIV is used for an in-deep analysis of the swirl flow regime inside the separator to gain a deeper insight into the dominant flow structures and the corresponding force field respectively. An extensive comparison of swirl number calculations is presented to evaluate the effects inside of the separation chamber. Additionally the flow field is used as a basis for a state-space-representation system of ordinary differential equations to calculate critical diameters based on an equilibrium force model. ISO 5011:2014 gas-solid-separation tests are presented for validation. Goal is to find correlation between the velocity field and the performance data of the uniflow cyclone for future prediction.

[1]Â Kofler, T., Pillei, M., and Kraxner, M., â€œInfluence of the Inlet Vane Geometry On the Uniflow Cyclones Performance (639f)â€. AIChE 2015 Annual Meeting, 2015

499601 Saponin Removal from Bitter Quinoa Ecotypes in Spouted BEDS (SB)

Abstract:

The removal of saponin from bitter quinoa grains (*Chenopodian quinoa* Willd.) by means of a Spouted Bed (SB) was experimentally studied. Basic SB performance characteristics and operation conditions were optimized on grains of various ecotypes of Quinoa Royal and their mixtures, grown in the Bolivian southern Altiplano.

Experimental runs were carried out using different laboratory-scale cylindrical-conical SBs of 7.5, 15 and 20 cm in diameter, nozzles of 1.4 â€“ 5 mm outlet diameter at static bed heights within 7.5 and 29 cm. Air was supplied by a 400 $Lmin^{-1}$ piston compressor, which included a 10 â€“ 300 $Lmin^{-1}$ rotameters and a digital air flow controller.

A two level factorial design was applied, including bed diameter, air linear inlet velocity (measured at the nozzle outlet diameter), bed height and ecotype as independent variables. Processing time was established at 30 min. Dependent variables were: residual saponin content and mass loss percentage of processed grains. The effect of process conditions on the primary nutrient content of the processed grains were evaluated and the morphology of both unprocessed and processed grains were also compared at micrograph scale.

Factors with larger effect on residual saponin content were air inlet linear velocity, reactor diameter followed by ecotype and bed height. The SB process achieved a saponin concentration of 0.00 - 0.02% in processed grains (below 0.12 %, which is the bitterness detection level) and a mass loss percentage < 7% using the following combination: air inlet velocity, 1200 ms^{-1} (1.4 mm nozzle diameter); 7.5 cm reactor diameter and 12.5 cm bed height. Primary nutrient contents (protein and lipids) in the treated grains increased with the outer layer removal, indicating that the optimum operation conditions do not damage the embryo (where the primary nutrient components are located) as observed at micrograph scale. Moreover, the process allowed total recovery of the removed saponin powder of high commercial value for diverse manufacture industries.

The kinetic studies showed that both, saponin removal and mass loss percentages occur at higher rates during the first period of 5 min and that both removal rates decrease afterwards. This behavior can be explained because of the fragility of the epispermâ€™s outer layer where the saponin is concentrated. The same kinetic studies showed a strong linear correlation between air inlet linear velocity and powder removal rates (measured at 5 min operation). There is also a strong linear correlation between air inlet linear velocity and saponin removal efficiencies as tested in different laboratory scale spouted beds.

These experiments show that the spouted bed dry process is an excellent technological alternative when applied to saponin removal from bitter quinoa grains.

491421 Influence of the Viscosity on the Velocity Propagation in Axial Direction in a Rietema Hydrocyclone

Abstract:

In the field of chemical or mineral industry hydrocyclones are widely used for solid-liquid separation [1]. In respect of geometrical similarity two well-known designs are dominating the field of hydrocyclones: The *Rietema* [2] and *Bradley* [3] hydrocyclones. For both types the geometric proportions are given. The main difference is the cone angle, which is 9° for *Bradley* hydrocyclones and 20° for *Rietema*hydrocyclones. Both designs mainly focus on water, which is the most common liquid media in hydrocyclone applications.

In the field of purification technology such as wastewater and sludge treatment, the rheological characteristics of the liquid media are significantly different from water [4]. As a result, the optimum geometrical proportions may not be sufficient for these applications. Therefore, the velocity propagation in the hydrocyclone is evaluated by PIV (Particle Image Velocimetry) measurements and CFD simulations in different axial positions.

For the rheological modelling of the sludge a mixture of water and glycerol with a viscosity of 6 mPas [5] is used. The results of measurements and simulations are compared with a water-only system. The hydrocyclone was designed and manufactured according to *Rietema*with a diameter of 100 mm and 500 mm in length.

For the water-only system the averaged $VelMag/VelMag_{max}$ is 0.86 at axial middle position of the cyclone while the velocity ratio is 0.53 for the system with higher viscosity. The viscosity has a strong influence on the axial propagation of the circumferential velocity which leads to a limited efficiency along the conical separation chamber of the hydrocyclone. In conclusion, both *Rietema* and *Bradley*geometrical hydrocyclone proportions may not be the optimum solutions for sludge processing. Further parametric studies on the cone angle and the cyclone length shall lead to optimization potential in the separation process.

[1] Svarovsky, L., Thew, M.T., Hydrocyclones â€“ Analysis and Applications, Kluwer Academic Publisher, Dordrecht, 1992.

[2] Bradley, D., The Hydrocyclone, Pergamon Press, London, 1965.

[3] Rietema, K., Performance and Design of Hydrocyclones, Chemical Engineering Science, 15, 298-325, 1961.

[4] Baudez, J.C., Markis, F., Eshtiaghi, N., Slatter, P., The rheological behaviour of anaerobic digested sludge, Water Research, 45, 2011.

[5] Haynes, W.M., CRC Handbook of Chemistry and Physics, 96th edition, 2015.

493030 Gas-Solid Flows with Tribocharging

Abstract:

Granular materials develop static charges when they come into contact with each other or other solid surfaces such as bounding walls, which is known as triboelectrification or tribocharging. Tribocharging may lead to safety hazards in extreme cases [1]; in other instances, it is believed to lead to undesirable outcomes such as sheeting in polymerization reactors [2]. The interplay between gas-particle flows and charge accumulation on particles is a topic of much interest. In this study, we probe the coupling between tribocharging in gas-solid flows via modeling and computations, aided by supplementary experiments.

We first study this coupling through CFD (Computational Fluid Dynamics) â€" DEM (Discrete Element Method) simulations of gas-particle flows, augmented by a finite-volume based particle-particle and particle-mesh type Poisson solver for the electric field [3] and Laurentie et al. model [4] for tribocharging. In the Laurentie et al. model, the charging tendency of particles is captured by effective work function difference between the contacting surfaces and the electrical field at contact. The effective work function is a lumped parameter that can be determined through experimental studies [4] or quantum mechanical calculations using the density functional theory [5]. In a recent study [6], we performed vibrated bed experiments and measured average charge on polyethylene (PE) particles at different controlled humidity conditions; DEM simulations of the same system were then performed to estimate the effective work function differences between the PE particles and the glass container at different humidity levels. CFD-DEM simulations of fluidized beds with these work function values captures the qualitatively different fluidization behavior seen at low and high humidity levels.

As CFD-DEM simulations are limited to relatively small systems and there is a need to examine the interplay of flow and tribocharging in large fluidized beds, we have also formulated and analyzed a kinetic-theory based Euler-Euler model for monodisperse particles with tribocharging. Towards this end, we have derived (i) the mean charge transport equation from the Boltzmann equation, assuming Maxwellian distributions for particle velocities and charges and allowing for conduction of mean charge through collisions in the presence of electric field; and, (ii) boundary condition capturing tribocharging at the wall. These models have been implemented in an open-source continuum physics software. Model predictions are then assessed through comparison with (i) hard-sphere Euler-Lagrange simulations and (ii) experimental studies.

References:

[1] T. Jones, J. King, J., Powder Handling and Electrostatics. Lewis Publishers Inc, (1991).

[2] G. Hendrickson G. Chemical Engineering Science, 61(4):1041â€"1064 (2006).

[3] J. Kolehmainen, A. Ozel, C. M. Boyce, and S. Sundaresan, *AIChE Journal*62, 2282 (2016).

[4] J. Laurentie,P. Traore, and L. Dascalescu, *Journal of Electrostatics*71, 951 (2013).

[5] S. Naik, S. Saurabh, G. Vipul, H. Bruno, Y. Abramov, W. Yu, and B. Chaudhuri, *International Journal of Pharmaceutics*491, 58 (2015).

[6] J. Kolehmainen, P. Sippola, O. Raitanen, A. Ozel, C. M. Boyce, and S. Sundaresan, submitted to Chemical Engineering Science.

501610 DEM Simulations of Flexible and Rigid Fibers for Describing Bulk Flow Behavior

Abstract:

Novel Discrete Element Method (DEM) simulations have enabled an investigation of the relationship between particle-level and bulk flow behavior for flexible and rigid fibers. Particle shape greatly affects the particle-phase stress; this is related to the degree of particle alignment induced by shearing. Particle shape also affects the collisional dissipation rate that increases linearly with increasing fiber aspect ratio. Flexible fibers introduce additional complexities such as fiber bending deformation that is characterized by the fiber bending stiffness and the fiber damping coefficient. The effects of individual fiber properties on bulk flow behavior and particle-phase stress have also been explored.

493115 Challenges in CFD Simulation of the Entire Circulating Fluidized Bed (CFB) Loop for Carbon Capture Process

Abstract:

Near-term applications of CO_2 capture from pre-combustion systems will likely involve physical or chemical absorption processes. However, these commercially available processes (e.g. SELEXOL) operate at low temperatures, imparting a severe energy penalty on the system and, consequently, their use could significantly increase the costs of electricity production. Therefore, development of high-temperature regenerative processes based on solid sorbents offers an attractive alternative option for carbon capture, at competitive costs. We have developed a regenerative high-temperature CO_2 capture process that is capable of removing more than 98% of CO_2 from a simulated water-gas-shift (WGS) mixture at IGCC conditions using highly reactive and mechanically strong MgO-based sorbents and a circulating fluidized bed (CFB) loop.

CFB loop-based reactors have the potential to be among the most important devices in the chemical and energy industries. CFB reactor ensures a continuous carbon dioxide removal process in a relatively compact unit using solid particles that makes it an excellent candidate for chemical looping of MgO-based sorbents for CO_2 capture and regeneration. Furthermore, computational fluid dynamics (CFD) provides an excellent approach to the reactors and entire CFB loop design in a systematic and economically feasible way. To use CFD to perform simulations of the CO_2 capture regenerative process, a model based on the multiphase flow dynamics governing equations taking into account the sorption/regeneration and the WGS kinetics is needed. Therefore in this work a CFD approach was used to describe CO_2 sorption and regeneration in the circulating fluidized bed loop reactor similar to the NETL carbon capture unit (C2U) experimental setup using an MgO-based sorbent.Â In this presentation we will discuss challenges involved in conducting numerical simulation (CFD) of the entire CFB loop. These challenges include developing fluid â€"particle drag force and Particle phase pressure and viscosity expressions which are applicable to entire range of solid concentration regimes from very dilute to packed moving bed, and significant computational time requirement to simulate three dimensional case of reacting system in such a complex geometry.

In addition, the key experimentally verified parameters needed for the CFD modeling of the sorption/regeneration and the WGS reaction rates and their dependence on the operating conditions (i.e., temperature, pressure, gas composition, catalyst/sorbent ratio, etc.) were also developed.

502488 Discrete Particle Modelling and Analysis of Gas-Solid Flow in Pneumatic Conveying

Abstract:

Pneumatic conveying is an important operation used in industries to transport bulk materials from one place to another. To achieve optimal design and control of this operation, it is very important to understand the complex behaviors of the gas-solid flow along the conveying pipelines. In recent years, many efforts have been made in this direction by means of the numerical approach of combining discrete element method (DEM) for particles and computational fluid dynamics (CFD) for the gas phase. This talk will present an overview of CFD-DEM studies of pneumatic conveying in our laboratory, focused on two aspects: model development and model application. The model development is discussed in terms of model formulation, new features and model validation. Then, the usefulness of the CFD-DEM approach is demonstrated through some case studies where the main findings at the particle scale are highlighted. In particular, the flow regimes and their transition are unveiled in terms of the flow behaviors of gas and solid phases, the forces acting on particles, and the energy dissipation among gas, particles and pipe wall. Also, the CFD-DEM simulation of the large-scale pneumatic conveying is demonstrated, where the transport distance is around one hundred meters and the pipelines consist of horizontal, vertical and inclined pipes, as well as bends. Finally, areas for future development are briefly discussed.

490263 Mesoscale Modeling of Multiphase Reactors: Theory and Applications

Abstract:

Although new catalysts and chemical technologies can be invented or patented in laboratory every year, process scaleup from laboratory to industrial application remains a troublesome or challenging issue. Successful cases are limited and risky, relying on the empirical correlations and the engineers whose knowledge and experience is acquainted through the long-term case study of previous well-established processes. It is generally acknowledged that the main technical problem, among others, is how to create an ideal transport environment for reactions and separation, and hence chemical reactions could be compatible with their carrier, i.e., the fluid flow, mass and heat transfer in multiphase reactors. A new angle to achieve a fundamental understanding and then seek efficient solutions of these classical problems is to reveal the mystery on mesoscales, i.e., the mesoscale transport phenomena and mechanisms relevant to bubbles, droplets and particles. Actually mesoscale problems are essential to a more fundamental understanding of momentum, mass and heat transfer in the classical study of transport phenomena, and to the mixing, residence time distribution and rate-limiting analysis in the chemical reaction engineering, yet they are beyond the scope of those classical textbooks of chemical engineering.

In this paper, we highlight a heuristic mesoscale modeling approach for multiphase reactor systems, starting from a conceptual Energy-Minimization Multiscale (EMMS) model and ending at the stability-constrained multifluid CFD model. While the stability condition determines the direction of system evolution, the stability-constrained CFD further describes the dynamics of structure evolution. Then this theory is applied to the simulation of gas-solid fluidization and gas-liquid bubble column reactors, with further extension to gas-liquid-solid three phase flow and stirred tanks. Several industrial applications on meso-scale modeling in liquid-solid polyethylene reactors and liquid-liquid emulsification systems will also be highlighted.

Taking the gas-liquid flow in bubble column reactors as an example, these reactors have found widespread applications in chemical and energy-related industries, such as coal liquefaction, waste water treatment and CO2 utilization, in view of their advantage in mixing, mass and heat transfer. CFD is playing important roles in the understanding of complex multi-phase fluid dynamics, phase holdup and bubble size distribution. However, the CFD simulation for gas-liquid flow based on Eulerian framework is reported to be sensitive to a number of constitutive equations. The accuracy or reliability of CFD prediction cannot be guaranteed and empirical correlations or parameters are always needed. Actually the CFD simulation is usually based on the Eulerian-Eulerian framework in which the complex meso-scale physics is averaged out and hidden in the so-called constitutive equations of two-fluid models, though the governing equations are rigorously derived from continuum mechanics or kinetic theory. The difficulty and challenging issue is therefore technically shifted to the constitutive models to describe the phase interaction forces (drag, lift and virtual mass forces) for momentum conservation equations, the turbulence coupling terms for turbulence modeling and the bubble breakup and coalescence kernel functions for population balance equations(PBM) of bubble number density. But physically the constitutive models relies on the understanding of mesoscale transport mechanisms.

In the new model, a stability condition is formulated as the minimization of the sum of two energy dissipations, reflecting the compromise of a liquid-dominant regime at which smaller bubbles prevail and a gas-dominant regime favoring the existence of larger bubbles. It supplies a mesoscale constraint for conservation equations, and a mesoscale perspective to understand the macroscale regime transition. With the zero-dimensional conceptual model, we obtain a jump change after a gradual increase of gas holdup with superficial gas velocity. The jump change corresponds to the regime shift from the homogeneous regime to the fully-developed heterogeneous regime, and the regime shift can be physical interpreted through the stability condition: transition of the minimum of the sum of two energy dissipations from one local minimum point to the other leads to the dramatic variation of structure parameters and finally gives rise to the regime shift. The model calculation for gas-liquid and gas-solid systems demonstrates the intrinsic similarity of the two systems: the system evolution at macroscale is driven by stability conditions.

Theoretically stability condition may offer closure laws for CFD simulation. While direct integration is difficult, we propose various simplified approaches which are able to derive new closure models for drag, bubble-induced turbulence and corrections for kernel functions of coalescence or breakage rates in population balance equations. The stability-constrained multifluid CFD model shows much advantage over traditional closure models. We also tentatively extended this model to gas-liquid-solid flows for slurry bubble columns. More details can be found the publications (Li et al., 2016; Yang et al., 2015, 2017; Zhou et al., 2017).

Reference

Yang, N.*, Xiao, Q., A mesoscale approach for population balance modeling of bubble size distribution in bubble column reactors, Chemical Engineering Science, 2017, in press.

Zhou, R., Yang, N.*, Li, J., CFD simulation of gas-liquid-solid flow in slurry bubble columns with EMMS drag model, Powder Technology, 2017, in press

Li, J., Ge, W., Wang, W., Yang, N., Huang, W., Focusing on mesoscales: from the energy-minimization multiscale model to mesoscience, Current Opinion in Chemical Engineering, 2016, 13, 10-23

Yang, N.*, Mesoscale transport phenomena and mechanisms in gas-liquid reaction systems. In: Guy B. Marin and Jinghai Li, eds., Advances in Chemical Engineering (book chapter), 46, 245-280.

503611 Reflection from Former Students

Abstract:

Several former students of Prof. Liang-Shih Fan will represent the over 150 Ph.D. and Postdoc advisees from the Fan group to reflect on their experiences interacting with Prof. Fan at both professional and personal levels. These former students, i.e. Prof. Alissa Park at Columbia University, Prof. Raymond Lau at Nanyang Technological University, Dr. Zhe Cui at Shell, Prof. Fanxing Li at NC State University, and Prof. Andrew Tong at the Ohio State University, came from various backgrounds and conducted doctoral research on a variety of subjects related to fluidization, particle technology, and reaction engineering. They will reflect on how Prof. Fanâ€™s influences have helped to shape their careers. Each of the students will identify the most important things he/she learned from Prof. Fan as well as their most exciting research findings from their ongoing research activities.Â

494158 New Classes of Organic Nanoparticles with Engineered Shape and Functionality: Particles with "Gecko Legs" and Environmentally Benign Antimicrobials

Abstract:

We will discuss the fabrication and applications of new classes of functional (bio)polymer particles, made by scalable processes of antisolvent-induced biphasic precipitation under shear. The process enables the formation of high surface area liquid structures, which can serve as templates for the formation of diverse classes of polymer and biopolymer nanoparticles and nanomaterials. The method was originally used to make dispersions of polymer rods that can act as "superstabilizers" of Pickering foams and emulsions. It also allows scalable manufacture of nanofibers and nanoribbons (*Adv. Mater.*, **27**, 2642, 2015). In this talk we will discuss two classes of novel particles made by this technique in our group. Modifications of the shear precipitation technique allowed making new types of sheet-like, ribbon-like and dendrimeric particles. The dendrimeric particles, in particular, constitute a novel type of material with extraordinary structure and properties. These particles are hierarchically structured, with a big branched corona of nanofibers spreading out in all directions. The nanofiber corona around these also called "gecko leg" particles endows them with extraordinary strong adhesion to almost any surface and to each other, and enables unique structure-forming abilities, adhesiveness and rheological characteristics. Secondly, we will present the making and properties of a new class of environmentally-benign nanoparticles with cores made of biodegradable lignin. These particles are fabricated by liquid shear process including water-only pH-jump precipitation. They can serve as highly potent microbicidal substitutes of common silver nanoparticles after being loaded with an optimal amount of silver in the form of adsorbed Ag^{+} ions (*Nature Nanotech.*, **10,** 817, 2015). The cationic surface modification of these particles facilitates their adsorption onto bacterial membranes and triggers a targeted release of active Ag^{+} ions. These environmentally-benign nanoparticles illustrate how green chemistry principles can be applied to design sustainable nanomaterials with increased activity and decreased environmental footprint.

500521 Cellulose Nanocrystal Templates for Transparent Conductive Films

Abstract:

There is great interest in using biorenewable materials for sustainable, next generation technologies. Here we present the use of cellulose nanocrystals (CNCs), derived from wood pulp, as a unique template for growing conductive silver nanoparticles (Ag NPs) as a potentai replacement for indium tin oxide and related transparent conductive materials. We demonstrate the potential to successfully nucleate and grow AgNPs on CNC templates in the form of a stable, optically clear and transparent aqueous colloid. Our approach to generating the Ag/CNC templated nanoparticles is unique compared to traditional metallic NP colloid growth schemes, in that we take advantage of starving the particle nucleation and growth processes to control NP size and resultant nanocomposite morphology. The co-integrated CNC templates serve as nucleation sites for NP growth. The CNC templating agent facilitates systematic alignment of conductive nanoparticles along its rigid backbone. Our results show the impact of CNC surface chemistry and functional group density on the CNC/Ag nanocomposite morphology, including Ag particle size/shape and population density on the CNC template.

We further present preliminary results to generate robust 3D percolating networks of CNC/Ag particles in the form of homogenous, optically transparent nanocomposite films and conformal coatings. Traditionally, electrical percolation requires exceeding a critical loading level with which particles are sufficiently close in a 3-D network to enable interparticle charge mobility for a conductive pathway. Our results demonstrate that the CNC/Ag colloids can be processed under ambient conditions into optically transparent and electically conductive films. Further, we shed light on the relationship of nanocomposite film structure/morphology, film formation approach, and measured electrical conductivity/optical transparency.

498896 Hydrophilic and Hydrophobic Functional Group Interactions on Organic Crystals Coated with SiO_2 Nanolayers By Physical Vapor Deposition

Abstract:

The purpose of this research work is to study the surface interactions of functional hydrophilic and hydrophobic forces with nanolayers of pharmaceutical additives coated on active pharmaceutical ingredient (API) [acetaminophen (APAP)] and excipient [lactose (Lc)] crystal surfaces. APAP and Lc crystals were developed in the lab and Atomic Force Microscopy (AFM) was used to investigate the effect of chemical nature of both excipient and API on adhesion forces. Physical Vapor Deposition (PVD) was used to coat the crystals with 10 nm, 20 nm and 50 nm thin layers of SiO_2. Uncoated and coated crystals were compared with respect to their adhesion forces. AFM probes were functionalized with hydrophilic and hydrophobic groups of 7-carboxy 1-heptane thiol and 1-decane thiol. It was found that the adhesion forces increased on both Lc and APAP crystal planes. Functionalized hydrophilic forces was found to be higher than the hydrophobic force on SiO_2 coated surfaces. Adhesion of functionalized groups was phenomenally affected by the chemical nature of crystals.

501592 Enzyme-Mimetic Antioxidant Luminescent Nanoparticles for Highly Sensitive Hydrogen Peroxide Biosensing

Abstract:

Hydrogen peroxide (H_2O_2) is an abundant molecule associated with biological implications and reacts with natural enzymes, such as catalase. Thus, H_2O_2 quantification constitutes a powerful tool for detection of disease biomarkers linked to enzyme-based assays. However, the optical H_2O_2 biosensing without organic-dyes in biological media and at low, submicromolar, concentrations has yet to be achieved. Herein, we rationally design biomimetic artificial enzymes based on antioxidant CeO_2 nanoparticles that become luminescent upon their Eu^{3+} doping. We vary systematically their diameter from 4 to 16 nm and study their catalase-mimetic antioxidant activity, manifested as catalytic H_2O_2 decomposition in aqueous solutions, revealing a strong nanoparticle surface area dependency. The interaction with H_2O_2 influences distinctly the nanozyme luminescence rendering them highly sensitive H_2O_2 biosensors down to 0.15 mM (5.2 ppb) in biologically-relevant solutions. Our results link two, so far, unrelated research domains, the CeO_2 nanoparticle antioxidant activity and luminescence by rare-earth doping. When these nanozymes are coupled with alcohol oxidase, biosensing can be extended to ethanol exemplifying how their detection potential can be broadened to additional biologically relevant metabolites. The biomimetic artificial enzyme developed here could serve as a starting point of sophisticated *in vitro* assays towards highly sensitive detection of disease biomarkers.

494092 Physicochemical Characterization of Curcumin-Polymer Nanoparticles and Behavioral Studies of Arthritic Pain in Rats

Abstract:

Physicochemical characterization of curcumin-polymer nanoparticles and behavioral studies of arthritic pain in rats

Sonal Mazumder[a*], Ashish K. Dewangan[a], Yamini Perumal[b], Naresh Pavurala[c], Kanwaljit Chopra[b]

[a] *Department of Chemical Engineering, Birla Institute of Technology and Science Pilani, Pilani Campus, Pilani – 333031, India*

[b] *University Institute of Pharmaceutical Sciences, Panjab University, Chandigarh – 160101, India*

[c] *Office of Pharmaceutical Quality, U.S. Food and Drug Administration, Silver Spring, Maryland – 20993, USA*

Rheumatoid Arthritis is a critical illness and is caused by systemic inflammatory disorder of the synovial membrane. Often the joints and cartilages are damaged, followed by loss of function. Curcumin (CUR), a plant-derived chemical and crystalline in nature is of interest in this study. Curcumin can effectively alleviate the symptoms of rheumatoid arthritis like tenderness and swelling of joints compared to regular drugs. However, its utility as a therapeutic agent is limited by its poor absorption, rapid metabolism and rapid systemic elimination. Formulation of nanoparticles has been proved very much promising to increase absorption and bioavailability, leading to effective drug administration. In our study, curcumin was molecularly dispersed in amorphous polymer and nanoparticles were produced by a rapid precipitation technique. Dynamic light scattering showed the particle size was 170 nm. Differential Scanning Calorimetry (DSC) results showed that the drug in the nanoparticles was in the amorphous state. Drug loading efficiency in the nanoparticles was 74 ± 2.4 %. The drug from the nanoparticle showed higher release compared to the pure as-received drug. Release of drug from the nanoparticles followed Korsmeyer-Peppas and Higuchi equations.

The affect of nanoparticles were studied in rats induced with complete Freund's adjuvant arthritis. Behavioral (pain) studies were conducted such as thermal hyperalgesia, mechanical hyperalgesia, mechanical allodynia, hematological assessment, radiographic analysis and various disease development parameters. Studies showed that nanoparticles have superior anti-inflammatory effects and could give faster relief from pain.

496927 Polyacid-Functionalized Gold Nanoparticles As an Amyloid-β Inhibitor Platform

Abstract:

Alzheimer's disease (AD) is a debilitating disease that is the sixth leading cause of death in the United States. Current understanding of the etiology of AD points to disruption of the amyloid-β (Aβ) aggregation pathway as a promising strategy for a potential cure. Functionalized gold nanoparticles (AuNPs) have gained attention as inhibitors of Aβ aggregation. We recently demonstrated that polyacrylic acid-coated (PAA) AuNPs are strong inhibitors of aggregation and proposed that this inhibition results from alterations in solution near the NP surface.

To further probe this hypothesis, spherical AuNPs were synthesized with two different average diameters (5 and 20 nm) and functionalized with PAA having two distinct degrees of polymerization (20 and 80). Using these two levels for these two parameters we obtain four unique AuNP samples. The effect that these functionalized AuNPs have on the aggregation of Aβ was assessed by incubating 10 μM Aβ monomer with 0 (control), 10 or 30 nM AuNPs in 40mM Tris-HCl (pH 8.0) containing 0 or 150 mM NaCl. The extent of aggregation was measured using the fluorescent dye thioflavin T (ThT) over the course of 72 hours. Subsequently, transmission electron microscopy (TEM) images were obtained for the Aβ species present. We have performed an analysis of the results from the two level-four factor full factorial design at varied NP and solution conditions.

We have correlated the results of this factorial analysis to the predictions made using a theoretical model of the polymers and solution near the NP surface. The PAA layer on the AuNPs was modeled with a self-consistent mean field theory that explicitly accounts for the size, shape, conformation and charge distribution of all molecular species present in the aggregation assay buffer. The conditions present in the experiments were used to parameterize the theoretical model. The theory consists of writing the Helmholtz free energy of the system in terms of the entropy and energy associated with all species. A conformational ensemble for the PAA molecules is generated using the rotational isomeric states model (RIS). Thermodynamic equilibrium expressions are derived for the density of each molecular species as a function of distance from the AuNP surface. These equations were solved numerically.

A similar self-consistent field model has been developed for Aβ itself. In this case, the chain molecule of interest is the protein and the conformational ensemble is generated from molecular dynamics simulations. Additionally, the model includes terms to account for the solvation of the protein surface, effectively capturing the hydrophobic properties of the amino acids constituting the molecule. Practical application of the framework involves constraining the centers of mass of the proteins at positions in space and solving the resultant system of thermodynamic equilibrium expressions. This yields a thermodynamic potential reflective of the relative configuration of Aβ molecules, allowing one to evaluate the potential of mean force for dimerization as well as the free energy landscape of the formation of oligomer structures. Moreover, all of this information is calculable in altered solution conditions evoked by the presence of PAA-coated AuNPs.

PAA-functionalized AuNPs displayed a remarkable ability to inhibit the aggregation of Aβ in a manner dependent on the NP properties. This relationship is sensitive to surface-solution interactions which are accurately captured with a combination of our self-consistent theories for PAA and Aβ. Surface modified NPs coupled with this robust and accurate molecular model represent a powerful platform to engineer nanotechnologies to modulate Aβ aggregation.

500150 Grindability of Torrefied Wood Pellets

Abstract:

Torrefied wood pellets can be used directly or co-fired with coal in an existing power plant, if they can be completely size reduced using the existing mill or coal pulverizes. The size reduction characteristics of torrefied pellets could be determined by grindability parameters. However, the generalized size reduction parameter or specific grinding procedure has not been developed yet for densified biomass materials. In this study, the effect of torrefaction temperature on the specific grinding energy of torrefied pellets was studied. The grinding energy followed an exponential or power trend rather linear trend over the torrefaction temperature. The applicability of three well-known grinding equations: Kick, Rittinger, and Bond was also investigated for torrefied and raw wood pellets. Rittinger's and Bond's equations were well matched with the experimental data (R^2=0.93-1). The HGI and ISO grindability indices, that are both based on the Rittinger's theory, were calculated for the torrefied pellets. Also, the bond work index (BWI), based on the Bond's theory were determined for each type of the pellets, and a correlation between HGI and BWI were proposed. The BWI decreased by an increase in torrefaction temperature contrary to HGI index. It decreased from 732 kWh t^{-1} for control wood pellets to 70 kWh t^{-1} for torrefied wood pellets at 290°C. Further studies are recommended for grinding through different mill types and different lignocellulosic materials.

494451 Effect of Media Shape on Impact Breakage of a Bed of Silica Sand Particles: An Attainable Region Approach

Abstract:

Different drop weight media were used to investigate single particle impact breakage characteristics of a bed of silica sand particles. This is done using different mass of silica bed and drop heights. The attainable Region technique, an analysis tool is used to optimise the amount of required size class. Results demonstrate that spherical drop weight leads to maximum amount of intermediate size class whilst the cubes provide a much courser product. It can be established that, the drop weight method coupled with the AR technique provides beneficial and practical way of evaluating the optimal impact breakage for size reduction processes.

495905 DEM-PBM Modelling of the Millling Process of Pharmaceutical Ribbons

Abstract:

Ribbon milling is a critical process step in roll compaction, which determines the properties of produced granules, in particular, the granule size distribution. It is well recognised that fragmentation during ribbon milling is governed by two mechanisms: abrasion and impact, depending on the milling speed. Prediction of the fragmentation process during ribbon milling will enable ribbon milling to be optimised. In this study, a predictive methodology for modelling size reduction processes in ribbon milling was developed by integrating two different modelling techniques, i.e. the discrete element method (DEM) and the population balance model (PBM). DEM was used to model fragmentation at the microscopic level, giving a detailed insight into the underlying breakage mechanism. In DEM modelling, virtual ribbons were created by introducing an appropriate interfacial energy using the cohesive particle model and the fragmentation process of the virtual ribbon during abrasion and impact was analysed. The fragmentation rate and the number and size of fragments (i.e. granules) resulting from the breakage of a ribbon were determined and the effects of milling speed and interfacial energy were explored. The DEM results for single ribbon fragmentation were then incorporated into the PBM so that the ribbon milling process can be modelled at the system level, from which the granule size distribution during ribbon milling can be obtained. Comparing with the data obtained from physical experiments, it was shown that the DEM-PBM approach can predict the ribbon milling process with careful consideration of the dominant fragmentation mechanism.

MODELING OF PARTICLE BREAKAGE AND DISPERSION IN A SLURRY FISCHER-TROPSCH REACTOR

Deekshitha Adapa[2], *Aruna C.M.*[2], *Udaya Bhaskar Reddy Ragula*[1,2]
[1]*Center of Excellence in Advanced Materials & Green Technologies, Amrita School of Engineering, Coimbatore-641112, Amrita Vishwa Vidyapeetham, Amrita University, India*
[2]*Department of Chemical Engineering & Materials Science, Amrita School of Engineering, Coimbatore-641112, Amrita Vishwa Vidyapeetham, Amrita University, India*

Introduction

The increase in demand for transportation fuels and the depletion of the fossil fuel reserves such as crude oil and natural gas motivated the production of transportation fuels such as petrol and diesel from renewable sources such as biomass. The production of transportation fuels from biomass is a two-step process. In the first step, the biomass is converted into synthesis gas (a mixture of carbon monoxide and hydrogen) and the synthesis gas is converted to liquid fuels via Fisher-Tropsch Synthesis (FTS) in the second step. The FTS process is represented generally using the reaction given below.

$$(2n + 1)\, H_2 + n\, CO \rightarrow C_nH_{2n+2} + n\, H_2O \qquad (1)$$

Where n is an integer typically varies from 1-20. This is also called as the carbon number.

The FTS occurs over a catalyst. The catalysts that are currently being used are iron, supported cobalt, nickel and ruthenium[1–4]. The ruthenium is often used as catalyst promoter for cobalt and nickel for FTS[4].

The different types of reactors commercially available for FTS are fixed bed, fluidized bed, and slurry bed reactors[5–8]. The FTS process is a bond formation reaction, it is highly exothermic. Since, the reaction is highly exothermic, the raise in temperature results in hot spots within the catalyst particle or in the catalyst bed[9]. These hot spots are undesirable, as they reduce the selectivity to liquids fuels produced. These hot spots can be eliminated by using efficient means of mixing. Among the commercially available reactors for FTS, slurry bed reactor offers efficient means of mixing. This reactor can offer effective temperature control, and hence higher selectivity to liquid fuels. To effectively remove the exothermic heat of reaction in a slurry reactor, a heat transfer fluid is passed through the tube bundle.

The slurry reactor for FTS is a three phase reactor. The slurry Fischer-Tropsch reactor is fed with a catalyst-liquid fuel slurry[10,11]. The synthesis gas is bubbled through the catalyst slurry from the bottom of the reactor. As the catalyst slurry progresses from bottom to the top of the reactor, the catalyst particles in the slurry undergo numerous collisions. These collisions are of three types: a) collision between the catalyst particles, b) collision of catalyst particles with the tube bundle, and c) collision of particles with reactor walls. These collisions result in particle breakage/agglomeration. The size and shape of the particles change as they undergo breakage or agglomeration. This change in size alters the interfacial surface area available for the reaction. It also changes the settling behavior of the particles and their residence time in the reactor.

The effect of catalyst particle size on both methane selectivity and liquid fuel selectivity was studied by Pendyala et al., [12].

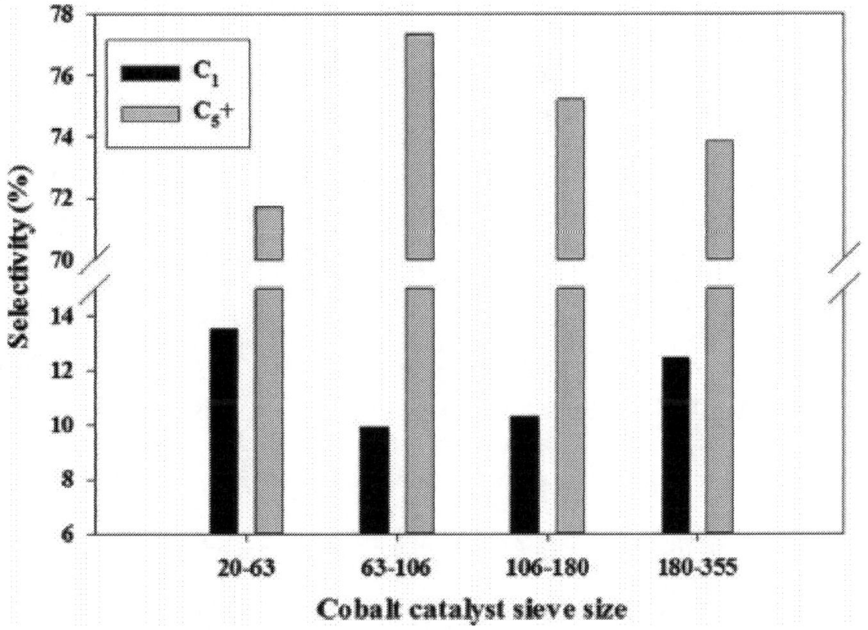

Figure 1: Effect of Cobalt Particle (sieve) Size on Selectivity of Liquid Fuels and Methane[12]

It is evident from Figure 1 that, the optimum particle size for maximizing the liquid fuel productivity is between 60 to 180 microns.

Due to the change in particle size, the dispersion of the catalyst particles in the slurry reactor will also get affected. It is important to tie the particle breakage with the particle dispersion, especially when the particles are suspended in a flowing fluid. To our knowledge there is no prior work on combining particle breakage and dispersion especially in catalytic systems.

This paper addresses the particle breakage and dispersion phenomena for an alumina supported cobalt FTS catalyst. The breakage and dispersion model are coupled and coded in MATLAB.

The objectives of the work are:

1. To develop a mathematical model for particle breakage of a Cobalt/Al_2O_3 Fischer-Tropsch catalyst.
2. Combine the particle breakage model with the particle dispersion model using Richardson-Zaki correlation for particle velocities for a catalyst with distributed particle size range.
3. Solve the models developed using finite difference scheme using MATLAB software.
4. Study the effect of fluid superficial velocity and catalyst loading on particle breakage and dispersion phenomena.

Modeling of Particle Breakage and Dispersion in Slurry Reactor

As the slurry moves upward along the length of the bed, the particles in the slurry undergoes several collisions, that is, collision with the other catalyst particles, with walls of the reactors and the tubes inside the reactor. These collisions result in breakage of the particles which results in smaller sized particles. After breakage, the settling behavior of catalyst particles changes and they get dispersed across the volume of the bed. This proposed method couples the phenomenon of particle breakage and particle dispersion phenomenon in a slurry bed reactor at FTS conditions.

The collision between the particles or collision of particles either with the tube bundle or with the reactor wall is a strong function of particle velocity. It is important to estimate the particle velocities of individual particle sizes present in a particular location of the slurry.

Particle Velocity Calculations

The catalyst particles present in the slurry bed reactor have a size range of 40 to 520 microns[12]. The slurry is fed in the upward direction, but the weight of the particles acts in the downward direction. The presence of multi-sized particles makes the scenario is analogous to hindered settling.

The particle velocity in the case of a hindered settling is given by the Richardson -Zaki model[13].

$$U_i = U_o + \sum_{i=1}^{n} U_{t_i} \varepsilon^{n_i - 1} C_i - U_{t_i} \varepsilon^{n_i - 1} \tag{2}$$

The Richardson-Zaki index can be evaluated from the correlation given by the following equation

$$n_i = 2.82 (d_{pi})^{0.2} \tag{3}$$

The particle dispersion co-efficient is found from the modified Peclet number given by equation (4)[14].

$$Pe_i = \frac{U_o d_{pi}}{\varepsilon D_i} \tag{4}$$

The Peclet number is calculated from the ratio of Froud number to Peclet number[1,2]

$$\frac{Fr_i}{Pe_i} = 7.9 \left(\frac{U_o - U_{mf_i}}{U_{t_i}} \right)^{2.141} \tag{5}$$

$$Fr_i = \frac{\rho_f}{\rho_p - \rho_f} \frac{U_o^2}{g d_{p_i}} \tag{6}$$

The equations (6) and (7) are calculated at the minimum fluidization conditions.

$$U_{mf_i} = \frac{\mathrm{Re}_{mf_i} \mu_f}{d_{p_i} \rho_f} \tag{7}$$

$$\mathrm{Re}_{mf_i} = \left[(33.7)^2 + 0.0408 Ga_i \right]^{0.5} - 33.7 \tag{8}$$

$$Ga_i = \frac{\rho_f (\rho_p - \rho_f) g d_{p_i}^3}{\mu_f^2} \tag{9}$$

The terminal settling velocity in the equation (4) is given by the Stoke's law.

$$U_t = \frac{2}{9} g R^2 \left(\frac{\rho_p - \rho_f}{\rho_f} \right) \tag{10}$$

Particle Breakage and Distribution Models

Weibull distribution

In recent times, the Weibull distribution model of statistics is used by researchers to model the particle survival under different scenarios. The model has successfully explained the probability of particle survival in case of crushing soil particles under the action of a compressive force, particle breakage when a tensile force is applied, to analyze the particle breakage phenomenon in a tri axial test, etc.[15]. In this work, the application of Weibull's model is extended to calculate the probability of breakage in a slurry bed reactor. The use of Weibull distribution demands a geometric similarity of the particles under study. Thus the catalyst particles are assumed to be spherical in shape. This model provides the probability of breakage as a function of stress applied on the particle.

Dynamic fragmentation model

The effect of impact can be calculated based on dynamic fragmentation given by Grady[16]. He developed a relationship for fragment size based on induced strain rate, fracture toughness, density and propagation velocity. The model has already been used successfully in predicting the breakage phenomena in vertical and horizontal impact crushers, ball mills etc.,

Weber model

The fragmented particle size along with the influence of hydrodynamic fluid behavior can be determined using Weber model[17]. The Weber number is proportional to the ratio between disruptive hydrodynamic force and stabilizing surface tension. Hence, the ratio between inertia force due to the penetrating velocity and surface tension. The model has addressed the interaction of molten metal with the coolants under the influence of thermal and hydrodynamic conditions.

Particle Breakage Probability

Weibull distribution model can be used to explain the breakage of brittle ceramics. This model gives the probability of survival as a function of the force applied[15].

$$P_s = \exp\left(-\frac{\sigma}{\sigma_0} \right)^m \tag{11}$$

Thus, the probability of breakage is given as

$$P_b = 1 - \exp\left(-\frac{\sigma}{\sigma_0} \right)^m \tag{12}$$

Where the dynamic stress induced on the particle is given by [18]

$$\sigma = \rho \times V_p \times U_p \tag{13}$$

Diameter of fragmented particles

Grady proposed the dynamic fragmentation model for particles breaking under the influence of an impact force. The diameter of the fragmented particles obtained after breakage is given by

$$D_0 = \left(\frac{4.472 \times Kl_c}{S_r \times V_p \times \rho} \right)^{\frac{2}{3}} \tag{14}$$

Where,

$$S_r = \frac{2 \times U_i}{d_{pi}} \tag{15}$$

The fragmentation characteristics can also be observed using the Weber number as a hydrodynamic parameter. The diameter of the broken particles can be calculated using Weber number[17],

$$We = \frac{\rho_p \times U_i^2 \times D_p}{\sigma_t} \tag{16}$$

$$\frac{D_0}{D_p} = 0.94 We^{-0.35} \tag{17}$$

Particle Dispersion

The flux at the inlet and exit are assumed to be zero. Thus the following inlet and exit boundary conditions are employed to compute the concentration of the particles.

$$D_i \rho_i \left(\frac{\partial w_i}{\partial z} \right) - U_i w_i \rho_i = 0 \tag{18}$$

The steady state force balance equation of the catalyst particles is given by [13]

$$\frac{\partial}{\partial z}\left[D_i \left(\frac{\partial C_i}{\partial z} - \frac{C_i}{\rho_m}\frac{\partial \rho_m}{\partial z} \right) - U_i C_i \right] = 0 \tag{19}$$

The equations (14) and (15) are descritized into a finite difference equation by using "central difference" for the second order differential terms and "forward difference" for first order differential terms.

Results and Discussion

The model developed was used to analyze the influence of various operating parameters on the size distribution of the catalyst. The parameters analyzed include:

- Weight hourly space velocity (whsv)
- Percentage of solids in the feed slurry
- Size distribution of catalyst in the feed slurry
- Superficial fluid velocity

For the simulation of the combined particle breakage and dispersion of particles in slurry FTS process, the FTS process parameters were obtained from Basha et. al.,[8] and the properties of alumina particles, for particle breakage calculations were obtained from Tromans [19].

Among the four different parameters mentioned above, the former two doesn't have any influence on the particle breakage phenomenon. As a result, the size distribution of the catalyst at the exit of the reactor remained unaltered.

The influence of superficial fluid velocity and the feed particle size distribution are addressed below:

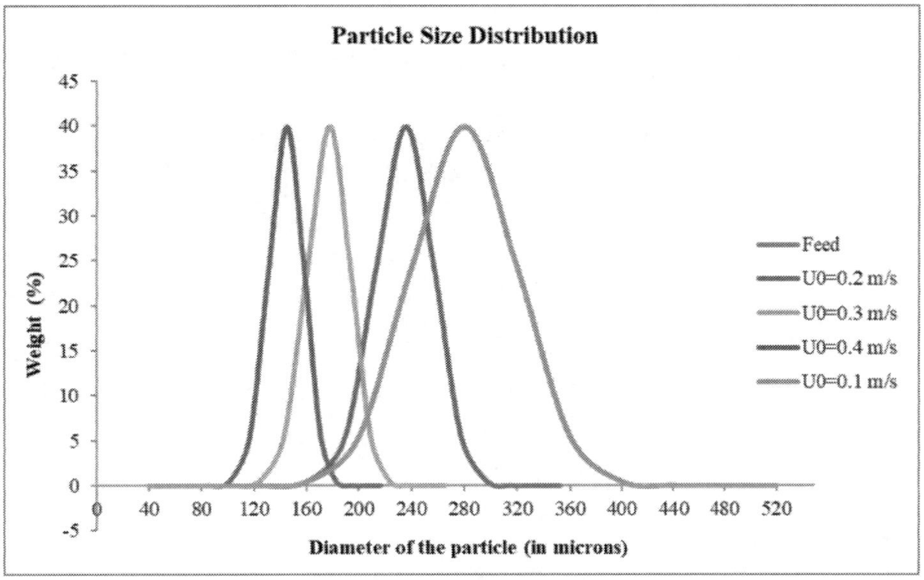

Figure 2: Effect of Feed Particle Size on Exit Particle Distribution - Normal Distribution of Feed Particle Size

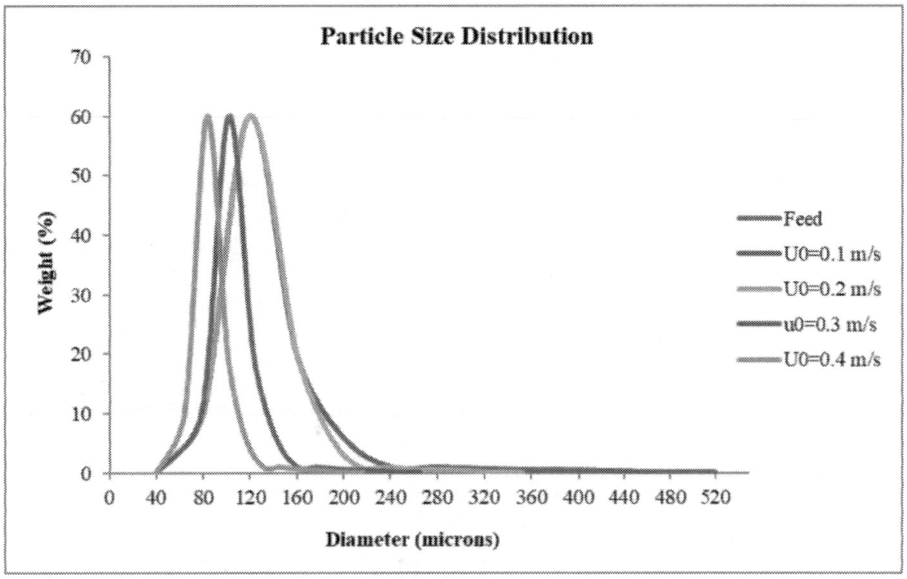

Figure 3: Effect of Feed Particle Size on Exit Particle Distribution – Feed Containing Higher Amounts of Fine Particles

The curves in the figure 2,3 and 4 presents the fact that the probability of breakage is very high in the case of larger sized particles. The feed with higher percentage of larger sized particle shows a very broad distribution of catalyst particle size at the exit of the reactor. Thus a normal distribution of catalyst size can prevent the entrainment of the catalyst to a greater extent, which occurs as a result of breakage of the catalyst particles.

The influence of superficial fluid velocity on the final particle size distribution at the reactor exit is studied. The figures 2 to 4 clearly depict the fact that the rate of generation of finer particles increases as the superficial fluid velocity increases. This is presented by the shift in the curve towards the finer end.

Note: There is no breakage of particles when the superficial fluid velocity is 0.1 m/s.

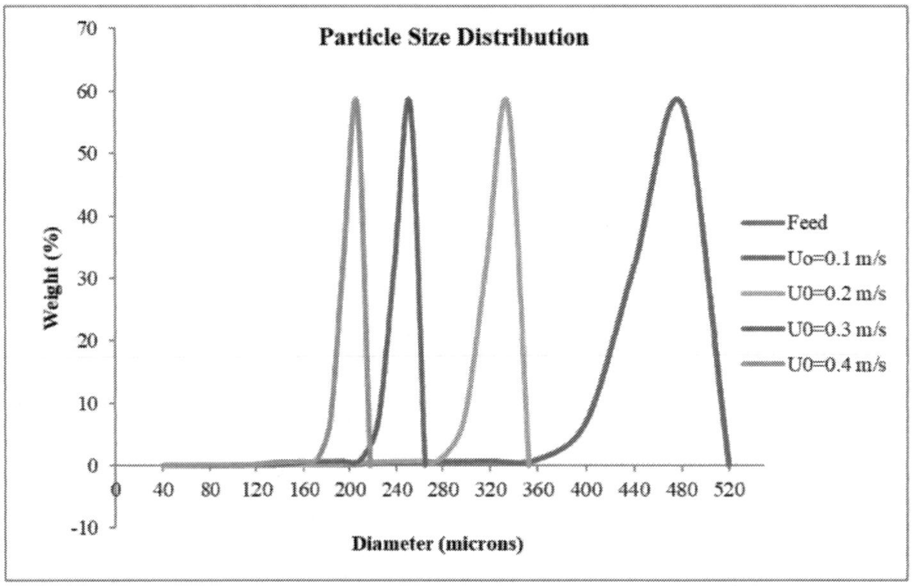

Figure 4: Effect of Feed Particle Size on Exit Particle Distribution – Feed Containing Higher Amounts of Large Particles

Table 1: Comparison of Mass mean diameter (in microns) for different feed size distributions

Feed Type	**Normal Distribution**	**Left skewed Distribution**	**Right Skewed Distribution**
Feed	280	137	456
U0=0.1 m/s	280	137	456
U0=0.2 m/s	234	133	323
U0=0.3 m/s	176	110	244
U0=0.4 m/s	144	90	199

The problem was also modeled using the Grady's fragmentation model. The results obtained showed that the model fails to address this scenario of particle breakage as it doesn't account for the solid fluid interactions which occur inside the reactor. The problem is thus addressed using the Weber model which gives due consideration to these interactions.

Nomenclature

d_{pi} Particle diameter of species i, mm

D_i Dispersion coefficient of particle species i, m^2/sec

C_i Fractional volume concentration of particle species i

ρ_m Bulk density of the bed, kg/m^3

U_i Particle velocity of the species i, m/sec

z Axial distance along the length of the bed, m

Fr_i Froude number of species i

U_{mf_i} Minimum fluidization velocity of species i, m/sec

ρ_f Density of fluid, kg/m^3

ρ_p Particle density, kg/m^3

g Acceleration due to gravity, m/sec^2

Re_{mf_i} Minimum fluidization Reynolds number for species i

Ga_i Galileo number for species i

μ_f Viscosity of the fluid, kg/(m.sec)

U_o Fluid superficial velocity, m/sec

Ut_i Terminal settling velocity of the species i, m/sec

ε Void fraction

n_i Richardson-Zaki index

R Radius of the particle, m

σ Stress applied on the particle, Pa

σ_0 Stress such that 37% of the specimens survive, Pa

Do Diameter of the particle after breakage (m)

K_{lc} Fracture toughness (Pa m^0.5)

S_r Induced strain rate (sec)

σ_t Surface tension (N/m)

We Weber Number

References

(1) van der Laan, G. P. *Kinetics, Selectivity and Scale Up of the Fischer-Tropsch Synthesis*; 1999.

(2) Sarup, B.; Wojciechowski, B. W. Studies of the Fischer-Tropsch Synthesis on a Cobalt Catalyst II. Kinetics of Carbon Monoxide Conversion to Methane and to Higher Hydrocarbons. *Can. J. Chem. Eng.* **1989**, *67*, 62.

(3) Yates, I. C.; Satterfield, C. N. Intrinsic Kinetics of the Fischer-Tropsch Synthesis on a Cobalt Catalyst. *Energy & Fuels* **1991**, *5* (1), 168.

(4) Zhao, S.; Nagineni, V. S.; Seetala, N. V.; Kuila, D. Microreactors for Syngas Conversion to Higher Alkanes: Effect of Ruthenium on Silica-Supported Iron–Cobalt Nanocatalysts. *Ind. Eng.*

Chem. Res. **2008**, *47* (5), 1684.

(5) Lee S.H, Kim J.H, L. J. G. Fixed and Fluidized Bed Water-Gas-Shift Reactor Using Syngas from Waste GAsification and Hydrigen Productionn Using the Same. US 8,313,700 B2, 2012.

(6) Hu, J.; Yu, F.; Lu, Y. Application of Fischer–Tropsch Synthesis in Biomass to Liquid Conversion. *Catalysts* **2012**, *2* (2), 303.

(7) Wright, M. M.; Brown, R. C. Distributed Processing of Biomass to Bio-Oil for Subsequent Production of Fischer-Tropsch Liquids. **2008**, 229.

(8) Basha, O. M.; Sehabiague, L.; Abdel-Wahab, A.; Morsi, B. I. Fischer-Tropsch Synthesis in Slurry Bubble Column Reactors: Experimental Investigations and Modeling - A Review. *Int. J. Chem. React. Eng.* **2015**, *13* (3), 201.

(9) Lualdi, M. *Fischer-Tropsch Synthesis over Cobalt-Based Catalysts for BTL Applications*; 2012.

(10) Wang, Y.; Fan, W.; Liu, Y.; Zeng, Z.; Hao, X.; Chang, M.; Zhang, C.; Xu, Y.; Xiang, H.; Li, Y. Modeling of the Fischer-Tropsch Synthesis in Slurry Bubble Column Reactors. *Chem. Eng. Process. Process Intensif.* **2008**, *47* (2), 222.

(11) Maretto, C.; Krishna, R. Modelling of a Bubble Column Slurry Reactor for Fischer-Tropsch Synthesis. *Catal. Today* **1999**, *52* (2–3), 279.

(12) Pendyala, V. R. R.; Jacobs, G.; Ma, W.; Klettlinger, J. L. S.; Yen, C. H.; Davis, B. H. Fischer-Tropsch Synthesis: Effect of Catalyst Particle (Sieve) Size Range on Activity, Selectivity, and Aging of a Pt Promoted Co/Al2O3 Catalyst. *Chem. Eng. J.* **2014**, *249*, 279.

(13) Juma, A. K. A.; Richardson, J. F. Segregation and Mixing in Liquid Fluidized Beds. *Chem. Eng. Sci.* **1983**, *38* (6), 955.

(14) Asif, M.; Petersen, J. N. Particle Dispersion in a Binary Solid-Liquid Fluidized Bed. *AIChE J.* **1993**, *39* (9), 1465.

(15) Fu, Y.; Tao, Z.; Hou, X. Weibull Distribution of the Fracture Strength of 99% Alumina Ceramic Reshaped by Cold Isostatic Pressing. *Ceram. Int.* **2014**, *40* (6), 7661.

(16) Srivastava, S.; Yazdchi, K.; Luding, S. Mesoscale Dynamic Coupling of Finite- and Discrete-Element Methods for Fluid-Particle Interactions. *Philos. Trans. R. Soc. A Math. Phys. Eng. Sci.* **2014**, *372* (2021), 20130386.

(17) Zhang, Z.-G.; Sugiyama, K.-I. Fragmentation of a Single Molten Metal Droplet Penetrating into Sodium Pool. IV. Thermal and Hydrodynamic Effects on Fragmentation in Copper. *J. Nucl. Sci. Technol.* **2012**, *49* (6), 602.

(18) Djordjevic, N.; Shi, F. N.; Morrison, R. D. Applying Discrete Element Modelling to Vertical and Horizontal Shaft Impact Crushers. *Miner. Eng.* **2003**, *16* (10), 983.

(19) Tromans, D. Crack Propagation in Britle Materials: Relevance to Minerals Comminution. *Int. J. Res. Rev. Appl. Sci.* **2012**, *13* (2), 406.

498940 Perspectives on Rotor Stator Wet Milling: Scaledown, Scaleup, Operations and Applications

Abstract:

Rotor stator wet milling is a versatile, scalable and portable method for manipulating the particle size and physical attributes of pharmaceutical APIs, intermediates and reagents, and can be utilized for a variety of applications. An overview of the theory for particle size reduction via wet milling will be provided, along with some general guidance on setup, operational parameters. Scaleup/scaledown issues (attrition, undesired annealing, reproducibility) and examples of novel applications will be presented by use of case studies, including the following: Particle size distribution modification by combination of terminal wet milling and heat/cool annealing; improving reaction kinetics and conversion efficiency by wet milling of a reagent to degranulate and expose fresh interfaces to solution phase; pushing the boundaries of rotor-stator wet milling with higher tip speed operations; and use of wet mills in sterile operations or for potent compounds, to reduce exposure risk and bioburden associated with dry powder processing.

499393 Crystal Wet Milling and Particle Attrition in High Shear Mixers

Abstract:

While wet milling of active pharmaceutical ingredients (API) and other crystalline substances has become an increasingly important focus of industrial application, there are few fundamental studies on which to base mill design, scale up and operating strategies. In this study, a systematic mechanistic framework is developed, and employed with definitive experiments performed in a Silverson L4R rotor-stator mixer, to investigate the effect of milling conditions, mixer geometry, and particle physical properties on ultimate particle size and particle breakage kinetics.

Based on the literature, it was reasoned that along with size and shape, stress intensity factor (or fracture toughness) and hardness are the important physical properties affecting breakage rate. To vary these properties, several crystalline materials were milled in anti-solvents (e.g. sugar in isopropyl alcohol) to allow discrimination between mechanical size reduction and solubility effects. Collision rates were varied by changing particle concentration; and coatings and small geometric changes were used to consider interactions with solid surfaces. Rotor speed and volumetric throughput were adjusted to independently vary energy input and mill head residence time. By varying operating conditions, mill head geometry and crystal mechanical properties, it is possible to discriminate among various crystal fracture (e.g., plastic vs. elastic deformation) and particle breakage mechanisms (e.g., fluid shear, particle-particle collisions, particle-blade collisions) to develop mechanistic relationships which aid in process scale up and serve as inputs to population balance models and other models for the evolution of the crystal size distribution.

Crystal mechanical properties were measured with a Hysitron TriboIndenter. Particle size distributions were measured by several methods. A Lasentec focused beam reflectance measurement (FBRM) probe and Particle View Microscope (PVM) were placed in the holding tank to continuously monitor particle size and suspension uniformity. Grab samples, acquired in the holding tank at well-defined time intervals, were analysed using a Horiba laser diffraction instrument and by microscopy with an automated image analysis technique. Based on the results appropriate diameters were defined to track the milling rate and ultimate size of the largest surviving particles as well as the characteristics of the fines produced via attrition from the parent crystals. CFD simulations were conducted to estimate the power draw for different rotor speed and rotor and stator gap clearance.

Classical grinding relations show a different dependency on particle size (e.g. proportional to size, surface area, volume, etc.) for the impact energy needed to fracture a particle. For particles colliding with a solid surface, Gahn and Mersmann (1997) and Ghadiri and Zhang (2002) have developed mechanistic models for fracture based on elastic and plastic deformation of the crystal, respectively. In this study, we also consider the case where both elastic and plastic response is important. This gives rise to a series of models that have their roots in the classical grinding relations, where the cohesive force resisting breakage can be identified and employed to construct and contrast correlations of practical utility. To this end, it is proposed that both macroscale (related to tip speed) and turbulent inertial subrange scale (related to power per mass) collision velocities can be responsible for breakage. This leads to a class of mechanistic models for the maximum stable crystal size and well as the breakage rate and production rate of fines. Model discrimination is based on comparison to the data.

The mechanistic theory is further exploited to provide breakage kernels based on probability of collision and collision rate theories. Application of the breakage functions to predict the attrition rate as well as their implementation within a Population Balance framework is discussed. The results of this study are not restricted to rotor-stator mixers; the scaling laws, non-dimensional correlations, and breakage functions developed here can be applied to mixing processes in which particle breakage occurs according to the same mechanism(s).

490676 An Investigation into the Performance of an Industrial-Scale Roll Mill

Abstract:

Prediction of mill performance is a difficult task. In order to predict the performance of a roll mill, it is useful to determine the selection and breakage functions of a mill for a particular bulk solid as a function of mill parameters such as roll gap, roll differential speed, and mill feed rate. This enables modeling of the grinding circuit under changing mill and feed conditions and estimation of the product particle size. In this investigation, a production-sized roll mill was fed tightly sized fractions of particles which yielded the selection function and daughter particle size distribution (PSD) directly. In addition, the mill was also fed a polydisperse PSD in order to investigate how well the exit PSD was modelled using the data from the tightly sized fractions. Lastly, the experimental breakage matrix was compared with the predictions of several breakage matrix models.

496755 A Novel Multi-Scale Modeling Approach for Simulating the Evolution of Particle Size during Dry Milling

Abstract:

A multi-scale modeling approach is formulated which combines the use of the discrete element method (DEM) and population balance modeling (PBM) to simulate the evolution of particle size distribution (PSD) in dry milling. DEM is used to elucidate the effect of the microdynamic environment on particle breakage behavior which is input into the PBM to predict the evolution of the PSD. While DEM and PBM can be used in combination at early milling times to simulate the evolution of the PSD, PBM must be used exclusively at prolonged milling times when particle numbers exceed the computational limit of DEM. As a major novelty, the proposed intermittent coupled DEMâ€'PBM approach uses a non-linear PBM, instead of a time-dependent PBM, which explicitly accounts for the effect of the microdynamic environment through a so-called effectiveness factor. The effectiveness factor accounts for the effect of the evolving PSD on breakage behavior and can be used to more accurately predict the PSD within the non-linear PBM framework. This study has demonstrated that a non-linear PBM, whose parameters were calibrated using DEM at early milling times, is capable of predicting the PSD evolution during prolonged milling and potentially obviates the need for computationally expensive DEM simulations which cannot practically simulate the whole milling duration.

495323 Using Magnetically Assisted Impact Coating (MAIC) to Improve Powder Flow Parameters

Abstract:

Many powders resist flowing and have poor basic flow properties. These poor properties manifest themselves in problems with pneumatic conveying, fluidization, and filling of dies. Powders can further exhibit problems upon storage, where consolidation and caking can further reduce the flow properties of a powder.

In order to provide more consistent processing and product quality, flow aids, such as fumed silica or magnesium stearate, are typically added to a powder to improve its flow and/or reduce its tendency to cake. These flow aids are typically added via a traditional blending method, such as v-blending, with the amount of flow aid being the only processing parameter that could be used to control changes in powder flow. Unfortunately, most processes have an inherent maximum amount of flow aid that can be added before product quality is affected. Frequently, the maximum is also governed by regulatory statues.

In a preliminary study [1], Magnetically Assisted Impact Coating (MAIC) was shown to improve flow properties of a model compound, 5 Î¼m citric acid, when compared with a traditional blending process, v-blending. The initial study was done with an FT4 Powder RheometerÂ® comparing fumed silica loadings of 0.5% and 2.0% (w/w), and showed that for a number of powder flow parameters, MAIC-processed material had better flow properties than that of v-blended material. This follow-up study verifies the original findings and expands it to include 0.0%, 0.5%, 1.0%, 2.0%, and 5.0% (w/w) silica loadings, and presents statistical comparisons between the flow properties of v-blended samples and MAIC-treated samples, including flow parameters not studied or well characterized in the initial study [2].

The result of the this study shows that MAIC improves flow properties of the model compound at all levels of flow aid addition, which allows for further improvement of flow properties or a reduction in the needed levels of flow aid for equivalent powder flow. MAIC was even shown to improve flow properties when no flow aid was added; the reason for this is still under investigation.

References

[1] Freeman, T; Hendrickson, W.; Bowman, C. R. *et al.* (2016, November). *Comparison of Magnetically Assisted Impact Coating (MAIC) with Traditional Mixing Techniques for the Addition of a Silica Flow Aid.* Paper presented at the meeting of the American Institute of Chemical Engineers, San Francisco, CA.

[2] Brockbank, K., Armstrong, B., Chandorkar, Y., *et al.*, *Understanding powder caking as a consequence of a range of mechanisms by means of powder rheometry.* 2015:33:102-108.

507484 Investigation of Humidity Effects on Electrostatic Behavior of a Small Cold Model Fluidized Bed

Abstract:

Introduction: Fluidized beds are being used in wide variety of industrial applications ranging from oil cracking to pharmaceutical powder manufacturing. The ambient humidity of the fluidizing gas can have a complicated role in the fluidization and can cause varying inter-particle phenomena, such as liquid bridging (LaMarche et al., 2016) in highly humid systems or lead to triboelectric charging in lower humidities (Park, 2002). Triboelectric charging of granular matter taking place at low ambient humidity (less than 60%) can cause particles to adherefluidizing column walls (Hendrickson, 2006), lead to particle clustering (Lee et al., 2015), and affect particleentrainment (Fotovat et al., 2017). We have studied a small cold model fluidized bed both experimentally and numerically to shed light on the system changes in humidities ranging from 0% to 60% relative humidity.

Methodology: In this work we considered a small rectangular fluidized bed with soda lime glass walls, nitrogen as the fluidizing gas, and spherical nearly monodisperse polyethylene particles. The extent of particle charging was adjusted by changing the humidity of the fluidizing gas. The experimental model was fitted with a pressure transducer that allowed us to measure the pressure drop across the bed. In addition, we employed particle image velocimetry (PIV) to investigate the particle velocities seen at the column wall to determine the extent which particles had adhered to the column walls. Particle charges were measured using a faraday cup method (Fotovat et al., 2017) that allowed us to determine the mean particle charge in the system. The measured particle charge was used as an input for a numerical model (Kolehmainen et al., 2017) to calibrate an effective work function value (Laurentie et al., 2013) that was used to adjust the extent of charging seen in the numerical model. The wall layer velocities and pressure drop obtained from the simulations were compared with the experimental results to validate the numerical model.

Results: The charging behavior of the system studied here showed non-monotonic relation between the relative humidity and the mean charge. Around 30% relative humidity the mean particle charge diminished; above this level it was negative; and below the 30% relative humidity the measured particle charge was positive. The measured mean charge correlated with the wall adhesion, and downward particle velocities were observed at 30% relative humidity that is consistent with our charge measurement. The numerical model employed was able to capture the pressure drop results for small particle charge levels (less than 10nC/g), but failed to reproduce the results for higher charge levels. This inconsistency was accounted for the numerical model only forming a monolayer of particles at the wall, while the observed wall layer thickness in the experimental model was multiple particles thick.

References:

[1]Fotovat, Farzam, Xiaotao T. Bi, and John R. Grace. "Electrostatics in Gas-Solid Fluidized Beds: A Review." *Chemical Engineering Science* (2017).

[2]Fotovat, Farzam, et al. "The relationship between fluidized bed electrostatics and entrainment." *Powder Technology* 316 (2017): 157-165.

[3]Hendrickson, Gregory. "Electrostatics and gas phase fluidized bed polymerization reactor wall sheeting." *Chemical Engineering Science* 61.4 (2006): 1041-1064.

[4]Kolehmainen, Jari, et al. "Triboelectric charging of monodisperse particles in fluidized beds." *AIChE Journal* 63.6 (2017): 1872-1891.

[5]LaMarche, Casey Q., et al. "Linking micro-scale predictions of capillary forces to macro-scale fluidization experiments in humid environments." *AIChE Journal* 62.10 (2016): 3585-3597.

[6]Laurentie, J. C., P. Traoré, and L. Dascalescu. "Discrete element modeling of triboelectric charging of insulating materials in vibrated granular beds." *Journal of Electrostatics* 71.6 (2013): 951-957.

[7]Lee, Victor, et al. "Direct observation of particle interactions and clustering in charged granular streams." *Nature Physics* 11.9 (2015): 733-737.

[8]Park, Ah-Hyung, Hsiaotao Bi, and John R. Grace. "Reduction of electrostatic charges in gas–solid fluidized beds." *Chemical Engineering Science* 57.1 (2002): 153-162.

507754 BubbleTree: A Rigorous Algorithm for Lagrangian Tracking and Statistical Analysis of Bubble or Cluster Motion within 3D Fluidized Bed Simulations

Abstract:

Bubbling dynamics and distribution of gas flow critically influence the performance of fluidized bed reactors and must be quantified accurately for their design and performance optimization. In this study, BubbleTree (BT) toolset is developed using python based Visualization Tool Kit (VTK) for analyzing 3D computational fluid dynamics simulation data and performing statistical analysis of bubbling dynamics. The algorithm has been rigorously validated using simulation data from pilot-scale bubbling bed simulations and, in addition to bubble properties such as their volumes, locations and velocities, this toolkit provides the capability to (a) track bubbles through coalescence and splitting events and (b) compute gas bypass (or throughflow) by integrating gas fluxes through bubble surfaces. Further, the applicability of this toolset is also demonstrated for tracking particle clusters in circulating fluidized bed simulations. BubbleTree will be integrated with MFiX CFD simulations allowing for *in situ* computation of bubble- (and cluster-) based physical model (e.g. sub-grid models) and enable the fundamental investigation of gas-solids flow interaction in complex reacting particulate flows.

486738 Interaction Between NO and Char in the Presence of Oxygen

Abstract:

Interaction between carbon and NO in the presence of oxygen

Wenxia Yan[1,2], Songgeng Li[1,2]and Cuigang Fan[1,2]

[1] Institute of Process Engineering, Chinese Academy of Sciences, Beijing, China

[2]University of Chinese Academy of Sciences, Beijing, China

NO_x emission from combustion of coal remain a problem of considerable interest. Deep reduction of NO_x has been required to meet more restrict environmental protection regulation in China. Use of carbonaceous materials as a reducing agent for NO removal has the advantages including the inexpensive carbon materials, their abundance and easy availability, very high reduction efficiency without using any catalysts and thus low cost. However, the interaction between NO and carbon is complicated and still under debate although much effort has been devoted in the past decades.

In this process, NO is reduced by carbon. CO, CO2 and N2 are formed. In this work, eleven coal chars are studied in a laboratory scale fixed bed reactor for NO removal. It is found out that the eleven coal chars exhibit different performance in terms of the selectivity of NO-char reaction in the presence of oxygen. Basically, it can be divided into two groups. One group of chars has the positive correlations of CO formation with NO reduction. The other group has completely the inversed trend. Various characterization techniques like XPS, BET and TEM have been employed to examine their physiochemical properties in attempt to give an explanation of this phenomenon. It is found that surface oxygen complexes play an important role. Further, the oxygen species are distinguished and their roles have been elucidated. Their textural and structural properties have also been discussed. Two different reaction mechanisms, i.e., surface carbon-oxygen complexes involved reactions and directly burn-off unoccupied surface, are put forward to explain different performance exhibited by the samples examined in this work.

Contact informationï¼š sgli@ipe.ac.cn

488463 Evaluation of 2MWe Oxy Circulating Fluidized Bed Boiler Performance

Abstract:

Many countries are investing extensively in research efforts to develop eco-friendly energy conversion technologies, due to the stringent environmental regulations and continuously increasing energy consumption with population growth. Among the clean technologies, CFB(Circulating fluidized bed) boiler, which is classified as a promising technology, has been investigated because of various fuel availability, distribution of uniform temperature and higher energy efficiency. In addition to CFB boier, there are other wayst to achieve eco-friendly energy conversion, such as oxy combustion, which is proposed to substitute air by oxygen in boilers to avoid dilution of the fuel gases with nitrogen in order to facilitate CO_2 handling for deposition, and the upgrading of steam quality which may help to increase electric conversion efficiency and to reduce CO_2 emission. Therefore, oxy-fuel circulating fluidized bed boiler with super-critical steam cycle has been developing in the world. In this study, the performance of 2MWe oxy-CFB boiler with super-critical steam cycle is evaluated using 1.5 dimensional model which had been based on IEA-CFBC model. Also, experimental results from an air-fired CFB boiler which has been operated in Korea Institute of Energy Research are used to confirm performance results from 1.5 dimensional model. The effect of oxygen concentration in the input to the oxy-CFB boiler on solid hold-up, temperature, pressure and heat flux in a riser has been determined.

488465 *In situÂ* Desulfurization Behavior Under Oxy-CFB Boiler Conditions

Abstract:

Owing to the continuously increasing use of high-end technologies, global energy consumption is also growing. In response to the price fluctuation of fossil fuels and stringent environmental regulations, many countries are investing extensively in research efforts to develop environment friendly energy conversion. Among the various technologies, Oxy-CFB(Circulating fluidized bed) boiler is recognized as separation of high purity carbon dioxide and a reduction of NOx emissions. In Oxy-CFB boilers, sulfur dioxide emissions have become major concern. To remove sulfur dioxide emissions in air fired CFB boilers, limestone has generally been used until now. Because operation conditions of oxy-CFB boiler, such as CO2 concentrations and temperature, are quite different from those of conventional air fired CFB boiler. According to the equilibrium of CaCO3 calcination reaction, indirect and direct desulfurization reaction might occur in oxy-CFB boilers. Also, recarbonization reaction could happen in oxy-CFB boilers. Therefore, direct desulfurization and recarbonization reaction of 4 different kinds of limestones(CaCO3 : 72-95wt%) were investigated by using thermogravimetric analyzer under oxy-CFB boiler conditions(CO2 concentration : 60-95%, Temperature : 600-1000â„ƒ). Also, reaction models based on the shrinking core model and volumetric model is proposed and evaluated to predict the behavior of limestone in oxy-CFB boilers.

489024 A Combined Experimental-Computational Study of Cohesive Powders Under Consolidation By Xrct Image Analysis and DEM Simulation

Abstract:

Understanding and predicting structural and dynamical characteristics of cohesive powder systems is a technical challenge in many powder technology processes and applications. Advances in experimental techniques such as X-ray computed tomography (XRCT) have enabled characterization of cohesive powder systems with sub-micron resolution. Combined with standard bulk powder characterization methods, this technique allows for mesoscopic structural characteristics of cohesive powder systems to be determined. In this study, XRCT and the Discrete Element Method (DEM) are used to characterize the mesoscale structure, under controlled consolidation conditions, of two cohesive (roughly spherical) powders with distinct material properties. The DEM simulations are based on a constitutive elasto-plastic model that makes use of the JKR theory to account for cohesive interparticle forces. The density and void space structure obtained from DEM simulation are compared to characterization results obtained from XRCT image analysis. The ability of DEM simulation and the contact model to accurately predict these characteristics is discussed.

490837 The Fabrication of Capillary Electrophoresis Microfluidic Chips with Metal Oxide Nanoparticles to Control Optical Properties

Abstract:

Microfluidic chips are commonly utilized to study the fluid flow through micron sized channels. These chips can be used in a variety of applications in the medical, environmental, and engineering fields. Typically, optical detection techniques are employed to verify performance and quantitative results, creating challenges in chip fabrication.

Hummingbird Nano (HBN) Inc. has developed a rapid and sustainable method for fabricating capillary electrophoresis microfluidic chips by molding channels using ferrofluid controlled by magnetic fields. The channels are molded in a highly functionalized, viscous monomer, dipentaerythritol pentaacrylate (SR-399). Polymerization through the application of UV light will create a solid polymer matrix surrounding the ferrofluid. The ferrofluid is then recovered to be reused, and micron sized channels are successfully embedded within the polymer.

Optical sensing devices can then be placed directly adjacent to the material at the point of manufacture. To avoid detection losses, the refractive index of the polymer can be matched to that of the fiber optic cable. Inorganic nanoparticle crystallites with known refractive indices and well-defined size distributions are used to control the refractive index of the nanocomposite chip material. Metal oxide nanoparticles can be synthesized with particle dimensions small enough to efficiently reduce light scattering, maintaining the monomerâ€™s transparent properties.

In collaboration with HBN, funded by NSF SBIR, a nanocomposite comprised of SR-399 and metal oxide nanoparticles are developed with the appropriate refractive index and material properties suitable for the manufacture of capillary electrophoresis microfluidic chips. The nanoparticles are synthesized, characterized, and then functionalized for dispersion in the monomer prior to polymerization. Transmission Electron Microscopy (TEM) and Energy-Dispersive X-ray Spectroscopy (EDS) are utilized to determine size, morphology, and composition of the nanoparticles.

491516 Extended Thin Film Electrocatalyst Structures Via Pt Atomic Layer Deposition

Abstract:

Platinum-nickel nanowires are a promising advanced catalyst material for the oxygen reduction reaction (ORR) in fuel cell applications. When synthesized by galvanic displacement of nickel nanowires (NiNWs), these extended surface electrocatalysts have demonstrated activity and durability superior to benchmark carbon-supported Pt catalysts. However, challenges in scale-up and reproducibility of this fabrication process have prompted investigation into atomic layer deposition (ALD) of Pt as a facile, scalable synthesis alternative for the production of Pt-NiNWs. Atomic layer deposition of Pt nanoparticles on NiNW substrate was performed in a packed bed reactor, using trimethyl(methylcyclopentadienyl)platinum(IV) ($MeCpPtMe_3$) as a metal source, and either H_2 or O_2 for ligand removal. Gases exiting the ALD reactor were monitored via an in-line mass spectrometer, which provided information on reaction mechanisms. Deposition of Pt was confirmed through inductively coupled plasma mass spectroscopy for a range of ALD cycles; oxidative ALD chemistry was found to yield greater Pt content at high cycle numbers. Transmission electron microscopy and elemental mapping techniques were also used to analyze the morphology and structure of ALD-modified NiNWs. The effects of post-processing treatments, such as H_2 annealing and acid leaching, were also studied. Electrochemical performance of ALD-synthesized catalysts in both half-cell and full-cell analyses was found to be comparable to that obtained with galvanic displacement catalysts, thereby demonstrating the efficacy of ALD as a fabrication method for extended thin film electrocatalyst structures.

493464 Effect of Cluster Size and Voidage on Gas-Solid Behavior Via CFD-DEM Simulation

Abstract:

Circulating fluidized beds are widely employed in industry for gas-solid contacting operations. In this systems, the riser is the core part, where usually the main reaction takes place. Our previous research showed that the particles clustering hindered the sufficient contact between the gas and the catalyst particles, which affected the reactions. However, most of researchers studied the impacts of a static cluster on reactions instead of a dynamic cluster, and the lifetime of cluster which may be the rate-controlled step was ignored. Thus, in order to investigate the conditions required to disperse a cluster, the present study investigates the dynamic structure of particle clusters using CFD-DEM model, in which the solids is obtained by the Discrete Element Method (DEM) applies Newton's laws of motion to every particle and the fluid phase is described by the local averaged Navier-Stokes equations that can be solved by the traditional Computational Fluid Dynamics (CFD).

Based on the research, the effects of cluster with different voidage (from 0.750 to 0.995) and size (5, 10 and 15 mm) on the gas-solid behavior were studied. The results show that with a decrease of voidage or with an increase of cluster size, the gas is harder to pass through the cluster and the cluster dispersed slowly, which means the lifetime of cluster is longer because of the stronger force between particles. Besides, as shown in Fig. 1, it is noted that cluster being eroded at the outer by the air flow and particles being scattered upward at beginning. Because air flow avoids the cluster which behaves as a rigid sphere at first, the velocity becomes a maximum at the sides of the sphere, this allows particles to more easily leave the cluster. Furthermore, at t=0.20 s, particles at the bottom of the cluster still have a low velocity and the cluster was not dispersed completely, which means the lifetime of cluster was too long to generate a poor contacting between the gas and solids phase and lead to severe process efficiency losses. However, at a very high voidage (i.e. ¦Å=0.995, show in Fig. 2), particles under the surface and at the side of the cluster have a relatively high velocity at t=0.05 s and the lifetime of cluster was much shorter, which means each particle moves under the influence of air flow because of the little force among particles.

t=0.00s t=0.05s t=0.10s t=0.20s

Fig. 1 Particle flow in different time at ¦Å=0.920

t=0.00s t=0.05s t=0.10s t=0.20s

Fig. 2 Particle flow in different time at ¦Å=0.995

Therefore, the large particle with low voidage clusters is deemed to have a long lifetime, which may affect the reaction and result in low yields of products.

494387 CPFD Simulation of Solids Residence Time and Mixing Behaviors in a Downer Reactor

Abstract:

A downer reactor is a relatively new type of circulating fluidized bed (CFB), in which the conversion and selectivity of reactions are closely related to the particle's residence time, thus the in-depth understanding of the particle's residence time and its mixing behavior is critical for the design, operation and scale-up of downer reactors. Computational particle fluid dynamics (CPFD) model, in which the fluid phase is calculated with a Eulerian computational grid and the solids are predicted using Lagrangian computational particles, provides an efficient approach to understand the comprehensive hydrodynamics in downer reactors. The objective of the present study is to systematically analyze the hydrodynamics and solids mixing behaviors in the downer using a modified gas-solid drag model based on the CPFD approach. The effects of operating conditions (i.e., superficial gas velocity, solids circulation rate) on the mixing behavior were addressed. The flowing conclusions were obtained. The comparison between the simulation results and the experimental data indicates that the CPFD method was capable to predict the hydrodynamics and solids mean residence time. The solids RTD featured a narrow peak with large axial Peclet numbers (Pe), indicating that solids flow approximate an ideal plug flow behavior. However, the solids mixing behavior reached higher levels of attainment to ideal plug flow patterns than that of the solids phase. Operating conditions exhibited a certain influence on the characteristics of gas-solid flowing and mixing behaviors in the downer. With the increase of solids circulation rate and the decrease of superficial gas velocity, particles reached the steady state faster and the radial distribution of solids holdup became more uneven, the mean residence time increased with the wider peak and lower peak value of particles RTD, and the Peclet numbers decreased, which mean that the mixing degree increased.

494856 Settling Rate of Agglomerates Consisting of Polydisperse Primary Particles By Brownian Dynamics

Abstract:

The dynamics and settling rate of nanoparticles dispersed in aqueous solutions is important for *in vitro* dosimetry in nanotoxicology, nanomedicine and stability of nanofluids. Settling rate measurements of sub-micrometer agglomerates are not trivial, contrary to millimeter-sized ones that can be easily tracked in a settling column.

Here, the mobility of such nanosized fractal-like SiO_2 agglomerates (Goudeli et al., 2016) in dilute suspensions is investigated and their settling rate is calculated by Brownian Dynamics (BD) simulations. The BD method is thoroughly validated and the agglomerate mobility diameter is calculated *ab initio* and compared with that obtained by well-known scaling laws for fractal-like agglomerates and settling rate experiments (Kasper et al., 1985). The effect of the constituent primary particle mean diameter and geometric standard deviation (polydispersity) as well as the agglomerate shape anisotropy on their settling rate is elucidated for the first time. For constant agglomerate mobility diameter, the BD-obtained settling rate increases with increasing primary particle size for agglomerates consisting of monodisperse primary particles as well as with increasing primary particle polydispersity when agglomerates consist of polydisperse spheres, but is hardly affected by the particle shape anisotropy due to their fast rotation and small mass.

Furthermore, an analytical expression for the settling rate of agglomerates consisting of lognormally distributed primary particles is derived. The common assumption of monodisperse constituent primary particles leads to underestimation of the agglomerate settling rate compared to the settling rate calculated from analytical expressions. This underestimation depends on the primary particle polydispersity and agglomerate mass mobility exponent D_{fm} (an indicator of the agglomerate structure). The proposed expression is compared also with experiments for the deposition rate of fumed SiO_2 agglomerates in aqueous suspensions (Spyrogianni et al., 2017).

References:

Goudeli, E.; Eggersdorfer, M. L.; Pratsinis, S. E. *Langmuir* **2016**, *32*, 9276-9285.

Kasper, G.; Niida, T.; Yang, M. *J. Aerosol Sci.* **1985**, *16*, 535-556.

Spyrogianni, A.; Sotiriou, G.A.; Brambilla, D.; Leroux, J.-C.; Pratsinis, S.E. *J. Aerosol Sci.* **2017**, *108*, 56-66.

496014 Particle Convection in Vibrating Bed

Abstract:

When particle transporting, the particle convection was observed in a bed caused by vibrating. The height of the bed in the vicinity of the center is larger than that of the periphery considerably. It is well known phenomenon as the particle convection. A simulation using DEM explains the phenomenon well.
Because the particle convection encourages the attrition, it is desirable to suppress when transporting. Suitable partition in the bed is discussed using DEM in this report.

496428 Atomically Deposited Sintering Aids: Assessing the Effects of Al_2O_3 Particle ALD on the Sintering and Performance of SOFC Electrolytes

Abstract:

Yttria-stabilized cubic zirconia (YSZ) is the most-common electrolyte material for solid oxide fuel cells (SOFC) due to its reasonable oxygen-ion conductivity and chemical stability across a wide range of environments. To achieve suitable ionic conductivity, YSZ electrolytes must be near theoretical density; this necessitates high sintering temperatures, often exceeding 1450 °C. This high sintering temperature limits the viability of low-cost one-step sintering during manufacturing and thus increases the cost of SOFC fabrication. Previous researchers have successfully lowered the required YSZ sintering temperature through incorporation of aluminum oxide (Al_2O_3) particles with the YSZ. While this technique has proven to be successful in reducing the sintering temperature, the presence of Al_2O_3 in the electrolyte can reduce YSZ ionic conductivity. In this work, we report on the use of atomic layer deposition (ALD) to precisely and conformally coat individual YSZ particles with the desired amount of Al_2O_3. Constant-rate-of-heating (CRH) experiments were conducted using a horizontal push-rod dilatometer to extract the activation energy of sintering in the initial stage and also gain mechanistic insights into the active diffusion mechanisms as a function of ALD film thickness. Transmission electron microscopy (TEM) and scanning electron microscopy (SEM) were used to characterize the as-deposited thin Al_2O_3 films and the effects of the Al_2O_3 films on grain growth and microstructure during densification. Finally, electrochemical impedance spectroscopy was used to characterize the ionic conductivity of dense YSZ electrolytes.

497518 Greener Ethylene Production Via Chemical Looping

Abstract:

Ethylene is commercially produced from ethane via high-temperature pyrolysis in the presence of diluting steam a.k.a. steam cracking. The process requires a high temperature (>1100 K) and a large steam generation load, making it highly energy intensive (17-21 GJ/tonne ethylene). The process also emits significant amount of NO_x and CO_2 and requires periodic shutdowns for coke burnout. Moreover, single-pass ethylene yield is limited by thermodynamic equilibrium. Oxidative dehydrogenation of ethane (ODH) has been investigated as a potentially more efficient approach. In ODH, ethane is partially oxidized into ethylene and water. The oxidation of hydrogen makes the ODH reaction exothermic, thereby reducing fuel consumption while overcoming the equilibrium limitations for cracking reactions. However, to allow economical product separation, conventional ODH schemes require the use of pure oxygen co-fed with ethane. The oxygen generation, which uses cryogenic air separation systems, is capital and energy intensive. We propose a chemical-looping ODH (CL-ODH) approach that can address these issues. In this scheme, a metal oxide based redox catalyst provides oxygen for the ODH reaction from its lattice. The oxygen depleted redox catalyst is subsequently oxidized in air, regenerating the catalyst and releasing the heat needed for the process. This allows built-in air separation with minimal parasitic energy loss and better temperature control.

We have identified, in our previous work, that Mg_6MnO_8, a mixed oxide with a cation deficient rocksalt structure, is an excellent model redox catalyst for ODH. When promoted with alkali salts, changes in the bulk and near surface properties of the Mg/MnO system help produce ethylene with exceptional selectivity by suppressing deep oxidation of ethane or ethylene. The facile combustion of hydrogen favored by the promoted redox catalyst leads to high ethylene yield and provides the heat required for the endothermic dehydrogenation reaction. In the current study, CL-ODH of ethane is modeled using ASPEN Plus® and is compared with a conventional stream cracking process. Results show that CL-ODH with 85% ethane conversion provides over 80% reduction in both the overall energy demand and CO_2 emissions. The exothermic nature of the regenerator and elimination of the steam requirement lead to major reductions in the upstream energy consumption. Kinetic modeling using CHEMKIN-PRO® is performed to determine the effect of gas phase and surface reactions on the CL-ODH product distributions. The proposed CL-ODH process in the context of circulating fluidized bed operations is also discussed.

498205 Control of Particle Structure and Size Distribution By Humidity

Abstract:

Control of Particle Structure and Size Distribution by Humidity

Georgios A. Kelesidis, Florian M. Furrer, Eirini Goudeli, Maximilian L. Eggersdorfer, Karsten Wegner, Sotiris E. Pratsinis

Particle Technology Laboratory, Institute of Process Engineering,

Department of Mechanical & Process Engineering, ETH Zurich, Switzerland

Nanoparticles with compact structure are attractive for biomedical applications, such as drug delivery and theranostics, due to enhanced bioavailability and light absorption in the near infrared spectrum (Sotiriou et al., 2014). Flame-made nanoparticles, however, typically are ramified agglomerates of primary particles with open structure and mass-mobility exponent, D_{fm}, of about 2.17 (Sorensen, 2011). Such agglomerates can restructure in the presence of humidity to smaller and more compact entities, as has been observed for soot (Ma et al., 2013). Here, this effect of agglomerate restructuring is investigated for flame-made silica aiming at small and compact nanoparticles with well-defined size distributions.

Water vapor produced by an evaporator is introduced into a diluted aerosol stream sampled above the flame, resulting in controlled saturation ratios, S, of 1.15 to 1.55. Water initially condenses on the particle surface and then is largely removed, as the humid aerosol flow passes through a series of two diffusion dryers. Combined Differential Mobility Analyser and Aerosol Particle Mass measurements are employed to determine particle structure and anisotropy by the D_{fm} and the mass-mobility prefactor, k_{fm}, respectively.

Silica agglomerates restructure more with increasing S resulting in more spherical and compact structures, corresponding to larger D_{fm} and smaller k_{fm}, respectively. The asymptotic morphology obtained after complete agglomerate collapse for S = 1.55 is characterized by a D_{fm} of 3.02 ± 0.11 and a k_{fm} of 0.26 ± 0.06, in good agreement with the D_{fm} of 2.79 and km of 0.63 measured from soot agglomerates restructured after water condensation and evaporation (Ma et al., 2013). The packing density, θ, is gradually increasing up to 19 % as the agglomerates become slightly more compact for S = 1.35. Further increase of S up to 1.55 results in silica agglomerate restructuring into compact, spherical entities with average θ = 0.28 ± 0.01, invariant of mobility diameter. The average θ is lower compared to that of soot (Zangmeister et al., 2014) due to the larger polydispersity of silica primary particles.

The compact structures attained with increasing S have smaller size, decreasing the average d_m from 107.6 nm in ambient conditions to 91.4 nm for S = 1.55. Larger agglomerates restructure more than smaller ones, resulting in narrower size distributions with smaller mobility-based geometric standard deviation,

$\sigma_{g,m}$. A gradual $\sigma_{g,m}$ reduction is obtained for increasing S, reaching a maximum 22.4 % decrease for S = 1.55. Thus, particle processing with humidity is a promising method to finely tune their morphology and size distribution.

References:

Ma, X.F., Zangmeister, C.D., Gigault, J., Mulholland, G.W., Zachariah, M.R., (2013) *J. Aerosol Sci.* **66** 209-219.

Sorensen, C.M. (2011) *Aerosol Sci. Technol.* **45**, 765-779.

Sotiriou, G.A., Starsich, F., Dasargyri, A., Wurnig, M.C., Krumeich, F., Boss, A., Leroux, J.C., Pratsinis, S.E. (2014) *Adv. Funct. Mater.* **24**, 2818-2827.

Zangmeister, C.D., Radney, J.G., Dockery, L.T., Young, J.T., Ma, X.F., You, R.A., Zachariah, M.R. (2014) *PNAS* **111**, 9037-9041.

499616 Transition Metal Oxide Powders Made from Flame Spray Pyrolysis for Li-Ion Batteries

Abstract:

Flame spray pyrolysis process provides a cost effective way to make metal oxide powders. When the solvent is a green chemical, the process could become a sustainable manufacturing process for materials production. In this presentation, we will show that transition metal oxide powders can be produced in a flame spray process with a green solvent as the metal oxide precursor. It is also shown that complex transition metal oxide powders, such as $Li(Ni_{1/3}Mn_{1/3}Co_{1/3})O_2$ and $Li(Ni_{0.8}Co_{0.15}Al_{0.05})O_2$, can be produced in this process. These powders can be controlled to a certain degree in terms of powder sizes and morphology, to make them suitable for use in Li-ion batteries. Detailed results of the powder synthesis, characterization, and battery testing will be presented.

499686 Drag Model Evaluation through Fluidized Beds and Free-Falling Particles

Abstract:

Drag models are evaluated for two different conditions: particles in a bed and accelerating particles. Some monodisperse drag models scale a drag coefficient by the voidage to predict the drag in a bed. For beds of non-spherical particles, the current models use a non-spherical drag coefficient with a spherical voidage scaling. Results from fluidized bed experiments with cylindrical particles are used to compare with the current non-spherical models. The experiments found that particle elongation can cause significant changes in behavior of the bed. Therefore, two-fluid simulations are performed with the current drag models to compare the flow behaviors. The models for the drag coefficient play a significant role in the accuracy of these predictions. One common agreement is that particle acceleration can impact the drag coefficient, but it is typically ignored in the models. Thus, simple free-falling particle experiments are performed with high-speed cameras to quantify the effect of particle acceleration.

NUMERICAL EVALUATION OF SOLID-LIQUID DRAG MODELS FOR A FLUIDIZED BED BIOREACTOR

Daniela M. Koerich, Gabriela C. Lopes
Chemical Engineering Department, Federal University of São Carlos (UFSCar), São Carlos, Brazil

Leonardo M. Rosa
Chemical Engineering Department, University of Blumenau (FURB), Blumenau, Brazil

Introduction

Human progress has promoted major technological challenges. The increasing generation of effluents, for example, requires efficient solutions in order to establish sustainable development. Various processes can be adopted for the treatment of organic effluents, including the Conventional Activated Sludge Systems (CAS) and its variants, UASB reactors, Membrane Biological Reactors (MBRs) and Sequencing Batch Reactors (SBRs) [22]. Anaerobic fluidized bed reactors have aroused greater interest since the 1980s. Fluidized bed bioreactors are characterized by the biocatalytic use of immobilized enzymes or microbial cells. The immobilization method most commonly used in wastewater treatment in a fluidized bed refers to the attached growth technique for microbial cells, and involving the growth of a biofilm (cell layers and excreted sludge) on the surface of each carrier particle during the course of fluidization. The growth of biofilm on the carrier particles changes its size, effective density, and surface properties – and, as a consequence, its fluidization characteristics [25]. The fluidization regime is maintained by the drag force associated with the upward movement of the effluent. The maintenance of fluidized beds requires a control of the superficial velocity of the effluent, which depends on the characteristics of the bioparticle – which, in turn, are altered as the microbial community evolves. Thus, in addition to the difficulties related to the maintenance of healthy cultures, as in other processes, there are also difficulties in maintaining the fluidized bed, which makes this one of the most complex processes available for wastewater treatment [22]. Due to the constant velocity along the height, the conventional anaerobic bioreactors can drag the particles out of the reactor (washout) [12], when these particles have their density changed due to the accumulation of biofilm on their surface. On the other hand, fluidized beds in tapered bioreactors (TBR) offer as an additional advantage the decrease of the ascending velocity along the height. This naturally prevents the bioparticle from being carried out of the bioreactor. To optimize the configuration of reactors, and consequently to improve their performance, it is essential to know the dynamics of the phases in their interior. In this sense, Computational Fluid Dynamics (CFD) offers several advantages. This technique is used to simulate phenomena involving moving fluids [4], and is able to simulate reactors in fluidized bed [10]. In these situations, numerical techniques have the advantage of establishing precise values for the process parameters [16]. With the application of CFD techniques, industrial equipment can be simulated using models based on thermodynamics, hydrodynamics and reaction kinetics. Processes can be visualized through simulation results before performing full scale experiments, saving time and reducing costs [23]. In the simulation of multiphase reactors, attention must be given to the modeling of the phenomena present. In gas-solid flows, for example, where the dispersed phase has a specific mass much higher than the continuous phase, the dominance of the drag force on the other interfacial forces is verified. The turbulence also plays a key role, as well as the interactions between particles, which can be modeled considering the Kinetic Theory of Granular Flow (KTGF) [1, 3, 20]. For gas-liquid systems, where the dispersed phase has a much lower specific mass than the continuous phase,

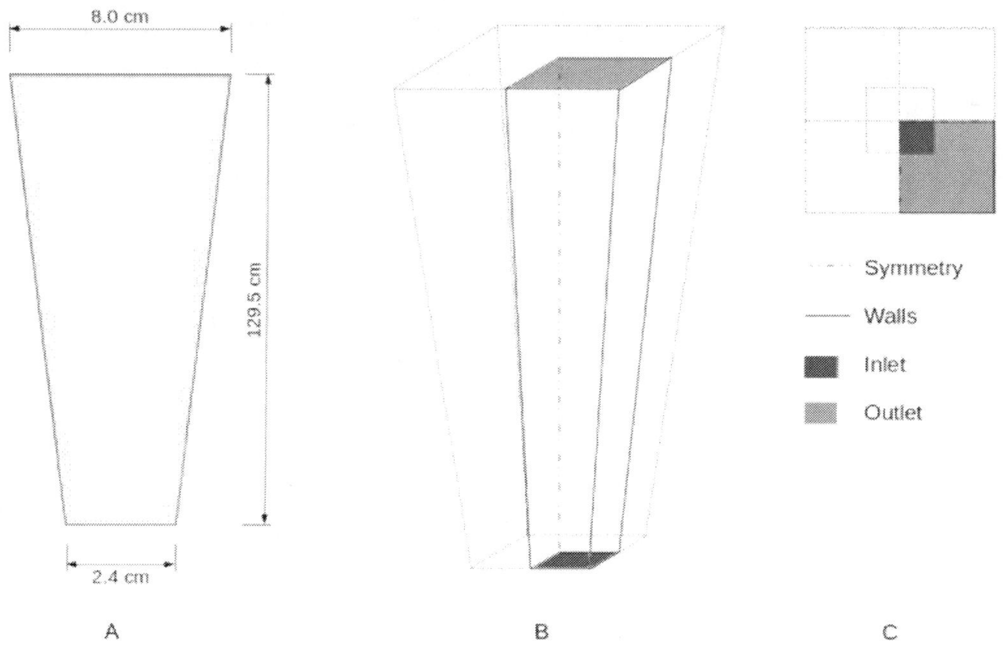

Figure 1: Tapered bioreactor dimensions and geometry.

also observed the influence of forces such as the lift and the virtual mass [17, 18, 19]. In solid-liquid systems, in which the properties of the phases are not so disparate, there is a lack of data regarding the magnitude of the acting forces. In simulations of these systems, solids were usually assumed to be spherical particles with the same diameter. However, in reality, biowaste particles may have unusual characteristics such as relatively large mean particle size, wide size distributions, and extreme shapes. The surface roughness of bioparticles can also affect the interfacial forces [8]. Such irregular particles need more research to correct the drag force and solid stress models for reducing uncertainty [26].

There is a lack in literature of numerical studies on solid-liquid flows. Thus, the present study aims to numerically evaluate the performance of different models for analysis of the interfacial forces acting on these systems. Experimental data from the literature was used to simulate different operating conditions of a fluidized bed bioreactor, using the OpenFOAM code.

Materials and methods

Bioreactor and operation conditions

The fluid dynamics was evaluated in a laboratory-scale fluidized bed tapered bioreactor, with a working volume of 4 L (Figure 1a). It has 1.295 m high, 0.024 m wide at the base, and an angle of 2.5° (1.25° on each wall) [24]. A two-phase (liquid-solid), transient, three-dimensional approach was considered to simulate 1/4 of the geometry, for which conditions of symmetry were applied in the faces (Figures 1b and 1c) so that it represented the complete bioreactor. The third phase (gas) was disregarded in the present study due to low influence on bed height as can be observed in the results presented by Wu *et al.* [24].

The bioparticles were composed of activated carbon coated by a biofilm layer and the substrate was a synthetic wastewater containing acetate [24]. The properties of the phases used in the simulations are presented in Table 1. For simulation purposes, it was considered that all bioparticles were spherical. The kinetic theory of granular flow (KTGF) was used to consider

Table 1: Phases properties

Phase	Density [kg/m^3]	Dynamic viscosity [kg/(m.s)]	Diameter [m]
Liquid	994.14	7.254×10^{-4}	–
Solid	1420.06	0	7.09×10^{-4}

(Adapted from Huang and Wu [9] and Wu *et al.* [24])

effects of collisions between particles, which modifies properties such as pressure, viscosity and granular temperature [6]. As initial condition a bioparticle mass of 0.2 kg was considered [24]. In all cases, the substrate is fed into the bioreactor base, with velocities of 0.0146 to 0.0783 m/s [24].

Mathematical modeling

The Eulerian-Eulerian model was used, considering a continuous phase (wastewater) and a dispersed phase (bioparticles). The continuity and momentum transport equations were calculated for each of the phases:

$$\frac{\partial}{\partial t}(\alpha_l\rho_l) + \nabla\cdot(\alpha_l\rho_l U_l) = 0 \tag{1}$$

$$\frac{\partial}{\partial t}(\alpha_p\rho_p) + \nabla\cdot(\alpha_p\rho_p U_p) = 0 \tag{2}$$

where α represents the volumetric fraction, ρ is the specific mass, and U is the velocity vector. The indexes l and p indicate the liquid and solid phases, respectively.

$$\frac{\partial}{\partial t}(\alpha_l\rho_l U_l) + \nabla\cdot(\alpha_l\rho_l U_l U_l) = \nabla\cdot\left(\alpha_l\mu_l(\nabla U_l + (\nabla U_l)^T)\right) + \alpha_l\rho_l g - \alpha_l\nabla p + M \tag{3}$$

$$\frac{\partial}{\partial t}(\alpha_p\rho_p U_p) + \nabla\cdot(\alpha_p\rho_p U_p U_p) = \nabla\cdot\left(\alpha_p\mu_p(\nabla U_p + (\nabla U_p)^T)\right) + \alpha_p\rho_p\vec{g} - \alpha_p G\nabla\alpha_p + M \tag{4}$$

where p and g are fluid pressure and gravity, respectively. G is the modulus of elasticity. M is the interfacial momentum transfer between the phases. The interfacial forces considered were of drag and lift (Equations 5, 6 and 7).

$$M = M_D + M_L \tag{5}$$

where the drag and lift forces are given by:

$$M_D = \alpha_p\frac{3}{4}C_D|U_r|U_r \tag{6}$$

$$M_L = \alpha_p C_L\rho_l U_r \times (\nabla\times U_p) \tag{7}$$

In equations 6 and 7, the coefficients C_D and C_L are estimated through the use of empirical correlations. Legendre-Magnaudet model was used for the lift force [15]. In the present study a correction to consider the effect of force on solid particles was implemented [14].

In the present study, the Ergun, Wen-Yu, Gidaspow and Syamlal-O'Brien drag models were used. It must be emphasized that these models were developed to predict the behavior of gas-solid flows.

Several models are available to consider the turbulent nature of the flow. Among these, the isotropic models tend to underestimate the turbulence in the reactor. However, the flow predicted by the k-epsilon model is qualitatively close to the flow predicted by an anisotropic model, requiring less computing power [11]. Thus, the k-epsilon model was used in this study. The kinetic theory of granular flow (KTGF) was used to consider the effect or particles collisions on the flow. In this model, there is an additional transport equation for the granular temperature.

Numerical procedure

For each simulation, a prescribed velocity was used in the reactor base, atmospheric pressure at the outlet and, non-slip condition in the walls. The solution of the equations set was done using the finite volume method, as available in OpenFOAM, version 4.1. A second-order interpolation scheme, and the PIMPLE algorithm for pressure-velocity coupling were used. Maximum residuals of 10^{-6} were considered as a convergence criterion. Adaptive time step was applied, in order to maintain the Courant number below 0.9. The simulations were performed on an Intel® Core^{TM} i5-2430M with 2.40 GHz processor.

The dimensions of the bioreactor were presented in Figure 1. For the elaboration of the numerical meshes was used the software "blockMesh". Four meshes were elaborated with different refining, with sizes of 1,170, 2,768, 5,625 and 14,700 control volumes. A growth rate of 1.33 was applied to the vertical axis in order to maintain the volumes approximately cubic throughout the entire height of the reactor. The grid convergence index (GCI) was used to estimate the discretization error [21], considering local and global variables. To determine the optimum mesh and simulation time, the highest velocity was used, of 0.0783 m/s. As all simulations started from an initial bed height, care was taken to guarantee that the same amount of bioparticles was present in all meshes, independent of the refinement used.

All simulations were performed on a mesh of 14,700 volumes of control to maintain GCI values between 10^{-4} and 10^{-5}, respectively, for fraction of bioparticles and pressure. The simulation time was 130 s, of which the last 80 s were used for the calculation of time-averaged values. In all cases, to obtain the fluidization height, the criterion used was the height corresponding to a maximum volume fraction of bioparticles equal to 0.05 [2].

Results and discussion

Lift influence on the bed height

In fluidized bed bioreactors, the bioparticles remain suspended in the fluid (fluidized) due to the drag forces associated with the upward movement of the wastewater [7, 22]. However, the lift force was also evaluated in order to determine its influence. Most of the models available for the lift force were developed for gas-liquid flows. The model considered here was Legendre-Magnaudet which was corrected, since the same authors had already established the necessary adjustment for their application in solids flows.

The inclusion of the lift force resulted in a change in the bed height at all flow rates considered (Figure 2), higher than in the obtained results disregarding this force. Observing this figure, it is noted that, at lower flow rates, the lift force has little influence on the bed height in relation to the heights obtained only with the drag force, of approximately 1.49% for the velocity of 0.015 m/s. On the other hand, at the higher flow rates the influence of the lift force was more representative, reaching a height up to 15.14% higher for the velocity of velocities above 0.078 m/s.

A possible explanation for this behavior can be observed in Figure 3. The lift force is characterized by being perpendicular to the flow, moving particles from regions with lower

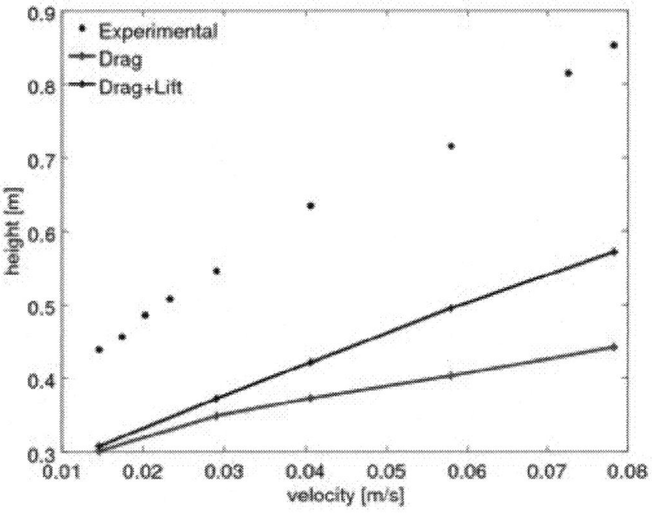

Figure 2: Lift influence on the fluidized bed height.

velocities to regions with higher velocities. As can be seen in Figure 3a, considering only the drag force, the bioparticles had a higher concentration along the walls. When considering drag and lift (Figure 3b), there was a radial redistribution of the bioparticles in the fluidized bed, which were carried to the center of the bioreactor and, consequently, raising the height of the fluidized bed.

Despite the influence of the lift force on the fluidization height results, the inclusion of this force was not enough to provide a good agreement between the simulated values and the experimental data.

Drag models evaluation

The Ergun, Gidaspow, Syamlal-O'Brien and Wen-Yu drag models were considered for the evaluation of the drag force. In all cases, the Legendre-Magnaudet model was included for the prediction of the lift force, due to its significant influence in the bed height prediction.

As can be seen in Figure 4, the Syamlal-O'Brien model provided slightly larger fluidizated bed heights than the other models. A possible justification for this variation would be the consideration of a group of particles for the calculations made in this model, as well as its estimative for the relative velocity between the particle and the surrounding particles group.

For the other models, Ergun was originally developed to estimate the behavior of fluidized beds, with particle concentration above 20% [6]. The Wen-Yu model has been extended to dilute beds with porosity above 80% and considers a single isolated particle [6]. Gidaspow established his model from the Ergun and Wen-Yu models, which considers each model according to its dilution range [5]. Further, Gidaspow made a correction to consider the effect of a set of particles in the fluid and act as a correction for Stokes's law for the terminal velocity of a single particle [6]. Thus, the Figure 4 shows that the fluidized bed is in dense regime, since the Gidaspow model estimated fluidization height results very close to the values obtained with the Ergun model. Wen-Yu's model was in a range between the Gidaspow / Ergun and Syamlal-O'Brien models.

From the results obtained it was possible to observe that none of the models accurately predicts the fluidization of bioparticles. The Syamlal-O'Brien model presented a more approximate prediction regarding the experimental data, with deviations between 27.11 and 30.01%.

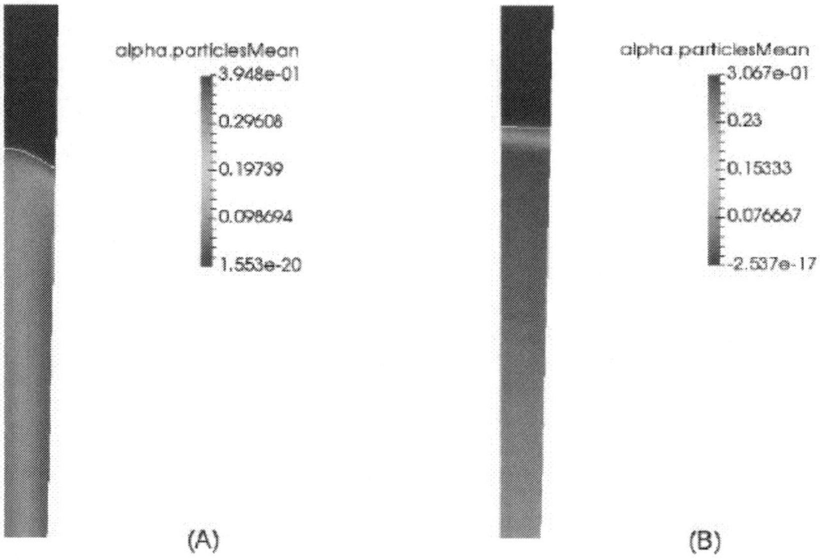

Figure 3: Bioparticles concentration (a) disregarding and (b) considering the lift force.

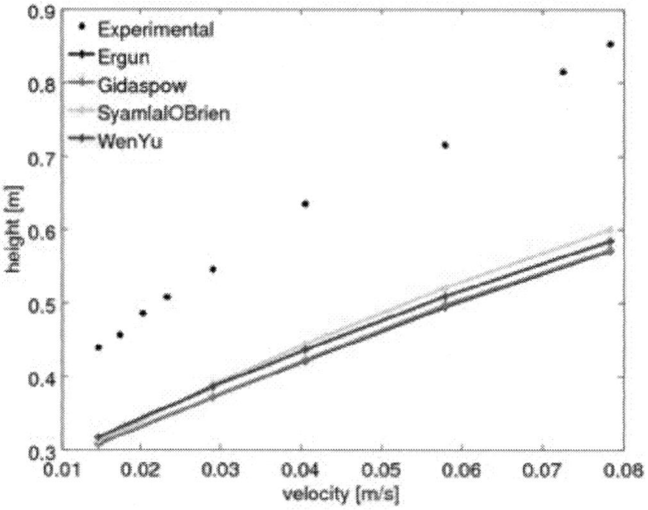

Figure 4: Fluidized bed height obtained using various drag models.

The Ergun and Gidaspow models exhibited a behavior very close to each other, and farther in relation to the experimental data, deviations between 30.51 and 33.63%.

Conclusion

In this study, a liquid-solid fluidized bed bioreactor was numerically evaluated. The influence of interfacial forces on the bioreactor prediction was first verified, comparing the results obtained with experimental values from the literature. It was observed that the lift force is important in the prediction of this system. Nevertheless, its inclusion was not sufficient to provide quantitative agreement with the fluidization curve obtained experimentally. Regarding the drag models evaluated, developed for gas-solid systems, these were found to be unsuitable for the liquid-solid system under the conditions evaluated in the study. Thus, it is emphasized the necessity to adapt the drag models available, for their use in liquid-solid fluidized beds such as those present in bioreactors.

Acknowledgements

The authors thank the National Council for Scientific and Technological Development (CNPq) for financial assistance, through project number 830795/1999-5.

References

[1] Arastoopour, H. (2001), "Numerical simulation and experimental analysis of gas-solid flow systems: 1999 Fluor-Daniel Plenary lecture," *Powder Technology*, v. 119, p. 59-67.

[2] Cornelissen, J. T.; Taghipoura, F.; Escudié, R.; Ellis, N.; Grace, J. R. (2007), "CFD modelling of a liquid-solid fluidized bed," *Chemical Engineering Science*, v. 62, p. 6334-6348.

[3] Crowe, C. T. (2000), "On models for turbulence modulation in fluid-particle flows," *International Journal of Multiphase Flow*, v. 26, p. 719-727.

[4] Fortuna, A. O. (2000), "Técnicas computacionais para dinâmica dos fluidos: conceitos básicos e aplicações," São Paulo: EDUSP.

[5] Gidaspow, D.; Seo, Y. C.; Ettehadieh, B. (1983), "Hydrodynamics of fluidization: Experimental and theoretical bubble sizes in a two-dimensional bed with a jet," *Chemical Engineering Communications*, v. 22, n. 5-6, p. 253-272.

[6] Gidaspow, D. (1994), "Multiphase flow and fluidization: Continuum and kinetic theory description," San Diego: Academic Press.

[7] Grady Jr, C. P. L.; Daigger, G. T.; Love, N. G.; Filipe, C. D. M. (2011), "Biological wastewater treatment," London: CRC press, IWA Publishing.

[8] Hermanowicz, S. W.; Ganczarczyk, J. J. (1983), "Some fluidization characteristics of biological beds," *Biotechnology and Bioengineering*, v. 25, n. 5, p. 1321-1330.

[9] Huang, J. S.; Wu, C. S. (1996), "Specifc energy dissipation rate for fluidized-bed bioreactors," *Biotechnology and Bioengineering*, v. 50, p. 643-654.

[10] Khan, M. J. H.; Hussain, M. A.; Mansourpour, Z.; Mostoufi, N.; Ghasem, N. M.; Abdullah, E. C. (2014), "CFD simulation of fluidized bed reactors for polyolefin production – A review," *Journal of Industrial and Engineering Chemistry*, v. 20, n. 6, p. 3919-3946.

[11] Koerich, D. M.; Rosa, L. M. (2016), "Numerical evaluation of the low Reynolds turbulent flow behaviour in a bioreactor," *International Journal of Simulation and Process Modelling*, v. 11, p. 66-75.

[12] Koerich, D. M.; Rosa, L. M. (2017), "Optimization of bioreactor operating conditions using computational fluid dynamics techniques," *Canadian Journal of Chemical Engineering*, v. 95, n. 1, p. 199-204.

[13] Launder, B. E.; Spalding, D. B. (1974), "The numerical computation of turbulent flows," *Computer methods in applied mechanics and engineering*, v. 3, n. 2, p. 269-289.

[14] Legendre, D.; Magnaudet, J. (1997), "A note on the lift force on a bubble or a drop in a low-Reynolds-number shear flow," *Physics of Fluids*, v. 9, p. 3572-3574.

[15] Legendre, D.; Magnaudet, J. (1998), "The lift force on a spherical bubble in a viscous linear shear flow," *Journal of Fluid Mechanics*, v. 368, p. 81-126.

[16] Maurina, G. Z.; Rosa, L. M.; Beal, L. L.; Baldasso, C.; Gimenez, J. R.; Torres, A. P.; Sousa, M. P. (2014), "Effect of Internal Recirculation Velocity in an Anaerobic Sequencing Batch Reactor (ASBR)," *Brazilian Journal of Chemical Engineering*, v. 31, n. 4, p. 895-903.

[17] Pang, M.; Wei, J.; Yu, B. (2010), "Numerical study of bubbly upflows in a vertical channel using the Euler–Lagrange two-way model," *Chemical Engineering Science*, v. 65, p. 6215-6228.

[18] Pang, M.; Wei, J. (2011), "Analysis of drag and lift coefficient expressions of bubbly flow system for low to medium Reynolds number," *Nuclear Engineering and Design*, v. 241, p. 2204-2213.

[19] Pourtousi, M.; Sahu, J. N.; Ganesan, P. (2014), "Effect of interfacial forces and turbulence models on predicting flow pattern inside the bubble column," *Chemical Engineering and Processing*, v. 75, p. 38-47.

[20] Rao, A.; Curtis, J. S.; Hancock, B. C.; Wassgreen, C. (2012), "Numerical Simulation of Dilute Turbulent Gas-Particle Flow with Turbulence Modulation," *AIChE Journal*, v. 58, n. 5, p. 1381-1396.

[21] Roache, P. J. (1997), "Quantification of uncertainty in computational fluid dynamics," *Annual Review in Fluid Mechanics*, v. 29, p. 123-60.

[22] Rosa, L. M.; Koerich, D. M.; Della Giustina, S. V. (2017), "The use of CFD in design and optimization of wastewater treatment units: A review," In: Organic waste: Management strategies, environmental impact and emerging technologies. Ed. Mark Collins, New York: Nova Science Publishers.

[23] Wang, X.; Ding, J.; Guo, W. Q.; Ren, N. Q. (2010), "Scale-up and optimization of biohydrogen production reactor from laboratory-scale to industrial-scale on the basis of computational fluid dynamics simulation," *International Journal of Hydrogen Energy*, v. 35, p. 10960-10966.

[24] Wu, C. S.; Huang, J. S.; Reiko, O. (2009), "Hydrodynamics of tapered anaerobic fluidized beds for metabolic gas production," *Chemical Engineering Journal*, v. 148, p. 279-289.

[25] Yang, W. C. (2003) "Handbook of Fluidization and Fluid-Particle Systems," New York: Marcel-Dekker, Inc.

[26] Yu, L.; Ma, J.; Frear, C.; Zhao, Q.; Dillon, R.; Li, X.; Chen, S. (2013), "Multiphase modeling of settling and suspension in anaerobic digester," *Applied energy*, v. 111, p. 28-39.

500491 Experimentally Validated Discrete Element Model to Predict Pharmaceutical Powder Flow at Different Humidity Conditions

Abstract:

Pharmaceutical powder flows are altered significantly based on the moisture contents and exposed humidity conditions. A Discrete Element Model (DEM) has been developed to predict the effects of humidity on powder flow, altering the cohesive forces based on granular bond numbers in three different hopper geometries. Experiments similar to the simulations are further performed for different pharmaceutical excipients at three different humidity conditions. To predict cohesive properties for powders, those did not flow under gravity; vertical vibration is introduced in both experiments and simulations. The equivalent DEM bond numbers are in good accordance with the experimental results. Pharmaceutical powders with lower moisture sorption show more variability in flow properties in comparison to cohesive powders having higher moisture contents. The equivalent DEM parameters suggest the same and are presently being implemented in a ring shear cell tester model to further quantify the flow variability at different humidity conditions.

500897 Heat Transfer in a Rotary Drum Using Experiments and Simulations

Abstract:

Heat transfer in a rotary drum using experiments and simulations.

Manogna Adepu, Shaohua Chen, Yang Jiao, Aytekin Gel, Heather Emady

Arizona State University, Tempe, AZ 85287

Granular material subjected to agitation is encountered in many practical applications of material processing (e.g. cooling, heating, granulation, and clinkering). Furthermore, many of these applications also involve heat transfer whereby solids face cold or hot surfaces and heat is exchanged not only between individual particles but also between the particles and external surfaces via the particle–particle or particle–surface contact. This work focuses on understanding the heat transfer mechanisms in the granular bed inside a rotary drum, one of the most commonly used process equipment. Experiments are performed using 4 mm diameter silica particles to investigate the granular flow and heat transfer mechanism inside a 3-inch radius and 3-inch long stainless-steel rotary drum. The drum rides on two titanium wheels, precluding the direct contact of the drum wall with the rollers. One side of the drum is closed with a transparent quartz window, which can handle high temperatures. Another side of the drum is closed with a sapphire window, compatible with an IR camera, specifically chosen to give a high transmittance to the infrared light. The setup is designed to handle up to 1000^0 C, making it possible to study all modes of heat transfer via conduction, convection, and radiation. The wall of the drum is maintained at a desired high temperature using heat guns placed around the drum. A drum tumbler is used to a rotate the drum and the temperature evolution in the particle bed is recorded using Omega thermocouples and a mid-wave infrared (MWIR) camera. All the modes of heat transfer are quantified under varying operation conditions to establish a strong understanding of the heat transport. For this, both experiments and CFD-DEM (using MFIX-DEM, an open source multi-solver suite) simulation techniques are used to analyze the thermal behavior. The digital image analysis (DIA) and the particle image velocimetry (PIV), are combined with IR thermography (DIA/PIV/IR technology [1]) to obtain comprehensive quantitative hydrodynamic and thermal data sets.

Reference:

[1]. Patil, A. V, Peters, E. A. J. F., Sutkar, V. S., Deen, N. G., & Kuipers, J. A. M. (2015). A study of heat transfer in fluidized beds using an integrated DIA / PIV / IR technique. Chemical Engineering Journal, 259, 90–106.

501591 Single Drop Impact on Heterogeneous Powder Beds

Abstract:

Wet granulation is a process by which agglomerates are formed using liquid to bind particles together. It is important in particulate applications like pharmaceuticals, detergents, agricultural chemicals and catalysts, with the purpose to achieve particle size enlargement, create dosage uniformity and improve the powder's flow properties. Specifically, the interaction between liquid and heterogeneous mixtures, where the powders have different wettabilities, has a lot of relevance in the pharmaceutical industry, where active ingredients are often relatively hydrophobic. Within granulation, single drop granule formation is of large interest, by revealing the fundamentals of the interaction between liquid and powders. Parameters like drop penetration time and granule formation mechanisms have been studied in single drop granulation to better quantify the drop impact and subsequent granule formation process. Studying the granule morphology and content uniformity can also help understanding the structure and property of granules, thus improving their quality. Therefore, it is novel to combine drop penetration time, granule formation mechanism, and granule characterization together to achieve a comprehensive map of the granulation process.

In this work, single drop impact of liquid on a static powder bed was studied to investigate the granule formation mechanism, droplet penetration time, granule content uniformity, as well as the characterization of granules (morphology and internal structure). Water was used as the liquid binder and two pharmaceutical powders, microcrystalline cellulose (MCC) and acetaminophen (APAP), were mixed to make heterogeneous powder beds of varying compositions. The complete drop impact and penetration was recorded with a high-speed camera. Three granule formation mechanisms identified previously [1] have occurred: Spreading, Tunneling, and Crater formation. At a low impact height, Spreading occurred for mixtures with an APAP amount of less than 25%, while there is a transitional regime from Spreading to Tunneling when the APAP amount increased from 25% to 50%. Tunneling dominates from 50% to 100% of APAP proportion. Crater formation occurred at a high impact height for the materials that exhibit Spreading at low impact height. The granule morphology and the internal structure were characterized by a prism method with image analysis [1] and micro-CT, respectively. The content uniformity was characterized via UV-vis spectrometry by measuring the UV absorbance of the APAP dissolved in ethanol. It is believed that the mean particle size of the powder bed is the predominant factor in influencing the granule formation mechanism, drop penetration time, and granule properties.

[1] Emady, H. N., Kayrak-Talay, D., Schwerin, W. C., & Litster, J. D. (2011). Granule formation mechanisms and morphology from single drop impact on powder beds. Powder technology, 212(1), 69-79

Drag Coefficients of Irregularly Shaped Particles

Fanhao Deng, **Xiaoling Chen** , Yongxing Zhang

Beijing Key Laboratory of Process Fluid Filtration and Separation, China University of Petroleum - Beijing, Beijing, China

Introduction

The Gas-solid two-phase flow is widely used in energy, environmental protection, chemical industry and other fields. Particles, both in physical and chemical processes of gas-solid two-phase flow, are non spherical in most case. The drag coefficient is one of the core parameters to reveal the interaction between particles and fluid mechanism. Thus, it has a very important significance to study non spherical particle drag coefficient for the development of new gas solid reaction process and improving energy efficiency. Most of the non spherical particle experiments are carried out in liquid medium, actually particles in many cases are moving in the gas medium. It will cause large error if apply the drag force characteristics of particles in the liquid medium to gas medium. In addition, it is difficult to accurately describe the irregular non spherical particles by single shape descriptors. Considering these problems, this paper use three shape descriptors to describe the shape of the irregular non spherical particles, and conduct experiment in gas medium to get the influence law of the shape of irregular non spherical particles to drag coefficient.

Research Methods

Image process method was employed to obtain particle shape descriptors using MATLAB. Figure 1 is an original image of iron ore particle, and figure 2 is the image of particle after processing. Three particle shape descriptors, including aspect ratio, polygonality and circularity, are obtained from the image processing. The major chord was defined as the maximum displacement between two pixels within the defined region, and the minor chord was defined as the line perpendicular to the major chord. The aspect ratio was simply defined as the ratio of the major chord to the minor chord. The polygonality of the particles was obtained by calculating the smallest bounding polygon with a set number of verticals. The circularity was defined as Square root of the ratio of largest inscribed circle diameter to smallest circumscribed circle diameter[2].

Figure 1 Original image of particle

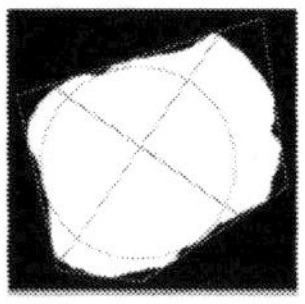

Figure 2 The image of particle after processing

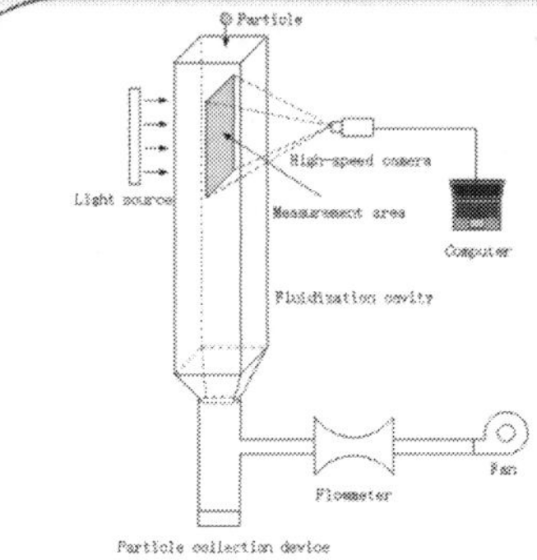

Figure 3 Schematic diagram of experimental device[1]

The experimental platform was set up to measure drag coefficients of irregular non spherical particles. The schematic diagram of experiment device was shown in figure 3. The falling process of irregular non spherical particles in a uniform high-speed turbulent flow was recorded by high-speed camera. Then, the coordinates of the centroid, instantaneous velocity, acceleration, frontal area, drag coefficient of the particles were calculated through MATLAB programming. The influence rules of particles shape and air speed on drag coefficient were studied by changing experimental conditions.

Results

In order to verify the feasibility and reliability of the scheme of the experiment, the drag coefficients for both spherical and non spherical particles determined from this test rig were compared with the data which are available in literatures, and the results as shown in figure 4 and figure 5. The maximum relative error of experimental value and standard value can be controlled within 10% in figure 4 and within 14% in figure 5,so the experimental values are in good agreement with the standard values.

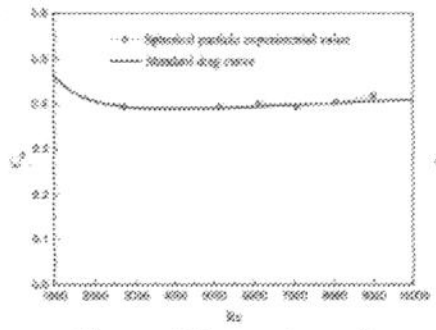

Figure 4 Comparison of experimental values and the standard spherical particle drag coefficient curve

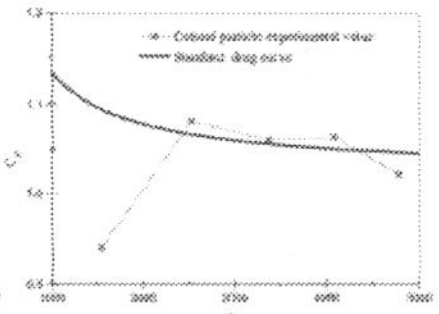

Figure 5 Comparison of experimental values and the standard cuboids particle drag coefficient curve

From figures 6 to figures 8, it is conclude that the three shape descriptor of particles have a great influence on drag coefficient. With the increase of the polygonality, the drag coefficient decreases first and then increases. With the increase of the aspect ratio, the drag coefficient decreases first and then increases. With the increase of the circularity, the drag coefficient decreases first and then increases.

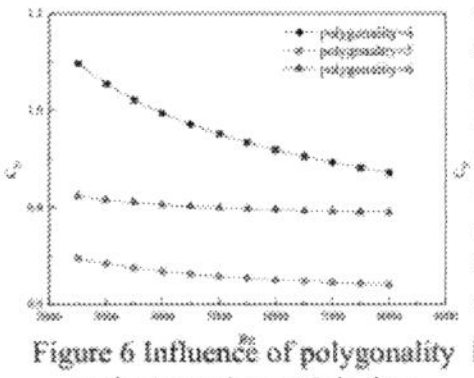

Figure 6 Influence of polygonality on the irregular particle drag coefficient

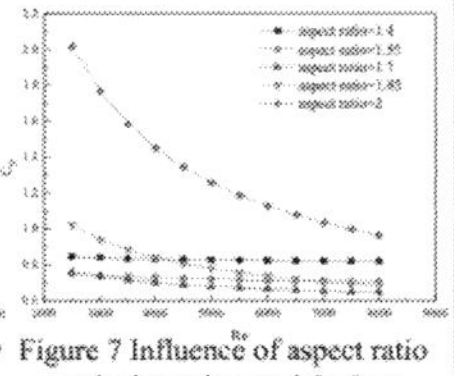

Figure 7 Influence of aspect ratio on the irregular particle drag coefficient

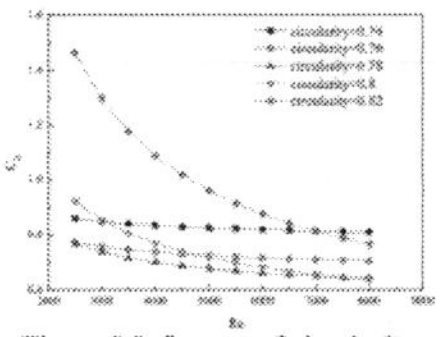

Figure 8 Influence of circularity on the irregular particle drag coefficient

The Empirical formula that I obtained was

$$\begin{cases} C_D = \frac{24}{\mathrm{Re}}\left(1 + C_1 \mathrm{Re}^{C_2}\right) + \frac{C_3}{1 + C_4/\mathrm{Re}} \\ C_1 = 3.406 P^{4.54} A^{16.09} C^{43.781} \\ C_2 = 2.795 P^{-2.976} A^{-5.847} C^{-18.484} \\ C_3 = 0.13 P^{6.356} A^{-1.603} C^{-5.792} \\ C_4 = -3.68*10^{-5} P^{-9.578} A^{3.009} C^{-111.248} \end{cases}$$

The range of factors are $2300<\mathrm{Re}<8700, 1.1<A<2.2, 0.7<C<0.85, P=4,5,6$.

Acknowledgements

The authors would like to acknowledge the financial support from National Natural Science Foundation (No.51504273) and Science Foundation of China University of Petroleum, Beijing (No.2462014YJRC 014)

References

[1] Ren B, Zhong W, Jin B, et al. Study on the Drag of a Cylinder-Shaped Particle in Steady Upward Gas Flow[J]. Industrial & Engineering Chemistry Research, 2011, 50(12):7593-7600.

[2] Williams K C, Chen W, Weeger S, et al. Particle shape characterisation and its application to discrete element modelling[J]. PARTICUOLOGY, 2014, 12(1):80-89.

Influence of flow hydrodynamics on pyridine synthesis reaction

Shuaishuai Zhou, Mengxi Liu, Chunxi Lu[*]
(College of Chemical Engineering, China University of Petroleum-Beijing, Beijing, China)

Abstract

In order to choose a proper reactor type for large scale pyridine synthesis, turbulent fluidized bed, fast fluidized bed and transport fluidized bed reactor are used. Based on experiment result, a model combines flow hydrodynamics and reaction kinetics is proposed to investigate the average catalysts activity by using different reactor types. The results show that the average catalysts activity is determined by catalysts circulation rates and superficial gas velocity. The average catalysts activity within experiment operating conditions are 1, 0.71 and 0.36 for transport fluidized bed reactor, fast fluidized bed reactor and turbulent fluidized bed reactor respectively. Empirical correlations for catalysts average activity prediction are obtained.

Introduction

Pyridine bases have been widely used to synthesis pharma and agrochemicals for its high chemical reactivity. Traditionally, pyridine bases were isolated from coal; the produced pyridine bases had high sulfur content, which means low quality. In 1920s, large demand of pyridine bases encouraged development of synthesis production technologies. Until now, the most successful synthesis method is the one using ammonia and aldehydes as reactants over ZSM-5 catalysts. Currently, research work mainly focus on catalysts optimization[1], catalysts deactivation[2], reaction mechanisms[3], reaction kinetics[4] for pyridine synthesis, while few address the influence of reactor types on pyridine synthesis from engineering aspect. Previous studies show that the fluidized bed reactor is better than fixed bed reactor for pyridine synthesis and the produce yield increased by 50% by using fluidized bed reactor[5]. For fluidized bed reactor, the flow patterns inside the reactor can range from turbulent fluidized bed to transport fluidized bed by changing the superficial gas velocity. No research is found to investigate the different flow patterns on pyridine synthesis reaction. The reaction rate is determined by the catalysts average activity and catalysts concentration. For turbulent fluidized bed, the catalysts concentration is high but the average catalysts activity is low owing to heavy back-mixing, while, for transport fluidized bed, the catalysts concentration is low but the average catalysts activity is high because of weak back-mixing[6]. The key is to find the balance between the catalysts concentration and the average catalysts when it comes to choosing the proper type reactors for pyridine synthesis.

Experiments

Experimental setup

The experimental setup is shown in Figure 1. To cover the turbulent fluidized bed flow pattern to transport fluidized bed flow pattern, two sets of experiment apparatus are used. The first one basically composed by a column with internal diameter of 0.014 m and 3.5 m in height, the second one basically composed by a prelift zone, riser reaction zone (0.084 m in internal diameter and 13.5 m in height), and the condenser.

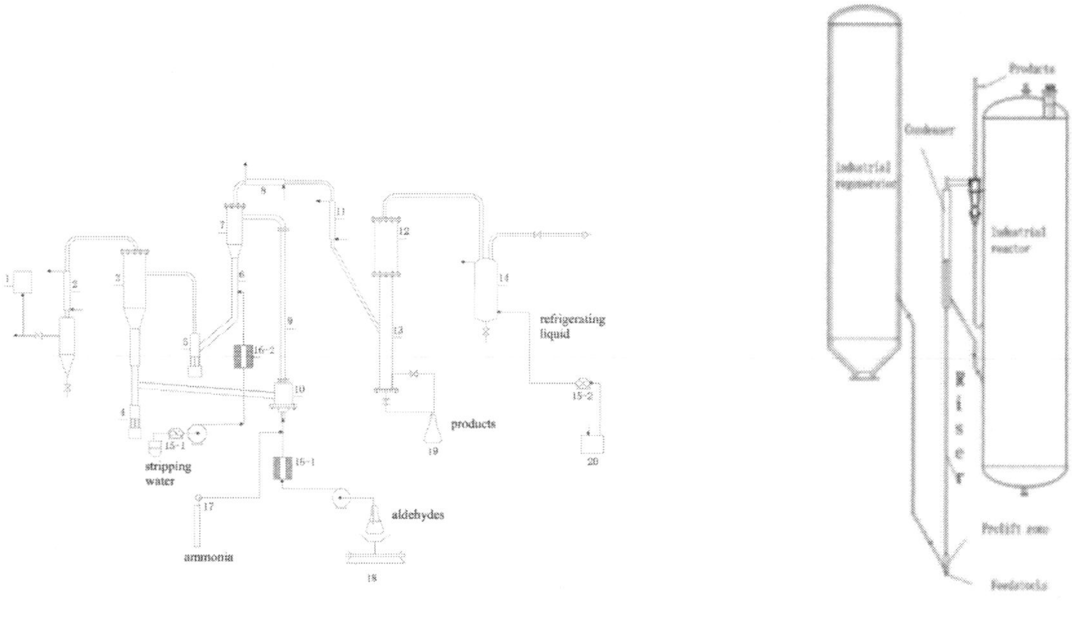

1. carbon dioxide detector; 2,8,11. condenser; 3. regenerator; 4,5. valve; 6. stripper; 7. separator; 9. riser; 10. bottom of riser; 12. pipe condenser; 13. gas-liquid separator; 14. liquid separation tower; 15-1~2. piston pump; 16-1~2. heater; 17. glass rotameter; 18. electronic balance; 19. product collector; 20. refrigerator;

Figure 1. Schematic diagram of experimental setup

Operating conditions

In this experiment, the reactants are acetaldehydes, formaldehydes and ammonia, the catalysts physical properties are shown in Table 1. The operating conditions in the experiment are given in Table 2.

Table 1. Properties of the used Catalysts

Properties	Value
Particle density (kg/m^3)	1300
Bulk density (kg/m^3)	930
Mean Particle size (μm)	70
Particle size range (μm)	30~90

Table 2. Operating conditions

Fast fluidized bed operating parameters variation range	
Parameters	Variation range
C_{A0} (mol/L)	0.0053~0.0056
u_g (m/s)	0.7~1.8
T_r (K)	723~798
P_r (kpa)	180~190
G_s (kg/m²/s)	11.9~22.1
Turbulent fluidized bed operating parameters variation range	
Parameters	Variation range
C_{A0} (mol/L)	0.0041~0.0056
u_g (m/s)	0.3~0.6
T_r (K)	723~798
P_r (kpa)	180~190
G_s (kg/m²/s)	11.9~24.2
Transport fluidized bed operating parameters variation range	
Parameters	Variation range
C_{A0} (mol/L)	0.0030~0.0056
u_g (m/s)	4.0~5.2
T_r (K)	711~757
P_r (kpa)	180~190
G_s (kg/m²/s)	10.0~79.0

Where, C_{A0} – the initial acetaldehydes concentration, mol/L;
u_g – the superficial gas velocity, m/s;
T_r – the reaction temperature, K;
P_r – the reaction pressure, kpa;
G_s – the catalysts circulation rates, kg/m²/s;

Experimental method

The catalysts circulation rates are calculated based on heat balance. The products are analyzed by gas chromatography ().

Experimental Results

Pyridine yield under different operating conditions

Table 3 is the produce yield under different operating conditions. Data 1 to 6 is the result in fast fluidized bed reactor, data 7-11 is the result in turbulent fluidized bed reactor and data 12 to 17 is the result from transport fluidized bed reactor. From the experiment results, it is hard to tell which flow pattern employs higher catalysts average activity. In that case, modeling analysis is required.

Table 3 Produce yield under different operating conditions

	C_{A0} (mol/L)	u_g (m/s)	T_r (K)	G_s (kg/m²/s)	y_P	y_{3P}	y_P+y_{3P}	y_c
1	0.0052	0.74	798	20.6	0.34	0.09	0.43	0.13
2	0.0052	0.74	798	22.1	0.41	0.11	0.52	0.15
3	0.0055	0.86	723	20.6	0.42	0.12	0.54	0.13
4	0.0054	0.88	748	20.6	0.41	0.14	0.54	0.13
5	0.0054	0.88	748	20.6	0.40	0.11	0.51	0.15
6	0.0054	0.88	748	20.6	0.44	0.14	0.58	0.16
7	0.0044	0.35	748	20.58	0.47	0.10	0.57	0.24
8	0.0044	0.35	748	20.58	0.42	0.09	0.52	0.26
9	0.0044	0.36	773	20.58	0.48	0.11	0.60	0.21
10	0.0044	0.36	773	20.58	0.46	0.10	0.56	0.24
11	0.0044	0.36	773	20.58	0.41	0.17	0.58	0.24
12	0.0054	4.31	728	17.89	0.38	0.13	0.51	
13	0.0054	4.35	736	19.15	0.33	0.12	0.45	
14	0.0052	4.43	742	20.37	0.29	0.11	0.40	
15	0.0052	4.65	748	22.73	0.28	0.12	0.40	
16	0.0051	5.34	757	28.72	0.39	0.14	0.53	

Modeling and Analysis

The modeling process illustrated here is for fast fluidized bed reactor, processes for turbulent fluidized bed reactor and transport fluidized bed reactor are similar.

Reaction Kinetics

There are few reports about the pyridine synthesis reaction kinetics. Reddy assumed that the pyridine synthesis reaction mechanism can be described by Figure 1, then based on micro reactor results and obtained the reaction kinetic parameters.

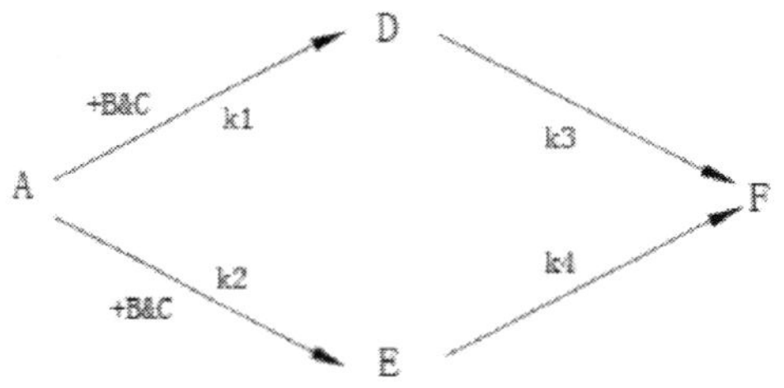

A: acetaldehyde B: ammonia C: formaldehyde D: pyridine E: 3-picoline F: coke

Figure 1. Reaction mechanism

As showed in Figure 1, the reactants aldehydes and ammonia convert into pyridine and 3-picoline at the same time with different rates, after that, the pyridine and 3-picoline convert into coke, the reactions are typical parallel-sequential reaction. The main reactions can be described by equations (1) ~ (2).

$$2CH_3CHO+HCHO+NH_3 \longrightarrow C_5H_5N+3H_2O+H_2 \quad (1)$$

$$2CH_3CHO+2HCHO+NH_3 \longrightarrow C_6H_7N+4H_2O \quad (2)$$

Equations (3) to (5) is the reaction rates expression given by Reddy.

$$-r_A = (k_1 + k_2)C_A^2 a \quad (3)$$

$$r_D = (k_1 C_A^2 - k_3 C_D)a \quad (4)$$

$$r_E = (k_2 C_A^2 - k_4 C_E)a \quad (5)$$

In this paper, the above reaction kinetics are adopted.

Solid phase flow hydrodynamics prediction model

The influence of solid phase flow hydrodynamics on reaction can be divided into two aspects. The first one is the catalysts concentration and the second one is the catalysts average activity. The model for catalysts concentration prediction can be classified into four categories according to its dimensions, which are zero dimension model, one dimension model, two dimensions model and three dimensions model. The three dimensions are axial direction, radial direction and circumferential direction. In this work, one dimension model is chosen to model the catalysts concentration at the following consideration.

1. Convenient to couple with reaction kinetics;
2. Take the catalysts concentration axial variation into consideration;

The one dimension models could be divided into two categories according to its theoretical bases: half empirical model and empirical model. Li-Kwauk combined experimental results and theoretical analysis and come up with the following equation to predict the catalysts concentration axial distribution in fast fluidized bed reactor[7]. However, the model parameters should be obtained by experiment, which limits its application.

$$\ln\left(\frac{\varepsilon - \varepsilon_a}{\varepsilon' - \varepsilon}\right) = -\frac{1}{Z_0}(z - z_i) \quad (6)$$

Bai assumed that the solid phase is pseudo-homogeneous and come up with the following equation to predict the catalysts concentration axial distribution in fast fluidized bed reactor[8]. The biggest problem for this model is that the drag coefficient is hard to predict.

$$\frac{d\bar{\varepsilon}}{dx} = \frac{\rho_p (1-\bar{\varepsilon})^2}{G_s^2}[F_D - (1-\bar{\varepsilon})\rho_p g - F_p] \quad (7)$$

Lu investigated large number of industrial fast fluidized bed data, and come up with the following empirical correlation to predict the catalysts concentration distribution in fast fluidized bed reactor. This empirical model has high prediction accuracy and easy to corporate with reaction kinetics. What's more, this model is obtained based on hot mode experiment data, and the predict result is trust worthy. In this paper, this model is chosen to predict the catalysts concentration axial distribution in fast fluidized bed reactor.

$$\rho = 1.923 u_g^{-0.915} h^{-0.347} G_s^{1.362} \tag{8}$$

Flow-Reaction model

In this paper the gas phase is assumed as plug flow. Based on that, the following equations are obtained to describe the Flow-Reaction model.

$$\frac{dC_{A0}}{dh} = \frac{(k_1 + k_2) C_A^2 \rho(h)}{u_g(h)} a \tag{9}$$

$$\frac{dC_{D0}}{dh} = \frac{(k_1 C_A^2 - k_3 C_D) \rho(h)}{u_g(h)} a \tag{10}$$

$$\frac{dC_{E0}}{dh} = \frac{(k_2 C_A^2 - k_4 C_E) \rho(h)}{u_g(h)} a \tag{11}$$

The catalysts average activity

The average catalysts activity is obtained by the following method. Since there is no report regarding the average catalysts activity in different flow patterns, in this paper, the catalysts average activity is first assumed to be 1, the corresponding prediction yield noted as y_i. Then the average catalysts activity is obtained by the following equation.

$$a_0 = y_e / y_i \tag{12}$$

Where, a_0– the average catalysts activity obtained by experiment;

y_e – the experimental pyridine yield + 3-picoline yield;

y_i – the predicted pyridine yield +3-picoline yield by assuming the catalysts average activity as unity.

For micro reactor the catalyst average activity can be taken as one, while for large scale reactor, the catalysts average activity is less than 1 owing to back mixing. The operating parameters that determine the backmixing are superficial gas velocity and catalysts circulation rates. In this paper, the relationship between catalysts average activity and operating parameters is given by equation (13).

$$a_0 = b_1 G_s^{b_2} u_g^{b_3} \tag{13}$$

By regression, the model parameters are obtained, b_1=0.14, b_2=0.586, b_3=0.129. The average catalysts average activity within our experiment operating conditions for fast fluidized bed reactor is 0.71. Similarly, for turbulent fluidized bed reactor, the model parameter are: b_1=0.10, b_2=0.476, b_3=0.215. The average catalysts average activity within our experiment operating conditions for turbulent fluidized bed reactor is 0.35. From equations (9) to (11), to have the same reaction rate for fast fluidized bed reactor and turbulent fluidized bed reactor, the catalysts concentration and the catalysts average activity should agree with the following equation.

$$\frac{\rho_{tb}}{\rho_{fb}} = \frac{0.14G_s^{0.586}u_{g_f}^{\ \ 0.129}}{0.10G_s^{0.476}u_{gt}^{\ \ 0.215}} \tag{14}$$

By solving equation (14), equation (15) is obtained.

$$\rho_{fb} = \frac{5u_{g_t}^{\ \ 0.215}G_s^{-0.110}\rho_{tb}}{7u_{gf}^{\ \ 0.129}} \tag{15}$$

By similar method, the catalysts average activity for transport fluidized bed reactor is near 1. The following equation is proposed to compute the corresponding catalysts concentration in transport fluidized bed reactor to guarantee the same reaction rate as in turbulent fluidized bed reactor.

$$\frac{\rho_{tb}}{\rho_{cb}} = \frac{1}{0.10G_s^{0.476}u_{gt}^{\ \ 0.215}} \tag{16}$$

$$\rho_{cb} = 0.10\rho_{tb}G_s^{0.476}u_{gt}^{\ \ 0.215} \tag{17}$$

Conclusions

The pyridine synthesis were carried out in different reactor types, which are turbulent fluidized bed reactor, fast fluidized bed reactor and transport fluidized bed reactor. The average catalysts activity were compared based on experiment results and model analysis, the results showed that the average catalysts activity is determined by catalysts circulation rates and superficial gas velocity. The average catalysts activity within experiment operating conditions are 1, 0.71 and 0.35 for transport fluidized bed reactor, fast fluidized bed reactor and turbulent fluidized bed reactor respectively. To have a high reaction rates, both catalysts concentration and catalysts average activity should be improved.

References

1. Antonyraj CA, Kannan S (2008), "Hantzsch Pyridine Synthesis using Hydrotalcites or Hydrotalcite-like materials as solid base catalysts," *Applied Catalysts: A General*, 338. 121-129.
2. Screedhar I (2011), "Optimization of vapor phase pyridine synthesis hindered by rapid catalyst deactivation," *International Journal of Chemical Reactor Engineering,* 6. 77-90.
3. Zhang X, Luo C, Huang D, Li A, Liu J, Chao Z (2013), "Reaction Mechanism of Aldehydes and Ammonia to form Pyridine bases," *Chinese Journal of Chemical Engineering*, 64. 2875-2882.
4. Reddy K, Screedhar I, Raghavan K (2011), "Kinetic studies on vapour phase pyridine synthesis and catalyst regeneration studies," *The Canadian Journal of Chemical Engineering,* 89. 854-863.
5. USA. 2807618 (1957), Reilly Tar & Chemical Corporation, invs: E. C. Francis, R. W. William..
6. Li J, Weinstein H (1989), "An experimental comparison of gas backmixing in fluidized beds across the regime spectrum," *Chemical Engineering Science*, 44. 1697-1705.
7. Li Y, Kwauck M (1980), "The Dynamics of Fast Fluidization," Fluidization (Edited by Grace J and Matsen J, Plenum Press, New York), 537-544.

8. Bai D, Zhu J, Jin Y, Yu Z (1997), "Novel Designs and Simulations of FCC Riser regeneration," *Industrial & Engineering Chemistry Research*, 36. 4543-4548.
9. Lu C, Wang Z (1991), "Research of industrial fast fluidized bed reactor (In Chinese)", Petrochemical Technology, 20. 695-699.

491256 Fluidized Bed Rheology I - Fundamentals

Abstract:

Recently Anton Paar released the Powder Cell for MCR Rheometers , this device is capable of high performance measurements with large ranges of shear rate and shear stress, as well as oscillation measurements for the first time in gas-fluidized beds. Being able to measure extremely small torques (+- 10 nNm) in a reproducible manner enables a proper rheological study of fluidized beds in the style of complex fluids.

Unsurprisingly the behavior of fluidized beds mirrors that of complex fluids (e.g. suspensions) in some regards, while still showing a large influence of the particulate matter itself but also presenting wholly unique phenomena.

Direct measurements of apparent viscosity, defluidization/deaeration rates in rotational as well as oscillatory manner will be discussed. Comparisons of multiple conventional concentric cylinder geometries for apparent viscosity measurement system based on the Stokesian surface viscosity relation will be demonstrated and discussed.

In this talk we will present multiple results from multiple Geldart groups and grain sizes/distributions as well as explore the underlying physical phenomena that govern the behaviour of fluidized and collapsed beds especially between incipient and full fluidization.

494503 Investigation of the Agglomeration Behaviors in Gas-Solid Fluidized Beds with Side-Wall Liquid Injecting

Abstract:

Investigation of the agglomeration behaviors in gas-solid fluidized beds with side-wall liquid injecting

Qiang Shi, Shaoshuo Li, Sihang Tian, Zhengliang Huang, Jingdai Wang, Yongrong Yang

State Key Laboratory of Chemical Engineering, Department of Chemical and Biological Engineering, Zhejiang University, Hangzhou, 310027, China

Abstract: Agglomeration exists in the wide range applications of fluidized beds, such as fluidized catalytic cracking, drying, solid fuel conversion, direct reduction of iron ore, biomass conversion, coal combustion and gas phase polymer production. The understanding of agglomeration mechanisms is critical to effectively control its evolution process. In gas phase olefin polymerization operated under condensed mode, liquid evaporation enhanced the heat removal efficiency, thereby the productivity. However, after being introduced to the bed, liquid adheres to the particles and forms liquid bridges among them which results in liquid bridge force. Meanwhile, as the polymerization is a highly exothermic process, particles soften if the surface temperature is high. Chain entanglements among particles form solid bridge, thereby the solid bridge force. Thus, under the actions of the two different strong interparticle forces, how does the bed hydrodynamics change ? And whatâ€™s the mechanism of particle agglomeration ? We have to get into insight into it so as to guide the operation of industrial fluidized bed reactors.

In order to mimic the exothermic process in an olefin polymerization fluidized bed reactor, we firstly designed a novel cold-mode fluidized bed based on induction heating method. Graphite particles coating with polyethylene wax on their surface were used as fluidized particles. Eddy currents formed on the surface of graphite particles generated heat and then heated up the particles. Polyethylene wax became sticky and formed solid bridge among particles after being heated. Liquid was injected into the bed through the side wall nozzle, and then adhered to the particles and formed liquid bridge among them. In this way, liquid bridge force and solid bridge force simultaneously existed in the fluidized bed. On this basis, we adjusted the liquid injecting rates so as to change the bed temperature and liquid contents among particles to investigate the dominant mechanisms of agglomeration under different interaction intensities of solid bridge force and liquid bridge force. Solid agglomerates (caused by solid-bridge force) sampled after experiments and liquid agglomerates (caused by liquid bridge force) deduced from bed pressure drop were used as characteristic variables to validate the mechanistic assumptions. Results showed that the interaction intensity of liquid bridge force enhanced with increasing liquid injecting rate, and more liquid agglomerates were formed till bed defluidized. Under low liquid injecting rate, liquid evaporation process dominated, leading to the decrease of bed temperature. The solid bridge force was weakened and the amount of solid agglomerates decreased. In this condition, gas came from liquid evaporation resulted in the increase of bubble diameter which was consistent with the variations of bed pressure fluctuations and standard deviations. However, further increasing liquid injecting rate, the enhancement of liquid bridge force and the formation of liquid agglomerates lead to the decrease of bubble diameter and the deterioration of bed fluidization quality, as agglomerates segregated in the distributor and formed dead zones. Heat accumulated in dead zones caused the overheating of particles surface and solid bridges were formed among them, so the amount of solid agglomerates increased. That is to say, dead zones caused by the segregation of liquid agglomerates created the acting environment for solid bridge force. Therefore, with increasing liquid injecting rate, the dominant mechanisms of agglomeration can be divided into three stages: solid bridge force dominated, dynamic equilibrium of liquid evaporation and liquid bridging, cooperated control of liquid bridge force and solid bridge force. Besides, we also investigated the effects of process variables and found that increasing heat releasing rate and the distance of the nozzle from distributor, the range of safe injecting rate of liquid widened.

Keywords: Induction heating; Liquid bridge force; Solid bridge force; Liquid evaporation; Cooperative control;

488037 Industrially Relevant Powder Characterisation Using a Uniaxial Powder Tester

Abstract:

During production, rapid and reliable evaluation of process intermediates is essential to ensure high efficiency and final product quality. It is therefore necessary that any powder characterisation methods utilised are simple and quick, and that the resulting data is easy to interpret. Whilst methods such as Angle of Repose (AOR) and Tapped Density can fulfil these criteria, they are often poorly repeatable, highly operator dependent and typically lack differentiation. These techniques may also be unsuitable for testing highly cohesive, and therefore typically problematic, powders. The recent development of an Advanced Uniaxial Powder Tester (AUPT) provides a tool with the ease and simplicity of the more traditional techniques with the added benefits of a robust, semi-automated and well defined operating procedure designed to deliver high levels of repeatability and significantly reduce operator variability.

The AUPT measures the uniaxial Unconfined Yield Stress (uUYS) - the stress required to fail a free-standing powder column which has been consolidated at a known Major Principal Stress (MPS). In addition to measuring uUYS, the AUPT also provides compressibility and bulk density evaluation. Adjustment of the MPS and/or consolidation time also enables assessment of the powder flow properties across a range of storage conditions.

In this study, 12 different powders representing a range of industries, including food and pharmaceutical materials, were evaluated using the AUPT. Each sample was tested at low, intermediate and high MPS values, with repeat measurements conducted for each data point. Clear and repeatable differences were demonstrated between the samples, especially when subjected to higher consolidation stresses. The highly repeatable compressibility, bulk density and uUYS values demonstrate that consistent consolidation was achieved at each MPS for all of the samples tested. Comparable results were also obtained between different users, demonstrating minimal influence of the operator on this technique.

A subset of the powders was also subjected to longer consolidation periods using off-line consolidation stations. These devices can be used to assess the impact of long storage periods, ranging from a few hours to several months, as well as the influence of elevated temperatures and relative humidity. This enables the effect of environmental conditions during long term storage of powders in IBCâ€™s, hoppers and silos to be accurately quantified. In this study, the samples were consolidated using a range of stress levels for time periods between 1 and 16 hours. Clear and repeatable differences were again observed between the samples with extended storage times shown to have a considerable impact on flowability of the powders.

The AUPT was therefore demonstrated to a reliable and sensitive instrument for both rapid evaluation and long term studies for a diverse range of materials from different applications, across a wide range of consolidation levels.

498179 Multiscale Modeling of Biomass Thermochemical Conversion in Fluidized Bed Reactors

Abstract:

Thermochemical conversion of biomass has recently gained a lot of interest as a potential renewable technology to provide both liquid and gaseous fuels from a variety of feedstock. However, design and scale-up of biomass thermochemical conversion reactors are mostly empirical processes, and require many pilot-scale reactor studies, which are both expensive and time consuming, and fail to provide the local and detailed quantitative information necessary to optimize the conversion process. Modeling and simulation tools are playing an increasingly important role for the design and optimization of the reactors, as they have the potential to provide detailed insight into the prevailing complex phenomena occurring during conversion. While significant progress has been achieved recently, predictive and reliable computational fluid dynamics (CFD) tools able to simulate the complex, reactive, multiphase flows found in fluidized-bed reactors are still lacking.

A major bottleneck in building predictive simulation capabilities for biomass thermochemical conversion is the adequate representation of the chemical processes, occurring both inside the biomass particle and in the gas phase. Particle-level chemical processes are coupled with intra-particle heat and mass transfer, occurring in a highly anisotropic porous structure. While intra-particle phenomena can affect the overall dynamics of the system, most CFD studies done at the reactor level neglect those and use a homogeneous sphere model to describe the biomass devolatilization. Secondary gas phase reactions are also most often neglected in reactor simulations, especially at the relatively low temperatures relevant for biomass pyrolysis processes.

This work aims at investigating the validity of those strong modeling assumptions, and characterizing quantitatively their impact on overall reactor performance. This is achieved by developing a one-dimensional intra-particle model for resolved simulation of single biomass particle devolatilization and a reduced model for the secondary gas phase reactions. The resulting models are integrated into a CFD solver specifically designed for the Lagrange-Euler simulations of multiphase flow. Simulations of pseudo-two-dimensional fluidized bed reactors are conducted with both the intra-particle and homogeneous particle models, and with and without secondary gas phase chemistry. A sensitivity analysis study of the results is performed to identify the controlling aspects.

494251 Reduced Order Particle-Scale Model for Biomass Pyrolysis and Gasification in Fluidized Bed Reactors

Abstract:

Thermochemical conversion of biomass is a viable route for the production of renewable fuels. Current technologies face challenges efficiency and scaling challenges, due in part to the lack of accurate process and reactor models. Computational Fluid Dynamics (CFD) can enable comprehensive simulations for optimizing reactor designs. One of the biggest challenges with reactive simulations, however, is the coupling of particle-scale phenomena with reactor-scale processes. Typically, global kinetic mechanisms are employed which are (a) feedstock specific and cannot be generalized and (b) do not capture key transient dynamics such as the evolution of primary pyrolysis.

This work is focused on (a) the fundamental investigation of biomass particle conversion and (b) the development of a reduced order model to enable robust coupling with reactor-scale processes. Towards this end, high-fidelity particle scale CFD simulations are conducted using bioSMOKE, developed at Politecnico di Milano. BioSMOKE captures the evolution and thermal degradation of arbitrarily-shaped woody biomass particles based on detailed kinetic mechanisms. Operating conditions are chosen typical to fixed and fluidized bed reactors: particle diameter in the range 0.1-5.0 cm, pyrolysis temperatures 800-1200 K and heat transfer coefficient 100-500 W/m^2-K. Using detailed statistics of the conversion phenomena, we show that the state of biomass and evolution of major volatile species (CO, CO_2, H_2, tar, etc.) can be parameterized using the mass fraction, Biot and pyrolysis numbers. This phase space is used to develop a reduced order model for biomass conversion which predicts the transient evolution of major species and char in excellent comparison with raw simulation data over the wide range of operating conditions investigated. The framework developed in this study allows for comprehensive characterization of particle scale phenomena within feasible computational time, without making any assumptions regarding the particle isotropy or reactivity whatsoever. The model can be easily incorporated into existing reactor-scale CFD simulations enabling robust simulations of fluidized and fixed bed reactors.

489677 Numerical Investigation of Solid Residence Time Distribution in a Circulating Fluidized Bed Riser for Catalytic Upgrading of Biomass Pyrolysis Vapors

Abstract:

Fast biomass pyrolysis is receiving more attention recently as it offers a promising technology to utilize renewable biomass resources for liquid transportation fuels. A circulating fluidized bed of catalyst material is commonly employed to upgrade the biomass pyrolysis vapor providing large throughput, excellent heat and mass transfer characteristics, and its controllable solid residence time distribution. The resident time distribution of solid catalyst is important to determine the product quality during the vapor phase upgrading process. Accurate prediction of catalyst residence time distribution is valuable for the design and scale up of reacting circulating fluidized bed systems. A multiscale Eulerian-Eulerian approach for modeling gas-solid flow in a circulating fluidized bed riser is used to model flow behavior and thus determine solids residence times. Several mesoscale drag models (e.g. filtered models) are employed and examined to consider their effect on particle clustering. Tracer particles are created from a labeled solid phase species and the solid residence time distribution is predicted by solving the solid phase transport equations for the labeled solids species. Subsequently, numerical predictions of pressure drop, solid volume fraction distribution, solid mass flux and solid resistance time distribution under different flow conditions are validated against of corresponding experimental data available in the literature. It is found that the flow hydrodynamics and solid mixing are well predicted.

490017 Thermal Desorption of Mercury from Coal Using a Draft Tube Spouted Bed

Abstract:

Thermal desorption of mercury from coal with a scalable, high throughput spouted bed desorber provides an effective method of pre-treatment for coal-fired power plants. A novel draft tube spouted bed desorber capable of processing a wide particle size distribution of pulverized coal was developed through sequential design iterations and a series of tracer visualization studies. Baseline experiments were performed for a conventional conical-cylindrical spouted bed and draft tube spouted bed. The mean particle residence time distribution and turnover rate were measured with variable particle sizes, fluidizing conditions, and reactor and draft tube geometries within a clear half-circular system. Effective heat and mass transfer rates were determined for a conventional spouted bed, straight draft tube, and modified draft tube geometry using surrogate pulverized coal. A self-contained, high-temperature spouted bed desorber test section was fabricated and demonstrated with bituminous, subbituminous, and lignite pulverized coal. Desorbed mercury species were continuously measured at temperatures between 150 °C to 350 °C, at several fluidizing gas flow rates. Several pulverized coal feeds with variable particle size distributions (between 10–500 μm) were fed into the spouted bed desorber under the range of operating conditions. The mercury thermal desorption rate for each coal type was determined as a function of particle size, temperature, and superficial gas velocity. Using the experimental data, process design and economics were evaluated for thermal desorption of mercury from a commercial scale coal-fired power plant.

490961 Particle Circulation Motions in an Evaporating Droplet

Abstract:

Particle deposition inside evaporating droplets has various applications including bioassays, spray drying, and microelectronics. In an evaporating particle-laden droplet, transport of particles influences the particle deposition pattern. Evaporative cooling of the droplets results in Marangoni currents near the droplet surface, and the Marangoni currents together with the internal edgeward flow form circulations inside the droplet. The direction of the Marangoni flow has been shown to change for various conditions. Hu and Larson (2005) found the direction of the Marangoni flow reverses at a critical contact angle. Ristenpart (2007) showed that the critical contact angle is related to the thermal conductivity of droplet and of the substrate respectively. Furthermore, Xu (2007) observed a stagnation point where the surface flow has opposite directions on either side. Recently, Barash (2015) showed that a multi-cellular flow exists inside the droplet by numerical simulation. In this work, we model the evaporating particle-laden droplet using the finite element method. Simulations show the variation of flow structures under various heating conditions, and can be used to predict deposition patterns.

486096 Fluidized Granular Activated Carbon (GAC) for the Mitigation of Membrane Fouling in Wastewater Treatment

Abstract:

The fluidization of Granular Activated Carbon (GAC) particles has received much attention in recent years as a promising method for mitigating membrane fouling particularly in membrane bioreactors (MBRs). In particular, the fluidized GAC particles are acknowledged to be beneficial in terms of adsorption of organic foulants, and mechanical scouring and surface shear of the membrane surface. Both lab-scale and pilot-scale studies have affirmed the efficacy of GAC fluidization in maintaining a depressed extent of membrane fouling in anaerobic fluidized bed - membrane bioreactors (AnFMBRs). More in-depth studies have also correlated the hydrodynamics of fluidized GAC particles with the effectiveness of membrane fouling mitigation. This study aimed at understanding the improvement of fine (micron-sized) particle critical flux in the presence of fluidized GAC via the Direct Observation Through the Membrane (DOTM) technique. The model foulant was a suspension of polystyrene particles (5 Î¼m sized), and the parameters investigated included GAC particle diameter (d_p), superficial liquid velocity (U_l) and thereby power requirement (P_r), and height along the vertically aligned membrane. Results indicate that: (i) fluidized GAC increased the critical flux by an order-of-magnitude relative to that with tangential liquid shear alone; (ii) the overall critical flux ($J_{critical,overall}$) expectedly increased with power input (P_r), but the relationship between local critical flux ($J_{critical}$) and power input (P_r) depended on the position along the membrane height; (iii) at the same P_r, although the larger GAC particles were more effective locally (i.e., higher $J_{critical}$) due to greater particle inertia, the smaller GAC particles were more effective overall (i.e., higher $J_{critical,overall}$) due to greater bed expansion enabling scouring of more membrane heights; and (iv) a higher power input (P_r) was required for sufficient bed expansion to enable more consistent critical flux ($J_{critical}$) values over the whole membrane height.

493340 Fluidized Bed Rheology II- Applications

Abstract:

Powder flow is a complex phenomenon and working with powders can be rather difficult when it comes to processing and storage. Powder handling is a common feature found in several industrial unit operations like mixing, storage, feeding, fluidization, compaction, blending etc. Most of these challenges are related to powder flow. It is very important to understand the flow of powders because most times it is this which dictates quality and efficiency in a powder process. Recently Anton Paar released the Powder Cell for MCR Rheometers, this device is capable of high performance measurements with large ranges of shear rate and shear stress, as well as oscillation measurements for the first time in gas-fluidized beds.

The closed nature of the cell makes even hazardous substances as well as sub-micron particles accessible in this fashion. Being able to measure extremely small torques (+- 10 nNm) in a reproducible manner enables a proper rheological study of fluidized beds in the style of complex fluids.

In this Talk we will show multiple applications of this technology in industrially relevant sector ranging from coating technology to fine chemicals.

Special attention will be given to the original application of fluidized bed: Fluid catalytic cracking and the possible repercussions for studying FCC Catalysts. Shear rate dependent apparent viscosity measurements correlated with pressure drop, oscillatory measurements and Elutriation rate for a variety of grain size distributions and compositions.

498935 Dry Mechanical Processing for Value-Enhanced Excipients

Abstract:

Dry mechanical processing for value-enhanced excipients

*Liang Chen, Xiaoyi Ding, Siqi Fan, Rajesh N. Davé**

Abstract

Excipients with good flowability, bulk density as well as compaction properties are desired for use in tableting since they play important roles in formulation development and processing, including, handling, mixing, feeding and compaction. However, it is often difficult to manufacture excipients that simultaneously possess all three desirable properties. In addition, most commercially available excipients that have excellent flow as well as compaction properties require multi-step processing involving use of solvents and subsequent drying, milling and sieving. The objective of this work is to examine the feasibility of using dry coating based surface modification of microcrystalline cellulose (MCC) using commercially available grades such as Avicel PH-101, PH-102 and PH-105, to produce engineered excipients having desired properties. Such an approach allows for preparing novel excipients without use of solvent and are expected to have lesser environmental footprint. Using a material sparing, model based dry coating methodology, as-received grades of MCC were dry coated with 1wt% silica R972P and Aerosil 200, along with selected cases using M-5P. The results indicated that as expected, the bulk density and flowability of dry coated MCCs were significantly improved, while there was an appreciable loss of compaction. To minimize the loss of compaction, attributed to decreased surface energy after coating, while maintaining improved bulk density and flowability, the effect of the type of silica was examined. Remarkably, use of Aerosil 200 led to significant improvements in bulk density and flowability with minimal reduction in compaction. The properties of the surface-engineered excipients were compared with as received and selected commercially available specialized pharmaceutical excipients. It was found that surface engineered excipients have a potential to achieve as good overall performance as any other materials in the market. Selected additional excipient materials are tested and overall, this work demonstrates feasibility for engineering excipients using dry processing instead of complex processes such as spray drying.

487823 Assessing the Impact of Dry Coating on the Surface Properties of Pharmaceutical Formulations

Abstract:

Dry coating of materials of pharmaceutical interest with nanopowders is recently gaining momentum as a promising formulation approach as it enables significant improvement of flowability under high relative humidity (RH) conditions, preventing, also, the caking upon storage under high RH.[1-3] Nevertheless, it remains a challenging and sophisticated process, where coating is, occasionally, accompanied by some undesirable phenomena.

This work proposes the use of vibrations for the efficient dry coating of materials of pharmaceutical interest with silica nanopowder.[4] The API or excipient of interest are placed in a jar, attached on a shaker, along with different loadings of silica nanopowder. The vibrations created by the shaker lead to the efficient coating. The coating efficiency is examined using SEM. The influence of particle habit and surface anisotropy on the efficiency of the coating was, also, investigated.

Inverse gas chromatography (IGC) was employed to provide an in depth understanding on the influence of coating on the surface properties of the particles. The surface energy of the coated particles, measured with IGC, decreases abruptly in the presence of relative humidity. For measurements conducted at RH>30 %, the surface energy measured was effectively the same. This suggests that the interactions were mainly driven by the presence of a water layer on the particles. Furthermore, IGC experiments were used to assess the roughness of the coated particles.[5] The measurements suggest that higher shaking times lead to a more rough area, associated with better flowability.

This study proposes an efficient, compact and moving parts-free dry coating process. It examines the influence of particle properties on dry coating and furthermore it employs vapour sorption tools to examine the roots of the improved behaviour of particles coated with nanopowders.

References

1. Yang J, Sliva A, Banerjee A, Dave RN, Pfeffer R. Dry particle coating for improving the flowability of cohesive powders. *Powder Technol.* **2005**, 158, 21-33.

2. Luo Y, Zhu J, Ma Y, Zhang H. Dry coating, a novel coating technology for solid pharmaceutical dosage forms. *Int J Pharm*. **2008**, 358, 16-22.

3. Zhang GZG, Ping T. Dry powder coating of silica nanoparticles on hygroscopic materials: Improving flow and reducing sensitivity to moisture. *Proceedings of 2013 AAPS Annual Meeting*. **2013**.

4. Polizzi M, Langdon BA, Beach LE, Mullarney MP. Applying dry powder coatings. *Pharmaceutical Technology*.

2011, 35, 94-102.

5. Burnett DJ, Heng JYY, Thielmann F, Garcia RA, Naderi M, Acharya M. Measuring surface roughness of pharmaceutical powders using vapor sorption methods. *AAPS Pharmaceutical Science Technology*. **2011**, 12, 56-61.

498398 Critical Material Attributes (CMAs) of Strip Films Loaded with Poorly Water-Soluble Drug Particles: IV. Surface-Modified Dry Micronized Drug Powders

Abstract:

Recent work has established polymer strip films as a robust platform for delivery of poorly water-soluble drug using their nanosuspensions. However, a simpler means of incorporating drug particles has yet to be demonstrated. Consequently, the feasibility of incorporating micronized fenofibrate (FNB), a model poorly water-soluble drug, via simultaneous micronization and surface modification with hydrophilic silica using fluid energy mill (FEM) was investigated. Milled silica coated FNB (MC-FNB) was prepared to faciliate direct incorporation into aqueous film precursor made using hydroxypropyl methylcellulose (HPMC-E15LV) as film former and glycerin as plasticizer. Films with as-received FNB (AR-FNB) and milled/uncoated FNB (MU-FNB) were prepared as control. The impacts of film precursor viscosity, low versus high shear mixers, and FNB surface modification on critical quality attributes (CQAs) of film product were assessed. Generally, high-shear planetary mixer resulted in better film CQAs; e.g., appearance (thickness uniformity, visible lumps/agglomerates), drug content uniformity (RSD%), and mechanical properties (ultimate tensile strength, elongation percentage, Youngâ€™s modulus). Use of MC-FNB particles was necessary to achieve superior film CQAs when commonly used lower-shear impellor mixing and low film precursor viscosity were used. The well dispersed hydrophilic silica particles conveyed a nano-scale surface roughness and helped to increase the wettability of poorly water soluble drug, which lead to a significant reduction in the tendency to form agglomerates and also had a positive effect on the dissolution of drug powders. Importantly, use of MC-FNB led to improved drug particle size recovery upon redispersion and fast, complete drug release. The results demonstrate feasibility and advantages of direct incorporation of dry surface modified-micronized poorly water-soluble drug powders in film manufacturing.

494730 Synthesis and Stability Study of Polymorphic Transformed Mannitol/ LB Agar Microcarriers for Dry Powder Inhalation

Abstract:

Spray drying is a widely used technique for the preparation of drug carriers as it can provide easily control of particle size, powder density and morphology. However, spray-dried microparticles are often in an unstable state. Most of the spray-dried microparticles have not reached an equilibrium state due to the short spray-drying processing time. In our previous work, novel rough mannitol/LB Agar microparticles were successfully prepared by polymorphic transformation method. The mannitol/LB Agar microparticles when used as carriers for dry powder inhalation has shown good drug delivery performance. In this study, the stability of the fully-transformed and fully-crystallized mannitol/LB Agar microparticles is enhanced with the use of agar as a stabilizer. As-prepared microparticles were characterized by X-ray diffraction spectra (XRD), differential scanning calorimetry (DSC), scanning electron microscopy (SEM), thermal gravimetric analysis (TGA) and Andersen Cascade Impactor (ACI).

The X-ray diffraction spectra (XRD) results indicate that Î´-mannitol was completely transformed to Î±-mannitol after immersing the spray-dried mannitol/LB Agar microparticles in hexane for a certain days. The scanning electron microscopy (SEM) shows that the spray-dried microparticles have smooth surface, while the roughness of the microparticle surface is increased after transformation. XRD indicates that the Î±-mannitol present in the transformed mannitol/LB Agar microparticles was stable for 33 months. SEM indicates that the small mannitol crystals in the transformed microparticles aggregated and self-assembled into nanorods. The transformed microparticles can maintain the rough-surface and spherical shape for 21 months. It is substantially longer than that of the microparticles without using agar as stabilizer, which can only maintain a stable shape for 6 months. Andersen Cascade Impactor (ACI) result shows that the fine particle fraction (FPF) of the model drug, Budesonide using the mannitol/LB Agar microparticles as drug carriers can also be maintained for 21-month.

2017 AIChE Annual Meeting
October 29 - November 3, 2017
Minneapolis Convention Center, Minneapolis, MN
Hilton Minneapolis, Minneapolis, MN

Understanding the Surface Chemistry, Surface Roughness and Wettability of Argon Plasma Treated Powder

Deepa Dixit[1], Shreya Bunk[1], Ramkrishna Rane[2] , Chinmay Ghoroi[1]
deepa.dixit@iitgn.ac.in, chinmayg@iitgn.ac.in

[1]DryProTech., Lab Chemical Engineering, Indian Institute of Technology Gandhinagar, Gandhinagar, Gujarat, India, 382355
[2]FCIPT, Institute for Plasma Research Centre, Gandhinagar, Gujarat, India

INTRODUCTION

Under humid conditions

Powder Bulk Properties

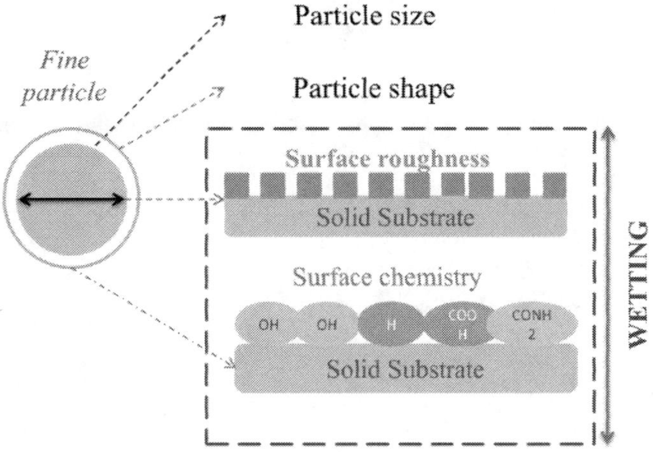

PARTICLE ENGINEERING

INTRODUCTION

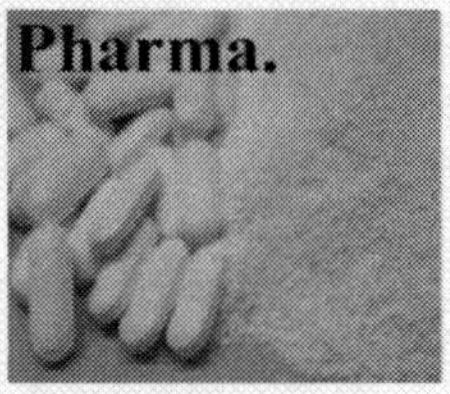

Under humid conditions

Particle engineering: ***<u>Wetting</u>*** *, flowability, dissolution, caking*

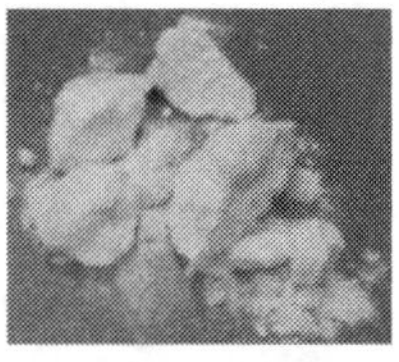

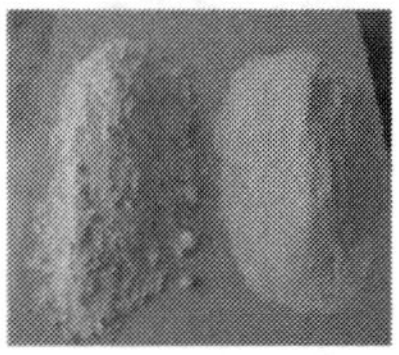

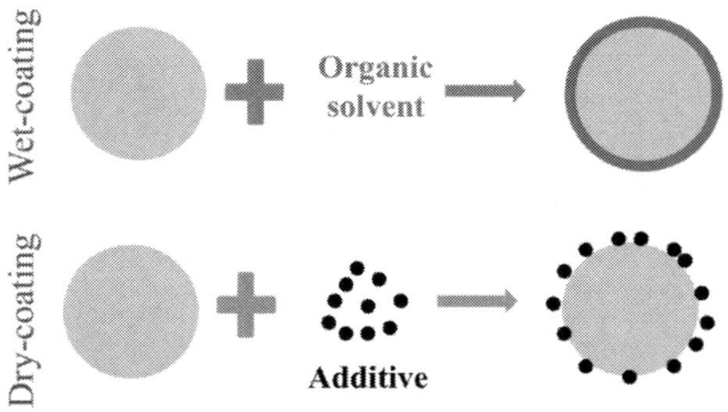

it is necessary to develop a process for the surface modification of fine powders which is environmentally benign, flexible, and economical.

MATERIAL & METHODS

Material
Corn Starch and Ibuprofen

Particle Size
Corn Starch 17 microns
Ibuprofen 90 microns
Used in pellet form* (10 mm dia.)

Operating Conditions
Plasma gas- Argon (Ar)
Voltage- 360-370 V
Current- 40-60 mA

Time- 5 min, 10 min, 20 min

Instrumentation diagram for the Setup at IPR

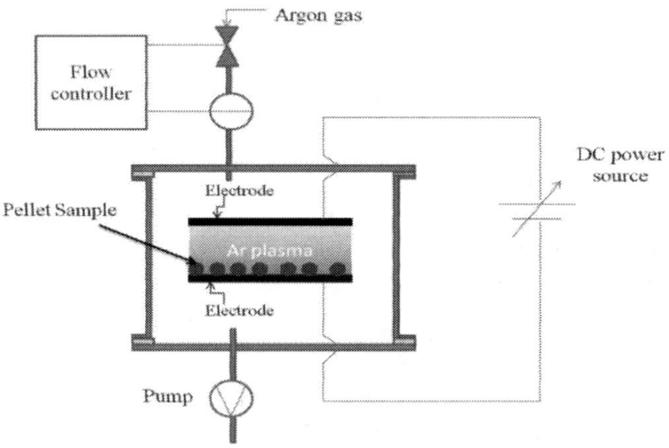

**The corn starch and ibuprofen pellets were prepared by compressing powder in a hydraulic pellet press at 5T and 2T pressure respectively (as they have different compactability).*

SURFACE MORPHOLOGY
Using FESEM

17AIChE
Annual Meeting, Minneapolis, MN

FESEM micrographs for treated CORN STARCH powders

Coral reef like nanostructures

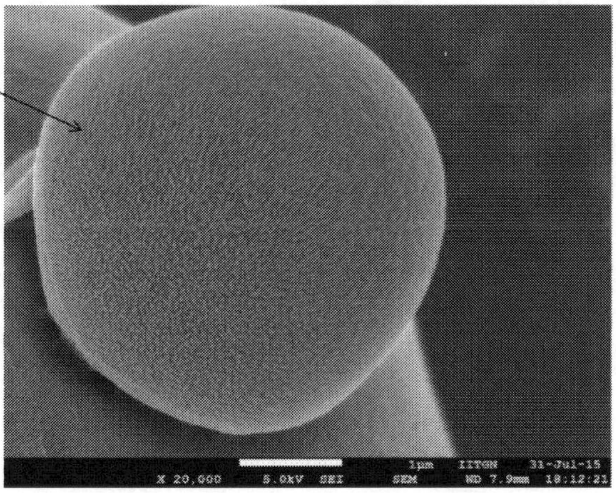

Material removal from the surface due to plasma active species bombardment

SURFACE MORPHOLOGY
Using FESEM

17AIChE
Annual Meeting, Minneapolis, MN

FESEM micrographs for CORN STARCH powders at low and high magnification

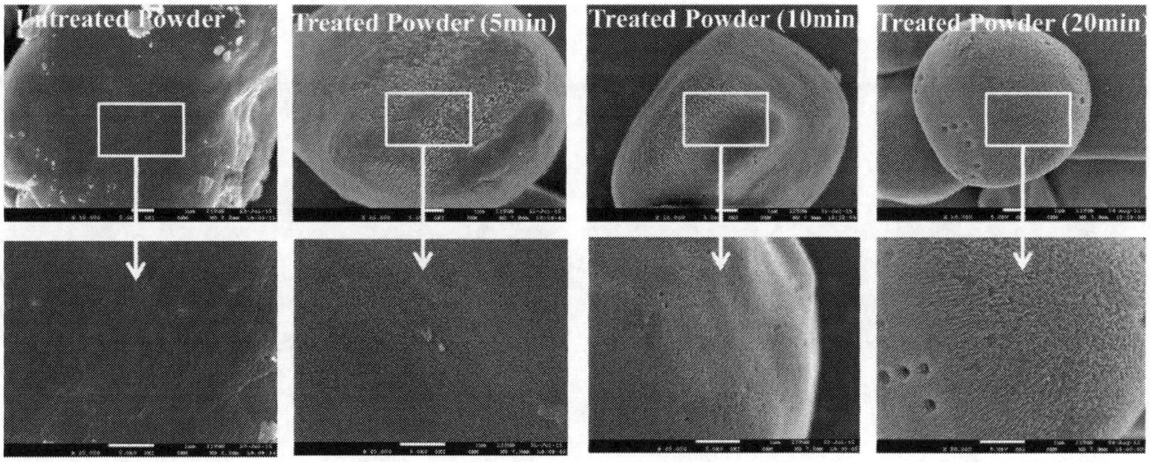

SURFACE TOPOGRAPHY
Using AFM

17AIChE Annual Meeting, Minneapolis, MN

Setup: Atomic Force Microscope ***Operating Mode:*** Tapping Mode ***Tip:*** Silicon nitride tip

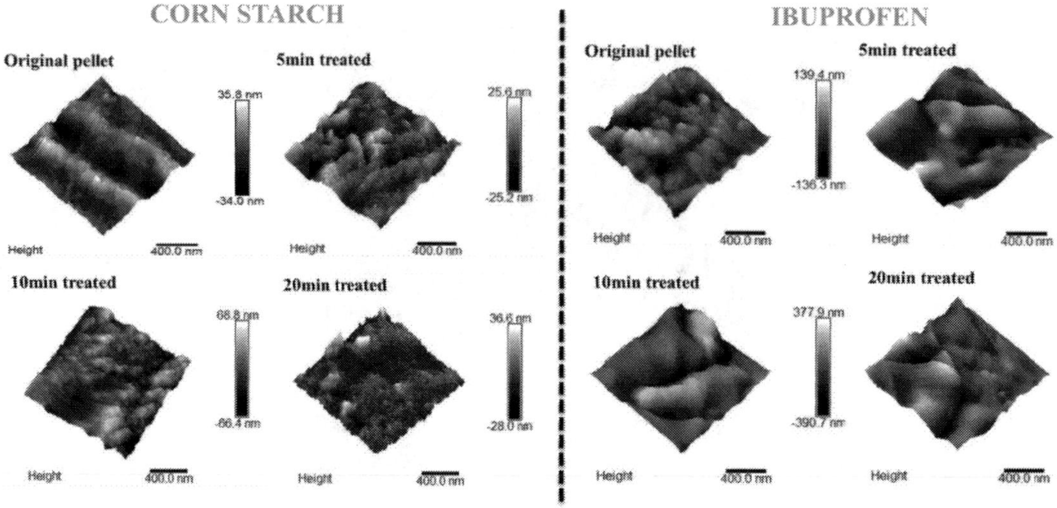

SURFACE TOPOGRAPHY
Using AFM

17AIChE Annual Meeting, Minneapolis, MN

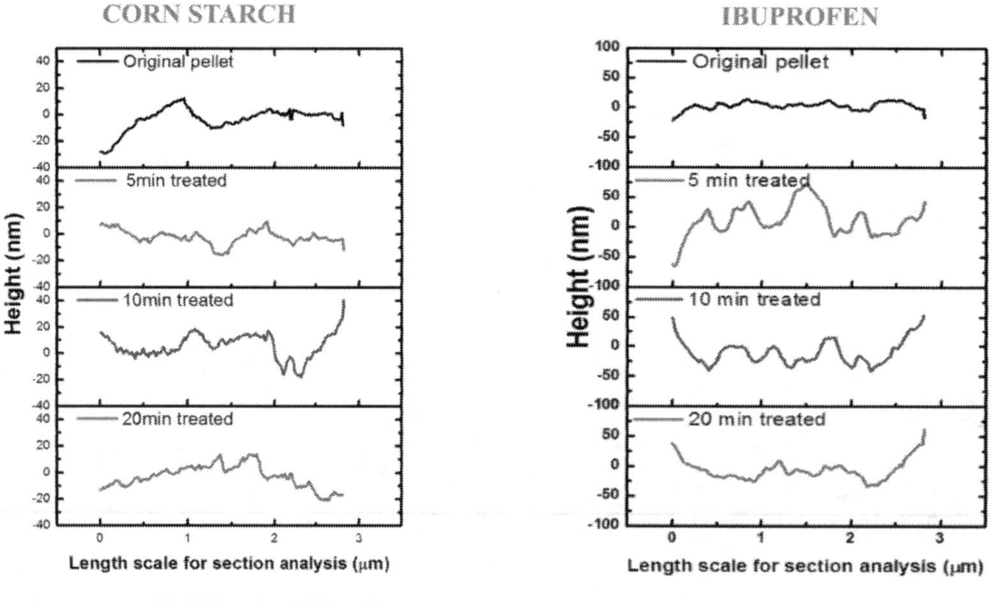

SURFACE TOPOGRAPHY

Using AFM

17AIChE
Annual Meeting, Minneapolis, MN

CORN STARCH

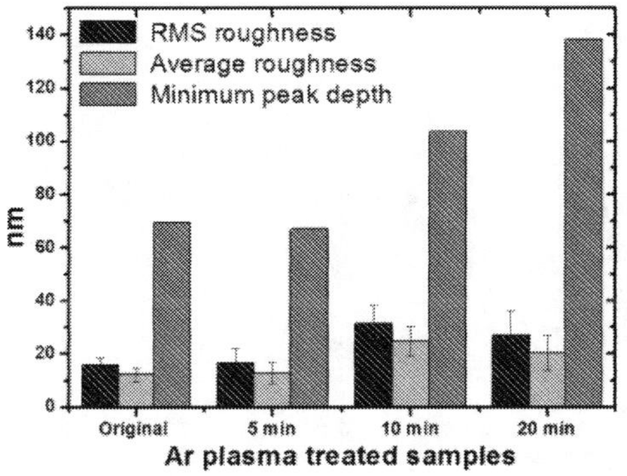

IBUPROFEN

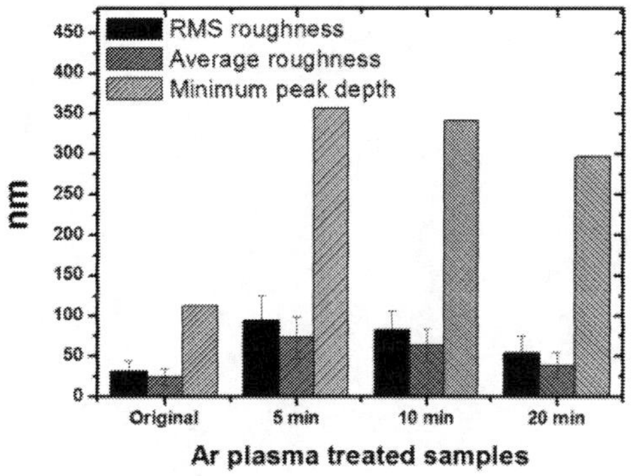

The change in scale of ibuprofen w.r.t. corn starch : *the greater the applied pressure (5T for cornstarch, 2 T for ibuprofen), the harder the pellets, the lesser the material etched out.*

The increase and decrease in roughness and depth suggest the building and damaging effect at the exposed surface.

SURFACE CHEMISTRY

Using XPS (*Wide scan spectra*)

17AIChE
Annual Meeting, Minneapolis, MN

(a) Original corn starch pellet, (b) 5 min treated corn starch pellet (c) 10 min treated corn starch pellet, (d) 20 min treated corn starch pellet

Treatment time	Mass % Carbon	Mass % Oxygen
0	59.26±0.51	39.60±0.45
5	53.75±0.41	45.23±0.38
10	53.35±0.46	44.90±0.43
20	52.12±0.71	45.35±0.67

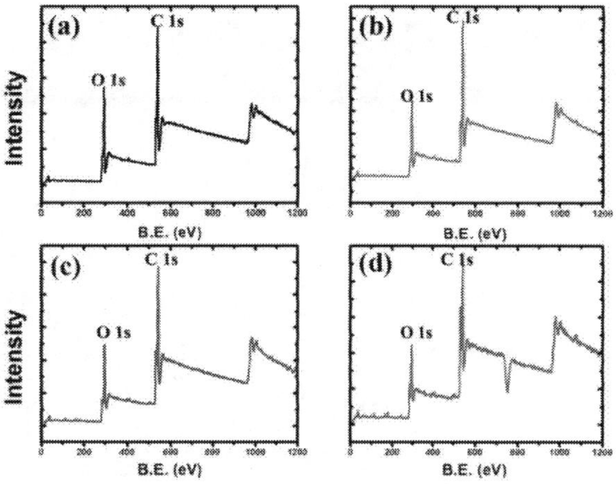

SURFACE CHEMISTRY
Using XPS (High resolution spectra)

High binding energy peak- Satellite peak

Indicating the presence of carbon atoms in different chemical state (bridging and non-bridging)

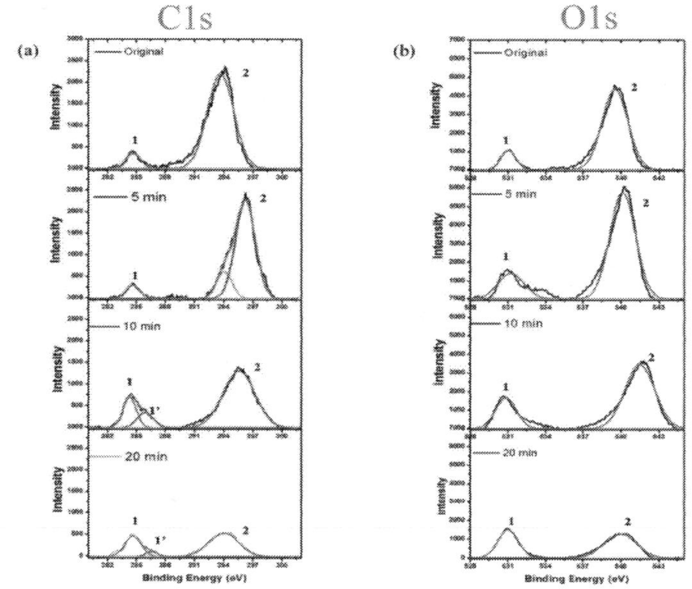

SURFACE WETTABILITY

Contact angle measurement using Sessile Drop Method

AR PLASMA TREATMENT FOR DIFFERENT TIME INTERVALS

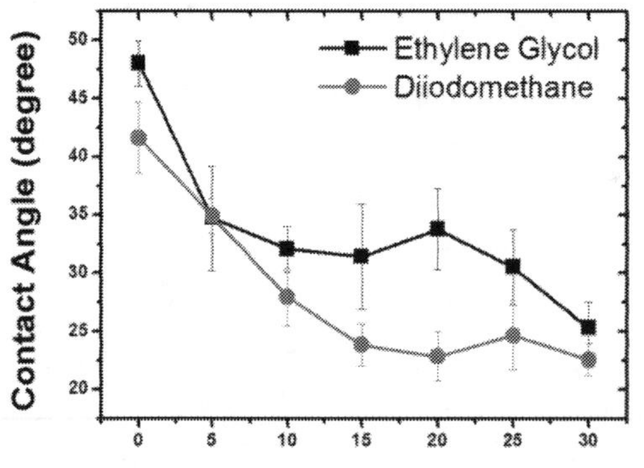

SURFACE WETTABILITY

Wenzel Model

$$\cos\theta_w = r\cos\theta_y$$

θ_y is contact angle for smooth surface
48° for corn starch and 68° for ibuprofen

θ_w is Wenzel's contact angle

'r' is roughness factor (determined using AFM)

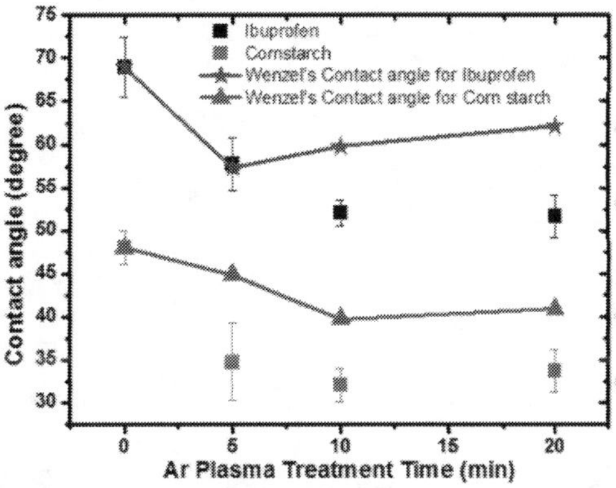

AR PLASMA TREATMENT FOR DIFFERENT TIME INTERVALS

SURFACE ENERGETICS
Using OWK method

$$\gamma_l^t(1+\cos\theta) = 2\left[\left(\sqrt{(\gamma_l^p\gamma_s^p)}\right)+\left(\sqrt{(\gamma_l^d\gamma_s^d)}\right)\right]$$

γ^d - Dispersive Surface energy
γ^p - Polar Surface Energy, θ – Contact Angle
$S_{l/s}$ – Spreading Coefficient, l & s are liquid and solid phase

CORN STARCH

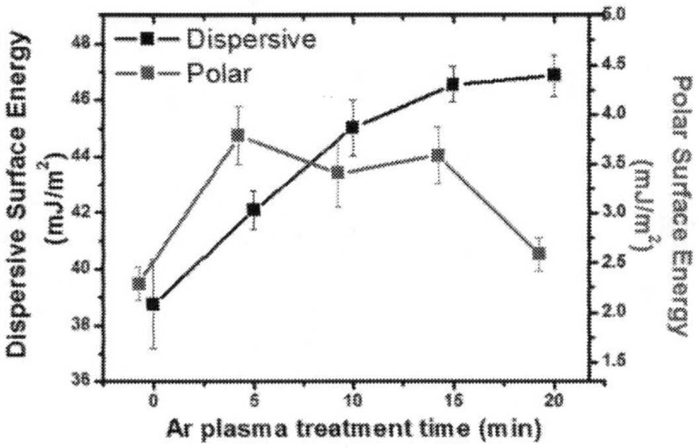

IBUPROFEN

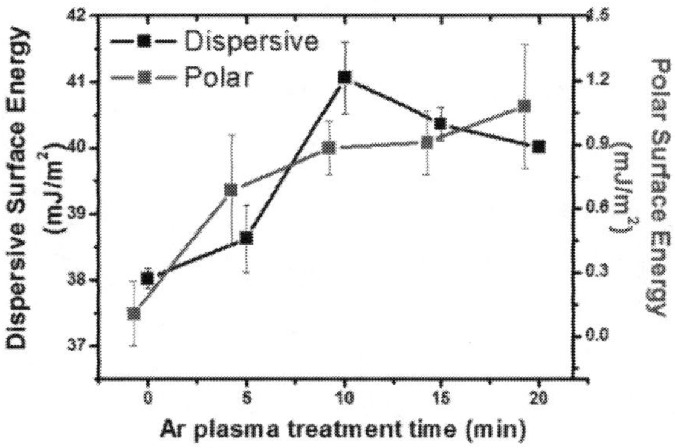

SURFACE ENERGETICS

Using OWK method

17AIChE Annual Meeting, Minneapolis, MN

<u>Spreading Coefficient:</u> $S_{l/s} = \frac{2}{\gamma_l^t}\left(\frac{\gamma_l^d \gamma_s^d}{\gamma_s^d + \gamma_s^d} + \frac{\gamma_l^p \gamma_s^p}{\gamma_l^p + \gamma_s^p}\right)$ <u>Work of Adhesion :</u> $W_a = \gamma_l^t(1+\cos\theta)$

CORN STARCH

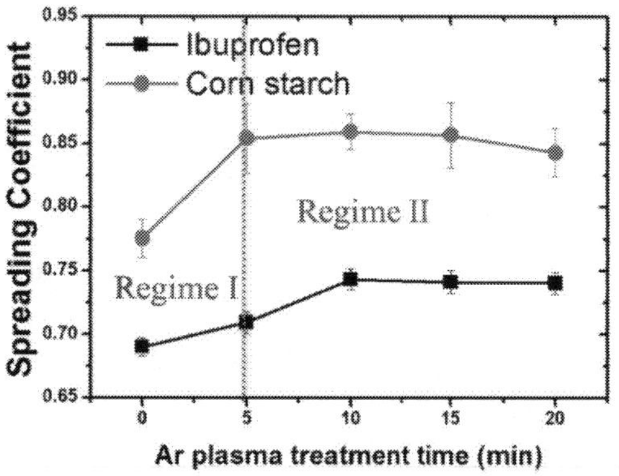

IBUPROFEN

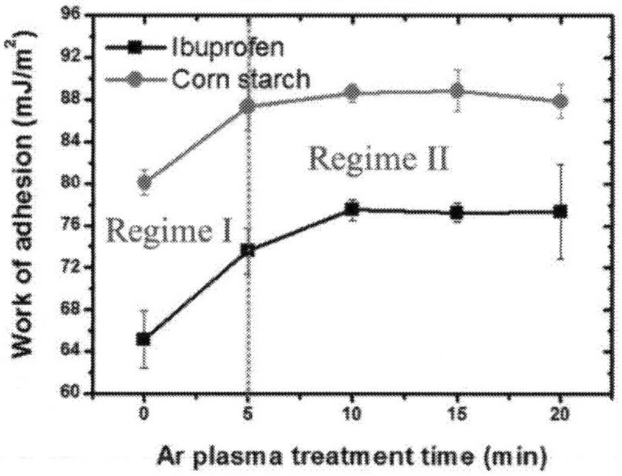

CONCLUSION

17AIChE Annual Meeting, Minneapolis, MN

- Study shows that Ar plasma significantly changed the surface topography and surface chemistry of the treated powders.
- First 5 min of exposure mainly cleans the surface. The increase in exposure time increases the nano-scale surface roughness depending on the nature of the material.
- ***Wetting characteristics***: Contact angle for corn starch and ibuprofen changes from 48° to 33° and 69° to 51° respectively.
- The regime I is dominated by the change in surface groups. In the regime II, spreading is mainly due to the combined effect of surface groups and nano-scale surface roughness.

ACKNOWLEDGEMENT

17AIChE
Annual Meeting, Minneapolis, MN

I would like to acknowledge the IIT Gandhinagar for providing the financial support to carry out this research work. I also acknowledge the FCIPT, IPR Gandhinagar for providing the experimental facilities.

I would like to thank my supervisor *Prof. Chinmay Ghoroi* , committee members and research groups for the support through out the work.

REFERENCES

17AIChE
Annual Meeting, Minneapolis, MN

1. *Daiki Murakami, Hiroshi Jinnai, and Atsushi Takahara "Wetting Transition from the Cassie–Baxter State to the Wenzel State on Textured Polymer Surfaces" Langmuir 2014, 30, 2061−2067.*
2. *Abraham Marmur "Wetting on Hydrophobic Rough Surfaces: To Be Heterogeneous or Not To Be?" Langmuir 2003, 19, 8343-8348.*
3. *Neil J. Shirtcliffe , Glen McHale, Shaun Atherton, Michael I. Newton "An introduction to superhydrophobicity" Advances in Colloid and Interface Science 161 (2010) 124–138.*
4. *Baldacchini T1, Carey JE Zhou M Mazur E «Superhydrophobic surfaces prepared by microstructuring of silicon using a femtosecond laser.» Langmuir. 2006 May 23;22(11):4917-9.*
5. *Sanjay Subhash Latthe, Annaso Basavraj Gurav, Chavan Shridhar Maruti, Rajiv Shrikant Vhatkar "Recent Progress in Preparation of Superhydrophobic Surfaces: A Review" Journal of Surface Engineered Materials and Advanced Technology, 2012, 2, 76-94.*
6. *Seung-Mo Lee1 and Tai Hun Kwon "Effects of intrinsic hydrophobicity on wettability of polymer replicas of a superhydrophobic lotus leaf" Journal of Micromechanics and Microengineering, Volume 17, Number 4.*
7. *Didem Oner and Thomas J. McCarthy "Ultrahydrophobic Surfaces. Effects of Topography Length Scales on Wettability" Langmuir 2000, 16, 7777-7782.*
8. *Stephen Michielsen and Hoon J. Lee Design of a Superhydrophobic Surface Using Woven Structures"Langmuir 2007, 23, 6004-6010.*
9. *W. Barthlott, C. Neinhuis "Purity of the sacred lotus, or escape from contamination in biological surfaces" Planta (1997) 202: 1–8*
10. *Edward Bormashenko "Historical perspective Progress in understanding wetting transitions on rough surfaces" Advances in Colloid and Interface Science 222 (2015) 92–103*

Thank you

495635 Effects of Baffle Configuration and Tank Size on Spherical Agglomerates of Dimethyl Fumarate in a Common Stirred Tank

Abstract:

To pave the way for technology transfer and scale up of the spherical agglomeration (SA) process for dimethyl fumarate, effects of the US, European and Kawashima type baffles and 0.5, 2.0 and 10 L-sized common stirred tank were studied. It was found that the particle size distribution varied significantly. However, the size-related properties such as dissolution profile and flowability of agglomerates from the same size cut after sieving could remain unchanged. The interior structure-related properties such as particle density and mechanical property of agglomerates upon baffle change and scale up from the same size cut were decayed and the agglomerates could become denser and stronger by prolonged maturation time. To maintain the same size distribution, agglomerates from any batch could have been separated and classified by sieving and then blended back together artificially by the desired weight% of each cut.

495419 UsingÂ Magnetically Assisted Impact Coating (MAIC) for Optimization of Powder Flow Characteristics

Abstract:

Many processes need to control powder flow in order to improve processing and maintain consistent product quality. Typical methods for control of powder flow involve the addition of flow aids, with few process handles except the amount of flow aid added to the powder. Typical flow aids used in this manner are fumed silica, a hydrophilic flow aid, and magnesium stearate, a hydrophobic flow aid.

Magnetically Assisted Impact Coating (MAIC) has been demonstrated as a technique for coating host particles without causing major changes to particle shape and size, which creates for a unique method for the application of flow aids [1]. This system relies on a process in which an oscillating magnetic field is used to accelerate and spin large magnetic particles, causing the magnetic particles to fluidize. The agitated magnetic particles impart energy to a substrate and its flow aid, and coating is then achieved as collisions occur between particles and between particles and the vessel wall [1].

MAIC has shown to improve flow properties of a model compound, 5 Î¼m citric acid, both with and without the addition of a hydrophilic fumed silica flow aid [2], and has been shown to improve the flow properties better than traditional mixing methods. In addition to changing the amount of flow aid added to a powder, the MAIC process has a number of different process parameters that can affect powder flow. The goal of this presentation is to offer a systematic analysis of the various process parameters, with a focus on the difference between batch [1] and continuous MAIC processing, residence time, intensity, and substrate effects. Additionally, a hydrophobic fumed silica will be compared with the original hydrophilic fumed silica with respect to their influence on powder flow properties. Powder flow measurements are done with an FT4 Powder RheometerÂ®, and include dynamic, bulk, and shear properties [3]. The study is ongoing, and the data are currently preliminary. Statistical comparisons of the flow properties will be presented.

References

[1] Ramlakhan, M., Wu, C.Y., Watano, S., et al., *Dry particle coating using magnetically assisted impaction coating: Modification of surface properties and optimization of system and operating parameters*. 2000:112:137-148.

[2] Freeman, T; Hendrickson, W.; Bowman, C. R. *et al*. (2016, November). *Comparison of Magnetically Assisted Impact Coating (MAIC) with Traditional Mixing Techniques for the Addition of a Silica Flow Aid*. Paper presented at the meeting of the American Institute of Chemical Engineers, San Francisco, CA.

[3] Brockbank, K., Armstrong, B., Chandorkar, Y., *et al*., *Understanding powder caking as a consequence of a range of mechanisms by means of powder rheometry*. 2015:33:102-108.

499796 Controlled Removal of Si-Rich Passivation Layer Using a Binary-Solid Fluidized Bed Reactor for Enhanced Carbon Capture and Storage Via Mineralization

Abstract:

Carbon capture, utilization, and storage (CCUS) is an important technology for our sustainable energy future. One of the combined CCUS approach is ex-situ carbonation of silicate minerals and alkaline industrial wastes such as steel slags. As magnesium and calcium silicates are chemically reacted with CO_2, value-added product can be produced in the form of thermodynamically stable solid carbonates. This is one of the most promising CO_2 sequestration technology because it can be integrated into various chemical processes and the overall process does not require the energy intensive solvent regeneration step. The main challenge of this technology is slow reaction rates associated with mineral dissolution, CO2 hydration and precipitation of solid carbonates. Our past studies showed that a mixture of Mg and Si-targeting ligands can significantly enhance the mineral dissolution, while whole-cell biocatalysts with carbonic anhydrase can improve the rate of CO2 hydration in the aqueous phase. While these advancements led to a better design of carbon mineralization process, the overall extent of reaction is still limited due to the development of Si-rich passivation layer on the surface of Mg and Ca-silicates. Thus, a binary-solid fluidized bed is designed to remove the Si-rich passivation layer during the mineral dissolution while bubbling CO2. The effects of the size and density ratios of mineral particles and in-situ grinding media are investigated in terms of the overall extent of reaction. A revised shrinking-core model is developed to describe how the thickness of the passivation layer impacts the dissolution behaviors of Mg- and Ca-bearing minerals.

501171 Copper Oxide-Based Oxygen Carrier for Chemical Looping Combustion of Methane: Investigation of Attrition Behaviors and Particulate Formation

Abstract:

Chemical looping combustion (CLC) process offers a potentially effective approach for CO_2 capture from carbonaceous fuel combustion via cyclic redox reactions of oxygen carrier particles in circulating fluidized bed systems. A potential challenge to CLC is the attrition of oxygen carrier particles, which can result in loss of materials and/or particulate emissions. In this study, attrition behavior and particle loss of a copper oxide-based oxygen carrier from a methane CLC process was investigated in a fluidized bed reactor. The aerodynamic diameters of most elutriated particulates, after passing through a horizontal settling device, ranged from 2 to 5 Î¼m. A notable number of submicron particulates were also identified. Cyclic reduction and oxidation reactions in CLC were determined to weaken the oxygen carrier particles, resulting in increased particulate emission rates when compared to oxygen carriers without redox reactions. The generation rate for particulates < 10 Î¼m was found to decrease with progressive cycles over as-prepared oxygen carrier particles and then reach a steady state. The surface of the oxygen carrier is also found to be coarsened due to a Kirkendall effect, which also explains the enrichment of Cu on particle surfaces and in small particles. It is therefore desirable to collect and reprocess small particles generated from chemical looping processes to reduce the loss of Cu. The current study also indicates that particulate control devices based on existing technologies may be required for methane CLC using Cu-based oxygen carriers. The lifetime of the Cu-based oxygen carrier examined is estimated to be ~5000 h.

498949 Chemical Looping Gasification (CLG) of Biomass Using a Novel Bimetallic [Copper-Ferrite] Oxygen Carrier

Abstract:

Chemical looping gasification (CLG) is a new novel technique that utilizes lattice oxygen of metal oxides in place of convention gasification agents like air, steam or pure oxygen to convert carbonaceous fuels into syngas. This method produces high quality syngas free from nitrogen dilution at a low cost. Unlike other traditional gasification technologies, the fuel and air in a CLG process do not mix. Development of an efficient and economic oxygen carrier is essential for the success of a CLG process. The objective of this study was to develop an oxygen carrier with optimum CuO and Fe_2O_3 composition supported on alumina for CLG and to investigate the use of CuO for the enhancement of an iron based oxygen carrier in biomass CLG. The synergetic properties of Cu-Fe mixed oxides make them a promising material for CLG. While copper has high reactivity, it has issues with agglomeration which is attributed to its low melting point. On the other hand, despite iron oxide being economically cheap and being able to withstand high temperatures, it has slower reactivity. In this work, the bimetallic oxygen carrier was developed using four different methods (Direct Decomposition, Dry Impregnation, Physical Mixing, and combination of Co-precipitation and dry impregnation). The method yielding a carrier with the best TPR-TPO performance (oxygen capacity) was chosen for fixed bed reactions. Biomass chemical looping gasification was investigated in a fixed bed flow reactor with inner diameter of 7 mm using Fe-Cu bimetallic oxide supported on alumina as an oxygen carrier. Fuel conversion was determined to be higher when Fe-Cu oxide was used compared to when only iron oxide was used as the oxygen carrier. Fresh and used oxygen carriers were characterized using SEM-EDS and XRD. The SEM results indicated agglomeration after one cycle of TPR-TPO when only CuO supported on alumina was used in place of a mixture of Fe-Cu oxides. Stability of the oxygen carrier was evaluated after several cycles of oxidation and reduction. It was confirmed that the bimetallic material has a longer lifecycle. XRD results indicated that the products of the bimetallic material reduction was Cu, Fe_3O_4 and FeO. XRD results also showed that the oxygen carrier can be regenerated to form a crystalline similar to fresh material.

510400 *Design and Operation of High Pressure and High Temperature Fluidization Systems for Chemical Looping Application*

Abstract:

Chemical looping technologies have evolved into a promising alternative for the efficient conversion of carbonaceous fuels to electricity and/or high value chemicals with minimal carbon emissions. These processes utilize an oxygen carrier solid material to indirectly supply oxygen to the fuel source. With the exponential growth of research and publications in this field, chemical looping has expanded to encompass power and chemical production with in-situ gas separation. The Ohio State University (OSU) has advanced the chemical looping concept in the development of 2 pilot-scale demonstration plants for coal combustion with CO_2 capture and gaseous fuel conversion to high purity H_2 as well as multiple sub-pilot test units for syngas generation from biomass, natural gas, and coal. In each of these processes, a moving bed reducer reactor is used for partial or full conversion of the carbonaceous fuels to CO/H_2 and CO_2/H_2O, respectively. A fluidized bed combustor reactor is used to regenerate the oxygen carriers to their original oxidation state. Interconnecting nonmechanical valves between reactors are used to control the solids circulation rate of the system for the mass balance of the redox reaction and heat balance for the system integration. This paper also summarizes the key developments of two chemical looping processes â€“ the 250kW$_{th}$-3MW$_{th}$ high pressure syngas chemical looping (SCL) pilot plant and the coal direct chemical looping (CDCL) 25 kW$_{th}$ sub-pilot unit and 250 kW$_{th}$ pilot plant. Key process design features and experimental results from over 1,500 hours of cumulative operation will be discussed. The counter-current moving bed reducer reactor design the CDCL and SCL process ensure nearly full fuel conversion to CO_2 with minimal solid circulation and capability of producing high purity H_2. During operation of the pilot plants, sliding mode control theory was adopted for modulating the aeration gas flow rate against the pressure fluctuations caused by the slugging bed combustor in order to maintain a constant solid circulation rate.

510401 *Chemical Looping Process for Post-Combustion Carbon Capture Using Calcium Hydroxide: Process Integration and Performance Results*

Abstract:

Calcium-based solid sorbents such as calcium oxide and calcium hydroxide can be used in a chemical looping scheme for the capture of carbon dioxide from flue gas streams, and both CaO and $Ca(OH)_2$ have been successfully demonstrated at the pilot-scale for CO_2 removal. Calcium-based sorbents have several process advantages when compared to other carbon capture processes such as amine scrubbing and oxy-combustion. They include being derived from a cheap and abundant mineral, limestone, possessing a high CO_2 carrying capacity (786 g CO_2/kg), potentially removing all acid gases produced, and having the ability to increase the net electricity generation of a power plant. If calcium oxide is the sorbent, then CO_2 is removed in a CaO-$CaCO_3$ cyclic loop in a two reactor system-Carbonator and Calciner. The Ohio State University has developed a process wherein calcium hydroxide is used as the solid sorbent for CO_2 removal in a $Ca(OH)_2$-$CaCO_3$-CaO cyclic loop using three reactors-Carbonator, Calciner, and Hydrator. Ohio State has demonstrated the simultaneous removal of CO_2 and SO_2 using calcium hydroxide at the 120 kW_{th}-scale and established its superior reactivity compared to CaO. In addition, a high-temperature steam hydrator using a bubbling fluidized bed has been constructed and tested. The $Ca(OH)_2$-$CaCO_3$-CaO process was simulated in Aspen Plus using the results obtained from experiments. The power plant type and level of CO_2 capture constrain the process integration options while also controlling the solids flow requirements and parasitic energy consumption. They are also the primary factors that affect the process performance. Ultimately, the performance results of the process is reflected in the cost of electricity, which is the single, most critical factor in assessing a post-combustion carbon capture process. The requirements and constraints placed on both a CaO-$CaCO_3$ and a $Ca(OH)_2$-$CaCO_3$-CaO cyclic loop and their effect on process performance at different levels of carbon capture for post-combustion carbon capture have been investigated.

484708 Experimental Study of a Riser-Based Carbon Stripper in a Solid Fuel-Chemical Looping Combustion System

Abstract:

Chemical looping combustion (CLC) has emerged as a promising combustion process with low-cost CO_2 separation. In fuel reactors of solid-fuel chemical looping combustion, the solid fuel gasification takes much longer time than the reduction of oxygen carriers. The slip of char particles into the air reactor will result in a decreased carbon capture efficiency. Therefore, carbon stripping is a critical process step which separates fuel particles from oxygen carrier particles. Carbon strippers were adopted in different reactor systems, including a 100 kW_{th} system in Chalmers, a 1 MW_{th} system in Darmstadt, and a 50 kW_{th} system in CSIC. Hongming et al. demonstrated that a riser-based carbon stripper can effectively separate fuel particles from oxygen carrier particles. Therefore, it is necessary to investigate the separation characteristics and technology cost of the riser-based carbon stripper in a three-fluidized-bed CLC system.

A three-fluidized-bed system was constructed based on Glickmanâ€™s simplified scaling ratios. The system is comprised of a fuel reactor, an air reactor and a carbon stripper in between. The fuel reactor is a bubbling fluidized bed; the air reactor is a riser which transports solid particles back into the fuel reactor, and the carbon stripper is riser-based. Plastic beads and ilmenite particles were used to simulate the fluidization behavior of char particles and oxygen carrier particles. Pressure profile, solid circulation rate, separation ratio and separation efficiency were measured and calculated to study the separation characteristics of the carbon stripper.

Long-term stable operations were achieved and results have proven the feasibility of the proposed three-reactor system. Solid circulation rate can be easily controlled by adjusting the fluidization gas velocity of loopseals. With an increase of gas velocity in the carbon stripper, more particles were elutriated back to the fuel reactor and the separation ratio and separation efficiency increased. With mixture feeding rates below 17 kg/m^2s, the separation efficiency increased from 71% to 97% as the gas velocity increased from 0.8 m/s to 1.6 m/s. Energy-cost ratios were calculated, and gas velocity higher than 0.8 m/s and mixture feeding rates lower than 17 kg/m^2s were preferred.

The separation characteristics of a carbon stripper is of vital importance to the operation of a three-fluidized-bed CLC system. The results of this paper reveals that carbon strippers can be used to increase the carbon capture efficiency and that riser is an economically feasible way to realize carbon stripping in a CLC system.

502989 Evolution of Powder Testers – Key Challenges and Opportunities

Abstract:

Powders have been utilised by mankind for many thousands of years but methods for the measurement of powder flow characteristics have really only existed in recent decades. It was not until the 1960's that engineers like Andrew Jenike, Jerry Johansson, Ralph J Carr Jr and Henry H Hauser started to develop more scientific and robust approaches to understanding and measuring powder flowability, with the development of several methodologies occurring within a few years. In 1964 Jenike released his ground breaking paper on the Storage and Flow of Solids, outlining his theory on the stresses generated in quasi-static and static regimes in hoppers and silos, and demonstrating the suitability of translational shear cells for measuring incipient flow during hopper discharge[1]. Carr [2] and Hasuner [3] published papers in in 1965 and 1967 respectively, describing the use of density changes as a function of consolidation to quantify powder flowability. Previously flow characterisation had been limited to techniques such as Angle of Repose (AOR), with Jenike stating that "it is only useful in the determination of the contour of a pile, and its popularity among engineers and investigators is due not to its usefulness but to the ease with which it is measured", although Carr disagreed, suggesting that AOR provided an indirect measurement of size, shape, porosity, cohesion, fluidity, surface area and bulk density of the material [2]. It was during this period that Jerry Johanson was first entering the field, completing his PhD under the supervision of Jenike in 1962 and joining Jenike in the formation of Jenike and Johanson in 1966.

As powder handling processes increased in complexity and manufactures moved to reduce costs and increase production output, the requirements for improved powder characterisation resulted in a flurry of further developments in the 1990's. The ATSM for measuring the flow properties of metallic powders using ball indentation was published[4] in 1990. Avalanching techniques were first proposed by the likes of Kaye in 1995[5] as a means of studying the rheological behaviour of powders. Meanwhile, Johanson and Schulze built on Jenike's work for determining flow behaviour in hopers and silos. In 1991 Johanson patented the Johanson Indicizers (Hang-up, Hopper Flow and Flow Rate)[6] whilst Schulze introduced the Ring Shear Tester in 1994[7]. Like the translational shear cell developed by Jenike, these methodologies derive, amongst other parameters, the Unconfined Yield Strength (UYS) of a powder as a function of the Major Principle Stress (MPS), however the Jenike Shear Cell required a highly skilled operator and took several hours to generate a Flow Function (FF), whilst the Johanson Indicizers and Schulze Ring Shear tester were much easier to use and reduced test times to a matter of minutes rather than hours. Other attempts had been made previously and during the same period to develop a Uniaxial Shear Tester, from Williams in 1970 who utilised a split mould [8] to Malby and the POSTEC Uniaxial Tester [9] in 1993, however these devices were never established commercially.

Since this period, advances in powder flow characterisation have remained steady. As engineers and scientists have better understood the need to measure powder flowability at the stress and strain rates found in modern processing equipment, methodologies have been developed which measure the powder under more dynamic conditions. In 2000 Freeman Technology developed their first Powder Rheometer[10], with Stable Micro Systems releasing a similar instrument around the same period. Freeman Technology further progressed the concept of stress/strain rate relevant measurements with the addition of the capability to introduce air, generating low stress aerated environments, and a shear cell for higher stress environments. Around this period, the measurement of the air pressure differential across the powder bed resulted in the introduction of the Permeability test on the Freeman FT4 Powder Rheometer, the Sevilla Powder Tester, patented in 1999 [11] and the Raining Bed Method (RBM), first proposed in 1967 but not developed fully until 2002 [12]. More recent developments based on the addition of air have included research by Tardos et al [13] investigating the fluidisation behaviour of powders in a Couette Device, and by Anton Paar who developed a Powder Cell which uses air to condition the powder prior to measurement. In terms of understanding flow in hoppers and silos the early work of Jenike is still as valid today as it was 55 years ago, however the desire to directly measure the UYS at lower stresses using simpler and lower costs technologies is evident, with techniques such as the Particulate Systems SSSpin Tester, developed by Kerry Johanson, the University of Leeds Environmental Caking Rig [14] and the Freeman Technology Uniaxial Powder Tester as recent additions to the field.

Going forwards, powder processing continues to grow in complexity, especially as industries such as the pharmaceutical sector move towards continuous manufacturing and emerging technologies such as 3D printing become established. Recognising the need to understand and measure powder behaviour under dynamic conditions, at low stress regimes, is driving research in terms of modelling and characterisation methods. Rationalising the physics of such environments is the key to a more fundamental understanding of these complex materials and is the next step in moving beyond Jenike's theory for quasi-static behaviour. The reality is this field is still very much in it's infancy, and there remains plenty of opportunity for current and future scientists and engineers to advance our understanding of bulk solids behaviour.

1. Jenike, A.W., *Storage and Flow of Solids*. 1964: University of Utah.

2. Carr, R.L., Evaluating flow properties of solids. *Chemical Engineering*, Vol. 72 No.(2) 1965 pp. 163-168

3. Hausner, H., Friction conditions in a mass of metal powder. *International Journal of Powder Metallurgy*, Vol. 3 No.(7) 1967

4. American Society for Testing and Materials, *Use of Automated Ball Indentation to Measure Flow Properties and Estimate Fracture Toughness in Metallic Materials*, in *ASTM 1092*. 1990, ASTM: Philadelphia. p. 188-208.

5. Kaye, B.H., J. Gratton-Liimatainen, and N. Faddis, Studying the Avalanching Behaviour of a Powder in a Rotating Disc. *Particle & Particle Systems Characterization*, Vol. 12 No.(5) 1995 pp. 232-236

6. Fayed, M. and L. Otten, *Handbook of Powder Science & Technology*. 2013: Springer US.

7. Schulze, D., Entwicklung und Anwendung eines neuartigen Ringscherätes (Development and Application of a novel ring shear tester). *Aufbereitungstechnik*, Vol. 35 No.(10) 1994 pp. 524-535

8. Schwedes, J., Review on testers for measuring flow properties of bulk solids. *Granular Matter*, Vol. 5 No.(1) 2003 pp. 1-43

9. Maltby, L. and G. Enstad, Uniaxial tester for quality control and flow property characterization of powders. *Bulk Solids Handling*, Vol. 13) 1993 pp. 135-135

10. Freeman, R., Measuring the flow properties of consolidated, conditioned and aerated powders — A comparative study using a powder rheometer and a rotational shear cell. *Powder Technology*, Vol. 174 No.(1–2) 2007 pp. 25-33

11. Castellanos, M.A., R.A. Ramos, and M.J.M. Valverde, *Dispositivo y procedimiento para medir la cohesion de medios granulares finos*. 1999, Google Patents.

12. Formisani, B., et al. *The bed support experiment in the analysis of the fluidization properties of fine*

solid. in *4th World Congress on Particle Techonology*. 2002. Sydney (Australia).

13. Langroudi, M.K., et al., An investigation of frictional and collisional powder flows using a unified constitutive equation. *Powder Technology*, Vol. 197 No.(1–2) 2010 pp. 91-101

14. Calvert, G., et al., A new environmental bulk powder caking tester. *Powder Technology*, Vol. 249 No.(0) 2013 pp. 323-329

500089 Predicting Segregation in Flowing Granular Materials

Abstract:

As granular materials flow they can segregate due to particle size or density differences. The degree of segregation depends on the flow conditions and the differences between the particle species. Decades ago, Dr. Jerry Johanson was a pioneer in directly addressing practical segregation problems, particularly in hoppers and bins. However, only more recently has a theoretical approach been suggested to predict segregation based on a continuum advection-diffusion-segregation equation for the concentration of each particle species. The model requires knowledge of the flow kinematics, which can come from experiments, simple models, or DEM simulations. The challenge is in defining the appropriate form of the segregation term in the equation. Several approaches have been proposed based on various models for particle percolation. One of the most successful is a relation for the segregation velocity based on the local shear rate, particle concentrations, and a semi-empirical parameter obtained from discrete element method (DEM) simulations for bi-disperse systems of particles. This approach can be applied to many flow geometries, particles of different sizes, densities, or shapes, and bidisperse, multi-disperse, or poly-disperse particle systems, provided that the particle segregation properties can be characterized. Recent research suggests progress in extracting segregation parameters from simple measurements so that even complex segregating materials can be modeled. The challenge lies in connecting these approaches to practical flow geometries such as bins and hoppers to provide a fundamental basis for methodologies that can be adapted to systems that are pertinent to industrial particle handling and processing.
Partially funded by NSF Grant No. CBET-1511450 and The Procter & Gamble Company.

491737 Making Solids Flow in Hoppers Using Passive Activation and Diamondback Hoppers

Abstract:

The secrets of Dr. Jerry R. Johansonâ€™s development of the famous Archbreaking Diamondback (ABD) hopper are revealed in this paper. Dr. Johanson combined his experiences during his doctoral studies at U of Utah with industry lessons and a very inventive mind to evolve the concept of â€˜passive activationâ€™ from first principles. The ABD is the direct result of passive activation. The paper will report how the ABD has revolutionized powder flow with applications and anecdotes about working with Dr. Johanson.

497436 The Impact and Use of Diamondback Hoppers in the Paper, Pulp and Biomass Industries

Abstract:

The early developments of using DiamondBackÂ® hopper technology for wood chips in a steam environment, and its impact on modern pulping technology will be discussed. The underlying needs for wood steaming and retention time control will be explained. The first successful application of the DiamondBackÂ® hopper in our industry opened a pathway to new developments, which have revolutionized the pulp and paper industry and its ability to build ever larger single line processing facilities, with improved quality and better financial results. The technology has lately been applied also in the fledging biomass treatment industry.Â Jerry Johansonâ€™s knowledge, understanding of the challenges, and commitment to the development was and is vital to its success.

498791 A Historical Perspective - the Impact of the Development of Bulk Solids Handling in Australia

Abstract:

This paper will present a historical account of the implementation of some of the work pioneered by Dr Johanson and the influence that work had in the development of bulk solids handling in Australia. It will look at the history of solids flow and examine some of the barriers and challenges that can be addressed in todayâ€™s world.

BULK SOLIDS FLOW THEORY THEN AND NOW

BY DR. J. R. JOHANSON OF JR JOHANSON INC.

ABSTRACT

This paper outlines Dr. Jerry R. Johanson's major contributions to bulk solids flow theory during the past sixty years and includes a new theory extension for viscous type solids.

THE BEGINNING

The bulk solids theory in the 1950's was only concerned with building bins strong enough to keep them from falling down. The existing German standards, DIN, were based solely on Janssen's silo loads of fifty years ago. Since silos still fell down with Janssen's silo loads [1], safety factors as high as five were used on the Janssen's loads. Dr. Andrew Jenike was unique in trying to mathematically describe solids flowing in bins and hoppers.

My association with him starting May 1958 was dumb luck. As a recently married student at the University of Utah, I needed a job to support our $125 monthly budget plus tuition and books. As I started my junior year, my job as lab assistant in the Mechanical Engineering Lab was just across the alley from a coal bin that was modified with a solid handling conveyor, a bin and a belt feeder. The lab director introduced me to the new Associate Research Professor, Dr. Andrew W. Jenike, for possible employment during the summer. During the interview, Dr. Jenike said, "If you are to work with me, you will need to apply to graduate school for a PhD." I needed work, so I went home and told my new bride, " I think I will go for a PhD." She replied, "How much longer will it take?" With tongue in cheek, I said, "three years." After some thought, she said, "Ok but no longer, and I want to start our family now." We didn't know that she was already pregnant with our first son Kerry.

Dr. Jenike had already developed a shear tester [2] capable of measuring the strength of bulk solids subjected to pressure and shear conditions in hoppers, and, with the aid of Dr. R. T. Shield [3], they formulated a mathematical model for bulk solids steady flow. This Effective Yield Locus (EYL) concept, (figure 1), provides the link between soil mechanics Yield Locus (YL), for insipient failure, and flowing bulk solids. Jenike also had developed the mathematical relationship between the stress at the abutments of an arch and the flow-no- flow criteria, which states that when the bulk solid's strength exceeds stress at abutments of the arch, the arch will be stable. Calculating the pressures in a flowing mass of bulk solids would be found in V. Sokolovskii's newly published book *"Statics of Soil* media" [4], available only in Russian and Polish. Fortunately, Polish is Jenike's native language, so we worked from his English translation. He excitedly came into my coal bin office one day talking about a Sokolovskii stress field where the stress increased linearly from the apex of a channel ($\sigma(\theta) = \gamma\ r\ s(\theta)$, radial stress) where ($r$,θ)are the

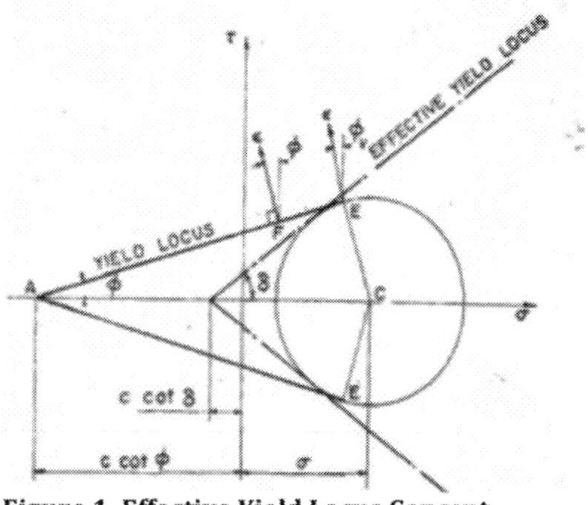

Figure 1, Effective Yield Locus Concept

radial coordinates. This replicated the results of tests in his garage a few years earlier. [5]. This radial stress has the potential of describing bulk solids flow in converging channels. We needed to prove that it does. I begin immediately calculating radial stress solutions for the range of usual hopper parameters, and using the general method of characteristics, showing that any conceivable boundary conditions always converged to the radial stress field [6,7].

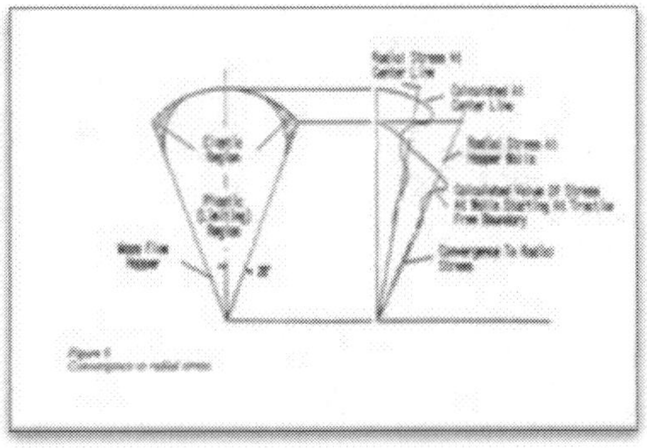

Figure 2, Convergence to radial stress

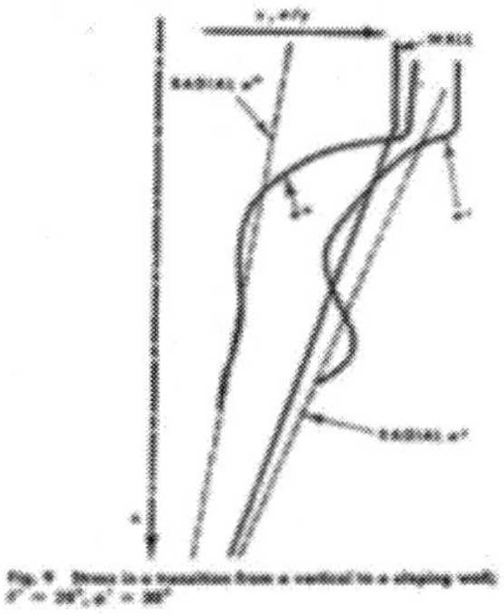

Figure 3, Convergence to radial stress

While calculating the radial stress, I discovered that conical axial-symmetric hopper solutions did not exist for certain hopper half angles θ and wall friction angles φ' (Figure 4). Could it be that I had s

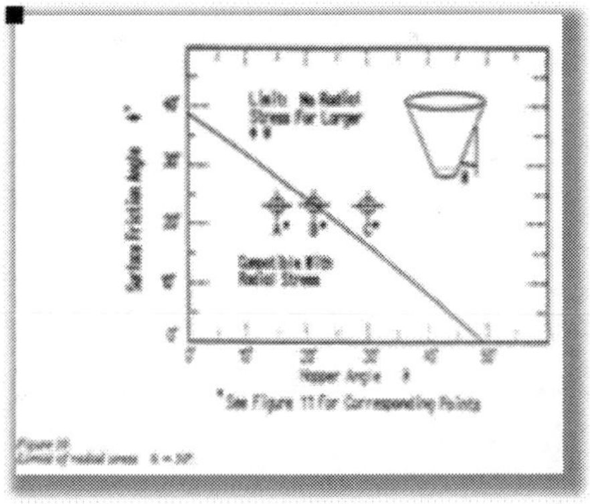

Figure 4, limits to radial stress

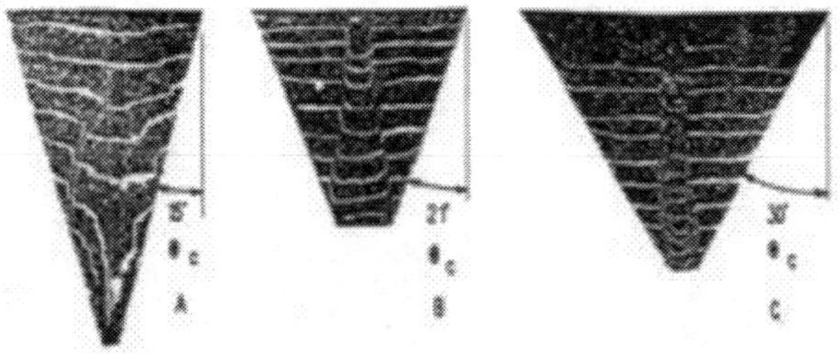

Figure 5, Flow in model conical hoppers

discovered why some hoppers flow only in the center. To verify this, I constructed three conical hoppers with different half angles that could be cut in two after flow to observe full three-dimensional flow (Figure 5). Indeed, flow at the hopper walls occurs only where radial stress is possible. I also successfully compared the radial stress slip lines (velocity characteristics) to the physical slip lines (hopper "A" Figure 5). At this time Dr. Jenike developed failure criteria for a deep well, later called a rathole by Hank Colijn who was the liaison for the Iron and Steel Institute, the organization that sponsored the bulk solids flow project. The work was almost finished.

PhD graduation came quickly after the birth of our second son, Bryan. Unfortunately, I confessed to my wife that I didn't think I could complete my thesis by June graduation. She replied, "Oh yes you will. You promised. I will help you. The boys and I will visit relatives in California. You will have the house to yourself. I will type your completed thesis when I return." It was an offer I couldn't refuse, especially since I had already accepted a job offer from U.S. Steel starting July 1, 1962.

EXTENSIONS OF BULK SOLIDS FLOW THEORY

At the US Steel Applied Research Center, working with Henk Colijn and the full resources of a large company, the possibilities were unbounded. I first wrote the script for the "Million Dollar Rathole" movie and associated paper [8] outlining the theories' practical application. While applying the theory to roll type briquetting presses for iron ore [9,10] and calculating theoretical insert placement in bins to expand flow regions [11,12,13], I approximated the shape of the stress characteristics as circles satisfying the boundary conditions (Figures 5 and 6). This unique method simplified the calculations, reducing the partial differential equations to ordinary differential equations.

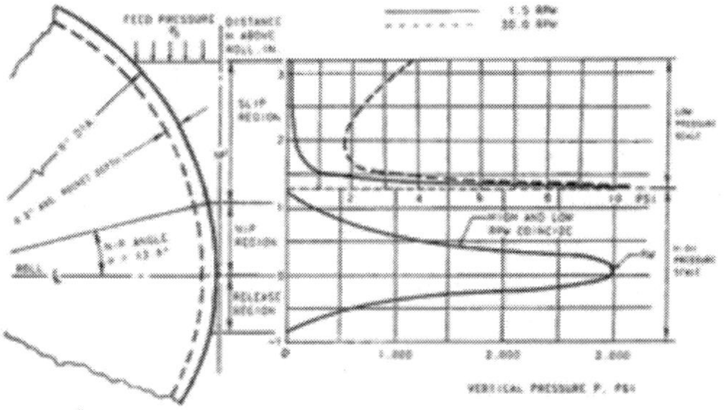

Figure 5, Roll press analysis

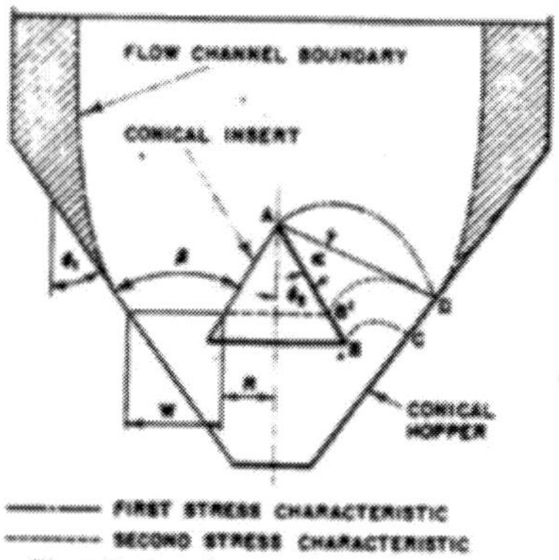

Figure 6, Insert placement analysis

In addition, I derived the limiting flow rate (Q) for converging hoppers [14].
$Q = \gamma A \sqrt{(B g/((1+m) \tan \theta))} \sqrt{(1- fc (1+m)/(\gamma B))}$, γ=bulk density, A=outlet area, B=outlet width or diameter, m= zero for slot outlet and 1 for conical outlet, θ= hopper half angle, g=gravitational constant, and fc= the bulk solid strength at the outlet. This theory includes acceleration and cohesive effects and was verified at the US Steel raw materials truck load-out hopper (Figures 7 and 8).

Figure 7, Truck load-out hopper

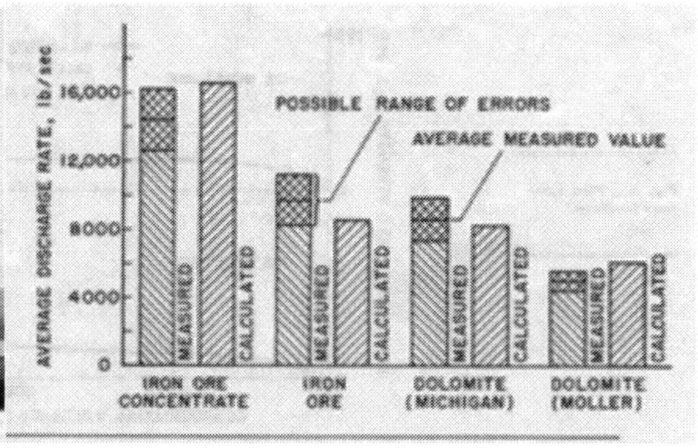

Figure 8, Measured and calculated flow rate

During my four years at US Steel, Jenike finished the University of Utah project and started consulting in Winchester, Massachusetts. I joined him July 1966 to form Jenike and Johanson Inc. He was working with finely ground anthracite, which floods, flushes, and flows like water. His mass flow hopper solution stopped the flushing but caused a limiting flow rate problem. He eliminated this by injecting a small quantity of air into the converging hopper. I had never worked with fine powders before. Our bulk solids flow theory now needed more research to include air effects. This seemed impossible since the solids flow partial differential equations are of a totally different type from fluid flow through porous media. The answer came to me. Extend the Janssen concept.

Janssen considered each horizontal grain layer in the silo subject to an average vertical stress σ_v a horizontal stress σ_h, and a shear stress τ at the slice edges where $\tau = \sigma_h \tan \varphi'$ such that σ_h/σ_v is a constant K. Consequently, the limiting silo wall stress is independent of K (Figure 9). Actually, K is different for different silo wall configurations, $K=(1-\sin\delta)/(1+\sin\delta)$ for diverging cylinders, K=0.5 for constant diameter cylinders, and K is about 2 for Converging cylinders or cones as calculated for radial stress solutions (Figure 9). Because σ_v must be continuous at the transition between the cylinder and the cone, a large peak wall stress occurs on the transition between the cone and the cylinder (Figure10) [15]. This extended Janssen became the basis of my two-phase flow calculations.

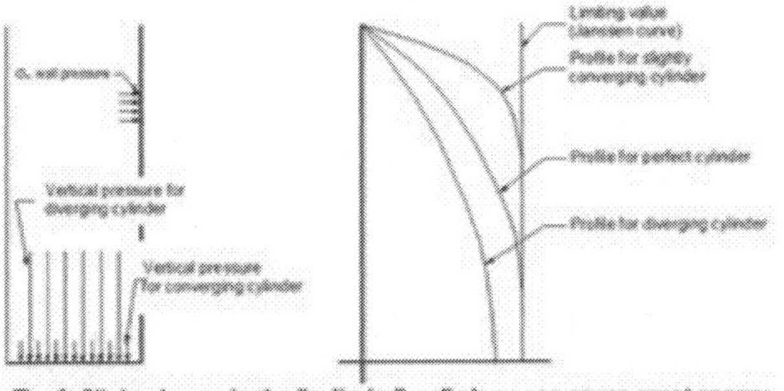

Fig. 9 Slight change in the "cylinder" wall slope can cause great energy dissipation. Vertical pressure can increase as much as 5X with a slightly diverging cylinder.

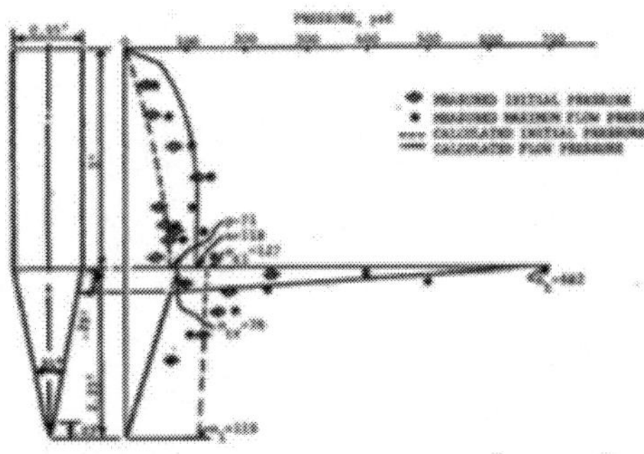

TWO PHASE FLOW

Using Janssen's slice concept with the revised body force BF= bulk density γ minus vertical gas pressure gradient dp/dx, solids continuity, gas continuity, and the gas flow through fine powders rule, $dp/dx=(U/K)\gamma$ where U is the superficial air velocity relative to the solid and $K = K_o (\gamma/\gamma_o)^A$, $\gamma = \gamma_o(\sigma/\sigma_o)^\beta$, where K_o, γ_o, A, σ_o, and β are measured materiel values, provides two phase flow solutions [16]. These include limiting rates resulting when gas is not injected (Figure 11), gas injection required to increase flow rate (figure 11), effectiveness of air blasters (Figure 12), Gas flow effects on slope stability (Figure 13), solids hang-up problems from gas injection in converging hoppers (figure 14), [18], and limiting flow rates in compressible packed beds. Later I extended the roll press theory to include entrained gas effects [17]

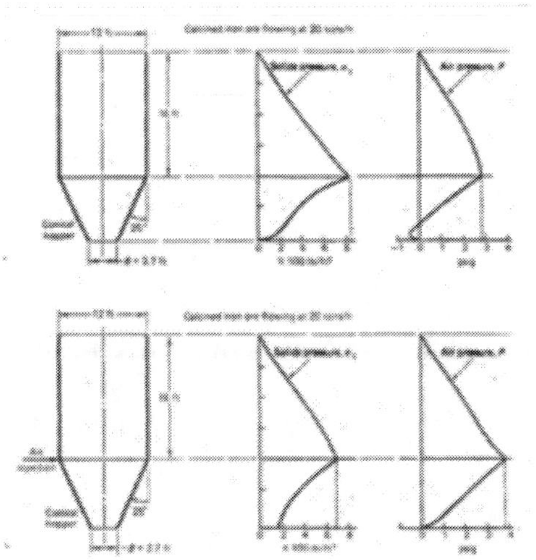

Figure 11, Gas pressures in hoppers

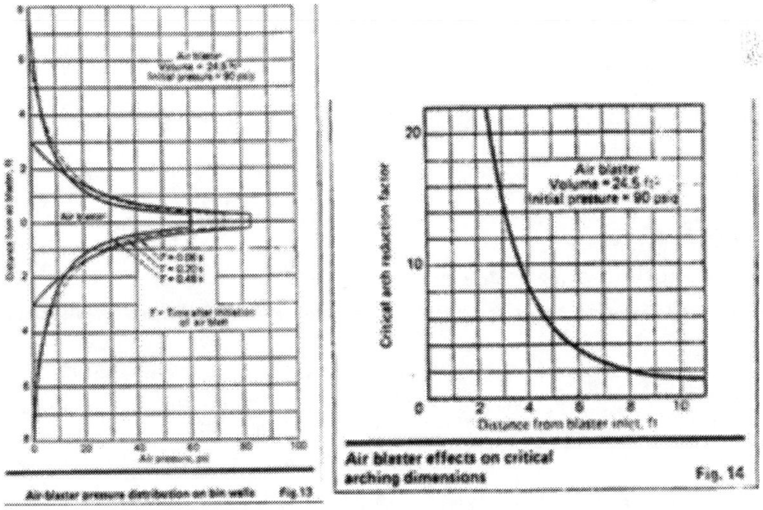

Figure 12, Air blaster analysis

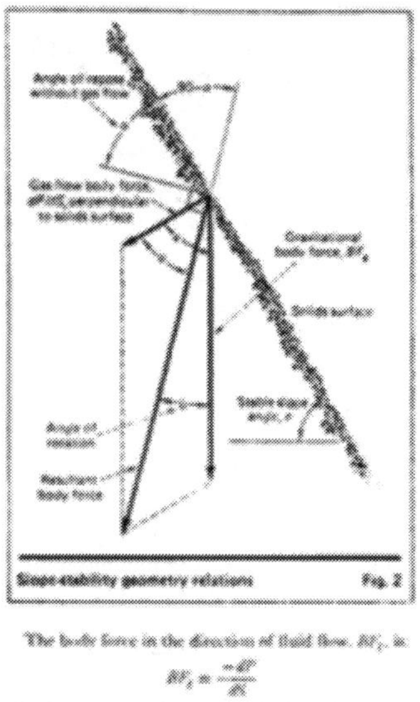

Figure 13, Slope stability

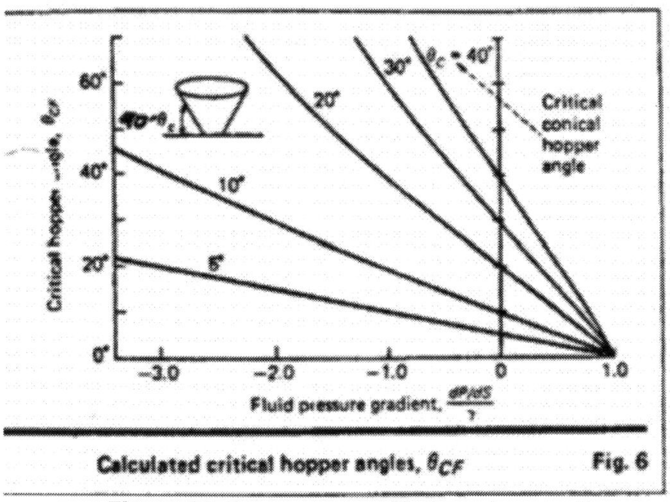

Figure 14, Gas pressure effects on hopper flow

RATHOLE ANALYSIS EXTENDED

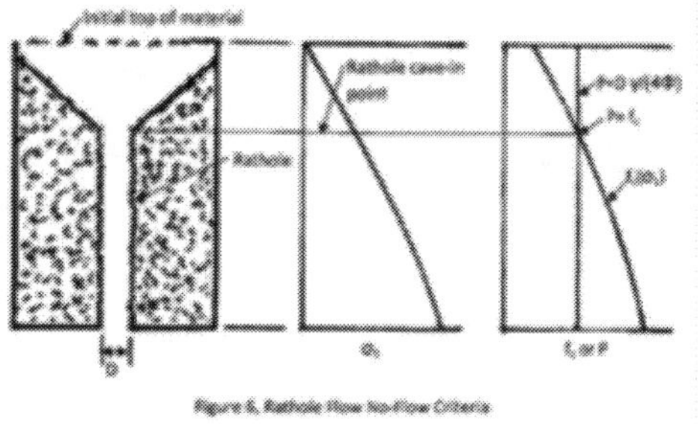

Figure 15, Rathole analysis expanded

The initial rathole failure concept ignored the consolidation history of the solids around the rathole. Using the consolidating pressures when the bin is initially filled to determine the solid's strength allows one to predict the position of the rathole formation in a non-mass flow flat bottom bin (figure 15). This concept is easily expanded non-mass flow hoppers also.

CONE IN CONE

The cone in cone concept came one night when I was awakened with the recollection of some unusual radial stress solutions I ran at the University of Utah involving an interior conical boundary. The limiting (θ, φ') allowable for these boundary conditions between the two cones was essentially the same as the limiting angle for a single conical boundary. This meant that conical hoppers with a cone in cone addition could have a half angle twice that of a single cone, consequently, ratholes can be eliminated in typical 60-degree hoppers [18].

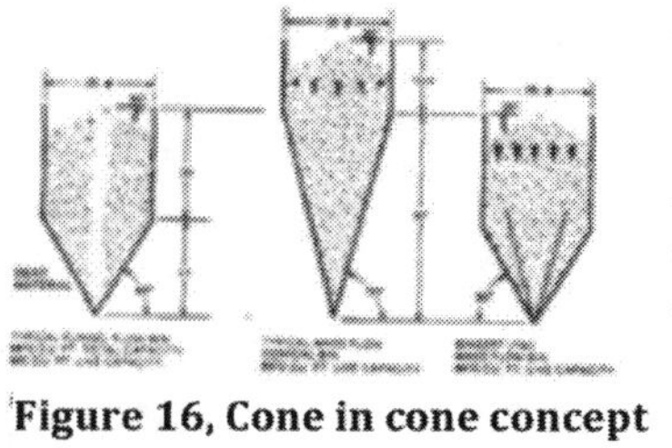

Figure 16, Cone in cone concept

Jenike retired in 1979 using an employ stock option plan. In 1985, my ownership became less than fifty percent; I was forced out of the company and formed JR Johanson Inc. This was good for me. I had to start over unencumbered by old ways. My first development was the uniaxial tester.

UNIAXIAL TESTER

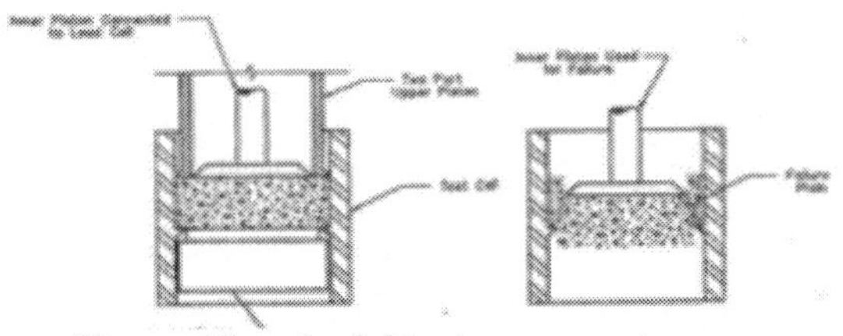

Figure 17, uniaxial tester concept

I needed a solids tester that produced accurate data in much less time than that required by the Jenike shear tester. The result shown here with a unique two piece top piston, and removable bottom plug that allows the flow function, fc verses σ_1, to be determined directly with only two tests. The tester is easily automated and the speed of the failure piston can be varied to measure the flow function at various strain rates [19]. Next was a hopper design taking advantage of the unrestricted hopper angle for plain strain (one-dimensional convergence).

ONE DIMENSIONAL-CONVERGENCE EXTENDED

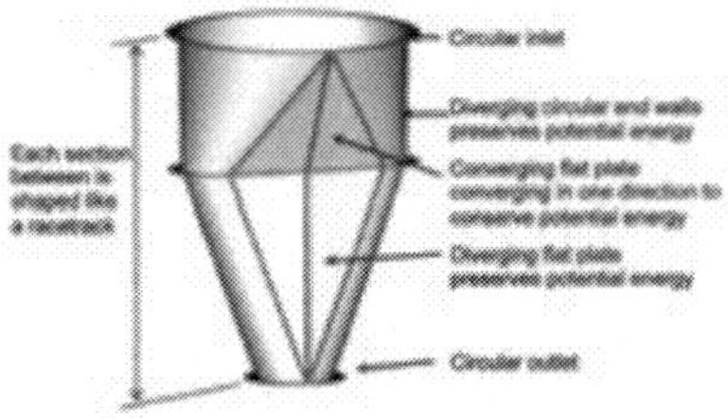

Figure 18, Diamond back hopper

Most significant part of this discovery was the realization that hoppers did not have to be infinitely long to achieve one dimensional flow conditions as long as the sidewalls diverged slightly [20] and the end walls were circular allowing side-to-side motion. Divergence minimizes the sidewall solids pressure and drag. The hopper angles are limited only by the material slide angle on the hopper walls or $(90-\theta)>\varphi'$ [22]. Note that φ' is the actual friction angle between the material and the hopper wall. The next discoveries were a new tester to apply stress uniformly to the sample [21] and oscillating hopper wall walls.

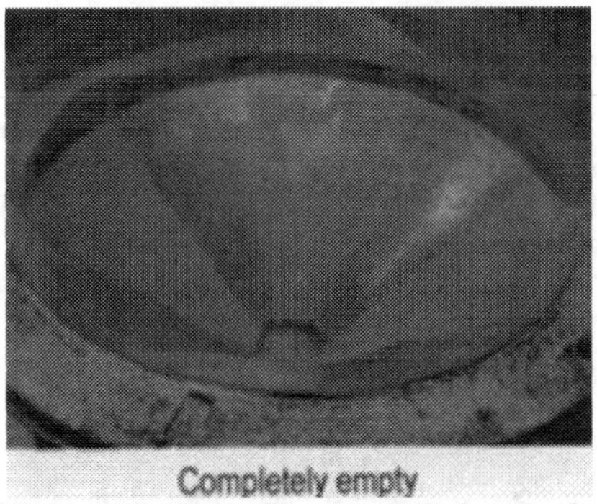

Figure 19, Mass flow in a 45-degree hopper

OSCILLATING HOPPER PLATES

In the 45-degree hopper pictured, the lighter surface plates, hinged at their top center and activated horizontally near the bottom, oscillate causing a zero vertical friction component during oscillation [22]. Activating a single plate eliminates arching and ratholling. Activating all plates simultaneously causes flow along all hopper walls including those not activated (mass flow). The activated plates of the configuration shown, effectively reduce the friction angle to half that when the lighter plates are not activated.

FLOW THEORY NOW

We now have the ability to size hopper outlets and feeders for highly viscous solids like oil sand, brown sugar and woodchips and thus avoid costly hang-ups associated with limiting flow rates simply by using the original EYL concept with effective cohesion c (Figure 1), and introducing a deviator stress $\sigma_d = c \operatorname{Cot} \delta$. The deviator stress is determined experimentally as a function of the strain rate SR imposed on the solids using the uniaxial tester. The geometrical relations of the new effective yield locus (figure 20) provide the following equations:

$\sigma_d = fcc\,(\sigma_1/2 - R)/(R - fcc/2)$, $R = (\operatorname{Sin} \varphi\,(\sigma_1 - fc/2) + fc/2)/(1 + \operatorname{Sin} \varphi)$, $\operatorname{Sin} \delta = R/(\sigma_1 - R - \sigma_d)$,

$\operatorname{Sin} \varphi e' = \tau'/(\sigma' + \sigma_d)$ where fcc is determined from the measured flow function where σ_1=fc, R is the stress circle radius for the σ_1 producing fc, φ is the internal friction angle, δ is the generalized EYL slope angle, and $\varphi e'$ is the effective wall friction angle used in the radial stress calculation results to determine critical conical hopper wall angles. The actual stresses can be calculated from radial stress simply by subtracting σ_d from the radial stress σ values (Figure 21).

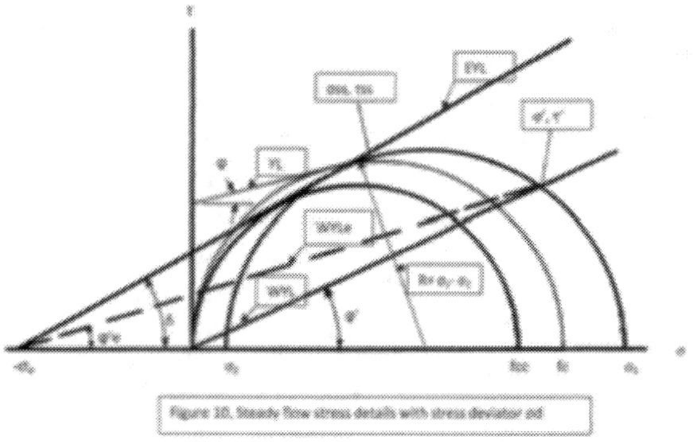

Figure 20 stress circle geometry with deviator σ_d

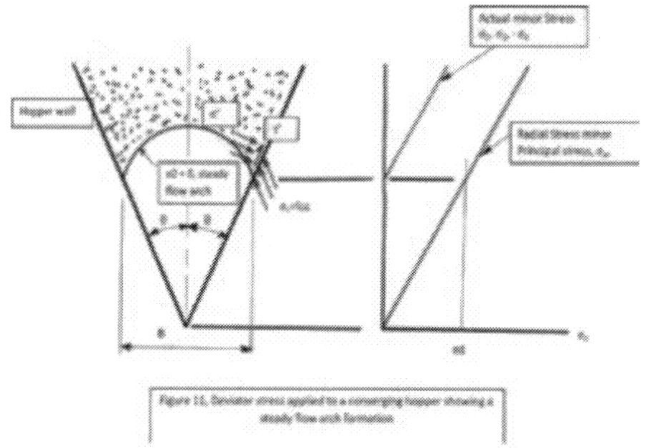

Figure 21 Arching using the deviator stress

A separation plane occurs where the actual minor principal stress $\sigma_2=0$ and the actual major stress σ_1=fcc. The radial stress mean normal stress σ at this failure condition is given by $\sigma = \gamma\ r\ s(\theta)$ where (r,θ) are the radial coordinates and $s(\theta)$ is the dimensionless radial stress. From the geometry (Figure 20, 21) σ = fcc /(2 Sin δ), r = B/(2 Sinθ). These equations combine, defining the critical arching dimension Bc = (fcc/Y) C where C=Sinθ/(s(θ) Sinδ). Typical C values are slightly less than 2 for conical hoppers and slightly less than 1 for one-dimensional hoppers. The average strain rate SR imposed on a bulk solid flowing at an average vertical velocity V in a hopper of half angle θ at a cross-sectional area A is given by SR=(dA/dt)/A = (dA/dx) (dx/dt)/A = (dA/dx) V/A = 2 V (1+m) tan θ / B where B is the hopper width or diameter. For the uniaxial tester, SR= V/H where V is the velocity of the failure piston and H is the height of the material above the tester bottom edge at the time of failure. These equations, then, make it possible to calculate limiting flow rates for vicious materials.

For example, the measured brown sugar (white sugar with molasses) test results illustrate this calculation. The measured flow functions (Figure 22), from the uniaxial tester, show fc increasing linearly with σ_1 and dramatically increasing with strain rate. This Figure 22 provides the critical fcc where σ_1=fcc for various strain rates. Bc(SR) (Figure 23) is calculate using these fcc values. Note that Bc increases dramatically with strain rate. To find the limiting flow rate at a hopper outlet, find the critical strain rate SR for the actual hopper outlet B (Figure 22). Use this SR to find the limiting velocity V=(SR B)/((1+m) 2 Tan θ) and consequently the limiting flow rate FR=Y A V. Note, with δ constant, φ decreases slightly with increased strain rate (Figure 24). These same calculations apply when there is no dependence on strain rate as indicated by the flow function and arching dimensions associated with SR=0.

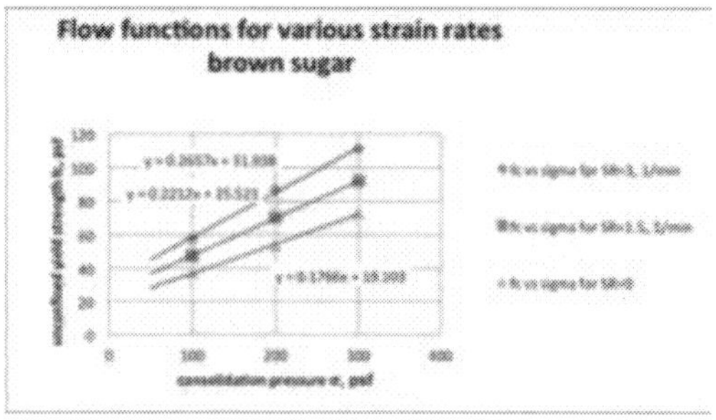

Figure 22, Flow functions for brown sugar

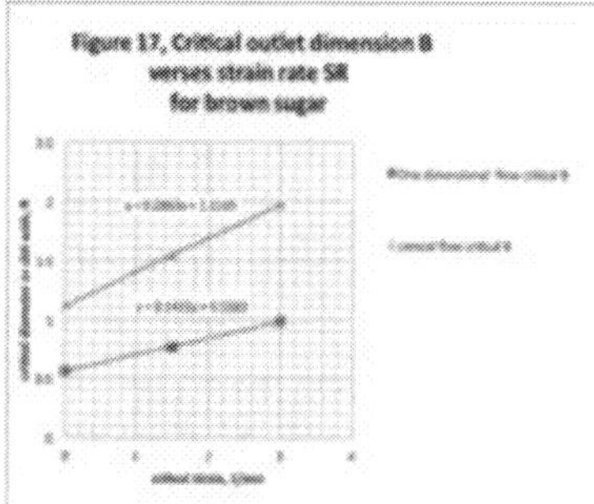

Figure 23, calculated outlet B

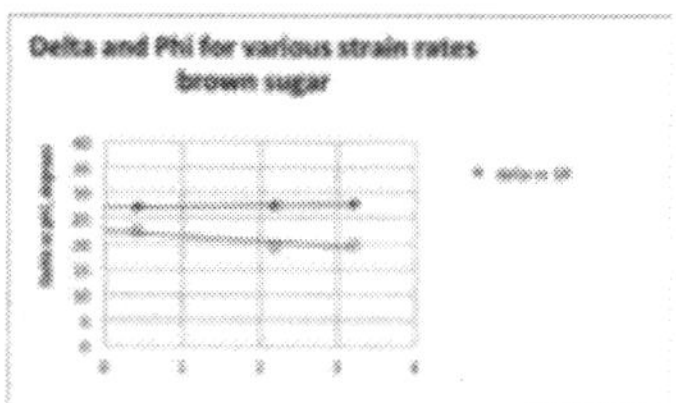

Figure 24, calculated δ

IMPLICATIONS TO NON VISCOUS FLOW

The new theory applied to non viscous bulk solids makes δ is a true constant and much less than previously thought (δ larger than 45 degrees is very unusual). Surface friction angles φe' are much less and the effective WYL approximate strait lines. As a result, the critical conical hopper half angles are larger. This explains why in practice, flow occurs at much flatter conical hopper walls than previously predicted. The modified arching equation also predicts slightly smaller arching dimensions.

THE FUTURE

Have all bulk solids problems been solved? The answer is a resounding no when you consider that: Steady state flow does not exist. Flow is always not symmetric. Bulk materials are not isotropic. Flow properties are changing during flow in processing vessels. Three-dimensional analysis with three-dimensional two-phase flow is required to predict process blow through instabilities. Elastic or rigid inclusions in flowing (plastic) regions) exist and must be included to predict non steady or oscillating flow patterns that can disrupt bulk solids chemical processing. It may take another sixty years but someone will solve these problems. There are still applications for one-dimensional convergence [23]

REFERENCES

1. Janssen, H.A., *Versuche Uber Geteridedruck in Silozellen*, Zeitschrift, Verein Deutscher Ingieure, 39 (August 31, 1895): 1045-1049
2. Jenike, A.W., P,J. Elsey, and R.H. Woolley, *Flow Properties of Bulk Solids*, Bulletin No. 95, Utah Engineering Experiment Station, University of Utah, Salt Lake City, Utah, (December 1958)
3. Jenike, A.W., and R.T. Shield, *On the Plastic Flow of Coulomb Solids Beyond Original Failure*, Journal of Applied Mechanics, (1959); 599.
4. Sokolovskii, V.V., *Statics of Earthy Media*, State Publishing House of Technical and Theoretical Literature, Moscow, Russia, (1954).
5. Jenike, A.W., *Flow of Bulk Solids*, Bulletin No. 64, Utah Engineering Experiment Station, University of Utah, Salt Lake City, Utah, (March 1954)
6. Johanson, J.R., *Stress and Velocity Fields in the Gravity Flow of Bulk Solids*, Bulletin No. 116 and PhD Thesis, University of Utah, Utah Engineering Station, University of Utah, Salt Lake City, Utah (1962).
7. Johanson, J.R., *Stress and Velocity Fields in the Gravity Field of Bulk Solids*, Journal of Applied Mechanics, Series E, 86, September 1964; 5-11.
8. Johanson, J.R. and H. Colijn, New Design Criteria for Hoppers and Bins, Iron and Steel Engineering, (October 1964); 85-105.
9. Johanson, J.R. *A Rolling Theory for Granular Solids*, Journal of Applied Mechanics, Series E, 1965, 32; 842-848.
10. Johanson, J.R., *Factors Influencing the Design of Roll Type Briquetting Presses*, Proceedings Ninth Biennial Conference of the International Briquetting Association, (September 1969; 17-31.
11. Johanson, J.R. *Use of Flow Corrective Inserts in Bins*, Journal of Engineering Industry, Series B, 88 (1966); 224-230.
12. Johanson, J.R. and W.K. Keysteuber, *Structural Support of Flow-Corrective Inserts in Bins*, Proceedings, AICE, Atlantic City, (September 1955).
13. Johanson, J.R., *A Placement of Inserts to Correct Flow in Bins*, Powder Technology, 1 (1967); 328-333/
14. Johanson, J.R., *Method of Calculating Rate of Discharge from Hoppers and Bins*, Transactions of Society of Mining Engineers, 232 (1965); 69-80.
15. Jenike, A.W., J.R. Johanson, and J.W. Carson, *Bin Loads, Part 2: Concepts*, Journal of Engineering for Industry, 99 (1973); 1-5.
16. Johanson, J.R., *Two Phase Flow Effects in Solids Processing and Handling*, Chemical Engineering (January 1, 1979); 77-86.
17. Johanson, J.R., *Predicting Limiting Roll Speeds for Briquetting Presses*, Proceedings, 13th Biennial Conference by Institute of Briquetting and Agglomeration, Vol. 13, (August 1973), 88-89.
18. U.S. Patent Number 4286,883, Blending apparatus for bulk solids (Sept, 1 1981
19. U.S. Patent Number 5,289.728, flow no flow tester, (March 1, 1994)
20. U.S. Patent Number 4,958.741, (Modular mass flow bin, (Sept. 25, 1990)
21. U.S. Patent Number 7,451.666 B2 Apparatus and test procedure for measuring the cohesive, adhesive, and frictional properties of bulk solids (Nov. 18, 2008)
22. U.S. Patent Number 6,086.307, (Hoppers with directionally applied relative motion to promote solids flow, (July 11, 2000)
23. U.S. Patent Number 8,821,697 B2 Reactor vessel having single convergence sidewall plates (Sept. 2, 2014)

501343 Colloidal Pd Nanoparticle Synthesis: The Effect of Ligand-Metal-Solvent Thermodynamics on Kinetics and Final Size

Abstract:

Metal nanoparticles are of great scientific interest due to their unique electronic, catalytic and optical properties. Colloidal metal nanoparticles of different sizes have been successfully synthesized by careful choice of precursors, solvents, and ligands. The roles of ligands and solvents in controlling the size are still not well understood. We have recently reported that surface capping by ligands as well as ligand-precursor coordination can affect the final size of Pd nanoparticles by modulating the surface growth and nucleation rates, respectively. This critical ligand role provides an opportunity for controlling the particle size by tuning the ligand interactions with both the metal precursor and nanoparticle surface. Herein, we report a study on how the thermodynamics of Pd coordination with phosphine and amine ligands in different solvents affect the kinetics of nanoparticles nucleation and growth, and consequently the final particle size. The kinetics were measured by in-situ small-angle X-ray scattering (SAXS) and nuclear magnetic resonance (NMR) was used to characterize the solvent-ligand coordination with the Pd precursors. The possible structures and free energies of ligand-solvent exchange reactions in solution were calculated using Density function theory (DFT). Additionally, we used isothermal titration calorimetry (ITC) to quantitatively compare the thermodynamic interactions between different ligands and palladium precursor and nanoparticle surface atoms by measuring the binding equilibrium constant, heat of binding and stoichiometry of binding. We show how the nucleation and growth rates, and final size are correlated with ligand-Pd(0) and ligand-Pd(II) binding equilibrium. The results give a better understanding of the ligand effect on synthesis kinetics and show the potential for tailoring the solvent-ligand-metal interactions to move closer towards predictive synthesis of metal colloidal nanoparticles.

489858 Switchable Surfactants for the Preparation of Monodisperse, Supported Nanoparticle Catalysts

Abstract:

Synthetic methods of preparing monodisperse, supported nanoparticles remain problematic. Organic surfactants are typically required to passivate the surface of nanoparticles to limit growth and control the size of the nanoparticles. Calcination is typically required following the deposition of the nanoparticles onto a catalyst support to remove the surfactant and activate the catalyst. In general, this calcination causes significant and unpredictable growth of the nanoparticle, undoing the monodispersity maintained during the synthesis. This undesired growth additionally leads to a decline in the catalytic activity of the nanoparticles due to a loss in surface area. We present the use of switchable surfactants (SwiS) for the preparation of monodisperse, supported nanoparticles. SwiS are molecules that undergo reversible, physical changes in response to an external stimulus. The SwiS can be "turned-on" to template nanoparticles during synthesis and then "turned-off" to release bare nanoparticles for deposition onto a catalyst support. The templating synthesis via a hypothesized reverse-micelle process allows for the production of monodisperse nanoparticles controllable by varying the size of the SwiS or the continuous phase. Nanoparticle size has shown to increase as the alkyl chain length of the SwiS is increased. The template is disassembled by reversal of the SwiS by simply heating the solution to drive off the CO_2 in the presence of a catalyst support. The deposition of bare nanoparticles allows for the formation of highly active supported nanoparticles that do not require calcination thus maintaining monodispersity as shown by analysis with transmission electron microscopy (TEM). Supported gold nanoparticles prepared using SwiS are more active in the hydrogenation of 4-nitrophenol than their traditionally prepared counter-parts while requiring less processing thus maintaining the size of the nanoparticle catalysts throughout preparation. Additionally, supported nanoparticles prepared with SwiS are significantly more resistant to sintering than their traditionally prepared analogs.

497875 Colloidal Iron Nanoparticles Provide Size-Control in the Catalytic CO Hydrogenation Reaction

Abstract:

Recent advances in synthetic nanotechnology to produce uniform metal nanoparticle catalysts and to use such "monodisperse" particles in catalytic reactions clearly allowed establishing correlations between the catalyst activity/selectivity patterns and the size and shape of metal catalyst particles. The rationale behind these efforts is the quest for high activity and selectivity in multi-path catalytic reactions, with little to no waste or polluting by-products. Among the various methods employed to prepare well-defined metal nanoparticles, those based on colloidal recipes have gained major importance. While the use of stabilizers ("capping agents", sometimes simply "ligands") to obtain stable colloidal dispersions with narrow particle size distribution is considered mandatory, there is ample evidence that these stabilizers cause alterations in the catalytic properties of the metal nanoparticles. On the other hand, an intriguing reaction for various green feedstock production on an industrial scale is the hydrogenation of carbon monoxide according to Fischer-Tropsch. Fe-, Co- and Ru-based catalysts have been most commonly investigated in the past for their capability to produce paraffins, olefins, and oxygenates with varying hydrocarbon chain length. Tuning these catalysts for high activity and selectivity to specific product classes is therefore of utmost interest. In this research we demonstrated that monodisperse Fe(0) particles in silica could be successfully prepared and tested in the catalytic CO hydrogenation (so-called Fischer-Tropsch process) without ligand effects to play any visible influence.[1] Different from all of the previous studies, which use iron oxide as precursor to prepare catalysts, we applied metallic (0) iron particles. This Fe(0) metal can be easily converted into iron carbide during Fischer-Tropsch synthesis since it is much more active than iron oxide. Palmitic acid and hexadecylamine used as capping agents in suitable ratios, were quantitatively eliminated through the reaction with hydrogen. We further reveal the occurrence of pronounced particle size effects with respect to catalytic activity and selectivity. In particular, an increase in the particle size causes increasing TOF (activity). TOF values of the same magnitude and particle size dependence have been found when relating the geometry-derived TOF to the active Fe(0) metal surface as determined through hydrogen-deuterium exchange. As to the selectivity signatures of Fe/MCF-17, methane and short-chain (terminal) olefin formation significantly increase with decreasing Fe particle size. Quite differently, both the long-chain C_{5+} olefin and (total) oxygenate slates show the opposite behavior. As to the C_{2+} paraffins, it seems that small particles are less efficient in chain lengthening than larger ones. According to XRD results, bulk Fe(0) of the as-prepared catalyst is chemically reconstructed into iron-carbides, e- and χ-Fe_5C_2, after reaction. The occurrence of iron-oxide phases, though not clearly detectable here, cannot be ruled out either. Reaction-induced nanoparticle sintering was limited according to TEM and microtome studies.

Summarizing, the methodological approach to study the catalytic CO hydrogenation over monodisperse Fe nanoparticles intercalated in the mesopores of silica provided new insights into the structure-sensitivity of the reaction and is anticipated to help design catalysts with specific selectivity signatures at high reaction rates.

1. V. Iablokov, Y. Xiang, A. Meffre, P.F. Fazzini, B. Chaudret, N. Kruse *ACS Catalysis* **2016**, *6*, 2496-2500

502479 Selective Dehydrogenation and Oxidation Reactions on Niau Alloys at the Single Atom Limit

Abstract:

Gold, while inert as a bulk metal, shows significant activity and selectivity when in nanoparticle (< 5 nm dia.) form[1]. Despite its high selectivity, monometallic Au is less active than commercial catalysts in reactions of industrial importance, such as the dehydrogenation of ethanol to acetaldehyde and the oxidative methanol-methacrolein esterification to methyl methacrylate. Therefore, the addition of a metal able to coordinate and activate unsaturated organic compounds (e.g. alcohols), such as Ni, could lead to improved catalyst reactivity. Atomically embedded Ni in Au surfaces would form an alloy at the atomic level (Single Atom Alloy), which allows us to reap the benefits of the high selectivity of Au coupled with the remarkable activity of Ni, thus obtaining better catalysts than their monometallic counterparts.

In the present work, we compare various NiAu structures, prepared as Ni-Au nanoparticles (NPs) supported on SiO_2, unsupported Ni-doped nanoporous Au (np-NiAu) and the industrially used Au@Ni core-shell catalyst[2]. The formation of highly dilute Ni-Au alloys is followed by CO-DRIFTS. Temperature Programmed Surface Reaction studies on ethanol dehydrogenation show 100% selectivity and noticeable stability for the NPs and np-NiAu materials. Kinetic measurements find the SAA NiAu to be more active than gold, and further show that the increased activity is induced by Ni, as the apparent activation energy values are lower for the SAA than for the monometallic Au samples. Comparison between Ni-Au NPs and the commercial catalyst was made via *in situ* ATR-IR spectroscopy combined with HPLC for the oxidative methacrolein-methanol coupling reaction in the liquid phase. The catalyst structure and composition was also followed by ICP, STEM/EDS and XAS.

Acknowledgements: This work was supported as part of the Integrated Mesoscale Architectures for Sustainable Catalysis, an Energy Frontier Research Center funded by the U.S. Department of Energy, Office of Science, Basic Energy Sciences, under award #DESC0012573.

References

1. M. Haruta, N. Yamada, T. Kobayashi and S. Iijima, *J. Catal.*, 1989, **115**, 301-309.
2. K. Suzuki, T. Yamaguchi, K. Matsushita, C. Iitsuka, J. Miura, T. Akaogi and H. Ishida, *ACS Catal.*, 2013, **3**, 1845-1849

488371 The Effect Pd Size on TiO_2 for Photocatalytic NO_x Removal

Abstract:

The effect Pd size on TiO_2 for Photocatalytic NOx Removal

Sotiris E. Pratsinis

Particle Technology Laboratory, Institute of Process Engineering,

Department of Mechanical & Process Engineering, ETH Zurich, Switzerland

Reducing the particle size of noble metals on ceramic supports can maximize noble metal performance and minimize its use (Flytzani-Stephanopoulos & Gates, *Annu. Rev. Chem. Biomol. Eng.* 2012, 3:545). Here Pd clusters onto nanostructured TiO_2 particles are prepared in one step by scalable flame aerosol technology while controlling the Pd cluster size from a few nanometers to that of single atoms (Fujiwara et al., *ACS Catalysis*, 2017, 6:1887). Annealing such materials at appropriate temperatures leads to photocatalytic NO_x removal in a standard ISO reactor up to10 times faster than that of commercial TiO_2 (P25, Evonik). Such superior performance can be attained by only 0.1 wt% Pd loading on TiO_2 (Fujiwara & Pratsinis, *AIChE J.*2017, 63:139-146).

Annealing these flame-made powders in air up to 600 °C decreases also the amorphous TiO_2 fraction and increases its crystal and particle sizes as observed by X-ray diffraction (XRD) and N_2 adsorption. The growth of single Pd atoms to Pd clusters on TiO_2 prepared at different Pd loading and annealing conditions was investigated by scanning transmission electron microscopy (STEM) and XRD. The single Pd atoms and clusters on TiO_2 are stable up to, at least, 600 °C for 2 hours in air but at 800 °C they grow into PdO nanoparticles whose fraction is comparable with the nominal Pd loading. Hence, most of Pd atoms are on the TiO_2 surface where at 800 °C they diffuse and coalesce. Diffuse reflectance infrared Fourier transform spectroscopy (DRIFTS) reveals NO adsorption on single, double, 3- and 4-fold coordinated Pd atoms depending on their synthesis and annealing conditions. The peak intensity of NO adsorption sites involving multiple Pd atoms is substantially lower in TiO_2 containing 0.1 wt% than 1 wt% Pd but that intensity from single Pd atoms is comparable. This indicates the dominance of isolated Pd atoms compared to clusters in Pd/TiO_2 containing 0.1 wt% Pd that match or exceed the photocatalytic NO_x removal at much higher Pd contents. Here, the number of isolated Pd sites in different Pd/TiO_2 catalysts is quantitatively compared.

492477 Exploring Fundamental Properties of Fe- and Mn-Promoted Rh Catalysts for Syngas Conversion to Higher Alcohols

Abstract:

As of today, large parts of the chemical industry rely on crude oil as a feedstock. However, in recent years dwindling supplies of this carbon source have led to the exploration of alternative feedstocks. A particularly promising candidate is syngas, a mixture of CO and H_2, which can either be produced from biomass or natural gas. Most commonly, syngas is either converted to methanol or to longer alkanes using the Fischer-Tropsch process, but the direct conversion of syngas to C_{2+} oxygenates would be highly desirable, due to their wide range of applications in chemical, fuel and polymer industries. A catalyst with particular promise in this regard is Rh. While monometallic systems show selectivity towards acetaldehyde, promoters, such as Fe and Mn, significantly influence selectivity. However, the atomic scale nature of the active site and its impact on the reaction network and selectivity has not been explored in detail.

In this contribution, we present our efforts in carefully controlling the synthesis and preparation of Fe- and Mn-promoted Rh catalysts supported on SiO_2. Specific amounts of either one or both promoters are deposited using controlled surface reactions and the impact of systematically varying the amounts of deposited Fe, Mn and Fe/Mn on the product selectivity is explored. This procedure reveals that both, Fe and Mn, increase the selectivity towards ethanol, an effect that is additive and leads to highest selectivity towards ethanol for a ternary system with molar ratios of 1:0.15:0.10 (Rh:Fe:Mn). In a subsequent step, the nature of the active sites is interrogated using a combination of XPS, CO-FTIR spectroscopy and density functional theory calculations. Under reaction conditions, Fe is found to be in a metallic state in a subsurface site. Mn on the other hand is present as an oxide on the 211 step edge. Finally, the implications of the structural changes due to the presence of the promoters on the reaction network and selectivities will be discussed.

The systematic study presented here is a step towards elucidating the effects and interactions of promoters on the selectivity of the studied catalysts. Especially the insights into the nature of the active site are a stepping-stone towards the computational exploration of the impact of various promoters on the activity, which is prerequisite for rational, modeling-based catalyst design.

500296 Controlling Energy Flow in Plasmonic Photocatalysis through the Design of Hybrid Plasmonic Nanostructures

Abstract:

It has been shown that photo-excitation of plasmonic metal nanoparticles (Au, Ag, and Cu) can induce direct photo-chemical reactions on the nanoparticles.[1,2] However, the widespread application of this technology in catalysis has been limited by the relatively poor chemical reactivity of noble metal surfaces. In this presentation, we present our recent work in which we conceptualized and designed hybrid nanostructures in which a plasmonic metal harvests the energy of visible light photons and selectively channels that energy into catalytically active centers on the nanostructure. To accomplish this, we developed a synthetic protocol to deposit a few monolayers of Pt onto plasmonic Ag nanocubes. This model system allows us to conclusively separate the optical and catalytic functions of the hybrid nanomaterial and analyze the interaction between the two functions. Through experimental and theoretical studies of the optical properties of these nanostructures, we show that the flow of energy is strongly biased towards absorption (i.e. excitation of energetic charge carriers) in the thin Pt shell. We demonstrate the utility of these nanostructures for photo-catalytic chemical reactions in the preferential oxidation of CO in excess H_2. The reactor studies conclusively show that photo-excitation of these nanostructures results in photo-chemical reactions on the Pt surface. We describe the fundamental physical reasons for the observed directed flow of energy and discuss how these discoveries impact the field of plasmonic catalysis.

1. Linic, S., Aslam, U., Boerigter, C. & Morabito, M. Photochemical transformations on plasmonic metal nanoparticles. *Nat. Mater.* **14,**567â€“576 (2015).

2. Christopher, P., Xin, H. & Linic, S. Visible-light-enhanced catalytic oxidation reactions on plasmonic silver nanostructures. *Nat. Chem.* **3,** 467â€“472 (2011).

491602 The Use of Genetic Algorithms to Predict the Crystal Structures of Energetics

Abstract:

Crystallography data of high energy density compounds such as TATB, TNT, HMX, and CL-20 are difficult to obtain experimentally; experimental crystallography techniques are expensive and hold many safety risks. The complex energy landscapes of these large energetic compounds make it difficult and time-intensive to predict stable crystal structures via computational means. Genetic search algorithms are regularly used in crystal structure prediction of other compounds by reducing dimensionality of the landscape to minimize computing costs and avoid local minima. We will present the use of an established genetic algorithm for the application of predicting the crystal structures of TATB, TNT, HMX, and CL-20. The results will be validated through comparison of experimental densities, lattice parameters, and x-ray diffraction data. This work establishes a methodology for predicting crystal structures of new energetics and novel energetic co-crystals.

496933 Investigation of Ignition in HMX Using the Henson-Smilowitz Reaction Model

Abstract:

Mesoscale simulations to study ignition behavior of an energetic material (HMX) due to hot spot formation by the shock-induced void collapse phenomena are performed using Eulerian Cartesian grid based solver (SCIMITAR3D). The chemistry is modeled using the Henson-Smilowitz(HS) multistep chemical kinetics mechanism. The HS model is implemented into the flow solver using the Strang splitting method. The phenomena of void collapse leading to formation of hot spots, triggering of chemical reactions and ignition are studied, taking into account the disparate time scales of each of the different physical mechanisms involved. The influence of heat diffusion on initiation of chemical reaction is also investigated. A parametric study based on different imposed particle velocities is performed to understand its dependency on the ignition behavior of HMX.

498658 Combustion Studies of Microscale Aluminum Functionalized with Perfluorotetradecanoic Acid

Abstract:

Micron aluminum (Al) particles are highly reactive when combined with a solid oxidizer, but not exceptionally reactive in an ambient air environment unless they are well dispersed. While enthalpy of oxidation for Al is high; many variables influence the viability of harnessing stored chemical energy within a single Al particle whose reaction is diffusion controlled. One way to enhance Al oxidation is to functionalize the particle surface with a condensed phase oxidizing agent. Fluorocarbons such as perfluorocarboxylic acids have been used to enhance Al combustion for nanoscale Al particles with some success. This study extends previous work toward synthesizing microscale Al particles with surface functionalized perfluorotetradecanoic acid (PFTD) ($C_{14}HF_{27}O_2$) and then characterizing reactivity. Samples were prepared with varying PFTD concentrations ranging from 0 to 25 wt. %. Equilibrium analysis including differential scanning calorimetry, DSC, and thermal gravimetric analysis, TG, reveals that 25 wt. % PFTD concentration is the minimum threshold to induce measurable exothermic pre-ignition reaction (PIR) behavior. In each sample, fluorinated species mostly evaporate upon PFTD decomposition but solid phase CF_2O_2H likely bonds with the Al particle surface. At higher heating rates (e.g., 100 °C/s) two stages of combustion were observed: the first associated with Al reaction with fluorine species from PFTD and a later stage associated with Al reacting with oxygen from the environment. Overall, functionalizing microscale Al particles with a fluorinated polymer facilitates its reaction in an air environment, even for small concentrations of PFTD coating (i.e., 5 wt. %).

494317 Fuel-Rich Metallic Energetics with Metal Fluoride Oxidizers

Abstract:

The importance of thermites over the years has risen due to the continuous improvement in its performance and tunability. The improvements in combustion and ignition behaviors were achieved through the particle size reduction, extending the contact area between fuel and oxidizer.Â Nanocomposite powders were prepared, in which particles remained in the micron-sized range, while containing nano-sized inclusions of individual metal and/or metal oxide phases. In the current work, the focus is to alter the properties of energetic thermite-like materials by changing the oxidizer to a metal fluoride, instead of the common metal oxides. The materials are prepared mechanochemically. The oxidizer; a meta-stable metal fluoride is ball-milled with a reactive metal fuel. Four composites, with fuel/oxidizer ratios fixed at 50-50 wt. % have been prepared and analyzed for the preliminary study; Al/BiF_3 (Equivalence ratio Ï• =9.8), Al/CoF_2 (Ï• =5.4), B/BiF_3 (Ï• =24.6), and B/CoF_2 (Ï• =13.4). The aluminum compositions so obtained are micron-sized composites with ca. 20-nm inclusions of the metal fluorides in the aluminum matrix. Experiments using an electrically heated filament showed that all prepared materials have very low ignition temperatures, with aluminum based ones around 480 °C while for the boron based ones, the temperatures are around 350 °C. Interestingly, all the prepared compositions are insensitive to electrostatic discharge except Al/CoF_2, which may be ignited by high-energy sparks. Thermal analysis of these materials indicates low-temperature reactions in both inert and oxidative environments. Aluminum samples have lower temperature exothermic reactions and no substantial mass loss in inert atmospheres. The Al/CoF_2 sample is more reactive of the two aluminum-based compositions. Boron samples have a mass loss of 25% when heated to 450°C in inert atmosphere suggestive of the formation of boron trifluoride gas. The ignition behavior of the prepared samples shall be studied further using laser and other ignition stimuli. Thermo-analytical kinetic studies of the materials in oxidative and inert atmospheres will be performed to explain and understand ignition behavior as well. Both as-prepared materials and their reaction products will be characterized by x-ray diffraction; their particle sizes will be measured to interpret the experimental observations.

494137 Nanocomposite Thermite Powders with Improved Flowability Prepared By Mechanical Milling

Abstract:

Nanocomposite thermite powders are of interest to develop variety of reactive parts and components. Manufacturing these components requires tailoring such properties of the thermite powders as their particle size distributions, particle shapes, and powder flowability. For example, an improved flowability is desired to use such powders as feedstock in additive manufacturing. Arrested reactive milling (ARM) offers a versatile and practical approach for preparing various nanocomposite thermites with fully dense particles, which will retain their structures and mixedness between reactive components while being stored, handled, and processed. However, ARM products usually have broad particle size distributions, rock-like particle shapes, and poor flowability. Here, ARM is modified to include an additional milling stage to tune the shapes and flowability of the prepared powders. Experiments are performed with aluminum-rich $Al{\cdot}Fe_2O_3$ thermites. After the initial nanocomposite thermite is prepared in a planetary mill, it is additionally milled at a reduced rotation rate, replacing milling balls with smaller glass beads, and adding different liquid process control agents. Powders with modified particle shapes and size distributions are obtained, which have substantially improved flowability compared to the initial material. Ignition temperature is obtained via filament ignition experiment performed in air. The rate of ignition and combustion are evaluated using electro-static discharge (ESD) experiment while combustion performance is tested using constant-volume explosion (CVE) experiment. A reaction kinetics study is performed at various heating rates using differential-scanning-calorimetry (DSC). The wire ignition set-up is modified with a copper plate mounted under the heated filament wire to capture igniting particles, thus enabling us to observe the products of partially burned individual particles for different prepared samples. In a modified ESD set-up, an aluminum foils is attached to also capture igniting particles at difference distances. The morphology of particles captured on foils are analyzed via scanning-electron-microscope (SEM) and compared for the same powders ignited using different heat sources (heated wire and ESD). The reactivity of the modified powders is not diminished; instead, an improved reactivity is observed for selected samples. Detailed results will be further discussed in the talk.

496988 Doping Boron with Iron for Better Combustion

Abstract:

Boron is a thermodynamically attractive fuel additive for propellants and explosives, however it has yet to reach its full potential because of its lengthy combustion times and ignition delays. This work focuses on increasing the rate of the heterogeneous combustion reaction on the surface of burning boron particles by introducing iron as a catalytic additive. Iron is readily oxidized and reduced, and is therefore expected to improve the surface reaction rate. Varying quantities of iron (0 â€“ 5 wt %) are introduced as a coating on boron agglomerates using iron pentacarbonyl as the iron source. The samples are prepared using 95% pure commercial boron and higher purity, 99%, boron powders in an inert reaction batch with dodecane and an organic surfactant. The prepared samples are characterized using SEM analysis to observe surface morphology and particle size distribution. Each sample is ignited in air using a CO_2 laser. Particle emission streaks of single burning particles are collected optically to determine burn times. Correlating statistical distributions of the measured burn times with the particle size distributions, an effect of the particle size on its burn time is determined for all materials. Flame temperatures are also recovered from the optical measurements. Combustion data will be compared to boron-iron composites prepared by mechanical milling to establish the effect of the preparation technique on the effectiveness of iron additive. The effect of purity of the starting material and iron dosing will also be discussed in detail.

498862 Synthesis and Reactive Characterization of Aluminum Combined with Aluminum Iodate Hexahydrate Crystals $[Al(H_2O)_6](IO_3)_3(HIO_3)_2$

Abstract:

Synthesis and Reactive Characterization of Aluminum Iodate Hexahydrate Crystals $[Al(H_2O)_6](IO_3)_3(HIO_3)_2$

A crystalline aluminum iodate acid salt has been synthesized from aluminum particles dissolved in iodic acid solution. The precipitate from solution is aluminum iodate hexahydrate $[Al(H_2O)_6](IO_3)_3(HIO_3)_2$, as confirmed by X-ray diffraction (XRD) analysis. The method of synthesis first dissolves iodine oxide in water, creating an IO_3^- solution with pH < 0.2. Aluminum nanoparticles are added to the IO_3^-solution and aluminum iodate hexahydrate (AIH) crystals precipitate. The bulk density of the crystalline AIH and Al composite is dependent on the initial water to aluminum concentration ratio during synthesis. Reactivity is characterized in terms of flame speed with measurements purposefully designed to capture less than 1% light emission, resulting in speeds as high as 3200 m/s for AIH + Al composite with a calculated density of 3.43 g/cc. AIH increases reactivity in Al energetic materials by replacing the Al_2O_3 passivation layer that naturally forms on Al particles with AIH. AIH is a reactive salt and the energy released form the formation of Al_2O_3 is significantly less than the energy released form the formation of Al_2O_3 from unreacted Al in the core of Al particles. Al_2O_3 acts as a heat sink and oxygen diffusion barrier and replacement of the Al_2O_3 passivation layer with AIH significantly reduces the apparent activation energy of Al oxidation. The reaction to form Al_2O_3 from AIH is extremely over oxidized (15.5 moles of extra oxygen for every mole of AIH) and replacement of the Al_2O_3 passivation layer with AIH increase the amount of available oxygen and reduces the distance between unreacted Al in the core and oxidizer. The combined effects of reducing apparent activation energy and reduced distance between fuel and oxidizer result in flame speeds as high as 3200 m/s in AIH mixtures.Â Reactivity is affected by the concentrations of AIH, Al, Î²-HIO_3, and HI_3O_8 of AIH mixtures. At an ER of 1.0, reactivity increases as concentration of Î²-HIO_3 and HI_3O_8 increase. When concentrations of Î²-HIO_3 and HI_3O_8 increase, AIH concentration decreases from 75wt% to 60wt%. The increase in reactivity at an ER of 1.0 could be a result of increased density from higher concentrations of iodic acids or could be a result of reduced concentration of AIH.

499762 Magnesium Diboride As a Fuel for Biocidal Formulations

Abstract:

Magnesium Diboride as a Fuel for Biocidal Formulations

Fidel D. Ruz-Nuglo, Hannah Dudak, and Lori J. Groven

Department of Chemical and Biological Engineering,

South Dakota School of Mines and Technology, Rapid City, South Dakota

Abstract

Magnesium diboride is typically considered a cheap and readily available ceramic material; however, it can also be used as a fuel in pyrotechnic applications. For example, Koch et al. recently demonstrated its use in infrared decoy flare formulations and Brusnahan et al. demonstrated its use as a green light emitter. This system is also interesting as a potential biocide where the limitations of aluminum can be overcome (more favorable aging and chemistry). In this work, we build upon our previous efforts focused on printable biocides and examine the role of magnesium diboride on the combustion and total iodine generation when combined with a suitable fluoropolymer binder and as fuel/oxidizer. It was observed that polymer content as low as 10% will result in printable formulations. Bomb Calorimetry under both inert (heat of reaction) and air (heat of combustion) atmosphere are investigated to shed light on its effect on the end products. Simultaneous differential scanning calorimetry and thermo-gravimetric analyses (DSC/TGA) were performed to explain the interactions between the polymer, iodate, and magnesium diboride. Rheological properties of the biocidal ink and combustion characteristics of the printed formulations will also be presented in detail.

499597 Fused Deposition Modeling of Biocidal Formulations

Abstract:

Fused Deposition Modeling of Biocidal Formulations

Fidel D. Ruz-Nuglo and Lori J. Groven

Department of Chemical and Biological Engineering,

South Dakota School of Mines and Technology, Rapid City, South Dakota

Abstract

Fused deposition modeling is a widely used technique for achieving additively manufactured materials. In recent years the technique has been used to print not only polymer filaments but highly solid loaded filaments where the inclusion content exceeds 70 wt%. A very good example is that of the metallized filaments where 80% aluminum or brass can be successfully printed. Therefore, it is expected that one could develop a highly solids loaded reactive filament and print if the polymer characteristics are favorable towards such. In this work, we demonstrate for the first time the development of a biocidal filament that can be printed with any off the shelf FDM system. The filament and subsequent printed materials are characterized for their aging/stability, combustion, thermal degradation profile, mechanical properties and ability for tailored iodine release.

490066 Computational Study of a MOF-Supported Single Site Ni Catalyst for Ethylene Dimerization

Abstract:

Metal-organic frameworks (MOFs) with ultrahigh porosity and large surface area have recently emerged as excellent heterogeneous supports for transition metal catalysts, either through functionalization of the linker or direct modification of the nodes. The mesoporous MOF, NU-1000, is of particular interest due to its chemical and thermal stability as well as its large pore size (~30 Å), which facilitates the diffusion of gaseous molecules. More importantly, the coordinatively unsaturated metal sites at the inorganic nodes provide specific sites for metal oxide species to anchor, enabling the synthesis of a highly dispersed and isolated single-site metal catalysts on the internal surface of NU-1000. Through atomic layer deposition, a highly dispersed Ni catalyst (Ni-NU-1000) was successfully synthesized; it shows good activity and stability for ethylene hydrogenation and was predicted to be a potential catalyst for ethylene oligomerization.(1,2)

In this work, we aim to study the structure of the Ni-NU-1000 catalyst and from there to identify the catalytic active sites for ethylene dimerization. In addition, the effect of the pore size on the selectity for 1-butene production have been studied. Furthermore, we have screened the first-row transition metals for their activity toward the ethylene dimerization. The goal is to provide physical insight into properties controlling the activity and selectivity for ethylene dimerization and then guide experimentalists to design a better catalyst for ethylene dimerization.

The current study provides physical insight into the properties of MOF supported metal clusters in controlling the activity and selectivity for dimerization of ethylene to produce 1-butene, which is a basic petrochemical building block and largely used as a co-monomer for the production of light weight polyethylene. As compared to homogeneous catalysts, the novel heterogeneous catalyst has many industrial advantages such as ease of product separation and catalyst recycling.

References

1. Li, Z.; Schweitzer, N. M.; League, A. B.; Bernales, V.; Peters, A. W.; Getsoian, A. B.; Wang, T. C.; Miller, J. T.; Vjunov, A.; Fulton, J. L. *J. Am. Chem. Soc.*, *138*, 1977 (2016).
2. Mondloch, J. E.; Bury, W.; Fairen-Jimenez, D.; Kwon, S.; DeMarco, E. J.; Weston, M. H.; Sarjeant, A. A.; Nguyen, S. T.; Stair, P. C.; Snurr, R. Q. *J. Am. Chem. Soc.*, *135*, 10294 (2013).

490094 Two-Dimensional Nanomaterials As Supports for Heterogenous Catalysis

Abstract:

Two-Dimensional (2D) nanomaterials have garnered remarkable interest recently due to their promising potential in versatile applications such as energy storage devices including lithium ion batteries (LIB) and capacitors. However, catalytic properties of 2D nanomaterials, especially for thermo-catalytic reactions, are rarely studied. The 2D structures may facilitate the characterizations of the supported nanoparticles and help determine the chemistry of active sites.

Water gas shift (WGS) reaction is widely used in industry to produce clean hydrogen for plenty of processes such as ammonia synthesis, methanol synthesis and hydrocarbon processing. Moreover, WGS is a useful probe reaction that can shed light on fundamental catalytic processes over considerable catalyst systems. In this presentation, we studied the effects of 2D nanomaterials as supports of noble metals for WGS reaction. Kinetics such as orders of reactants cast light on absorption and coverage of the reactants. Scanning transmission electron microscopy (STEM) as well as X-ray absorption fine structure (XAFS) were employed to identify the nature of active sites.

496895 Mitigating the Oxidative Deactivation on a Multi-Faceted Fe Catalyst: A Multi-Scale Model from First Principles

Abstract:

One promising method for addressing the growing need for diverse, sustainable energy sources is bio-oil produced via the fast-pyrolysis of biomass. However, these bio-oils have high concentrations of oxygenated compounds, resulting in fuel quality issues[1] that require catalytic hydrodeoxygenation (HDO), thereby reducing the overall oxygen content and improving fuel quality. Bimetallic, doped iron, catalysts have demonstrated synergistic behavior that contributes to a more cost-effective and longer-lasting HDO catalyst[2-6]. In order to improve the performance of such bimetallic iron catalysts, the oxidative deactivation mechanism that limits these catalysts must be determined.

While the addition of dopants to Fe catalysts exhibits an overall synergistic interaction, further investigation into the behavior of oxygen on a doped catalytic Fe grain is required to minimize the use of precious metals while maximizing its synergistic properties[2,4,7]. Here, we model the oxidative deactivation of a full catalytic Fe grain by exposure to O_2 as well as taking into account both coverage and facet effects on the kinetic behavior. Additionally, as the application of external electric fields has demonstrated favorable effects on other catalytic processes, such as methane steam reforming, the effect of a range of electric fields were also investigated[8-11]. Such a realistic, multi-faceted model of a doped Fe grain is an effective tool for optimizing the use of precious metal dopants and taking advantage of the effect of an electric field. These realistic, predictive models help to bridge the gap between experiment and theory, streamlining research efforts and accelerating progress.

In order to build the multi-faceted catalytic Fe grain model, the adsorption of oxygen was calculated on both clean and doped Fe(100), Fe(110), and Fe(111) over a range of surface coverages. The effect of an electric field on oxygen adsorption on these three Fe facets was studied over a range of ±1.0 V/Å, examining each available adsorption site over a range of coverages. Surprisingly, we found that the undoped Fe(111) facet favors a high oxygen coverage even under low O_2 exposure and all electric field strengths (see Figure 1). Results indicate that the undoped Fe(110) surface is the most strongly affected by the electric field over the coverage range, while the undoped Fe(100) surface is the least effected. As seen in Figure 1, the undoped Fe(100) and Fe(111) surfaces have the strongest adsorption energy, resulting in high oxygen coverage upon exposure to O_2. Preliminary work on Pd doped Fe(100) with varying Pd concentration has shown that increasing the surface concentration of Pd significantly decreased the adsorption strength of oxygen, which is consistent with other studies on the adsorption of benzene and phenol on Fe doped surfaces[12,13]. These results suggest that the primary function of Pd is the disruption of oxide formation due to the aversion that oxygen has for Pd on the Pd/Fe(100) surfaces. Consequently, the degree of catalyst deactivation due to the formation of a surface oxide likely decreases with the addition of Pd to an Fe surface, thereby promoting HDO with minimal deactivation of the active Fe surface.

The three Fe facets studied here were then used to construct a model of a multi-faceted grain surface, incorporating the anisotropic effects of the surface orientation on kinetic characteristics[14]. Such a model accounts for the effect of pressure, temperature, electric field, dopants, and surface orientation on the adsorption/desorption and oxidative deactivation behavior. This doped Fe catalytic grain model can be used as an experimentally predictive, useful tool for optimizing methods in which HDO catalysts are designed to reduce oxidative deactivation while minimizing the use of costly precious metals.

Figure 1: Equilibrium oxygen coverage on a multi-faceted catalytic iron grain at 500 K,1.5 Pa, and an electric field of -1.0 V/Å.

References:

1. Wang, H., Male, J. & Wang, Y. Recent Advances in Hydrotreating of Pyrolysis Bio-Oil and Its Oxygen-Containing Model Compounds. *ACS Catal.* **3,** 1047–1070 (2013).

2. Hong, Y., Zhang, H., Sun, J., Ayman, K. M. & Hensley, A. Synergistic catalysis between Pd and Fe in gas phase hydrodeoxygenation of m-cresol. *ACS Catalysis* (2014). doi:10.1021/cs500578g

3. Nie, L., de Souza, P. M., Noronha, F. B. & An, W. Selective conversion of m-cresol to toluene over bimetallic Ni–Fe catalysts. *Journal of Molecular …* (2014). doi:10.1016/j.molcata.2013.09.029

4. Sun, J. *et al.* Carbon-supported bimetallic Pd–Fe catalysts for vapor-phase hydrodeoxygenation of guaiacol. *Journal of Catalysis* **306,** 47–57 (2013).

5. Sitthisa, S., An, W. & Resasco, D. E. Selective conversion of furfural to methylfuran over silica-supported Ni Fe bimetallic catalysts. *Journal of Catalysis* (2011). doi:10.1016/j.jcat.2011.09.005

6. González-Borja, M. Á. & Resasco, D. E. Anisole and guaiacol hydrodeoxygenation over monolithic Pt–Sn catalysts. *Energy & Fuels* (2011). doi:10.1021/ef200728r

7. Hensley, A. J. R. *et al.* Enhanced Fe 2O 3Reducibility via Surface Modification with Pd: Characterizing the Synergy within Pd/Fe Catalysts for Hydrodeoxygenation Reactions. *ACS Catal.* **4,** 3381–3392 (2014).

8. Che, F., Ha, S., McEwen, J. S., Hensley, A. J. & Zhang, R. Elucidating the field influence on the energetics of the methane steam reforming reaction: A density functional theory study. *'Applied Catalysis B, Environmental'* (2016). doi:10.1016/j.apcatb.2016.04.026

9. Che, F., Gray, J. T., Ha, S. & McEwen, J.-S. Improving Ni Catalysts Using Electric Fields: A DFT and Experimental Study of the Methane Steam Reforming Reaction. *ACS Catal.* (2016). doi:10.1021/acscatal.6b02318

10. Che, F., Ha, S. & McEwen, J. S. Hydrogen Oxidation and Water Dissociation over an Oxygen-Enriched Ni/YSZ Electrode in the Presence of an Electric Field: A First Principles-Based Microkinetic …. *Industrial & Engineering Chemistry …* (2017). doi:10.1021/acs.iecr.6b04028

11. Che, F., Gray, J. T., Ha, S. & McEwen, J. S. Catalytic water dehydrogenation and formation on nickel: Dual path mechanism in high electric fields. *Journal of Catalysis* (2015). doi:10.1016/j.jcat.2015.09.010

12. Hensley, A. J. R., Zhang, R., Wang, Y. & McEwen, J.-S. Tailoring the Adsorption of Benzene on PdFe Surfaces: A Density Functional Theory Study. *J. Phys. Chem. C* **117,** 24317–24328 (2013).

13. Hensley, A., Schneider, S., Wang, Y. & McEwen, J. S. Adsorption of aromatics on the (111) surface of PtM and PtM 3 (M= Fe, Ni) alloys. *RSC Adv.* (2015). doi:10.1039/C5RA13578H

14. McEwen, J. S., Gaspard, P., de Bocarmé, T. V. & Kruse, N. Oscillations and Bistability in the Catalytic Formation of Water on Rhodium in High Electric Fields. *J. Phys. Chem. C* **113,** 17045–17058 (2009).

485927 Heterogeneous Dephosphorylation Using Ceria Nanocatalysts: Identifying the Active Site and the Rate-Determining Step

Abstract:

As an essential element in living systems, phosphorus is crucial for cell division and growth, energy storage and conversion, respiration, photosynthesis, and other biological processes. Phosphorus is found naturally in soil, but concentrations can fall to low levels depending on geographic conditions, and artificial phosphorus fertilizers are demanded to sustain the growth of crops. Before the early 1900s, most of the worldâ€™s phosphorus was derived from animal wastes. Today the vast majority (nearly 80%) of phosphorus in fertilizers comes from phosphate rocks, which are primarily harvested in the remote Western Sahara region. It is predicted that the production of phosphate rocks will reach its peak before 2040 and the reserves will be completely depleted by the end of this century. Therefore it is imperative to develop innovative and sustainable methods for production of phosphorus from renewable sources.

One solution toward this sustainability challenge is to extract phosphorus from phosphorylated biomolecules (e.g., phospholipids and nucleic acids) and recycle it for fertilizer production. This is conventionally done via fermentation, which produces struvite, a precipitate that can be processed to make phosphorus fertilizers. Alternatively, a more robust approach is catalytic dephosphorylation. By hydrolytic cleavage of the phosphate ester bond, free phosphate anions can be released from biomass and agriculture wastes, which can then be captured and regenerated as chemical streams for further applications. Previously studies of cerium oxide (CeO_2) nanocrystals for biological applications (e.g., cancer therapy, pharmacology and toxin mitigation) have shown that these nanomaterials can function as artificial phosphatases and catalyze the dephosphorylation of nucleic acids, peptides, DNA and RNA under ambient conditions. Thereby CeO_2 is believed to be a promising catalyst for dephosphorylation and recovery of phosphorus from biomass and organic wastes

Here we report on catalytic dephosphorylation for recovery of phosphates from organic and biological molecules. Ceria (CeO_2) nanocrystals were synthesized with shape control and applied as â€œartificial phosphatasesâ€ to cleave the phosphate ester bond in para-nitrophenyl phosphate (p-NPP) and release free phosphate anions in aqueous solutions. After a fundamental study on this model reactant, CeO_2 nanocrystals were also applied to feedstocks to phospholipids and nucleic acids. The dephosphorylation reaction was studied on the CeO_2 nanocrystals at various temperatures to evaluate the dependences of rate constant, activation energy and recyclability on the particle shape. The structure-property relationship established in these studies suggests that the oxygen vacancies on the surface of CeO_2 are the active sites for dephosphorylation. Adsorption studies of poisoning species allow for the proposal of a potential rate-determing step in this reaction.

487048 Surface Reconstruction of Pd Concave Icosahedra: Atomic-Scale Mechanisms from First-Principles and Experiments

Abstract:

Concave surfaces exhibit a large density of high-index facets, such as steps, edges, and kinks. These structural features offer sites of high surface energy, and low coordination number[1]. Consequently, they have shown improved catalytic activities toward a number of reactions[2], including formic acid electro-oxidation[3], and Suzuki coupling reaction[4]. A common aspect of this diverse set of reactions is that they take place at low temperatures (<100 °C), which does not compromise the structural integrity of the as-synthesized concave structures. However, a myriad of structure-sensitive catalytic processes take place at much higher temperatures. Therefore, acquiring an atomic-level understanding of the mechanisms by which the concave surface structures transform into more common surface facets at elevated temperatures is of paramount interest. Here, we focus on the atomic surface reconstruction of Pd concave icosahedra to regular icosahedra. Applying self-consistent periodic density functional theory (PW91-GGA) to an atomistic model of a concave icosahedron edge, we quantify activation energy barriers for key steps responsible for the surface reconstruction, including step deformation and adatom diffusion. Our calculated energetics are surmountable at the temperatures at which the reconstruction experimentally takes place, as observed by *in situ* high-resolution transmission electron microscopy (HRTEM). Combining experimental and theoretical methods, we offer insight into the mechanism of surface reconstruction of Pd concave icosahedra to regular icosahedra, which offers an opportunity to rationally design thermally-durable concave nanoparticles.

References

1. Ruditskiy A, Peng HC, Xia YN. Shape-Controlled Metal Nanocrystals for Heterogeneous Catalysis. *Annual Review of Chemical and Biomolecular Engineering, Vol 7.* 2016;7:327-348.

2. Zhang JA, Langille MR, Personick ML, Zhang K, Li SY, Mirkin CA. Concave Cubic Gold Nanocrystals with High-Index Facets. *Journal of the American Chemical Society.* Oct 2010;132(40):14012-14014.

3. Huang XQ, Zhao ZP, Fan JM, Tan YM, Zheng NF. Amine-Assisted Synthesis of Concave Polyhedral Platinum Nanocrystals Having {411} High-Index Facets. *Journal of the American Chemical Society.* Apr 2011;133(13):4718-4721.

4. Jin MS, Zhang H, Xie ZX, Xia YN. Palladium Concave Nanocubes with High-Index Facets and Their Enhanced Catalytic Properties. *Angewandte Chemie-International Edition.* 2011;50(34):7850-7854.

496857 Enhanced Catalysis By Optical Nanoantenna Reduced on Monolayer Transition Metal Dichalcogenide

Abstract:

Enhanced Catalysis by Optical Nanoantenna Reduced on Monolayer Transition Metal Dichalcogenide

D. Keith Roper[a,b*], Jeremy R. Dunklin[c], Gregory T. Forcherio[d], Alexander O'Brien[a]

[a]*Department of Chemical Engineering, University of Arkansas, Fayetteville, AR 72701*

[b]*Microelectronics and Photonics Graduate Program, University of Arkansas, Fayetteville, AR 72701*

[c]*National Renewable Energy Laboratory, Golden, CO 80401*

[d]*Army Research Laboratory, Adelphia, MD 20783*

Heterostructures of two-dimensional (2D) transition metal dichalcogenide (TMD) and optical nanoantenna (NA) offer optoelectronic inducibility with superior electron mobility and gate tunability. This enables enhanced photocatalytic activity of interest for chemical and biological catalysis, energy, sensing, desalination, and nanoelectromechanical systems. However, characterization of interrelated electro-optical and thermal effects at TMD-NA heterointerfaces is predominantly empirical. Simulation and its integration with microscopic and spectroscopic analysis of TMD-NA has been limited to date by complexity and convolution of effects, particularly for characterizing dynamic interactions at nanometer (nm) scales. Compact, multi-scale, integrated analysis of optical, electronic and catalytic effects could identify extraordinary features to guide design and implementation.

This work compared simulated vs. measured optical, electronic, and spectrocopic properties of TMD-NA for photoreductive catalysis. Heterostructures of monolayer (1L) TMD and NA were self-assembled via exfoliation and redox chemistry (Fig 1). Nanometer- and femtosecond-resolved electron energy loss spectroscopy (EELS) was used to simulate and measure low-energy NA plasmon modes, damping and electric near fields at heterointerfaces (Fig 2). Discrete dipole approximation (DDA) was used to analyze transmission ultraviolet-visible (TUV-vis) extinction spectra (Fig 3). Heterostructures of 1LTMD-NA were tested as catalyst electrodes for hydrogen evolution reaction using linear sweep voltammography.

Transmission electron microscopy (TEM; Fig 1) and x-ray photoelectron spectroscopy (XPS) showed NA were reduced directly onto 1LTMD flake edges forming covalent Au-sulfur bonds, the first such reported result. EELS (Fig 2) showed enhanced plasmon damping attributable to direct electron transfer at the catalytic 1LTMD-NA heterointerface. DDA (Fig 3) revealed plasmon and exciton features enhanced in far field extinction spectra of 1LTMD-NA heterostructures. Photocatalytic reaction rates measured for these novel 1LTMD-NA were over 10-fold higher than those reported for other TMD-based electrodes.

This integration of novel experimental, computational, microscopic and spectroscopic results leads to a new paradigm for improved photocatalysis at 1LTMD-NA interfaces. These new tools and data support improved design of 1LTMD-NA photocatalysts and their implementation in a variety of chemical, biological, energy and water systems.

[1] J.R. Dunklin, D.K. Roper et al., in preparation. **[2]** G. T. Forcherio, J.R. Dunklin, D.K. Roper et al., in submission. **[3]** J.R. Dunklin, D.K. Roper et al., in preparation.

DEVELOPMENTS OF COAL FIRED POWER GENARATION PROCESSES IN JAPAN AFTER THE FIRST OIL CRISIS (1973)

Shigekatsu Mori, Emeritus Professor of Nagoya University, Nagoya, Japan

Introduction

Forty-four years have passed during my research contribution with coal conversion technologies. It is great honor to present this paper on the developments of various coal fired power generation processes in Japan. Additionally, recent Japanese situations of power generation system are also briefly introduced

1. Atmospheric Pressure Type Bubbling-bed Fluidized-bed Boilers

Most of Japanese coal conversion projects were started after the first oil crisis. Until 1973, main power generation plants in Japan had been oil fired plants and therefore the replacement of oil with coal were strongly requested at this point. Then our Japanese chemical engineering group started projects to develop the fluidized-bed coal combustion process with desulfurization by lime-stone. Fundamental stages were passed through in a short term and the national project was started immediately. Both 20t/h pilot plant and also 160t/h (50 MW) demonstration plat operations were successfully achieved and then 350MW oil fired power generation plant was replaced by coal fired FBC plant in1995. To keep same power generation capacity as the original oil fired plant, stacked up structure was newly adopted as the coal fired FBC boiler plant as shown in Fig.1.

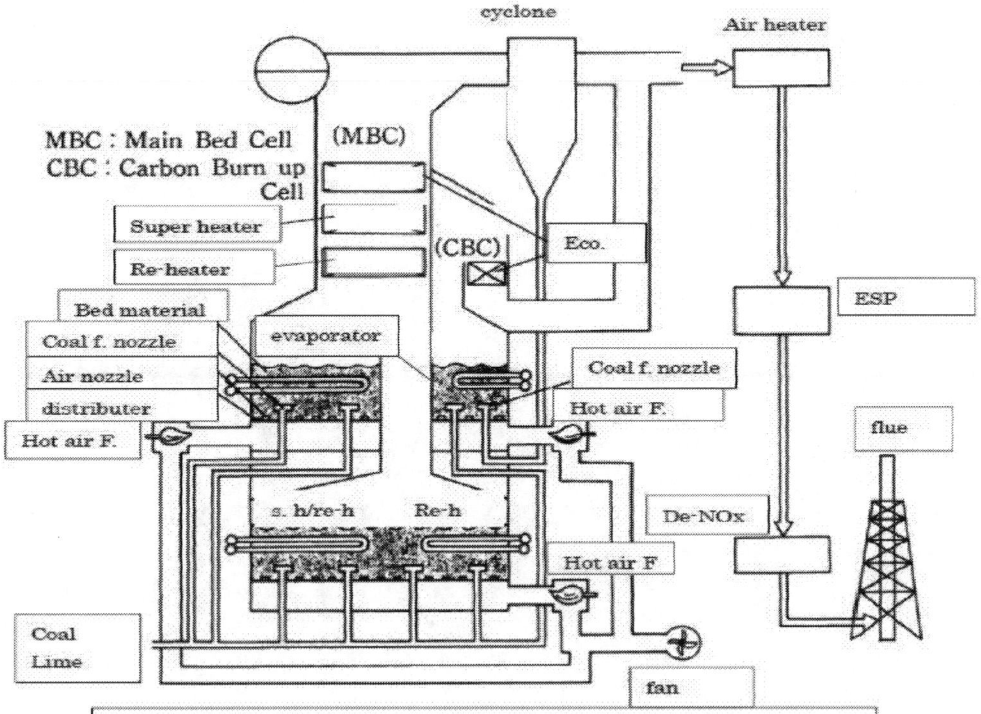

Fig.1 350MW Takehara No2 AFBC Boiler of J-Power

Excellent wasted gas treatment technologies were also established and then three single values were achieved, such as SOx and NOx concentrations in flue gas were less than 10 ppm and particulate concentration was less than 10 mg/m3.

However, around the same time, the global warming problem generated by carbon dioxide emission had become the significant worldwide social problem. Then fossil power efficiency of power generation plants had been emphasized. Since the bed temperature of A-FBC plants were limited less than 1273K, and their steam conditions were limited at the sub critical level, their power generation efficiency were limited less than 40 %. Then the development program of A-FBC coal fired power generation plant was suspended. However, bubbling fluidized bed combustion technologies have been adopted for relatively small-scale incineration plants of municipal wastes. Recently, FBC plants have been also widely expanded as power generation processes of woody biomass in Japan.

2. Development of pulverized coal-fired power generation process

In 1960's, a few pulverized coal-fired power generation plants had been operated and their maximum unit capacity were 250Mw and their steam condition were sub-critical, then their fossil power efficiency(LHV) were 37 to 39%. After the energy crisis, pulverized coal fired power plants were thoroughly developed. As shown in Fig.2, the super critical condition was achieved and their efficiency were 39 to 40% in 1983. Their maximum unit capacity was reached to 500Mw in 1981 and 700Mw in 1988. In 1996, the ultra-super critical condition was achieved and their efficiency were 41 to 43%. Their maximum unit capacity was reached 1000Mw in1990 and now it has been 1050Mw.

More important achievements of their developments were establishment the of air-pollution control processes. Such as, the electrostatic precipitator was adopted as the dust collector, the wet lime plaster method was adopted as the desulfurization process and the dry ammonia selective catalytic reduction was adopted as the denitrification process.

Fig.2 Efficiency of Pulverized Coal Fired Boiler in Japan

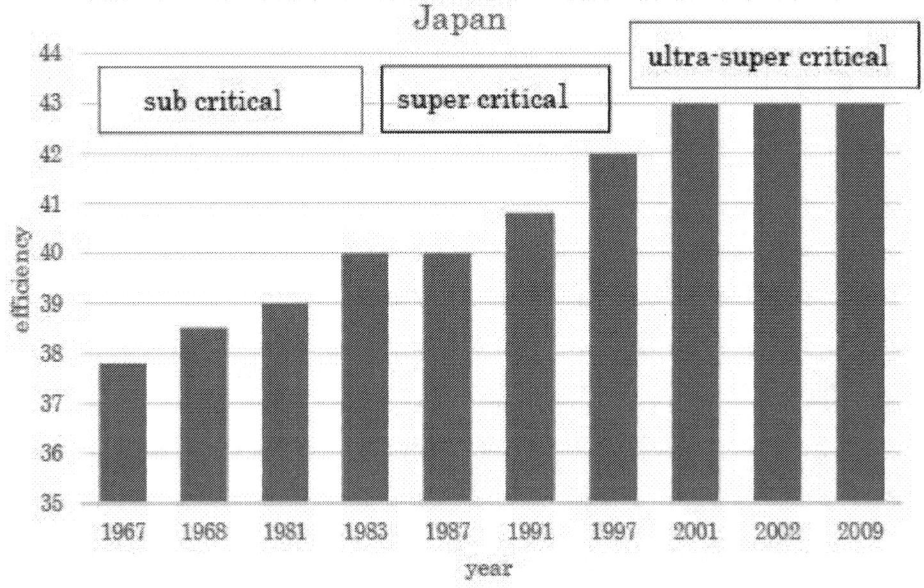

Although their attractive advantage had been its cheapest power generation cost, the most important principal problem of the coal power generation process has been its emission of carbon-dioxide gas. Developments of improved technologies of emission of carbon-dioxide have been still most important issues for any thermal power generation processes in these days.

3.Gas-turbine combined cycle power generation process of liquified natural gas

In 1984, gas-turbine combined cycle power generation process was adopted as the power generation process of natural gas in Japan. The combined cycle was constituted with gas turbine generation process and waste heat recovery steam boiler system. This engineering achievement had been extremely effective to introduce the power generation process of imported liquified natural gas(LNG) from overseas. The thermal efficiency of the gas turbine combined cycle was determined by the gas turbine inlet temperature. In 1984, the gas turbine inlet temperature was 1473 K and its thermal efficiency (LHV) was 48 %. Turbine inlet temperature was limited by heat resistant properties of turbine blades and then aggressive research works on the material development of turbine blades had been continued. Then turbine inlet temperature had been increased and then thermal efficiency had been raised up, such as; turbine inlet temperature was 1633 K and its thermal efficiency(LHV) was 49% in 1994, turbine inlet temperature was 1773 K and its thermal efficiency(LHV) was 58% in 1996, thermal efficiency(LHV) was reached 60% in 1998, turbine inlet temperature was 1873K in 2011 and its efficiency(LHV) was 62%.

4.Entrainment type coal gasification process for integrated combined cycle

Just after the successful operation of gas-turbine combined cycle power generation process of LNG, developments of integrated coal slagging gasification combined cycle(IGCC) were started. Two types of coal slagging gasification processes were developed in Japan, such as the air blown slagging coal gasification and the oxygen blown slagging coal gasification. Dry pneumatic conveying systems of coal were adopted in both plants.

4.1. Integrated air blown slagging coal gasification combined cycle[1)]

Based on the fundamental research with 2t/d process-development unit during 1983 to 1995, 200t/d pilot plant of air blown slagging gasification combined cycle was operated in 1998 to 1995. 1700t/d and 250Mw demonstration unit shown in Fig.3 was designed and constructed. Then this demonstrated operation from 2001 was successfully finished in 2012. From 2013, commercial operation had been started as 250MW Nakoso No.10 plant. Specification of this plant are mentioned as follows, such as: capacity is 250MW gross, coal consumption is approx.1700t/d, gas turbine is 1473K class(50Hz). Efficiency gross is 48%(LHV) and net is 42%(LHV). Flue gas properties are achieved as follows, such as; particulates < 4 mg/m^{3}N, $SO_x < 8$ppm, $NO_x < 5$ppm

4.2. Integrated oxygen blown slagging coal gasification combined cycle[2)].

Based on the operation experience of 10t/d test plant (HYCOL plant) during 1990-1993, 150t/day of 2.5MPa pilot plant (EAGLE plant) shown in Fig.4 was constructed and operated in 2002- 2013. This entrainment slagging coal gasifier had two stage oxygen blown gasification burners in the single gasification room. Slugged coal ash was quenched in the bottom quencher by water and discharged. Since this plant was oxygen blown gasification process, both nitrogen and CO_2 concentration in the product gas was limited and then CO and H_2 were enriched.

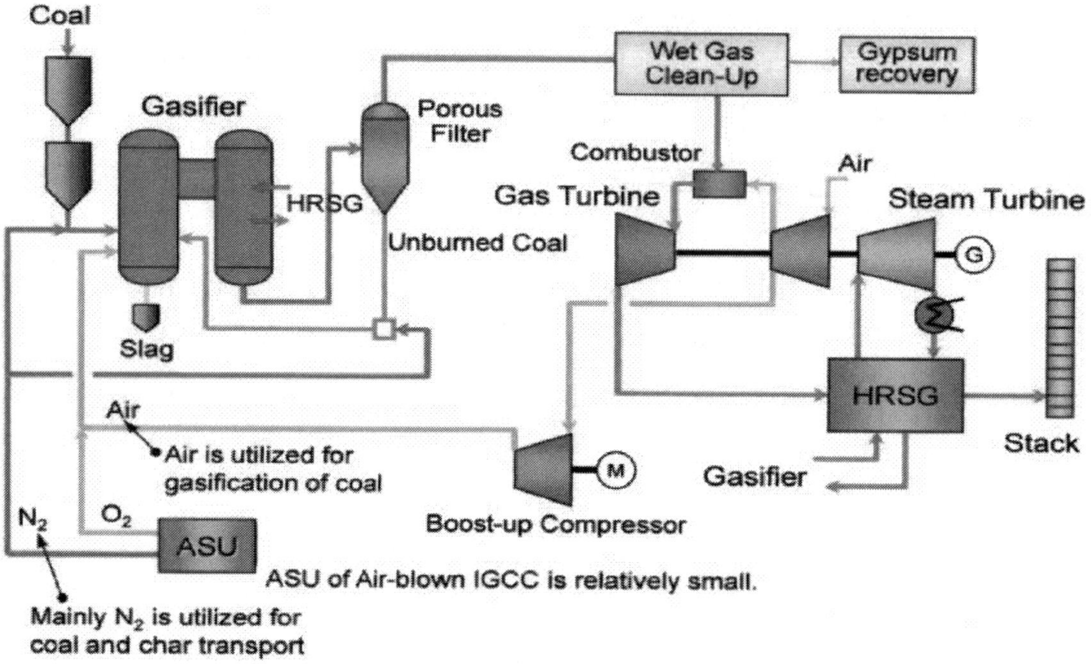

Fig.3 Schematic Diagram of NAKOSO IGCC Plant

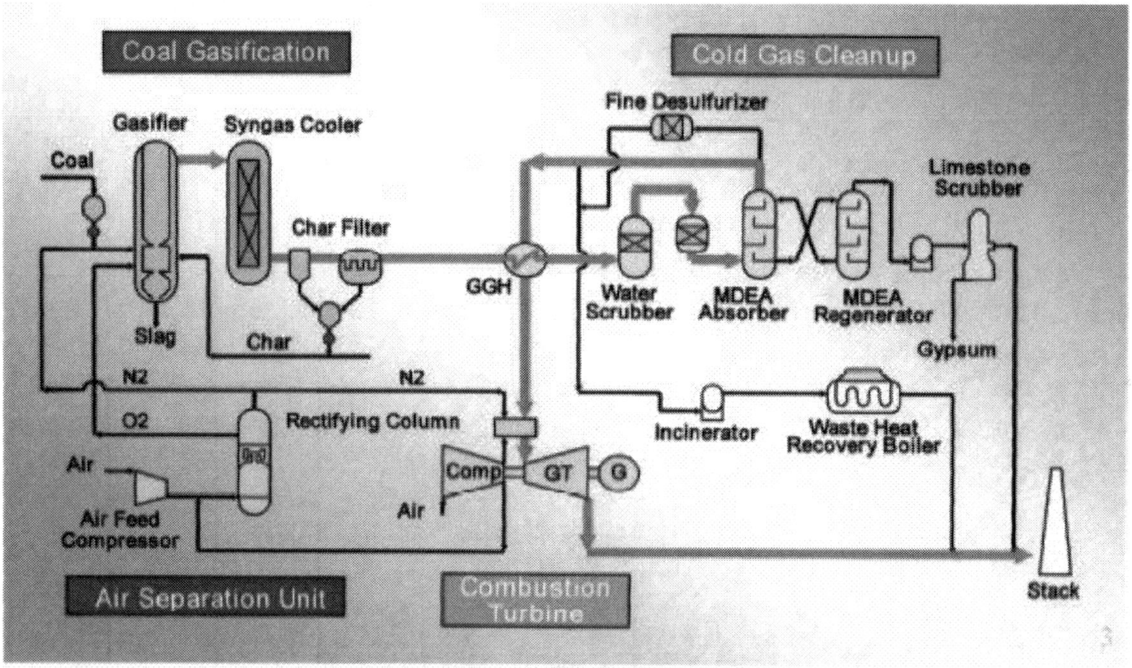

Fig. 4 EAGLE Plant of 150t/d Coal Gasification Pilot Plant

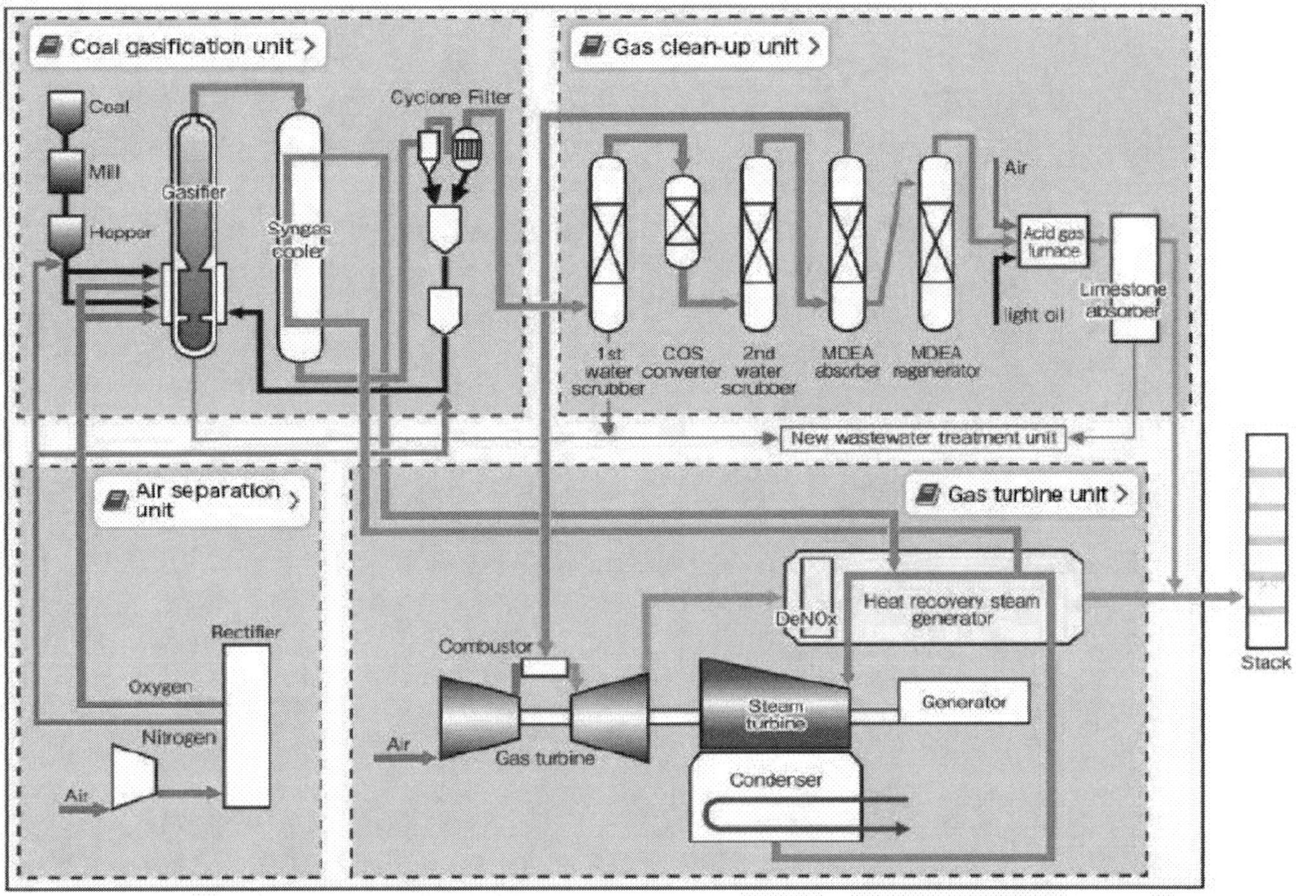

Fig.5 Osaki-Coolgen Oxygen-blown IGCC demonstration unit

Very low emission level of product gas was achieved, such as, SO_x , Halogen and Ammonia were kept less than 1ppm and dust was kept less than $1mg/m^3N$.

Based on these excellent operation experience, 1180 t/d, 166Mw demonstration plant (Oosaki-Coolgen plant)as shown in Fig.5 was constructed and gas purification test facilities were configured with CO_S conversion and removal unit and also with H_2S adsorption unit. The demonstration and semi-commercial operation was started in 2017. As the second stage, the demonstration of CO2 removal and recovery operation will be also challenged to develop the IGFC (Integrated Coal Gasification Fuel Cell Combined Cycle) and also CCS(Carbon Dioxide Capture and Storage). This Process is also applied to hydrogen production unit from coal directly.

5. Recent situations of Japanese power generation system

Here, resent situations of Japanese power system are introduced briefly. In Fig.6, trends of crude oil price(WTI) and Japanese electricity fee during these 44years after the oil crisis. Both oil price and electricity fee sharply increased during 1973~1980 and gradually decreased until 2000. Japanese electricity fee was continuously decreased further until Lehmann shock.

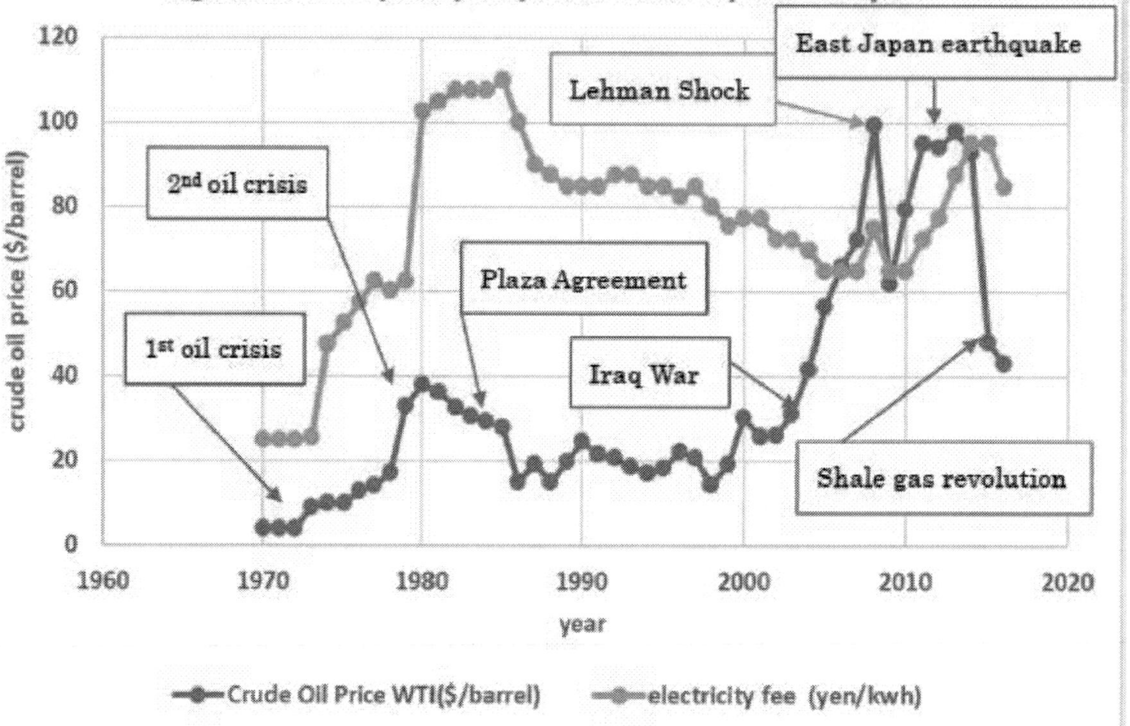

Fig.6 Crude oil price(WTI) and Electricity fee in Japan

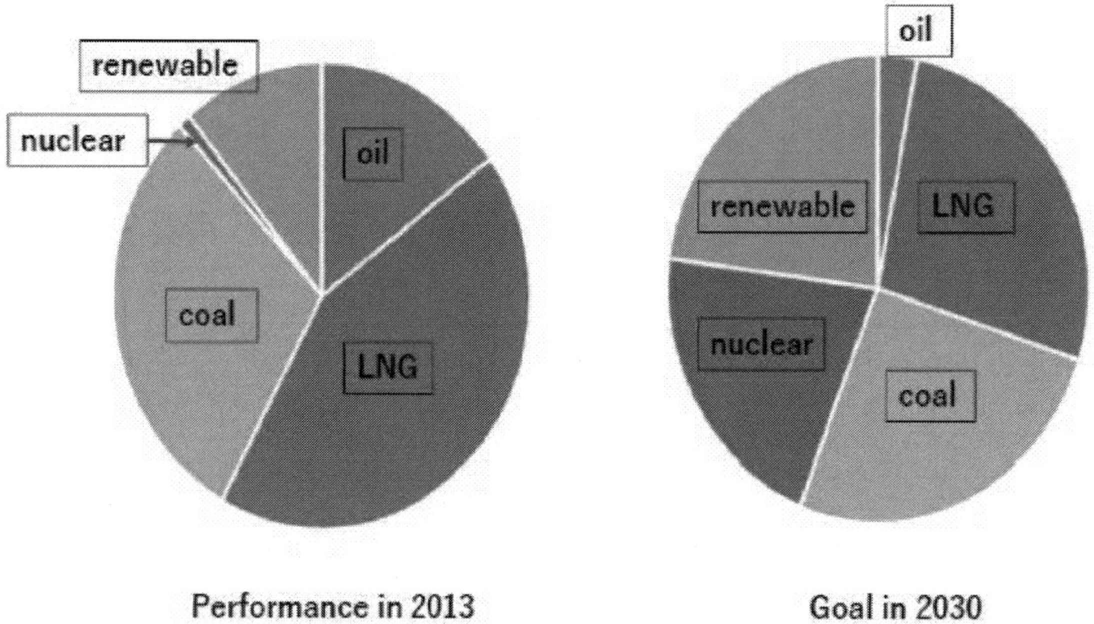

Fig.7 Power supply configuration in Japan in 2013 and its goal in 2013

The oil price was sharply increased again after Iraq War till Lehman shock. Both oil price and the electricity fee were increased again for the reason of East Japan Earthquake and the accident of Fukushima nuclear power plants. However, oil price was sharply decreased again recently by the effect of the shale gas and oil revolution.

In Japan,16 nuclear power plants were presented and total their capacity was 44,264MW, however Fukushima nuclear power plants were completely destroyed by the great tsunami caused by East Japan Earthquake disaster. All Japanese nuclear plants were shut down temporarily and then only three plants have been re-operated now. Therefore, the power supply configuration was affected directly as shown in Fig.7. Although Japanese government has promoted their restart and also determined the Japanese power supply configuration in 2030 as shown in Fig.7, major peoples have been disagreed with their re-start and full operation.

Worldwide uncertainty factors make the future energy situation extremely unclear as mentioned in this paper, then it is very difficult to clearly predict the optimum energy system in the world.

As shown in Fig.6, the world supply-demand relationship of energy resource such as oil, LNG and coal are extremely important factors in Japan since all of these resources are completely imported from foreign countries.

The importance of CO_2 reduction for the protection of the green-house problem have been well recognized and following policies are considered to be effective,

1)nuclear power generation
2)reduction of CO_2 from thermal power generation plant
3)renewable energy
4) import of hydrogen from foreign countries

However, any effective policy has not been adopted yet in japan.

Acknowledgments

Author express great gratitude to AIChE and Society of Chemical Engineering of Japan and also to Professor LS Fan and Professor Nobusuke Kobayashi for their great contribution on the organization of this session.

References

1) Nunokawa, M.: Paper presented in International Conference on Power Engineering (2013)
2) Aiso, K.: Paper presented in Subcommittee of power generation from coal, 9th China-Japan Forums on energy conservation and environment (2015)

494901 Selected Topics in Fluidization Fundamentals and Fluidized Bed Applications â€“ a Presentation Honoring Prof. Shigekatsu Mori

Abstract:

This presentation is especially prepared to celebrate Prof. Shigekatsu Moriâ€™s career long accomplishments in fluidization and fluid-particle systems. In this presentation, selected topics in fluidization fundamentals, including the minimum fluidization velocity and the bubbling bed properties as well as its modeling options in a gas-solid fluidized bed will be reviewed. The presentation will also cover several selected fluidized bed applications in the energy and environmental areas, including a fluidized bed technology which uses various sorbents to capture heavy metals at high temperatures during coal combustion or waste incineration. Finally, a potential fluidized bed technology which uses microwave energy to convert nanoparticle-embedded plastic waste into activated carbon capable of responding to magnetic field will also be introduced.

491050 Fluidized-Bed Drying Process Based on Self-Heat Recuperation Technology

Abstract:

A novel fluidized-bed evaporator for seawater desalination and biomass drying processes was proposed based on self-heat recuperation technology, in which all the heat of process streams is recirculated inside the system. Process simulation results showed that the energy consumption of the proposed process can be reduced to one-quarter than of a conventional process. However, in the case of high moisture content several problems are caused such as scale deposition on the surface of the heat exchanger, agglomeration of fluidizing particles, etc. In the present study, for the development of the drying process using the fluidized bed, evaporation experiments in a fluidized bed were conducted and the agglomeration behavior of the fluidized particles with seawater feed was investigated.

498902 The Role of Pressure Balance in Nonmechanical Device Design

Abstract:

Nonmechanical devices used for both control or automatic operation are increasingly being used in chemical processes. However, for these devices to operate successfully, conducting an accurate pressure balance is critical.

The reason for an accurate pressure balance and how the pressure balance affects nonmechanical device operation will be presented for two types of devices â€" a nonmechanical L-valve used for control of the solids, and a loop seal for automatic control of the solids.

In the first case, how a misunderstanding of how the L-valve works led to a non-working system will be detailed. For the second case, an application of the loop seal that is essentially 180-degrees different that most loop seals are designed will be discussed. Both system problems and applications will be discussed using detailed pressure balances to illustrate the problems/solutions.

岐阜大学
GIFU UNIVERSITY

2017 AIChE Annual Meeting
October 29 - November 3, 2017
Minneapolis, USA

Particle Technology Forum
Special Session: Celebrating Prof. Mori's Career Long Accomplishments
Wednesday, November 1, 1:55pm-2:15pm

573e
Carbpon Fiber Reclamation from CFRP Waste

Gifu University, Japan
Senior Professor
Hiroshi MORITOMI
moritomi@gifu-u.ac.jp

1

岐阜大学
GIFU UNIVERSITY

Contents

1. Introduction
2. Materials and methods
3. Results and discussion
(1) Saving energy
(2) Strength of rCF and rCFRP
- **1) Carbonization (pyrolysis)**
- **2) Two-stage thermal treatment**
- **3) Volume fraction of rCF**
- **4) Fiber length**
- **5) Fiber strength**
- **6) Ratio of rCF to vCF**
- **7) Resin**

4. Conclusions

2

CFRP demand

CFRP : Carbon Fiber Reinforced Plastic

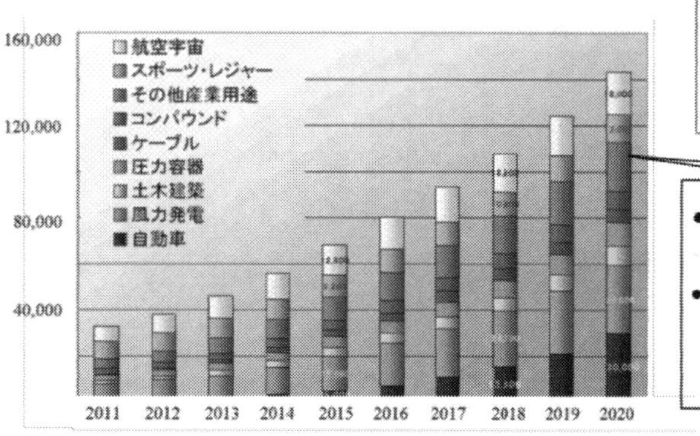

Carbon fiber

- Low density
- Light weight
- High strength
- Expensive

- Production in 2020 will be 3.5 times in 2012
- Increasing demand is for industrial application (wind power and vehicle, etc.)

Demand forecasting of CF production ⇨ If 10% of total virgin CF is recycled, the recycling market will be 15,000 ton/year CF.

- Higher energy consumption for virgin CF production
- Increasing of waste CFRP

Necessary to establish a recycling technology

3

Recycling technologies

Japan

Tech	Thermal	Thermal	Solvent	Super-critical	Sub-critical	Super heated	Catalytic	Electrical	Super heated
Organiz.	CFRI	CF Associ	Hitachi Kasei	Shizuoka U	Kumamoto U	JFCC	Shinshu U	Hatinohe U	Doshisha U
Type of resin	All	All	Ester	Ester	Ester	All	All	All	All
Recover	CF (Long & short) Gases	CF (Milled)	CF (Long) Chemicals	CF Resin compounds	CF Chemicals	CF	CF	CF	CF
Capacity	1,800t/y	1,000t/y	12t/y	Lab	Lab	30t/y ?	Lab	Lab	Lab

Overseas

Country	UK		Germany		USA	
Technique	Thermal (Milled)	Micro-wave Fluidized Heated fluid	Reforming?	Thermal	LP thermal LP fluid HT fluid	Micro-wave
Organization	ELG(Recycle Carbonfiber)	University of Nottingham	Fraunhofer	CarboNEXT	Adherant Technologies.	Firebird Advanced Materials
Capacity	1,200t/y	PDU?	Lab	1,000t/y	Pilot	Pilot

4

岐阜大学 GIFU UNIVERSITY

Contents

岐阜大学 GIFU UNIVERSITY

Present reclamation process

Recovering carbon fiber as "long fiber"

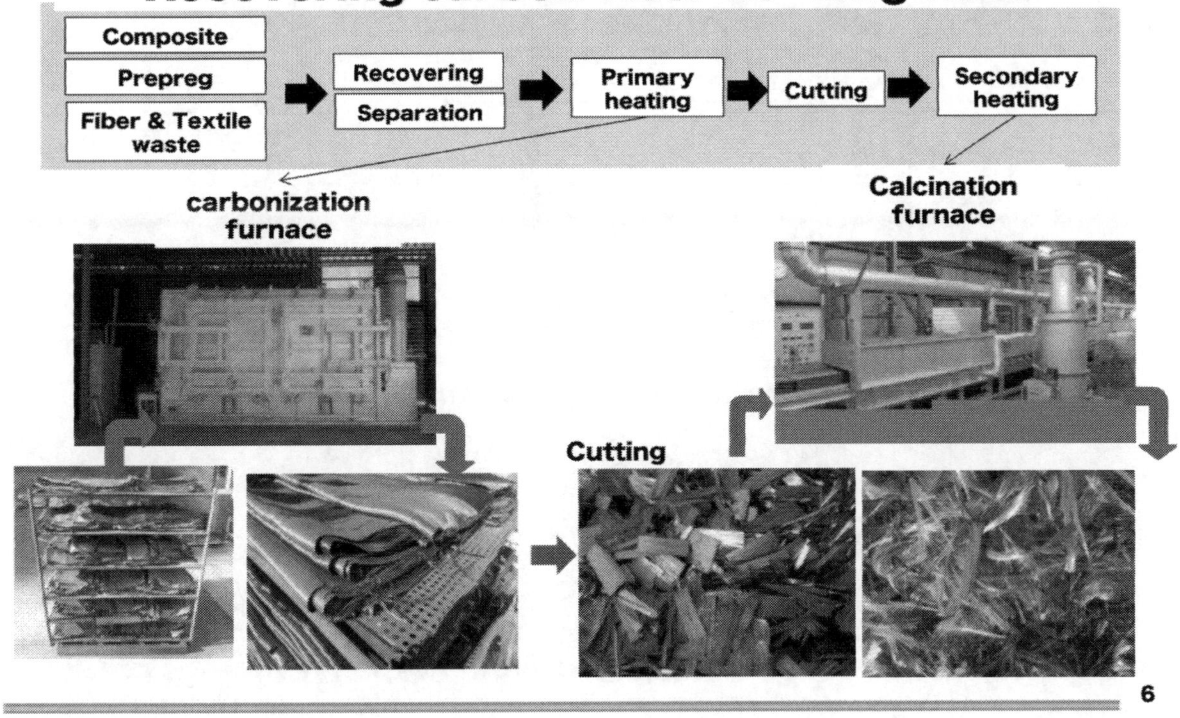

Heating up by kerosene until starting the resin decomposition

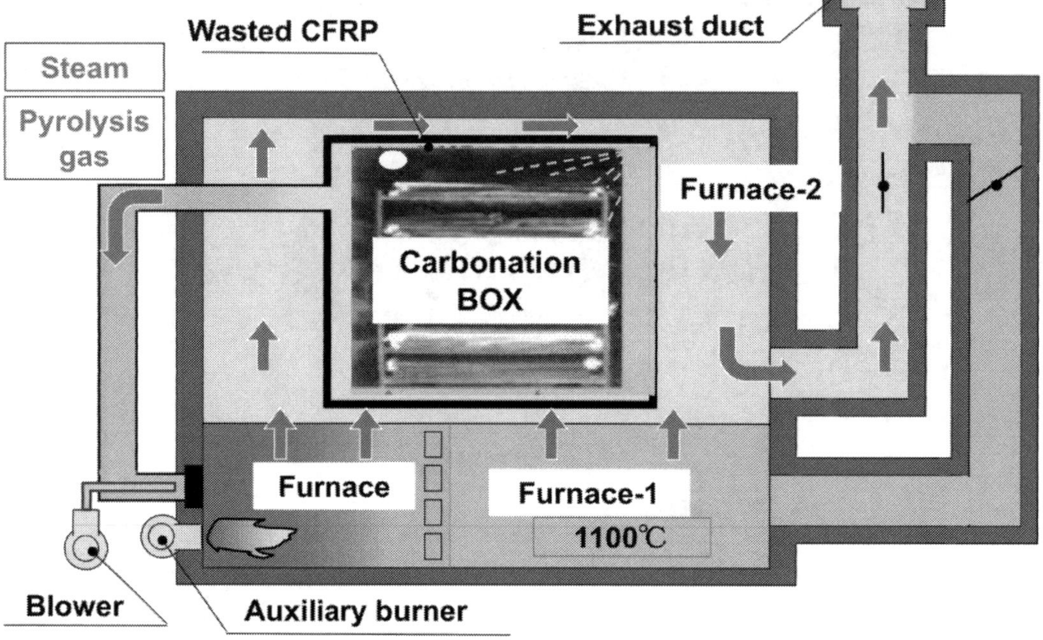

7

Secondary furnace (Lab)

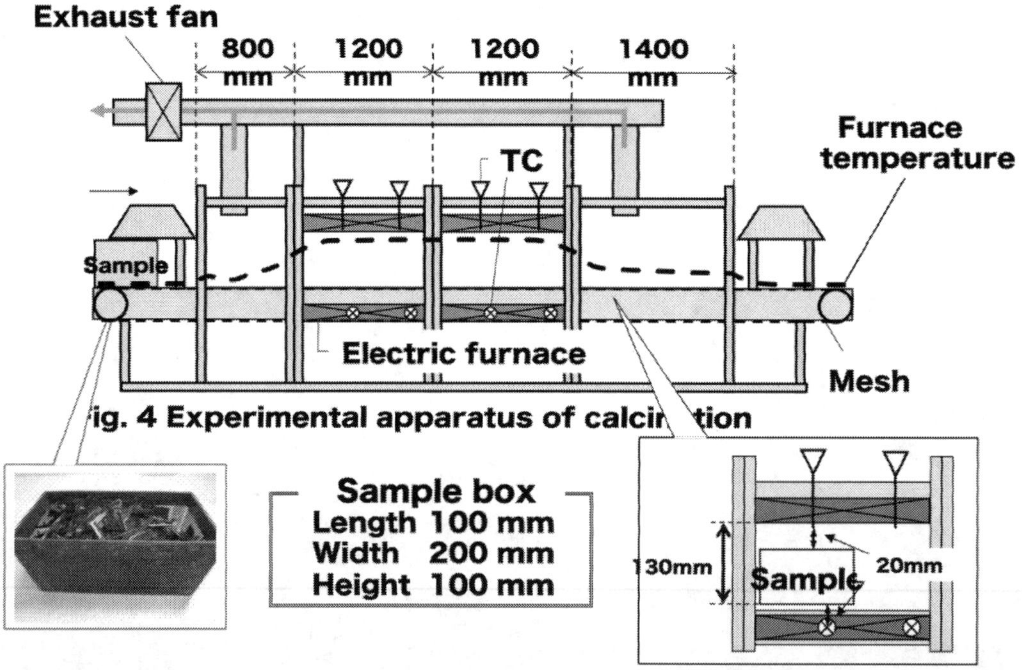

Fig. 4 Experimental apparatus of calcination

Enlarged view of experimental equipment

8

岐阜大学 GIFU UNIVERSITY

Two-staged thermal treatment Process

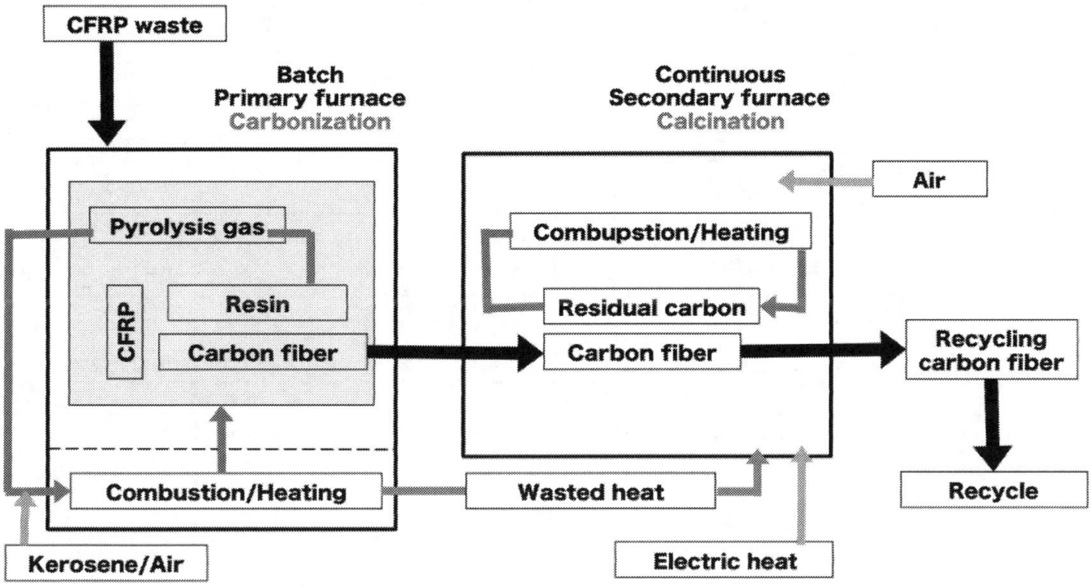

9

岐阜大学 GIFU UNIVERSITY

Contents

1. Introduction
2. Materials and methods
3. Results and discussion

(1) Saving energy

(2) Strength of rCF and rCFRP

1) Carbonization (pyrolysis)
2) Two-stage thermal treatment
3) Volume fraction of rCF
4) Fiber length
5) Fiber strength
6) Ratio of rCF to vCF
7) Resin

4. Conclusions

10

岐阜大学 GIFU UNIVERSITY

Operating conditions and results of carbonization process (CYT-1700)

	Test-1	Test-2	Test-3
Carbonization furnace	CYT-1700 (a unit operation)	CYT-1700 (two units consolidated)	CYT-1700 (two units consolidated)
Inside box	Usual	High temperature box Input/Output	High temperature box Input/output
Charging temperature	100°C	250°C	300°C
Superheated steam condition	200mL/min 0.3Mpa	400mL/min 0.3Mpa	400mL/min 0.3Mpa
Setting furnace temperature	650°C	650°C	650°C
Carbonization time	195min	180min	150min
Heating time to 650°C	30min	15min	10min
Energy consumption	17.3MJ/kg-CF	15.6MJ/kg-CF	12.6MJ/kg-CF

Table 2 Energy consumption of carbonization and calcination processes

	Initial		Present	
	Carbonization (CYT-1700)	Calcination (30m kiln)	Carbonization (CYT-1700)	Calcination (30m kiln)
Feeding	201kg	23.5kg/h	150 kg	35.3 kg/h
Energy consumption	18.2MJ/kg-CF	6.50 MJ/kg-CF	12.6 MJ/kg-CF	2.75 MJ/kg-CF

Ref: 24.6MJ/kg-CF at 2015

11

岐阜大学 GIFU UNIVERSITY

Saving energy & low cost

Recycling carbon fiber	Pyrolysis (The Japan Carbon Fiber Manufacturers Association)	Pyrolysis (Carbon Fiber Recycle Industry)	Solvent Extraction (Hitachi Chemical)
Recovering energy [MJ/kg-CF]	46.0	14.6 → 8.5	63.0 (Distillation 38.0)
Materials	Carbon fiber	Steel	Aluminum
Production energy [MJ/kg]	286	23.0	12

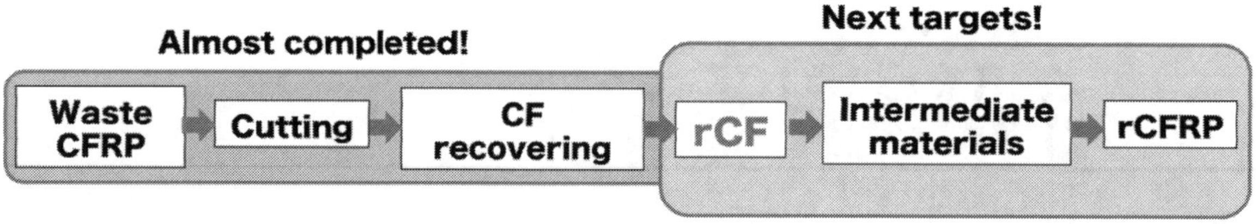

12

Contents

Surface observation with secondary heating

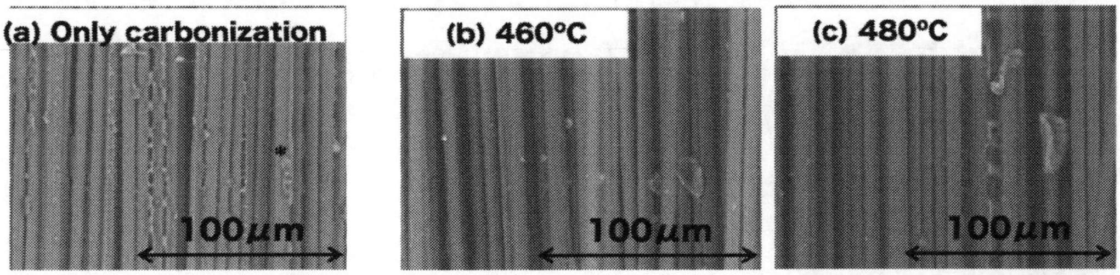

Carbonized CFRP and residual carbon

	Temperature [°C]	Carbonized CFRP [wt%]	Residual carbon [wt%]
(a)	—	100	11.6
(b)	460	93.9	5.5
(c)	480	92.5	4.1
(d)	500	9.06	2.2
(e)	550	87.4	-1.0

** Residual carbon = Carbonized CFRP - 88.*

岐阜大学 GIFU UNIVERSITY

Examination of tensile strength

Weight method

Sample	Recycled CFRP (SampleA)	Virgin CF (SampleA)
1-step Temp	25-600 °C , 210min (400 and 500 °C, 30 min hold)	Non
2-step Temp	460-500 °C 130 min	
Length	90 mm	90 mm

Single fiber

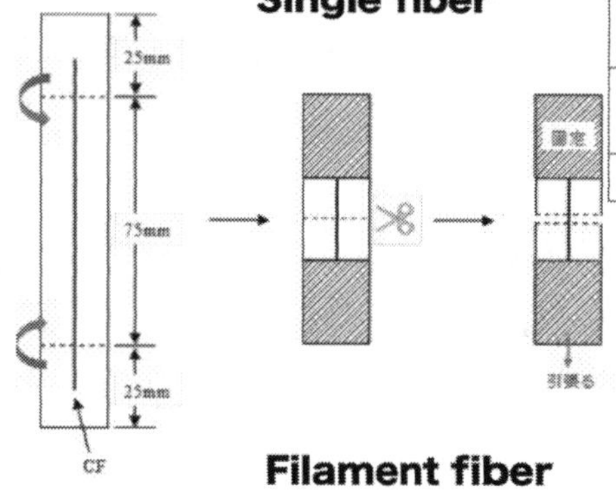

Filament fiber

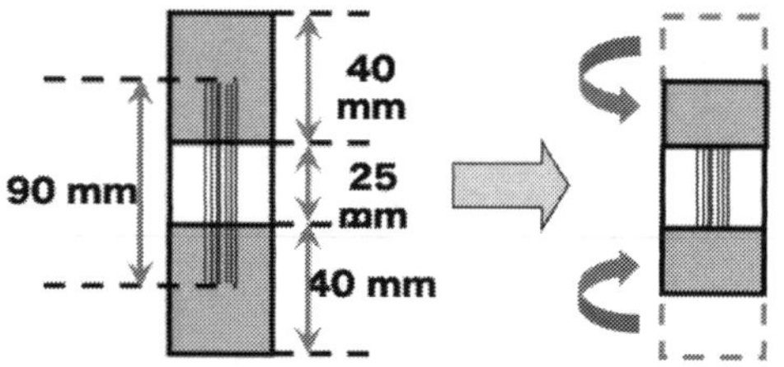

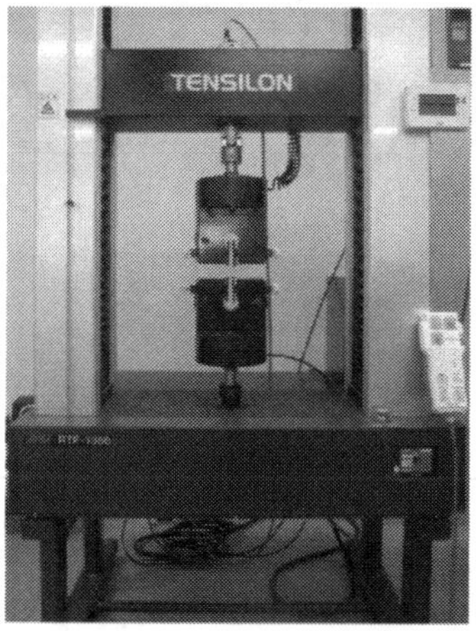

15

岐阜大学 GIFU UNIVERSITY

Tensile strength(1)

Effect of carbonizing temperature

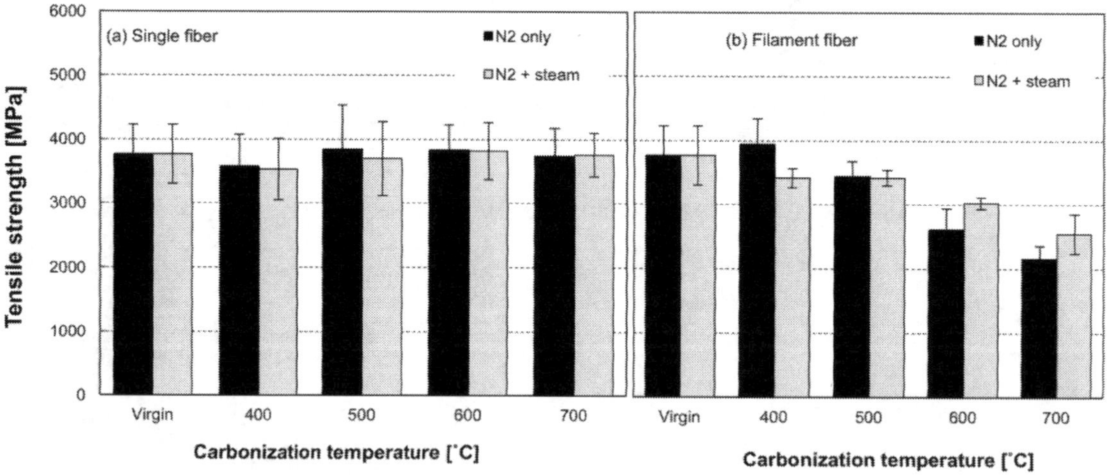

JIS R 7606

16

岐阜大学
GIFU UNIVERSITY

Tensile strength(2)

Effect of Calcination temperature

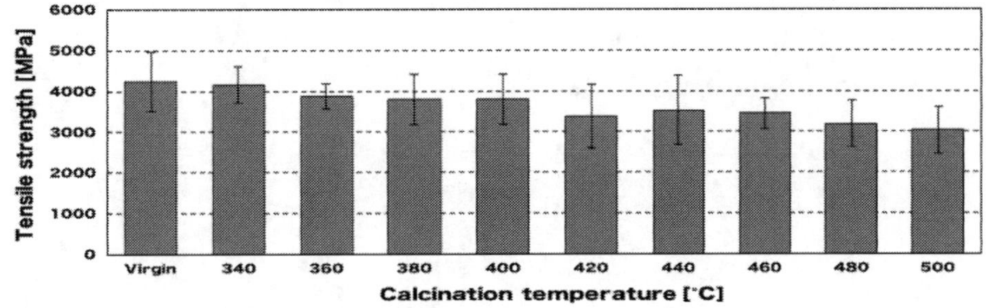

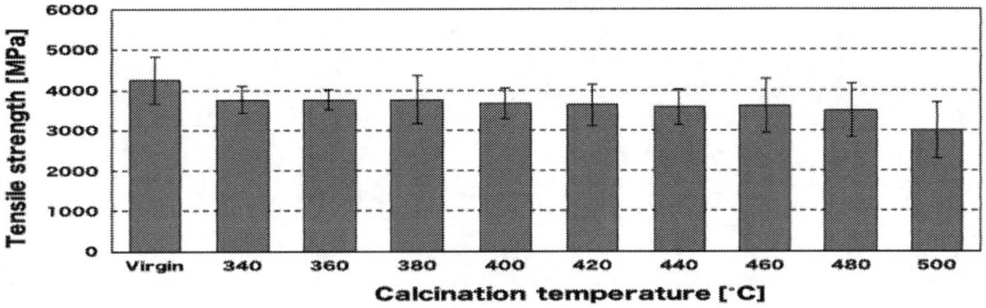

17

岐阜大学
GIFU UNIVERSITY

Contents

1. Introduction
2. Materials and methods
3. Results and discussion
(1) Saving energy
(2) Strength of rCF and rCFRP
 1) Carbonization (pyrolysis)
 2) Two-stage thermal treatment
 3) Volume fraction of rCF
 4) Fiber length
 5) Fiber strength
 6) Ratio of rCF to vCF
 7) Resin
4. Conclusions

18

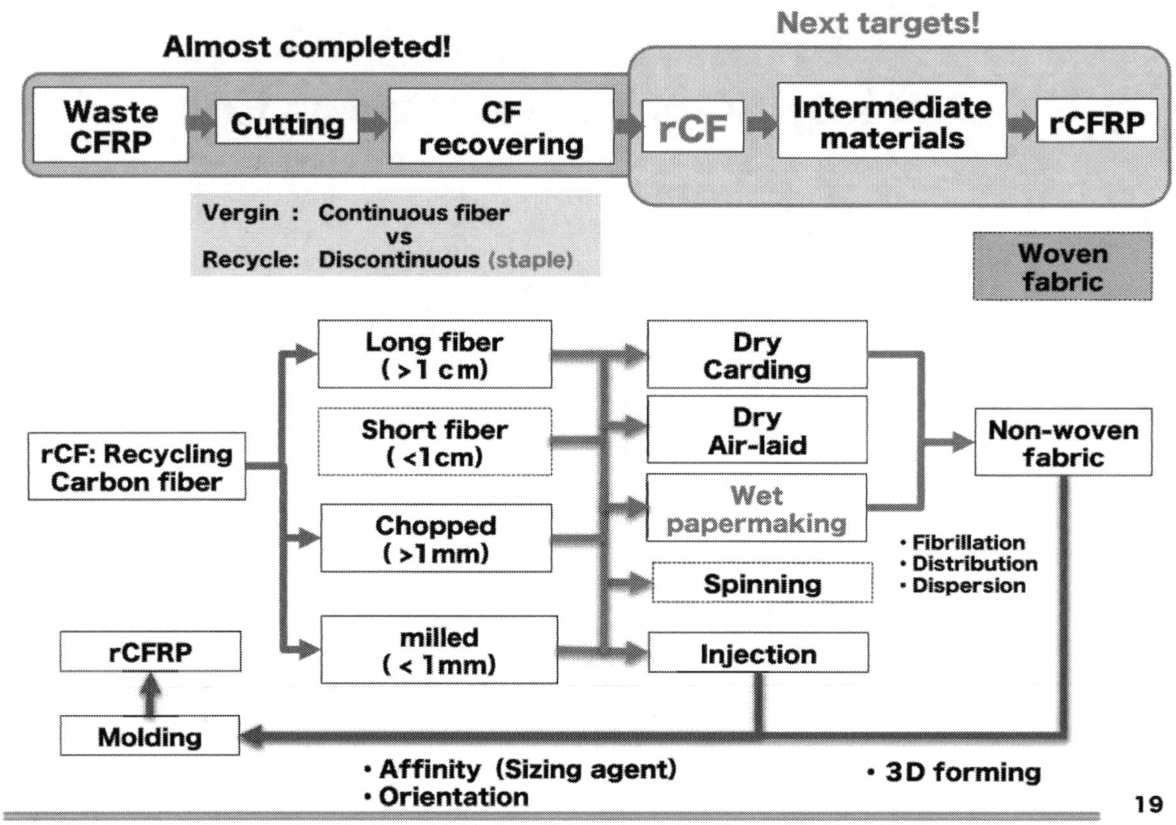

19

Tensile strength(3)

Effect of volume fraction of rCF

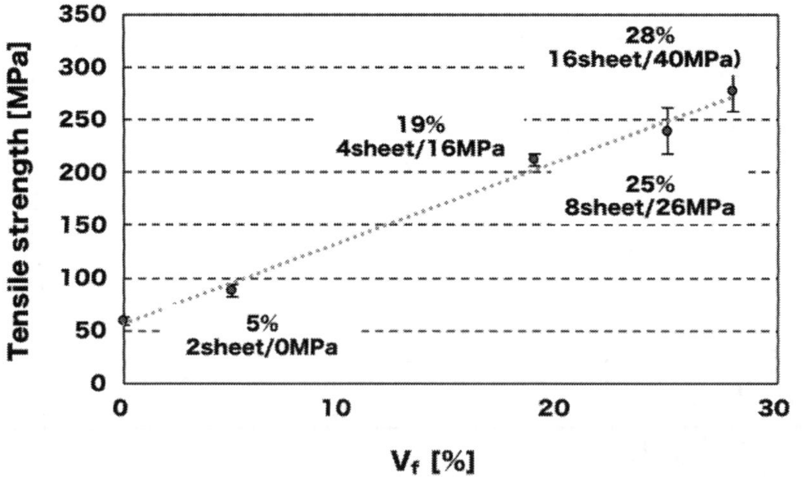

20

Tensile strength(4)

Effect of fiber length

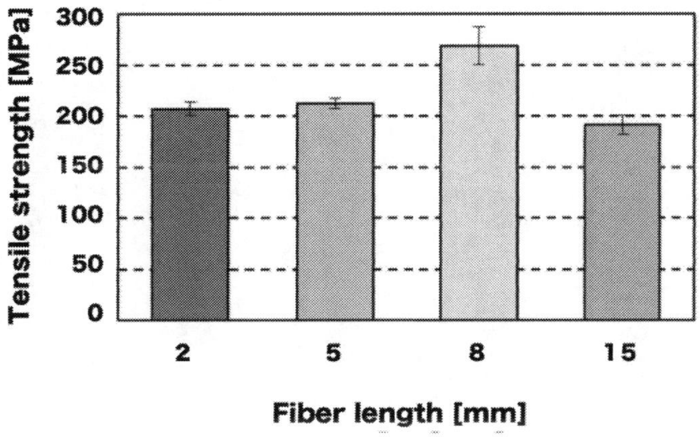

21

Tensile strength(5)

Effect of fiber strength

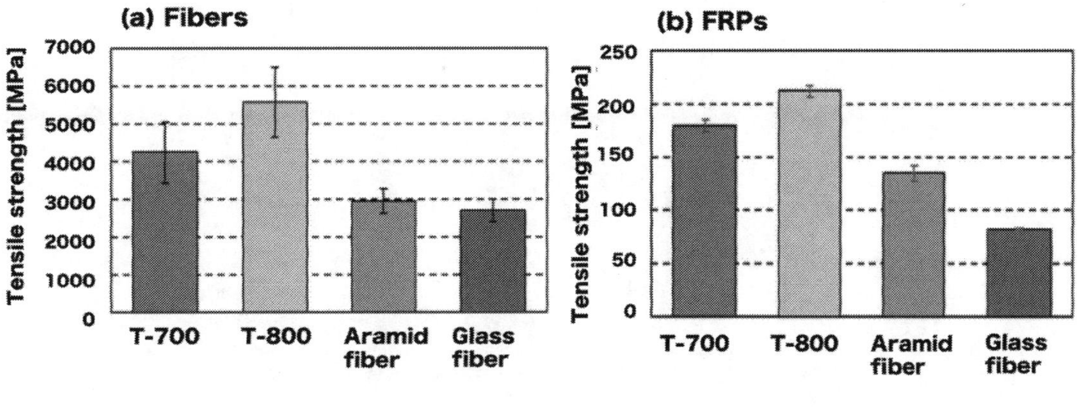

22

岐阜大学 GIFU UNIVERSITY

Tensile strength(6)

Effect of rCF ratio in paper making

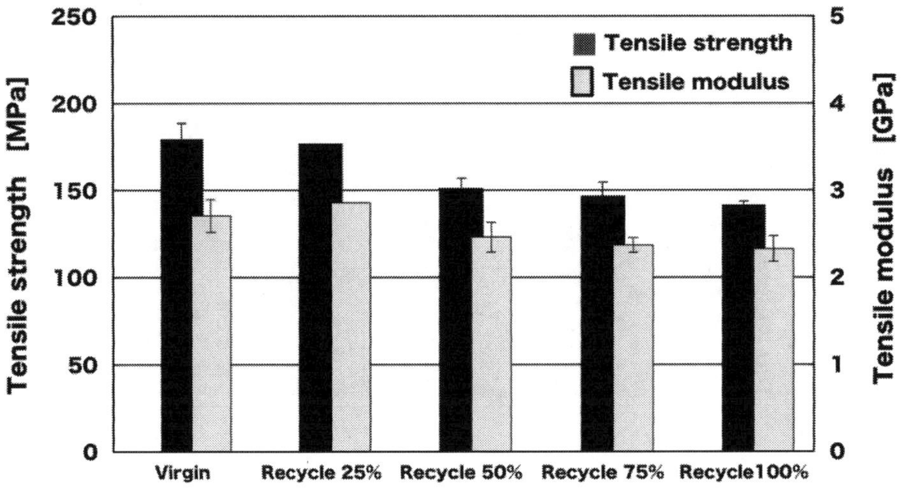

23

岐阜大学 GIFU UNIVERSITY

Tensile strength(7)

Effect of layer structure of vCF and rCF sheets in paper making

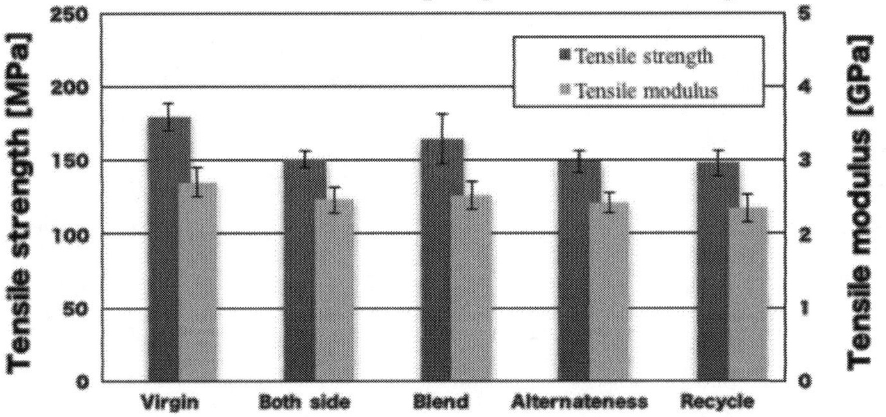

Total 8 sheet
Both side: rCF_4 sheet and vCF_2 sheet x 2
Blend: Mixed sheet with rCF and vCF_1 sheet x 8
Alternateness: Layer_rCF_sheet by layer_vCF_sheet x 4
Recycle: rCF_1 sheet x8

24

岐阜大学 GIFU UNIVERSITY

Tensile strength(8)

Effect of virgin cross sheet and rCF sheets in paper making

Tensile modulus

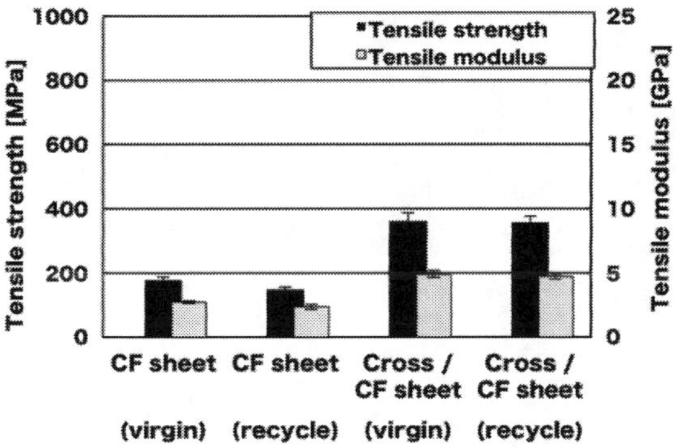

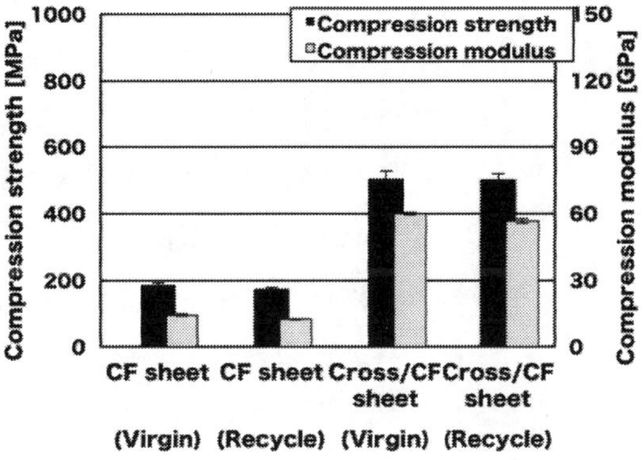

Total 8 sheet
Both side: rCF_6 sheet and cross_1 sheet x 2

岐阜大学 GIFU UNIVERSITY

Tensile strength(9)

Effect of reinforcement of rCF for different resins (TSP and TP)

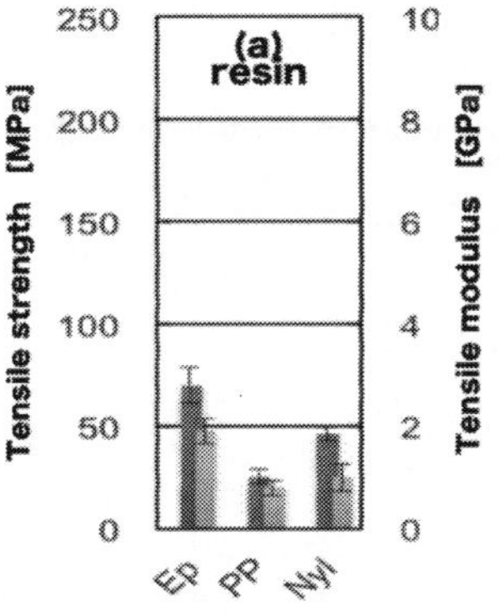

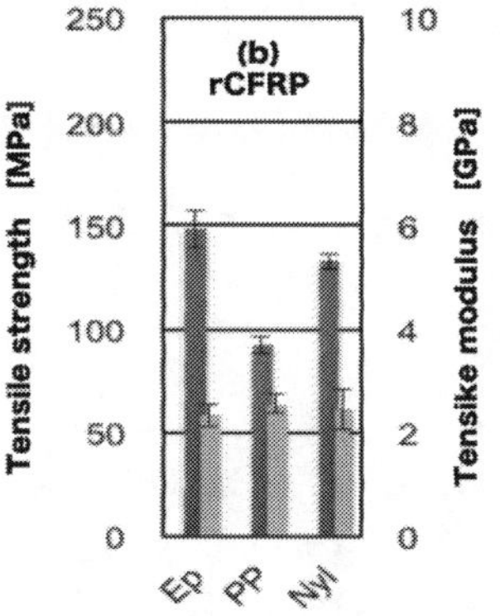

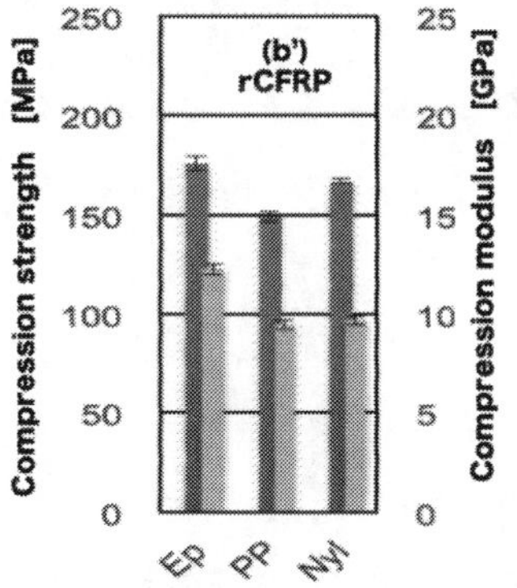

Ep: Epoxy resin(TSP)
PP: Polypropylene resin (TP)
Nyl: Nylon resin (TP)

岐阜大学 GIFU UNIVERSITY

Contents

1. Introduction
2. Materials and methods
3. Results and discussion
(1) Saving energy
(2) Strength of rCF and rCFRP
   1) Carbonization (pyrolysis)
   2) Two-stage thermal treatment
   3) Volume fraction of rCF
   4) Fiber length
   5) Fiber strength
   6) Ratio of rCF to vCF
   7) Resin
4. Conclusions

27

岐阜大学 GIFU UNIVERSITY

Conclusions

Establishing high efficient thermal treatment method for recycling carbon fibers from CFRP

- Achieved less than 10MJ/kg -CF of fuel consumption
- Achieved more than 80% strength of virgin CF
- Cleared the non-woven rCFRP characteristcs

Acknowledgement

This paper is based on results obtained from a project subsidized by the New Energy and Industrial Technology Development Organization (NEDO). We would like to express the gratitude here.

28

岐阜大学
GIFU UNIVERSITY

Thank you for your warm support!

Senior Professor
Hiroshi MORITOMI
moritomi@gifu-u.ac.jp

29

499412 A Nature-Inspired Approach to Aid the Understanding and Improve the Performance of Fluidized Beds

Abstract:

Gas-solid fluidized beds remain fascinatingly complex, despite many decades of research and industrial use. Chosen for their desirable mass and heat transfer properties for processes as diverse as drying, chemical production, waste-to-energy, fuel combustion and gasification, the complex fluid and particle dynamics are notoriously difficult to model, leading to a predominance of empirical approaches for scale-up and design.

Our research has focused on ways to structure fluidized beds, imposing more uniformity in mixing and bubble patterns, so that their behavior becomes more predictable, and they become easier to operate and scale-up. Biology provides us with wonderful examples of efficient, scalable hierarchical transport networks, such as the airways of the lungs and the vascular network, while regular sand ripples on dunes and beaches form despite the complexity of granular dynamics. This has inspired, respectively, the use of a fractal branching fluid injector for bubbling fluidized beds to facilitate scalability and promote more uniform mixing, and the pulsation of the gas flow to turn a chaotic bubble flow into a periodic bubble pattern.

An unexpected side-effect to our most recent work on pulsed fluidization is that it provides fundamental insights into the mesoscopic physics of fluidization, and helps to validate computational methods. Experimentation with a variety of particles and operation parameters allows us to survey the conditions for which regular bubble patterns appear. Coupling computational fluid dynamics (CFD) simulations with discrete element modeling (DEM) for the particles, helps us to understand how the patterns form, by comparing the numerical results with the experimentally observed patterns. As the patterns are sensitive to the drag law and particle friction, they also serve as an excellent fingerprint to investigate the physics of fluidization more generally.

496459 Combustion of Mg-S and Zr-S Reactive Nanocomposite Powders Heated to Ignition at Different Rates

Abstract:

Recently, it was observed that the burn rates of nanocomposite aluminum-based thermite powders prepared by arrested reactive milling change substantially depending on the heating rate achieved during their ignition. Micron-sized powder particles ignited by passing through a focused CO2 laser beam are heated at about 106 K/s. Such particles burn during several ms, which is slightly longer than the burn times of the similarly sized particles of pure aluminum. The burn times are reduced to hundreds of microseconds for same powder particles ignited by an electro-static discharge (spark), for which the heating rate is estimated to be close to 109 K/s. It was proposed that the much higher burn rates observed for the particles heated very rapidly are caused by essentially volumetric heterogeneous reaction occurring at all interfaces between reactive components within the nanocomposite structure. Thus, the nanocomposite structure is preserved upon ignition, at least partially. Conversely, when the heating rates are lower, the nanocomposite structure is lost upon ignition, so that the combustion is controlled by the external surface of the molten particle. In this work, the concept described above is tested for a different type of nanocomposite materials, mixing metal as a fuel and sulfur as an oxidizer. Such materials have high reaction enthalpy and generate potentially biocidal combustion products. Reactive nanocomposite powders were prepared by arrested reactive milling using elemental sulfur and ball milling it separately with two different metal fuels, zirconium and magnesium. Milling conditions are identified for each composition to prepare fully dense composites with components mixed on the nanoscale. As in previous work, the powders are ignited using a CO2 laser beam and electrostatic discharge. Emission traces of the ignited materials are captured using a single PMT filtered at the 568 nm wavelength. Time-resolved spectra of the ignited materials are also taken using a 32-channel spectrometer covering the visible range of wavelengths. Partially burned materials are captured and examined using electron microscopy. Results will be presented and discussed in this talk.

490854 Effect of Flow Conditions on Burn Rates of Magnesium and Magnesium-Containing Reactive Material Particles

Abstract:

In practical applications, such as propellants, explosives, and pyrotechnics, powders of metals and metal-based reactive materials burn in turbulent flows. However, in most experiments aimed to determine the particle burn rates, metal particles are burning in quiescent gases. Recently, it was shown that the burn rates of aluminum, titanium, boron, and several metal-based reactive materials are affected by flow conditions; higher burn rates are observed in turbulent flows compared to those in quiescent gas or laminar flow. The effect is different for different materials. In this work, the effect of flow conditions on combustion of magnesium and magnesium-containing reactive materials is studied. Magnesium commonly burns with a vapor phase flame forming away from the particle surface, so that the effect of flow conditions is expected to be particularly strong. The materials used in this project are a fine spherical Mg powder and mechanically prepared Al-Mg alloys and Mg-S composite powders. All the powders are fed into a focused beam of a CO2 laser, where they are ignited and burned in air. Different flow patterns are produced using a circular flow air knife, through which the particles are passing. The light emission pulses produced by burning particles are recorded using two photomultipliers equipped with interference filters at 700 and 800 nm. In addition, the emission pulses are recorded using a 32-channel spectrometer covering visible part of the spectrum. Finally, time-integrated spectra are measured from multiple burning particles. The optical measurements are interpreted to recover the flame temperature as a function of the particle size and time-dependent temperature traces for individual particles. Combustion products are also captured and examined using electron microscopy.

This paper will present results of experiments and discuss them in terms of mechanistic understanding of combustion of magnesium and magnesium containing reactive materials. This understanding is expected to serve as a foundation for a model describing combustion of different reactive material particles in different flow configurations.

494737 Quantitative Description of the Biocidal Effectiveness of Combustion Products of Iodine-Bearing Reactive Materials

Abstract:

Quantitative description of the biocidal effectiveness of combustion products of iodine-bearing reactive materials

Song Wang[1], Mirko Schoenitz[1], Sergey A. Grinshpun[2], Edward L. Dreizin[1]

[1]New Jersey Institute of Technology, Newark, NJ 07102, USA.

[2]University of Cincinnati, Cincinnati, OH 45267, USA

**Corresponding Author Email: dreizin@njit.edu*

Recent research efforts focused on developing new metal-based iodine bearing reactive materials (RMs). The materials are prepared mechanochemically; thermodynamic calculations predict that they generate I_2 gas among other combustion products. This work is aimed to develop a phenomenological model describing biocidal effectiveness of these recently developed RMs. The model exploits experiments, in which aerosolized spores of Bacillus anthracis surrogates (B. atrophaeus and B. thuringiensis var kurstaki) passed through an enclosed flow system, where they were briefly (sub-second duration) exposed to elevated temperatures and combustion products of different RM powders injected in a small, air-acetylene flame. The flame location varied relative to the tube through which the bioaerosol passed, producing conditions with distinctly different temperature and iodine concentration profiles. Reference experiments were performed with only the pilot air-acetylene flame and with the flame with injected particles of pure aluminum without added iodine. In addition, different lengths of the tube section where the combustion products mixed with bioaerosol were heated, to achieve different times of exposure for the microorganisms to the iodine-bearing gas. The flow rate carrying the bioaerosol varied as well to further alter its exposure time to the combustion products. An inactivation factor IF was defined as an inverse fraction of microorganisms surviving the exposure. It was obtained from experiments at different experimental conditions and different RMs. A higher IF was observed for elevated temperatures; an additional significant increase in the IF was observed when iodine-bearing RMs were injected in the flame. An increase in the exposure time also led to higher IF values. A computational fluid dynamics (CFD) model was used to describe the flow, temperature, and iodine concentration profiles for the experimental setup. The iodine release was described as a function of the burn rate, specific for each RM powder, which was measured in separate experiments. Microorganisms were modeled as individual 1-μm diameter particles with the density of water injected in the flow system. To describe the biocidal effect quantitatively, an exposure function, $F_{e,cT}$, for each microorganism was introduced as

where integral is taken for the entire time, t, while the microorganism is passing through the system, T and c are

respectively temperature and iodine concentration obtained from the CFD calculations, and E_1, E_2, and k are adjustable parameters representing the activation energies for purely thermal inactivation, for inactivation in iodine-bearing environments, and pre-exponent, respectively. Values of $F_{e,cT}$ for all microorganisms were calculated enabling one to produce an analog of the experimental IF function, considering a fraction of microorganisms for which the exposure function exceeds a certain threshold. Comparing the experimental data with the model, it was found that E_1 and E_2 can be taken as 5 and 10 kJ/mol, respectively, while k=25 m^3/mol to describe a large set of the available experiments with Al·B·I_2as an RM powder. Further work focuses on describing the effect for other iodine bearing RM powders.

Keywords: Agent defeat, Reactive materials, Biological aerosols

499113 Process Scale-up for Production of Bis-Tetrazol-Amine and *N,N'*-Bis-(1*H*-tetrazol-5-yl)-Hydrazine

Abstract:

Two nitrogen-rich materials, namely 5,5'-bis(1H-tetrazolyl)amine (BTA) and 5,5'-hydrazinebistetrazole (HBT) were found to be excellent burning rate modifiers in nitrocellulose based propellant formulations. These two nitrogen-rich energetic materials were also found to be compatible with nitrocellulose and nitrated ester plasticizers according to the STANAG 4147 guidelines. These results demonstrated that pursuing the characterization of the impact of these materials on propellants further was worthwhile. However, in order to do so, large quantities of materials are required. A scale-up of the synthesis of BTA and HBT to the pilot scale in a 30 L CSTR was undertaken. The scale-up of the synthesis of BTA proved easy to perform and followed the previously published synthesis protocols for quantities at the gram scale. The enthalpy of mixing of adding acid to the reactive media proved to be a non-issue and was the only exothermic step of this reaction. The scale-up of the synthesis of HBT however proved more challenging. The synthesis of HBT involves the synthesis of an intermediary, sodium 5,5'-azotetrazolate pentahydrate (NaZT). Due to volume constraints, concentrations of reactants were increased for the synthesis of NaZT. This proved a challenge due to the exothermal nature of the step of adding one of the reactants to reactive mixture in the CSTR reactor. Isolating the synthesized NaZT also proved challenging due to the high amount of water needed to be removed to isolate the product in satisfactory yields. To circumvent this, it was decided to minimize the amount of water to be evaporated by reusing the reactive media for the subsequent synthesis step after filtering out the MnO_2 by-product resulting from the synthesis of NaZT. This permitted a significant reduction of the amount of water to be evaporated. However, this choice of re-using the reactive media from the synthesis of NaZT for the subsequent synthesis of HBT proved challenging for two reasons. Firstly, the synthesis of HBT requires an initial pH of around 7 to be performed in a timely manner which required the neutralization of the reactive media as the synthesis of NaZT requires a highly alkaline media. This was an unknown as to our knowledge, this has never been attempted as fully isolating NaZT was not a challenge at the laboratory scale. Secondly, the addition of acid to neutralize the reactive media could incur the degradation of NaZT. The second challenge was tackled by carefully adding the neutralizing acid at a flow rate of 30 $mL \cdot min^{-1}$ and by isolating part of the NaZT through partial recrystallization of the product prior to the neutralization step. The final step of the synthesis of HBT which was performed with ease in the laboratory also proved more challenging at the pilot scale. Exothermal effects from the addition of magnesium to the CSTR which were a non-issue at the laboratory scale proved to be of a high importance in the 30 L CTSR during the first synthesis at the pilot scale. As a result, the heat release associated with each reaction step were measured experimentally. Both reactions for the synthesis of HBT are performed at high temperatures. A batch process design taking advantage of the exothermal nature of part of the synthesis is proposed to minimize the energy requirements for the synthesis of HBT. A batch process design for the synthesis of BTA is also presented.

499378 Investigation of the Processability of Tetrazole Polyelectrolytes As Binders for Nitrogen-Rich Composite Propellants

Abstract:

Poly(5-vinyltetrazole) is a brittle energetic polymer of little interest to industrial or military applications due to its poor mechanical properties and low processability. However, a weakly acidic tetrazole moiety makes it an interesting starting point to formulate energetic polyelectrolytes with very different physical properties from their neutral precursor: the thermal, physical and rheological properties of each polyelectrolyte vary depending on the nature of the electrolyte present with the tetrazole ion and must be studied accordingly. In this work, several concentrated 5-vinyltetrazole polyelectrolyte solutions were studied to compare their processability at different temperatures; the most promising polyelectrolytes were extruded, foamed and compounded with various energetic compounds to study their thermal stability, mechanical properties, and compatibility for use as energetic binders in water-processable composite propellants.

Commercially available high molecular weight polyacrylonitrile (PAN, >100kDa) was used as the preferred starting point for the synthesis of poly(5-vinyltetrazole) (PVT) to increase the molecular weight of the resulting polymer, improving its mechanical properties. The process used to synthesize PVT is described in the literature and takes place in either N,Nâ€™-dimethylformamide or water in the presence of a catalyst (ammonium chloride or zinc chloride, respectively) and sodium azide. The resulting neutral poly(5-vinyltetrazole) is ground into a coarse powder and converted to a variety of energetic polyelectrolytes in water: it can be neutralized by alkali metals and organic bases to yield 5-tetrazolate polymers or protonated by strong acids to yield 5-tetrazolium polymers.

Through this method, previously known basic polyelectrolytes (e.g. triaminoguanidinium, ammonium polyvinyltetrazolates) as well as mostly undocumented polyelectrolytes (e.g. polyvinyltetrazolium perchlorate and nitrate) were prepared and their properties reported. The thermal stability of these samples was studied by thermogravimetric analysis (TGA) and differential scanning calorimetry (DSC). Glass transition temperatures were evaluated through Dynamic Mechanical Analysis (DMA). Rotational rheometry tests were also conducted at different temperatures to observe the thermoplastic behavior of concentrated polyelectrolyte solutions. Sample molding was completed using a heated press, foamed samples were obtained using small amounts of azobisisobutyronitrile (AIBN) as a low-temperature foaming agent. Chemical compatibility with energetic compounds, oxidizers and possible plasticizers was assessed by comparing thermal stability (DSC) of mixtures to each of their respective components.

Overall results indicate that compounds using various tetrazole polyelectrolytes as binders can be used to tailor materials with desired properties over a wide temperature range. A limited number of these compounds in concentrated solutions also show thermoplastic behaviors deemed interesting for automated manufacturing, offering new possibilities for water-processed propellants.

499430 Tuning of Energetic Material Microwave Enhancement through Micro/Nanostructure

Abstract:

Dynamic techniques to control the combustion of energetic materials are desired to enable *in* situ tuning of energetic output (e.g. solid rocket motor thrust and pyrotechnic emission). Previous work has demonstrated that microwave irradiation of aluminized ammonium perchlorate composite propellant flames of propellants doped with readily ionizing alkali earth metals (e.g. sodium in form of sodium nitrate, $NaNO_3$), can result in the enhancement of atmospheric pressure flame structure, bulk flame temperature, and burning surface regression rate without significant effect on equilibrium performance.[1–3] It has found that microwave energy is primarily absorbed by the flame structure rather than the condensed phase propellant and microwave irradiation is proposed to primarily result in enhancement through (1) plasma formation within the flame and (2) dielectric absorption in hot aluminum oxide features of aluminum particle diffusion flames, resulting in enhanced aluminum particle combustion. In these processes, plasma formation is precluded by sodium ionization and microwave energy absorption of free electron populations within the flame, and high energy loss to aluminum oxide features within aluminum flames are expected to be a result of the exponential temperature dependence of the aluminum oxide loss tangent at propellant-relevant temperatures.[4] Together, these effects enhance heat feedback to the burning surface of the propellant, resulting in enhanced regression rate. In consideration of the overall flame structure of a composite propellant, the aluminum droplet diffusion flame, with a local motor-pressure flame envelope temperature of ~3800 K[5] and hot aluminum oxide product stream having a temperature bounded by the oxide boiling temperature, is the highest temperature feature of the composite solid propellant flame. As such, it is expected that local, high temperatures of aluminum diffusion flames and direct aluminum diffusion flame heating, which may occur prior to buildup of critical plasma-level electron populations, may enhance free electron populations and result in reduction in both plasma kernel formation times and the minimum required dopant concentration.

To investigate plasma formation near aluminum particle combustion, we target sodium ionization to aluminum diffusion flames within burning propellants through development of $Al/NaNO_3$ mechanically activated composites. Ball milled $Al/NaNO_3$ composites are characterized via scanning electron microscopy and x-ray energy dispersive spectroscopy, X-ray diffraction, and calorimetry. Microwave-enhancement of atmospheric pressure burning rate, kernel formation time, and minimum required dopant concentration are investigated using 900 W continuous wave and microsecond-pulsed (100 kW peak power) S-band irradiation of propellant flame structures within resonant cavities. Propellants containing balled milled $Al/NaNO_3$ are compared to propellants of the same formulation having physically mixed ingredients. The degree of combustion enhancement and flame structure is observed using high speed video, two-color pyrometry, and UV-VIS spectroscopy, and time-resolved diode measurements of propellant flame microwave absorption to explore the ability to optimize microwave plasma enhancement through micro/nanostructure control of energetic materials within the model system of a composite solid propellant.

Bibliography

[1] S. Barkley, K. Zhu, J. Lynch, M. Ballestero, J. Michael, and T. Sippel, in *AIAA Aerosp. Sci. Meet.* (2016), pp. 1–9.

[2] S.J. Barkley, K. Zhu, J.B. Michael, and T.R. Sippel, in *55th AIAA Aerosp. Sci. Meet.* (2017), pp. 1–8.

[3] S.J. Barkley, K. Zhu, J.B. Michael, and T.R. Sippel, in *52nd AIAA/SAE/ASEE Jt. Propuls. Conf. AIAA* (2016), pp. 1–11.

[4] W.H. Sutton, Am. Ceram. Soc. Bull. **68**, 376 (1989).

[5] W.E. Price, in *Fundam. Solid-Propellant Combust.*, edited by K. Kuo (AIAA, New York, 1984), pp. 479–513.

499959 Burning Rate Control of Energetic Materials with Thermally Switchable Microwave Properties

Abstract:

The dynamic control of energetic material burning rate remains an elusive goal. Previous work has demonstrated that continuous/pulsed microwave irradiation of doped ammonium perchlorate composite propellants can enhance flame structure, bulk flame temperature, and increase propellant regression rates at atmospheric conditions.[1–3] In these efforts, increases in propellant burning rate were found to be in part a result of dielectric absorption in hot aluminum oxide combustion features of aluminum droplet diffusion flames and subsequent increase in heat feedback to regression surfaces. Energy deposition to hot metal oxide features are hypothesized to be a result of the exponential temperature dependence of aluminum oxide dielectric loss at propellant flame temperatures,[4] where loss is due to aluminum oxide electron promotion to a valencene shell, resulting in 3000% more efficient energy deposition at 1500 K. While in these studies energy depositions was observed in burning aluminum agglomerates within the bulk of the propellant flame, more efficient combustion enhancement may be realized by localization of temperature-dependent deposition directly to an energetic material burning surface rather than utilization of indirect enhancement through thermal radiation feedback. Additionally, rather than exploiting thermal runaway effects of a combustion product (e.g. aluminum oxide), it may be possible to effect regression surface energy deposition through incorporation of materials in to energetics that have thermally switchable microwave properties.

To investigate the ability to control burning rate through microwave energy deposition directly to an energetic material regression surface, we formulate AP composite propellants containing nanostructured/nanoscale additives of (1) ball milled Al/MoO_3 nanostructured energetics or (2) graphene oxide nanoscale platelets. Ball milled Al/MoO3 nanostructured composite particles are hypothesized to effect microwave energy deposition to/near the burning surface through rapid ignition and production of aluminum oxide and flame temperatures in excess of the bulk propellant flame temperature. Separately, graphene oxide (GO) upon heating to propellant burning surface temperatures is converted from its microwave-reflective form of GO to microwave-absorptive reduced-graphene oxide (r-GO) structures[5–7] In order to explore the ability to enhance propellant burning rates, experiments will be conducted with continuous wave, 900 W and microsecond-pulsed (100 kW peak power) S-band microwave radiation propagated within a resonant cavity. Burning rate enhancement and near-burning surface flame temperatures are measured using high speed video and two-color pyrometry, respectively. Condensed phase dielectric properties of propellants containing Al/MoO3, GO, and r-GO as well as forward/reflected microwave power measurements are used to explore microwave absorption of the condensed phase. Additionally, the microwave ignition delay of GO and r-GO structures and the use of microwave energy as a deflagration ignition source using r-GO burning surface chars are explored. The prognosis for use of microwave energy deposition coupled with thermally tuned energetic material nanostructures as a means for use in energetic material ignition/ extinguishment/ reignition as well as regression rate throttling applications is discussed.

Bibliography

[1] S. Barkley, K. Zhu, J. Lynch, M. Ballestero, J. Michael, and T. Sippel, in *AIAA Aerosp. Sci. Meet.* (2016), pp. 1–9.

[2] S.J. Barkley, K. Zhu, J.B. Michael, and T.R. Sippel, in *52nd AIAA/SAE/ASEE Jt. Propuls. Conf. AIAA* (2016), pp. 1–11.

[3] S.J. Barkley, K. Zhu, J.B. Michael, and T.R. Sippel, in *55th AIAA Aerosp. Sci. Meet.* (2017), pp. 1–8.

[4] W.H. Sutton, Am. Ceram. Soc. Bull. **68**, 376 (1989).

[5] O. Akhavan, Carbon N. Y. **48**, 509 (2009).

[6] M. Han, X. Yin, L. Kong, M. Li, W. Duan, L. Zhang, and L. Cheng, J. Mater. Chem. A **2**, 16403 (2014).

[7] L.N. Wang, X.L. Jia, Y.F. Li, F. Yang, L.Q. Zhang, L.P. Liu, X. Ren, and H.T. Yang, J. Mater. Chem. A **2**, 14940 (2014).

501214 Field Switching of Monopropellant Burning Rate: Dielectrophoretic Control of Nitromethane Thermal Conductivity Using Field Aligned Carbon Nanomaterials

Abstract:

Field Switching of Monopropellant Burning Rate:

Dielectrophoretic Control of Nitromethane Thermal Conductivity Using Field Aligned Carbon Nanomaterials

A. R. Lawrence, T. R. Sippel

Department of Mechanical Engineering

Iowa State University, Ames IA

Submission to AIChE Fall Meeting

Session: Energetic Materials

October 29 - November 3, 2017, Minneapolis MN

Abstract

Control of propellant burn rates is vital to optimizing energetic material utility. Single-walled carbon nanotubes (SWCNT's) have a uniquely high, anisotropic thermal conductivity in the axial direction, and recent nanofluids research has shown that the addition of carbon nanotubes to liquids can increase the thermal conductivity of the mixtures by 300%.[1] When a non-uniform electric field is applied, CNTs will align due to dielectric forces (i.e. dielectrophoresis). Aligning the CNT's creates a thermally and electrically conductive path through the mixture, and can be affected in a highly efficient manner using low duty cycle fields, as typical alignment times in water are 100 times shorter than Brownian motion unalignment.[2] In separate energetic material research, it has been found that the addition of other nanostructured additives, such as fuctionalized graphene sheets, increased the burning rates by more than 200%.[3]

Taken together, these two bodies of research suggest that this alignment method could be utilized to obtain efficient, high-speed switching of the deflagration of liquid propellants--aligning the SWCNT's perpendicular to the burning surface would illicit a high speed burning mode, while a parallel alignment would result in near-normal combustion. We explore dielectrophoretic switching of burning rate in liquid nitromethane doped with single-walled carbon nanotubes burning at motor pressures under the influence of kHz-rate modulation of 300 kV/m static fields aligned both parallel and transverse to the burning surface. Burning rates are measured using high speed video and thermal

conductivity will be measured using hot wire technique. The effects of dopant type, concentration, pressure, and alignment direction are explored.

Bibliography

1. E. S. Choi, J. S. Brooks, D. L. Eaton, M. S. Al-Haik, M. Y. Hussaini, H. Garmestani, D. Li, and K. Dahmen, J. Appl. Phys. **94**, 6034 (2003).

2. K. Bubke, H. Gnewuch, M. Hempstead, J. Hammer, and M. L. H. Green, Appl. Phys. Lett. **71**, 1906 (1997).

3. J. L. Sabourin, D. M. Dabbs, R. A. Yetter, F. L. Dryer, and I. A. Aksay, **3**, 3945 (2009).

490842 Numerical Simulation and Experimental Study of a Micro Circulating Fluidized Bed

Abstract:

Numerical simulation and experimental study of a micro circulating fluidized bed

Yupeng Xu[a,*], Jordan Musser[a], Tingwen Li[a,b], Balaji Gopalan[a,c], Rupen Panday[a,d], Jonathan Tucker[a,e], Greggory Breault[a,d], William A. Rogers[a]

a. National Energy Technology Laboratory, Morgantown, WV 26505, USA

b. AECOM, Morgantown, WV 26505, USA

c. West Virginia University Research Corporation, Morgantown, WV 26506, USA

d. REM Engineering Services, Morgantown, WV 26506, USA

e. West Virginia University, Morgantown, WV 26506, USA

Abstract

Circulating fluidized beds (CFB) are widely employed in industry for a wide variety of gas-solid contacting operations. Over the past few decades, CFB hydrodynamics have been studied using various experimental techniques and by numerical simulations. Recently, several efforts have modeled full-loop CFBs using the two-fluid model (TFM) whereby gas and solids are treated as interpenetrating continua. However, this technique only resolves solid motion at the computational cell level. With the advancement of high-performance computing, higher fidelity models, like Computational Fluid Dynamics-Discrete Element Method (CFD-DEM) which tracks the trajectory of individual particles, have become increasingly popular as research tools. This is because particle-scale information including residence time, collision forces, and dispersion intensities are readily available for detailed analysis of complex flow phenomena.

In this first of its kind work, a small scale, full-loop circulating fluidized bed is investigated using physical experiments and CFD-DEM simulations conducted with the U.S. Department of Energy's Multiphase Flow with Interphase eXchanges code, MFIX-DEM. Experimental and simulation results are compared using pressure drops data, solids circulation rate and solids standpipe inventory height. Additionally, the performance of different drag laws is examined.

Fig.1. Direct comparison between simulation and experiments

492358 Attrition Prediction and Reactive Jet Cup Testing of Oxygen Carriers for Chemical Looping Combustion

Abstract:

Novel technologies for chemical productions and carbon capture utilizing circulating fluidized bed (CFB) reactor configurations have garnered much attention in recent years. An example of these technologies is the chemical looping combustion (CLC) scheme. CLC provides an increase in power efficiencies for lower cost, while providing low energy penalty carbon capture, than traditional combustion or gasification processes that additionally utilize CO_2 capture. Chemical looping combustion uses metal oxide particles often termed â€œoxygen carriersâ€ in a cyclic redox mode using two separate reactors termed the fuel and air reactors, respectively. Uncertainty lies in the chemical loopingâ€™s economic feasibility, which is partially due to the required solids makeup rate. Attrition due to impact and wear affect the solids makeup rate needed in CLC systems. Additionally, an oxygen carriers structure can be severely affected by the extreme conditions it experiences during operation. This includes high temperatures (800-1000°C) and shrinking/expanding of the particles due to the reduction/oxidation of the oxygen carrier. This can lead to additional change in the oxygen carriers structure such as crack formation and pore structural changes which can affect the overall oxygen carrier resistance to attrit.

This work investigates attrition of oxygen carriers for CLC on two fronts: (1) development of a simple spreadsheet model to help predict attrition expected in different CFB systems and (2) design and operation of a high-temperature, reactive jet-cup to help validate a portion of the model. The spreadsheet model uses developed models for predicting attrition rates in various portions of CFB units including: fluidized beds, cyclones, standpipes, impactors, risers, and corners such as L-valves. The model uses inputs based off the material properties (hardness, wear constant, impact constant, and fracture toughness), gas properties, flows, temperature, and the dimensions and number of each configuration encountered in the CFB system. An example modelling case using hematite as an oxygen carrier is performed using the design aspects for the $50kW_{th}$ chemical looping reactor at the National Energy Technology Laboratory (NETL). Additionally, the model can be used to perform a sensitivity analysis on key configurations and variables that could affect attrition the most. Additionally, the design of a reactive jet cup will be discussed as a novel determination of how thermal and chemical stresses undergone by oxygen carriers affect the overall attrition rates of the oxygen carrier. Investigation of a hematite oxygen carrier will be presented at a room temperature attrition test to show best case scenario for attrition of the oxygen carrier as well as under hot temperatures under both inert and reactive cases. For the reactive cases, the oxygen carrier is tested under fluidized bed conditions in methane/air redox schemes with nitrogen as a diluent. After specific reduction and oxidization cycles, attrition studies using the jet portion of the reactor is conducted to test the attrition rate observed.

502418 Agglomerate Number: Experimental Validation of the Energy-Based Dimensionless Group in Entrainment of Cohesive Particles

Abstract:

Cohesion significantly impacts entrainment behavior, and predicting entrainment rates for cohesive particles has been problematic for decades. Understanding the macroscopic (many-particle) cohesive behavior via the microscopic (particle-level) interactions is critical for improved predictive capabilities. For systems dominated by brief contacts (e.g., dilute systems), the Agglomerate number (Ag) is the relevant dimensionless group. Ag is defined as the ratio of the critical kinetic energy required for agglomeration to occur to the characteristic (fluctuation) kinetic energy available to overcome agglomeration. Previously, the appropriateness of Ag in collision-dominated (relatively dilute) systems has been shown theoretically for a myriad of systems, including risers, via collapse of the fraction of agglomerates versus Ag. In this work, a comprehensive set of agglomeration measurements of mildly cohesive particles in a dilute riser is compared to Ag for the first time. These local, in situ, non-intrusive measurements include mean particle and agglomerate velocities, relative velocity of particles and agglomerates, agglomerate diameters, fraction of agglomerates, and solids volume fraction. Experiments were designed to test a wide range of Ag values by varying the degree of cohesion via relative humidity (numerator of Ag) and different sized particles (denominator of Ag). Although not all dimensionless groups in the system could be controlled, e.g., the ratio of riser diameter to particle diameter, decent collapse of dimensionless measurements and Ag is observed. Overall, the presented results demonstrate increased agglomeration at high Ag and decreased agglomeration at low Ag.

496990 Experimental Analysis of a Vortexing CFB for Process Intensification via High-G Flows

Abstract:

EXPERIMENTAL ANALYSIS OF A VORTEXING CFB FOR PROCESS INTENSIFICATION VIA HIGH-G FLOWS

Michael Bobek[1,2], Jingsi Yang[1,2], Justin Weber[1], Frank Shaffer[1], Ronald W. Breault[1]

[1]*Nattional Energy Technology Laboratory, Morgantown, WV, USA 26507*

[2]*Oak Ridge Institute for Science and Education, Morgantown, WV, USA, 26507*

Fluidized beds are used in many industries for their favorable characteristics of good solids mixing, high rates of heat and mass transfer, and large throughputs. In the never-ending challenge of increasing efficiencies one avenue is through process intensification, specifically this work focuses on high-g flows, flows with the centripetal force typically on the order of 10 times the force of gravity. This operating regime provides intensified gas-solids contact through higher mass transfer, heat transfer, and gas throughput. This work focuses on an experimental analysis of a cold flow vortexing circulating fluidized bed. Through mapping the pressure distribution in the riser and utilizing high speed cameras to track particle behavior the insights into the behavior of the system were made and compared to barracuda CFPD modeling of the system. The figure below shows the high-speed video particle tracking, with results from the video analysis calculating particle velocities in the vortex of around 19 m/s and particle velocities of 1 to 2 m/s as they leave the recirculation section.

502422 Experimental Investigation of Horizontal Air Jets in Semi-Circular, Gas-Solid Fluidized Bed

Abstract:

Experiments were undertaken at PSRI with the intent to provide a validation benchmark problem with characterized uncertainties for the coupled computational fluid dynamicsâ€"discrete element method (CFD-DEM) modeling approach. In this talk, we will focus on the first material studied: 6 mm plastic beads. For this material, the particle count in the system is roughly 60,000, making it amenable for direct comparison with CFD-DEM using todayâ€™s computational resources. The Geldard Group D particles are well characterized; particle size, shape, coefficients of restitution and kinetic friction are measured. The test section is a semi-circular bed approximately 12 inches in diameter. Additional fluidization is provided by two horizontal air jets which penetrate the unit near the flat, front face. The opposing jets are operated at one of two vertical locations. Moreover, uncertainties in each of the experimental parameters are quantified for use in formal uncertainty quantification (UQ) analyses. During operation, the bed pressure drop is measured with pressure taps in and above the bed. Additionally, high speed video (HSV) imaging of the front face is taken to record the dynamics of the particles in and near the jet regions. The HSV images are processed with an in-house particle tracking algorithm and time- and spatially- averaged. By producing a map of the particle Froude number (using 2-D velocity magnitude), we trace isolines of $Fr = 0.15$ which are used to calculate the jet penetration depths. In this manner, we report a quantity that can be compared in a straightforward manner with both CFD-DEM and TFM simulations. We show further how the interaction of the jets varies with separation distance by extending the jets into the domain. Finally, we will conclude by previewing additional materials studied in this unitâ €"specifically, with smaller diameter particles and thus higher particle counts, we hope this validation set will grow with the increasing power of the CFD-DEM technology itself.

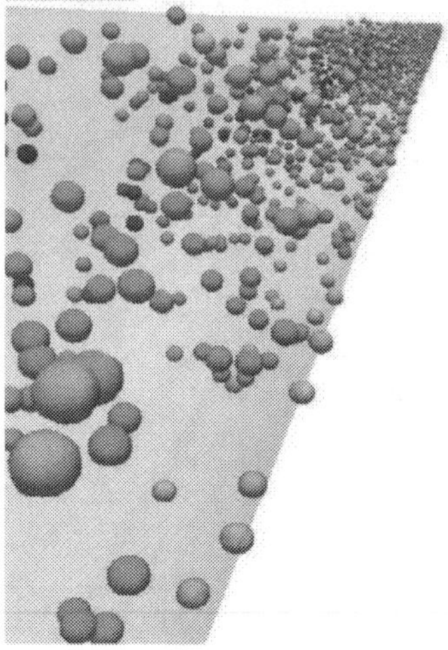

On the Euler-Lagrange simulations of a pilot scale CFB riser: experimental unit CFD-based design, flow structures and particle forces analysis

Utzig, J. [1]; Martignoni, W.P. [1]*; Souza, F.J. [2]; Meier, H.F. [1]

[1]University of Blumenau, Brazil
[2]Federal University of Uberlandia, Brazil

CFBs Simulation

- There are many ways to model the particle flow, either by Eulerian and/or Lagrangian approach:
  - Examples: Kinetic Theory of Granular Flow (KTGF), Discret Element Method (DEM), Multiphase Particle-in-Cell (MP-PIC)...
- Gas-solid phenomenology:
  - Phase interactions - clustering (Capecelatro et al., 2015)
  - Chemical reactions (Lopes et al., 2011)
  - Non-spherical particles (van Wachem et al., 2015)
- Riser geometry:
  - Exit geometry: C, L and T-shape (Shi et al., 2015)

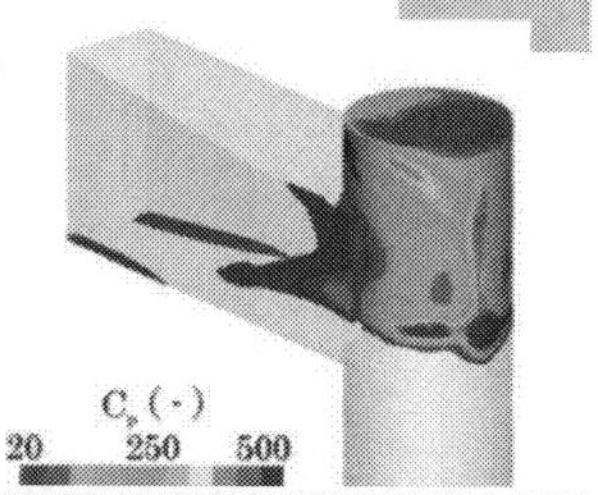

lfclvv-furb.com

Objective of this work

Given the gas-solid turbulent riser flow and the experimental data acquired by Utzig (2016) (previous work in this session), the present study aims to explore and predict the upward flow at the Pilot Unit of Riser and Cyclones, using an E-L point-particle approach.

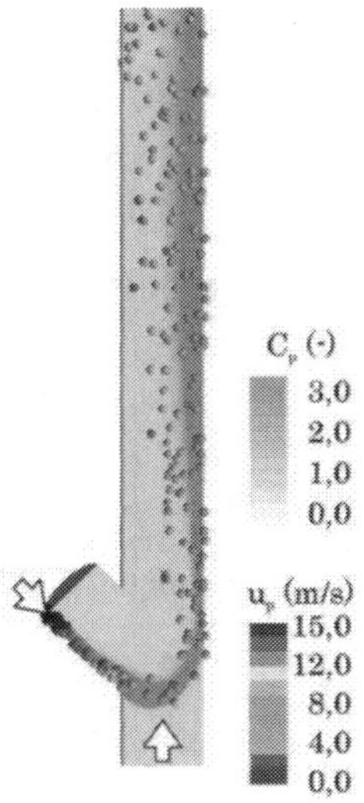

Pilot Unit of Riser and Cyclones

- Riser: 12m tall, 0.1m i.d.
  - None gas distributor
  - T-shape outlet

Operating Conditions

	$U(m/s)$	$\eta(kg_p/kg_g)$	$G_S(kg_p/m^2s)$
1		0,50	4,144
2	7,0	0,75	6,216
3		1,00	8,288
4		0,50	5,920
5	10,0	0,75	8,880
6		1,00	11,840

Fluid phase:
Athmospheric air
$\rho_p = 1.2\ kg/m^3$
$\mu = 1.8\ 10^5 Pa.s$

Particle phase:
Glass beads
$d_{p,32} = 72\ \mu m$
$\rho_p = 2500\ kg/m^3$

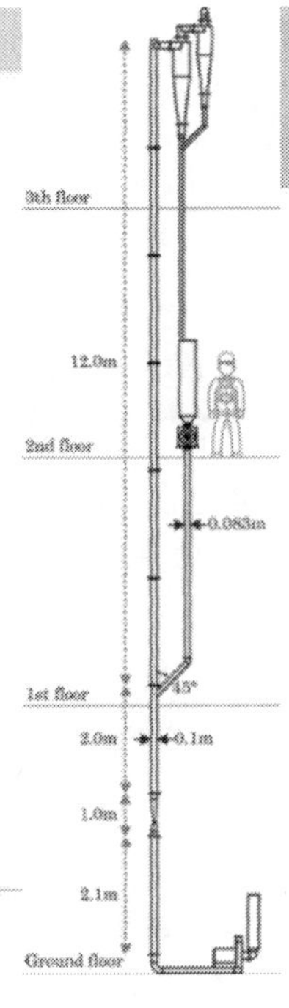

Simulation geometry

- The whole riser is the flow domain, as well as the Venturi flowmeter, the particles downcomer and the T-shape riser exit
- Two meshes will be tested: 318,952 nodes and 768,816 nodes; $y^+ \approx 30$
  (feasible due to computational cost)

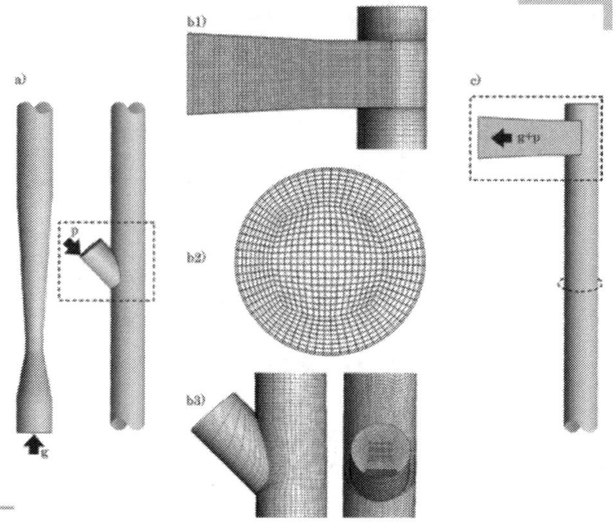

Mathematical Model

- Eulerian-Lagrangian Approach
- Gas-phase model: Continuity and incompressible, newtonian Unsteady Navier-Stokes Equations (URANS)

$$\frac{\partial(\rho u_i)}{\partial x_i} = 0 \quad , \qquad \frac{\partial}{\partial t}(\rho u_i) + \frac{\partial}{\partial x_j}(\rho u_i u_j) = -\frac{\partial p}{\partial x_i} + \frac{\partial}{\partial x_j}\left[\mu\left(\frac{\partial u_i}{\partial x_j} + \frac{\partial u_j}{\partial x_i}\right)\right] + \rho g_i + S_{p,Ui}$$

with the Two-Layers k-ε Model

$$\frac{\partial(\rho k)}{\partial t} + \frac{\partial(\rho u_j k)}{\partial x_j} = \frac{\partial}{\partial x_j}\left[\left(\mu + \frac{\mu_t}{\sigma_k}\right)\frac{\partial k}{\partial x_j}\right] + P - \rho\varepsilon$$

$$\frac{\partial(\rho\varepsilon)}{\partial t} + \frac{\partial(\rho u_j \varepsilon)}{\partial x_j} = \frac{\partial}{\partial x_j}\left[\left(\mu + \frac{\mu_t}{\sigma_\varepsilon}\right)\frac{\partial \varepsilon}{\partial x_j}\right] + C_{\varepsilon 1}\frac{\varepsilon}{k}P - C_{\varepsilon 2}\rho\frac{\varepsilon^2}{k}\rho\varepsilon$$

the turbulent viscosity

$$\mu_t = \lambda_\varepsilon \mu_{t,padrao} + (1 - \lambda_\varepsilon)\, \mu_{t,2camadas} \qquad \mu_{t,padrao} = C_\mu \rho \frac{k^2}{\varepsilon} \qquad \mu_{t,2camadas} = \rho C_\mu l_\mu \sqrt{k}$$

(other closing models will be omitted)

and the k-ω SST Model

$$\frac{\partial \rho k}{\partial t} + \frac{\partial}{\partial x_i}(\rho k u_i) = \frac{\partial}{\partial x_j}\left[\left(\mu + \frac{\mu_t}{\sigma_k}\right)\frac{\partial k}{\partial x_j}\right] + \widetilde{G}_k - \rho \beta^* k \omega$$

$$\frac{\partial \rho \omega}{\partial x_i} + \frac{\partial}{\partial x_i}(\rho \omega u_i) = \frac{\partial}{\partial x_j}\left[\left(\mu + \frac{\mu_t}{\sigma_\omega}\right)\frac{\partial \omega}{\partial x_j}\right] + G_\omega - \rho \beta_i \omega^2 + 2(1 - F_1)\frac{\rho \sigma_{\omega,2}}{\omega}\frac{\partial k}{\partial x_j}\frac{\partial \omega}{\partial x_j}$$

with the turbulent viscosity

$$\mu_t = \frac{\rho k}{\omega}\frac{1}{\max\left[\frac{1}{\alpha^*}, \frac{SF_2}{0{,}31\omega}\right]}$$

(other closing models will be omitted)

Finally, the particle source term (two-way coupling)

$$S_{p,Ui} = -n\left\langle m_p\left[\frac{du_{pi}}{dt} - \left(1 - \frac{\rho}{\rho_p}\right) g_i\right]\right\rangle$$

- Particle (parcel) motion model: **trajectory, linear and angular momentum equations**

$$\frac{dx_{pi}}{dt} = u_{pi}$$

$$m_p \frac{du_{pi}}{dt} = m_p \frac{3\rho C_D}{4\rho_p d_p}(u_i - u_{pi}) + F_{si} + F_{ri} + \left(1 - \frac{\rho}{\rho_p}\right) m_p g_i$$

$$I_p \frac{d\omega_{pi}}{dt} = T_i$$

Gas-phase fluctuations sinthetized by Langevin equation (turbophoresis)

$$u_i = U_i + u'_i\,, \quad u'_{i,n+1} = R_{p,ui}\, u'_{i,n} + \sqrt{\frac{2}{3}k(1 - R^2_{p,ui})}\,\chi\,, \quad R_{p,ui} = R_L \times R_{E,i}\,, \quad R_L = \exp\left(-\frac{\Delta t}{T_L}\right), \quad T_L = 0{,}24\,\frac{2}{3}\frac{k}{\varepsilon}$$

Sommerfeld (2001)

Stochastic particle collisions model (four-way coupling)

$$\kappa_{col} = 0,25\pi \left(d_p + d_{p,fict}\right)^2 \left\|u_p - u_{p,fict}\right\| C_p \;, \qquad \Psi = 1 - e^{(-\kappa_{col}\Delta t)}$$

Oesterlé and Petitjean (1993)

with the Saffman force $F_{si} = 1,615 d_p \mu \sqrt{Re_s} C_{ls} \left[(u_i - u_{pi}) \times \omega_i\right]$

Mei (1992)

Magnus force $$F_{ri} = \rho d_p^3 \frac{\pi Re_p}{8 Re_r} C_{lr} \frac{\left[\Omega_i \times (u_i - u_{pi})\right]}{|\Omega_i|}$$

Rubinow e Keller (1961)

+ Lun and Liu (1997) coefficient;

the torque $$T_i = C_r \frac{\rho d_p^5}{64} |\omega_i| \omega_i$$

Rubinow e Keller (1961)

+ Dennis, Singh e Ingham (1980) coefficient;

and the Schiller and Naumann (1935) drag coefficient.

Besides that, the wall roughness $\alpha_p^+ = \alpha_p^- + \Delta\gamma\xi$

Sommerfeld and Huber (1999)

Numerical Methods

- The in-house *UNSCYFL3D v2.2* code was used:
  - Finite Volume Method based
  - Solves unstrutured three-dimensional grids
  - Collocated arrangement and Rhie-Chow interpolation scheme
  - Discretizations:
    - Time: three-levels method
    - Advection: second order Upwind
    - Diffusion: centered scheme

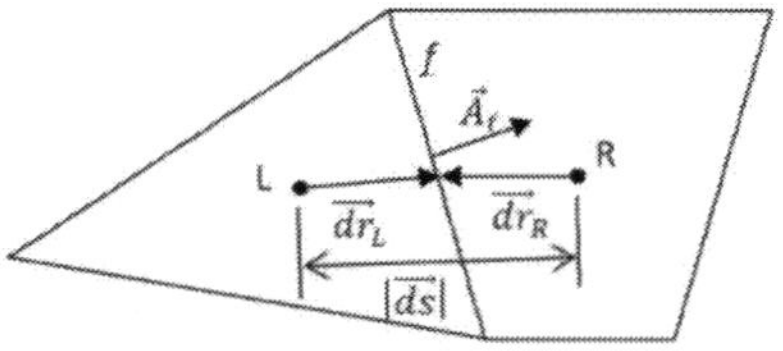

Boundary Conditions

- Gas-phase model:
  - Uniform velocity profile at the Ventury flowmeter bottom: $U = 7m/s$ and $U = 10m/s$; 5% turbulence intensity and lenght scale equals 0.07 riser diameter
  - No-slip condition at the wall
  - Relative pressure prescribed equal to zero at the riser outlet
  - $\Delta t = 10^{-4}s$
- Solid-phase model:
  - 250,000 – 350,000 computational particles/parcels fed uniformly at the Y side inlet, each of them with $U_p = 15m/s$
  - Perfectly elastic collisions
  - $\Delta t = 10^{-4}s$

- Pressure-Velocity coupling: SIMPLE algorithm
- Biconjugate gradient and Algebric Multigrid methods
- Particle tracking through unstructured grids: the method proposed by Haselbacher et al. (2007)
- Anallytical solutions of the particle motion ODEs

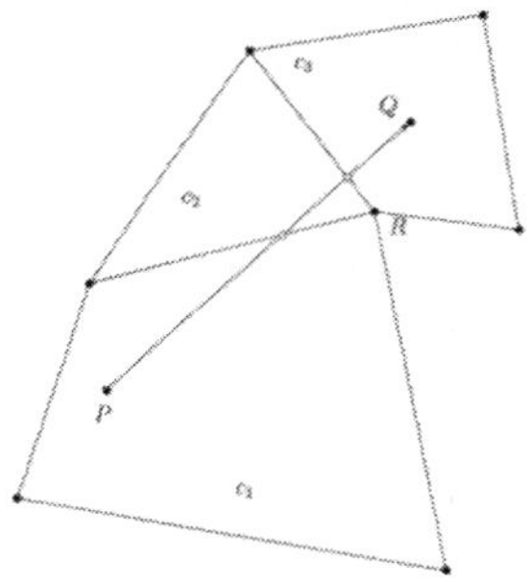

- Spatial discretization verification: Grid Convergence Index - GCI (Roache, 1994)
  - Based on Richardson Extrapolation Method

$$GCI_{ref} = F_S|E_{ref}| = F_S\left\|\frac{f_2 - f_1}{1 - R^m}\right\| \qquad GCI_{esp} = F_S|E_{esp}| = F_S\left\|\frac{R^m(f_2 - f_1)}{1 - R^m}\right\|$$

Results: GCI

- The spatial discretization sensibility was tested for the Case 1
- Radial profiles were taken at 5.5m
- The finer grid results are close to the extrapolated ones
- 718,816 nodes was selected

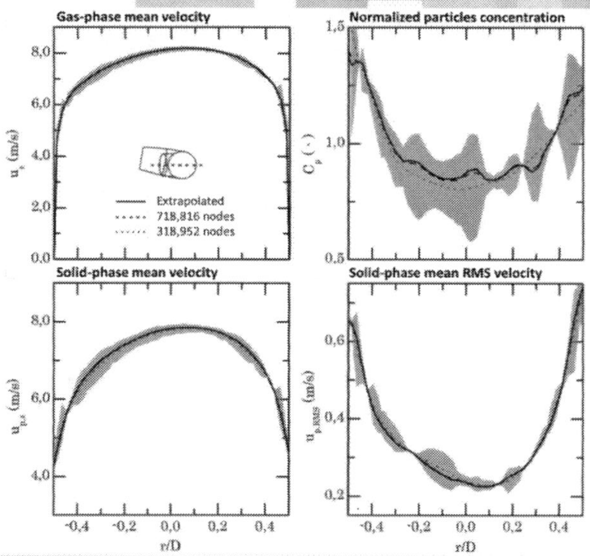

Results: Particle-model analysis

Caso 6: U=10m/s, η=1.0

Particle collisions:

- Negligible effect on axial mean velocity and RMS velocity
- Particle concentration: high axial momentum in the two-way coupling

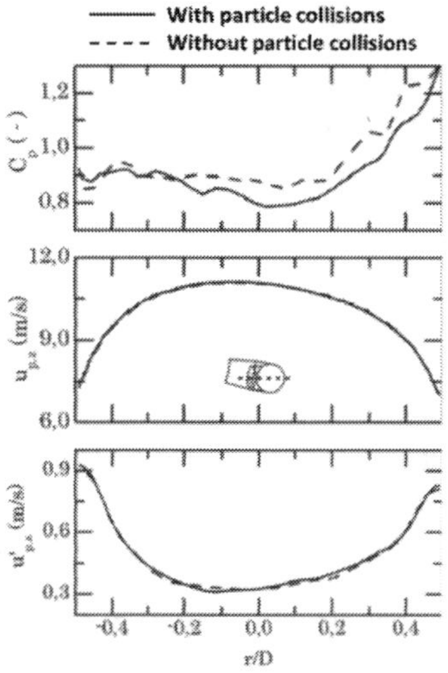

Wall roughness:

- Effect of highly rough wall on particle concentration: not expected behavior

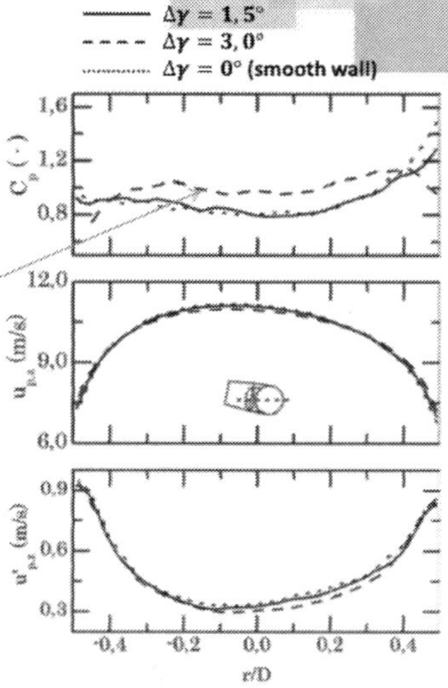

Caso 6: U=10m/s, η=1.0

Results: Particle-model analysis

Particle forces:

- Negligible effect on axial mean velocity and RMS velocity
- Particle concentration tendency mantained

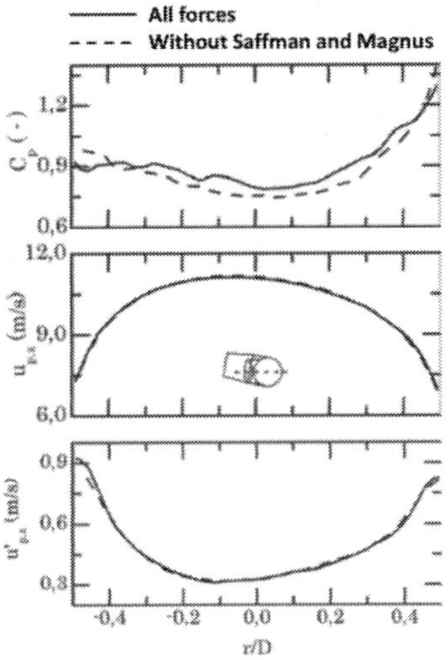

Turbophoresis effect:

- Turbulence fluctuations have large influence on particles distribution

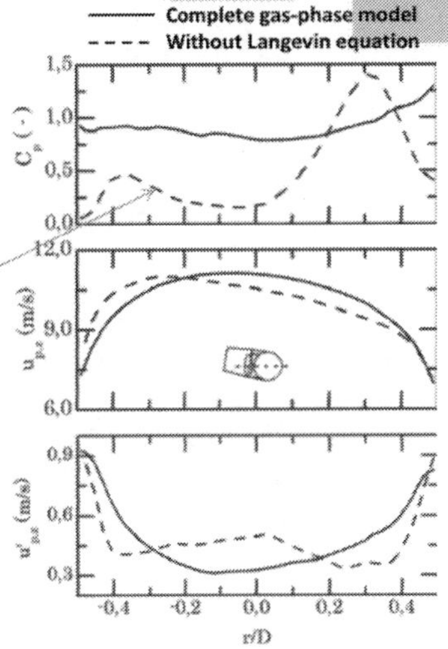

LFC LVV FURB

lfclvv-furb.com

Results: Validation (C_p)

- Numerical simulations: flat profiles that don't predict the core-annulus particle flow structure
- Better prediction of the two-layers k-ε model, mainly at riser center
- Possible numerical reasons: poor grid resolution near the wall and low computational particles number, since the turbophoresis was already predicted (Utzig et al., 2014)

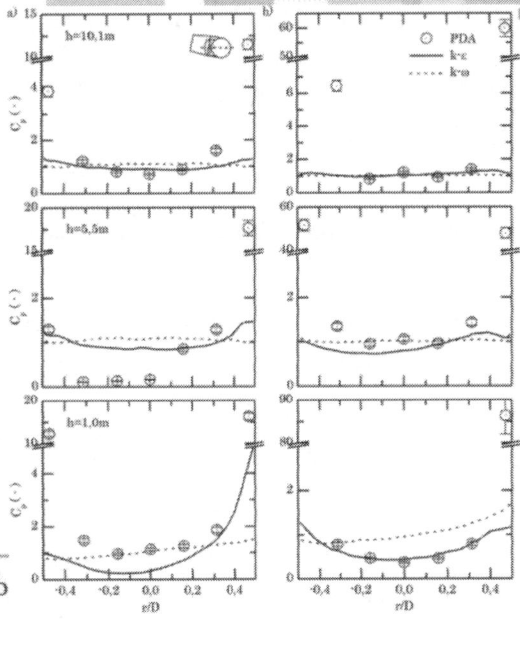

LFC LVV FURB

lfclvv-furb

Results: Validation (u_p)

- Similar predictions by the both turbulence models
- Better validation at the riser center
- Near the wall, due to the concentration error, there is lower momentum transfer between gas and particles, therefore, overprediction of particles velocity

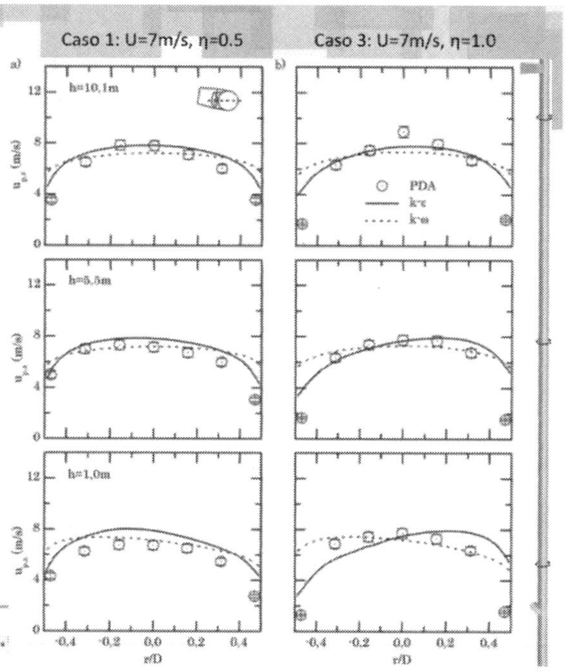

LFC LW FURB lfclvv-furb.

Results: Validation (u'_p)

- High particle velocity fluctuation, mainly at top and bottom riser regions (feed and exit effects), subpredicted by both models in any mass loadings
- Better prediction of k-ω SST model compared to the two-layers k-ε model

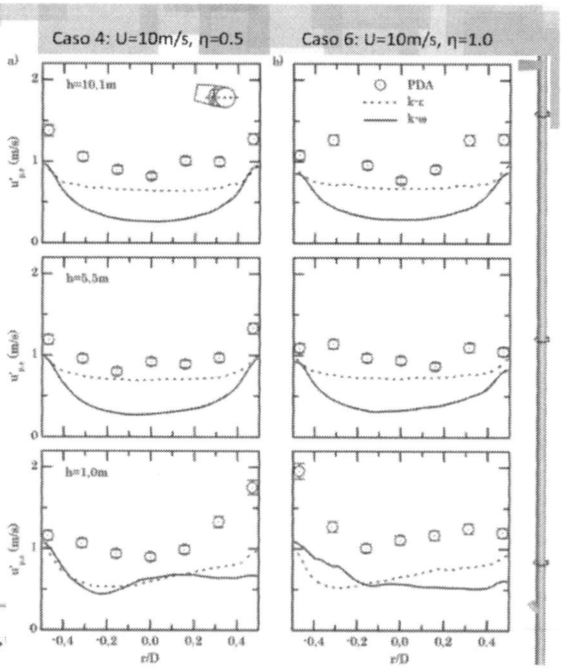

LFC LW FURB lfclvv-furb.

Conclusions

- The gas-particle flow on a pilot scale CFB riser was predicted by an Eulerian-Lagrangian point-particle, four-way coupling model
- A model sensitivity test shows that wall roughness and gas-phase velocity fluctuations given by the Langevin equation (turbophoresis effect) have remarkable influence on particle dynamics
- Good mean axial particles velocity validation, subprediction of particles velocity fluctuation and core-annulus structure not predicted

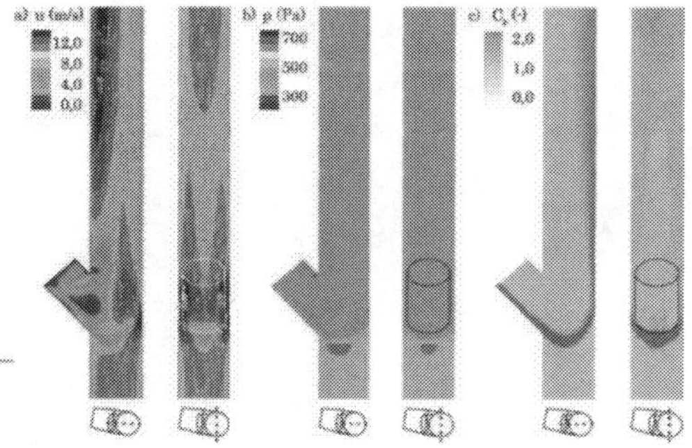

References

CAPECELATRO, J.; DESJARDINS, O.; FOX, R. O. **On fluid-particle dynamics in fully developed cluster-induced turbulence**. Journal of Fluid Mechanics, v. 780, p. 578-635, 2015.

DENNIS, S. C. R.; SINGH, S. N.; INGHAM, D. B. **The steady ow due to a rotating sphere at low and moderate Reynolds numbers**. Journal of Fluid Mechanics, v. 101, p. 257-279, 1980.

LOPES, G. C. et al. **Three-dimensional modeling of fluid catalytic cracking industrial riser ow and reactions.** Computers and Chemical Engineering, Elsevier Ltd, v. 35, n. 11, p. 2159-2168, 2011.

LUN, C. K. K.; LIU, H. S. **Numerical simulation of dilute turbulent gas-solid flows in horizontal channels.** International Journal of Multiphase Flow, v. 23, n. 3, p. 575-605, 1997.

MEI, R. **An approximate expression for the shear lift force on a spherical particle at finite Reynolds number**. International Journal of Multiphase Flow, v. 18, p. 145-147, 1992.

OESTERLE, B.; PETITJEAN, A. **Simulation of particle-to-particle interactions in gas-solid flows**. International Journal of Multiphase Flow, v. 19, n. I, p. 199-211, 1993.

RUBINOW, S. I.; KELLER, J. B. **The transverse force on a spinning sphere moving in a viscous liquid.** Journal of Fluid Mechanics, v. 11, p. 447-459, 1961.

SCHILLER, L.; NAUMANN, A. **A drag coefficient correlation**. Z. Ver. Deutsch. Ing., 1935.

SHI, X. et al. **Effects of the riser exit geometries on the hydrodynamics and solids back-mixing in CFB risers: 3D simulation using CPFD approach**. Powder Technology, Elsevier B.V., v. 284, p. 130-142, 2015.

SOMMERFELD, M.; HUBER, N. **Experimental analysis and modelling of particle-wall collisions.** v. 25, p. 1457-1489, 1999.

SOMMERFELD, M. **Validation of a stochastic Lagrangian modelling approach for inter-particle collisions in homogeneous isotropic turbulence**. International Journal of Multiphase Flow, v. 27, p. 1829-1858, 2001.

SOUZA, F. J.; SILVA, A. L.; UTZIG, J. **Four-way coupled simulations of the gas-particle flow in a diffuser**. Powder Technology, Elsevier B.V., v. 253, p. 496-508, 2014.

VAN WACHEM, B. et al. **Modelling of gas-solid turbulent channel flow with non-spherical particles with large Stokes numbers**. International Journal of Multiphase Flow, Elsevier Ltd, v. 68, p. 80-92, 2015.

UTZIG, J.; SOUZA, F. J.; MEIER, H. F. **A Numerical Analysis of the Turbophoresis in a Turbulent Gas-Particle Flow.** In: Volume 1C, Symposia: Fundamentals Issues and Perspectives in Fluid Mechanics; Industrial and Environmental Applications of Fluid Mechanics; Issues and Perspectives in Automotive Flows; Gas-Solid Flows: Dedicated to the Memory of Professor Clayton T. Crowe. ASME, 2014. p. V01CT18A011.

UTZIG, J. **Dinâmica das Partículas em Leito Fluidizado Circulante.** PhD Thesis, Federal University of Uberlandia, Brazil, 2016.

Thank you!

Acknowledgements: CNPq Conselho Nacional de Desenvolvimento Científico e Tecnológico; BR PETROBRAS

Questions or suggestions:
Jonathan Utzig, DSc
jutzig@furb.br

Visit Blumenau, – Brazil!

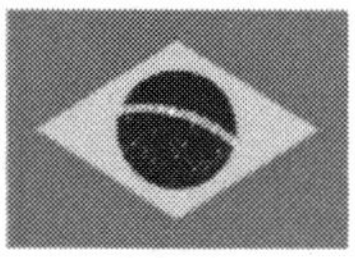

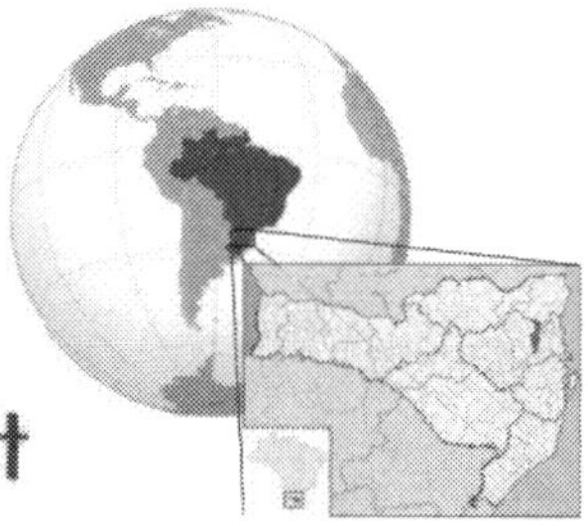

Oktoberfest BLUMENAU

LFC LW FURB lfclvv-furb.com 21

498943 Particle Phase Behavior Experimental Analysis on a Pilot Scale CFB Riser Using Phase Doppler Anemometry

Abstract:

The particle phase concentration and diameter distributions are quite important data of gas-solid multiphase flows to modeling validation. Measurements of the relationship between them and solids velocity can be useful to the fluid dynamics understanding and analysis of the relevant phenomena impacting on particles flow. Mainly in circulating fluidized beds (CFBs) applications, where commonly a very large number of particles are present, an appropriate definition of the mathematical modeling is imperative. In this way, experiments were carried out in three axial locations of the Pilot Unit of Riser and Cyclones: near the solids inlet, in the middle riser section, and near the top riser section. Phase Doppler Anemometry, a high accuracy, minimally intrusive technique, was used to measure the solids axial velocity, particles concentration and particle diameter, in 17 points of each cross section. 5,000 measurements were acquired in triplicate at each point, under turbulent flow conditions and dilute particles loadings (Gs=4 â€“ 12kg/mÂ²s) since the study of typical riser phenomena under these conditions is one of the objectives. Results show the relationship between particles velocity and concentration, as well as particles diameter. An asymptotic behavior is found relating solids velocity and concentration: low concentration allows for high velocity, although very high concentrations with low particles velocity are found along the riser. Negative particles velocities were found mainly at the bottom region regardless of particle diameter. The particles diameter and velocity variations are presented by means probability density distributions that show a prominent occurrence of inertial particles near the wall while smaller particles are more present in the center of the riser. An unexpected axial segregation of particles diameter was also found.

Keywords: Circulating Fluidized Bed. Phase Doppler Anemometry. Particles concentration.

492321 Non-Intrusive Characterisation of Particle Cluster Behaviours in a Riser through Electrostatic and Vibration Sensing

Abstract:

Particle clusters are important mesoscale flow structures in gas-solid circulating fluidised beds (CFBs). An electrostatic sensing system and two accelerometers are installed on the riser of a CFB test rig to collect signals simultaneously. Cross correlation, Hilbert-Huang transform (HHT), *V*-statistic analysis, and wavelet transform are applied for signal identification and cluster characterisation near the wall. Solids velocities are obtained through cross correlation. Non-stationary and non-linear characteristics are distinctly exhibited in the Hilbert spectra of the electrostatic and vibration signals, and the cluster dynamic behaviours are represented by the energy distributions of the signal intrinsic mode functions (IMFs). The energy distribution shifts towards larger scales (lower frequencies) with the solids flux due to the enhanced formation and motion of particle clusters, while shifts towards smaller scales with the superficial gas velocity because of the homogenised flow field. The cycle feature and main cycle frequency of cluster motion are characterised through *V*-statistic analysis of the vibration signals. Consistent characteristic information about particle clusters is extracted from the electrostatic and vibration signals. Furthermore, a cluster identification criterion for electrostatic signals is proposed, including a fixed and a wavelet dynamic thresholds, based on which the cluster time fraction, average cluster duration time, cluster frequency, and average cluster vertical size are quantified. Especially, the cluster frequency obtained from this criterion agrees well with that from the aforementioned *V*-statistic analysis. Results from this work provide a new non-intrusive approach to the characterisation of cluster dynamic behaviours and their effects on the flow field.

491644 Changing the Mass Flow Limit - What Bin Designs Will Minimize Segregation and How Do We Change the Mass Flow Limit to Optimize Segregation Prevention

Abstract:

Mass flow is often needed in hoppers to prevent stable rathole formation, mitigate segregation, control residence time, and enhance blending in process equipment. Traditional theory assumes that stresses in the hopper section are a linear function of the span of the hopper. However, there are two common situations when this assumption proves untrue: when the hopper is connected to a vertical section or with a stand-alone hopper, when the area of concern is simply flow near the top of the hopper. In these situations, the traditional mass flow analysis fails to predict reality.

Todayâ€™s process engineer has come to look at mass flow as the cure-all for all flow problems. By definition, mass flow means flow at the walls of the vessel as material is discharged. There are many velocity profiles that will induce flow along equipment walls. But, this flow profile may not be what is required by the process. For example, to minimize changeover times it is necessary minimize the velocity profile across the process equipment during discharge. Typical mass flow limit maintains a 3 to 1 velocity ratio across the hopper. This may not be enough to minimize changeover of product from the bin. In this case, the real parameter to design to is the sharpness of the residence time distribution in the process equipment. Further, a material sensitive to caking requires sufficient inter-particle movement within the critical time constant of caking to break bonds between all of the particles. Applying a plug flow (uniform velocity design) in this material would be a mistake. In this case, it is necessary to maximize the velocity profile without generating regions in the bin where the velocity profile becomes flat. Mass flow is best, but this must be a mass flow condition that has some optimal velocity profile. Alternately, when a layered segregation pattern exists in a bin, then mass flow is desirable. But, the mass flow must induce enough velocity profile to remix the layers upon discharge. If the segregation pattern is purely radial, then mass flow with a nearly uniform velocity profile is required to mix radial concentration distributions upon exit of the material. Therefore, proper hopper design is not so much a question of mass flow as it is a question of getting the right residence time distribution or velocity profile for the task at hand.

Current design charts always give just the mass flow limit. Some give the velocity ratio in the hopper. None present the mass flow limit to obtain the proper residence time distribution and none express the mass flow limit as a means to minimize segregation. This paper will look at redefining the mass flow limit for the process specific case, and providing guidance to assist design engineers in selecting the right hopper for the task at hand - or inducing the right gas or stress gradients through external means to utilize an existing bin and change the mass flow limit.

501041 Multilayer Granular Segregation in Discharging Cylindrical Hoppers

Abstract:

Multilayer granular segregation in discharging cylindrical hoppers

Manogna Adepu, Shaohua Chen, Yang Jiao, Aytekin Gel, Heather Emady

School for Engineering of Matter, Transport and Energy, Arizona State University, Tempe, AZ 85287

Particulate systems play a significant role in many industries, like pharmaceuticals, polymers, mineral processing, munitions, energy, food processing, environment, and agriculture. The discharging of granular materials from a hopper is a critical topic of industrial importance, and the discharge flow rate from hoppers is of continuous research interest. Granular materials typically consist of particles with a distribution of sizes, shapes, and densities, which may induce segregation of the material upon handling. Since the product quality is often contingent on maintaining blend homogeneity, segregation of granular materials is undesirable for many solids handling processes. Therefore, it is important to examine the causes and extent of segregation in various systems.

This work uses experiments to investigate the segregation of polydispersed silica particles during hopper discharge. The variables of interest here include the size distribution of particles and particle friction. A 3D printed lab scale hopper with four size ranges (diameters of 1.0-1.2 mm, 1.4-1.7 mm, 2.0-2.2 mm, and 3.7-4.0 mm) of silica particles is used for the study. The hopper is filled with the four layers of particles, starting with the fine layer on the bottom and finishing with the coarse layer on the top. The particles are discharged and collected transiently in equal volumes. Sieve analysis is done to analyze the weight fractions of all four layers. Preliminary results show that a central flow channel is formed, and the top layer begins to deflect and forms a â€˜Vâ€™-shaped surface. The incline formed promotes segregation via percolation whereby the fine layers shift downward into the inert or slow-moving material below, while coarse particles tend to roll down the incline toward the hopper centerline. As a result, the coarse layers start to discharge. Since the velocities are greatest near the centerline, the fines-depleted material that accumulates near the hopper centerline discharges quickly. As discharge continues, fines-rich material near the hopper walls is discharged, resulting in a peak at the end of the segregation profile, which is consistent with previous two-layer hopper segregation studies [1]. The same batch of particles will be subjected to etching, thereby increasing the surface roughness. Those particles will be used to investigate the friction effects on segregation.

Reference:

[1]. Ketterhagen, William R., et al. "Granular Segregation in discharging cylindrical hoppers: a discrete element and experimental study." *Chemical Engineering Science* 62.22 (2007): 6423-6439

498899 DEM Investigation of Adhesive Mixing of Fine and Coarse Particles: Dynamics of Collisional Mixing and Attachment

Abstract:

When small number of fine particles are mixed with coarser particles, they form ordered mixtures or what may be called adhesive mixtures. In this work, discrete element method (DEM) simulations are carried out to characterize the adhesive mixing process. The dynamic processes such as deagglomeration and adhesion of fine particles are investigated by analyzing fine-fine (FF) and coarse-fine (CF) particle contact numbers along with the mixing quality. The FF and CF contact numbers are used to quantify the deagglomeration and adhesion processes, respectively. The DEM simulation results showed that FF contact number decreases first and then reaches equilibrium; while CF contact number and mixing quality increase first and then reach equilibrium. Interestingly, the number of fine particles per coarse particle follows lognormal distribution. The time scales for equilibrium FF contact number and mixing quality are comparable, indicating that deagglomeration is the dominant factor for achieving a uniform adhesive mixture. This time scale increases with a larger ratio of fine particles and a higher surface energy. Collisional mixing dynamics provide a deeper insight into the adhesive mixing process. The results indicate that increasing surface energy of fine particles leads to worse mixing quality because the fine particles with high surface energy create such strong agglomerates that cannot be easily broken by the collision. Collision rate can dominate the mixing quality with the prerequisite that the collision energy is greater than the corresponding detachment energy of fine particles agglomerates. The experimental results validate that the modeling is suitable to describe the adhesive mixing process.

500899 Improved Pharmaceutical Blend Content Uniformity Due to Reduced Agglomeration of Dry Coated Micronized Drug Powders

Abstract:

Content uniformity of low dose blends with fine active pharmaceutical ingredients (API) is adversely impacted due to API agglomeration caused by high powder cohesion. In this work, dry coating using high-intensity vibratory mixing is employed to reduce API cohesion. This leads to a dramatic decrease in granular Bond number resulting in reduced agglomeration as predicted by contact models, thus greatly improving blend content uniformity (CU). Micronized acetaminophen (mAPAP) (~10 µm), a model API, was dry coated with nano-silica R972P (20nm), and mixed with Avicel 102 as a model excipient. The amount of silica was varied from 0 to 2.74 wt %, corresponding to theoretical surface area coverages (SAC) from 0 to 100 % respectively. Bulk density, unconfined yield strength, and dispersive surface energy results indicated dry coating with 0.27 to 1 wt % silica was adequate for API property enhancement. Sieving of mAPAP illustrated the reduction in mAPAP agglomeration, necessary for improved CU after dry coating, corroborating model based predictions. The predictions were corroborated by excellent CU for 3, 5, 10 wt % API loaded blends. The CU with dry coated mAPAP was significantly lower and within the acceptable range as compared to control blends without silica, as well as those with silica added during blending. Compared to theoretical prediction, actual CU was higher unless API agglomerate size distribution obtained via sieving was taken into account. Further investigations are ongoing to properly quantify powder agglomeration before/after mixing. In return, this will allow for better understanding of the powder agglomeration/CU relationship with an eye toward proper prediction of blend CU. Overall, cohesion reduction by dry coating is shown as a promising approach for improving content uniformity of cohesive API blends.

488271 Segregation of Fragile Granular Materials

Abstract:

Segregation of granular mixtures of different size particles has wide implications in various industrial processes such as hopper and bin filling and discharge, blending in rotating tumblers, or chute transferring. Segregation in these industrial processes often cause loss of process efficiency or product quality due to particle spatial inhomogeneity. Recently, a continuum approach [1] combining the advection, percolation, and diffusion mechanisms of granular flow has been developed to quantitatively model percolation-driven particle size segregation. This model has demonstrated the ability of accurately predict particle segregation in a range of flow geometries, processing conditions, and particle size distributions. However, the model is limited to non-fragile particles, whereas in many industrial applications, particles such as coals, food stuffs, or agglomerated chemicals and pharmaceuticals can break during processing or transportation, which can result in segregation or impact segregation dynamics. In this research, we develop a theoretical model considering particle attrition in segregating granular flow by incorporating a population balance model of size reduction into the continuum segregation approach. Inclined chute flow is used as a model flow to examine this segregation-attrition model for an initially monodisperse granular material that can break into either a bidisperse granular mixture or a polydisperse granular mixture.

[1] Fan et al. â€œModelling size segregation of granular materials: the roles of segregation, advection and diffusionâ€ Journal of Fluid Mechanics, 741, 252-279, 2014.

487707 Lubrication in Continuous Tubular Powder Blenders

Abstract:

Granular mixing, based on the scale of mixing, can be generally classified as macro-mixing and micro-mixing. Macro-mixing is defined as the macro scale transport of particles or particle assemblies from one spatial co-ordinate to the other. It is responsible for most reduction in variance in unmixed powder streams. Micro-mixing is defined as mixing at the scale of a single particle. Micro-mixing is always associated with a change in the properties of individual particles [1].

The role of micro-mixing becomes particularly important in the presence of shear-sensitive ingredients, such as magnesium stearate, or other lubricants. Magnesium stearate, generally present as thin nanoparticles, is known to coat larger shear insensitive excipients and active ingredients [2]. This coating reduces the bonding ability of particles, and consequently makes it more difficult to process the material into tablets [3]. Magnesium stearate is also hydrophobic in nature and the coating of active ingredient particles by magnesium stearate results in increase in their hydrophobicity and consequently decrease in the dissolution rate of the active ingredient [4, 5]. Also observed is the improvement in packing efficiency of the coated materials (increased bulk density) and improved flow properties [3]. It is known that the coating extent of the shear sensitive material and consequently its associated effects are directly proportional to the total amount of strain experienced by the material. The greater the strain, the more pronounced is the extent of coating and consequently more pronounced are the associated effects [2-5]. Close attention is thus paid to the extent of lubricant mixing in traditional batch manufacturing to ensure that the lubricant is not subjected to excessive strain levels during the blending operation. Typically, shear-sensitive ingredients are added in a second stage, after the mixing of other ingredients, to minimize the strain they experience. Therefore, the bulk ingredients are blended for a few hundred revolutions in the blender before the lubricant is added to the system, and the lubricant is generally mixed for a few tens of revolutions.

This work presents a comparison between the amount of strain experienced by material in a typical batch blending process and a continuous blending process, specifically, in the presence of shear-sensitive ingredients, such as magnesium stearate. Findings show that the total amount of strain experienced by the in-process material in a batch tumbling blender is greater compared to that in a continuous paddle blender, under typical operating conditions. The higher strain experienced in batch processing necessitates the addition of shear sensitive materials, such as magnesium stearate, at a later stage in the blending operation, resulting in two-stages of ingredient addition. This may not be necessary in continuous operations due to the gentler nature of the process. It is also hypothesized that the material experiences a more uniform degree of shear in a continuous blender compared to a batch blender. Lastly, the effect of strain rate on blend lubricity is quantified. It was observed that the rate of strain addition does not have an impact on the lubricity of the final blend. Whereas the lubricity of the blend is only dependent on the total strain experienced by it.

References

1. Muzzio, F., Mixing Mechanics: Practical powder blending: Micromixing, in Powder and Bulk Engineering. 2016.

2. Pingali, K., et al., Mixing order of glidant and lubricant â€“ Influence on powder and tablet properties. International Journal of Pharmaceutics, 2011. 409(1â€“2): p. 269-277.

3. Mehrotra, A., et al., Influence of shear intensity and total shear on properties of blends and tablets of lactose and cellulose lubricated with magnesium stearate. International Journal of Pharmaceutics, 2007. 336(2): p. 284-291.

4. Llusa, M., et al., Measuring the hydrophobicity of lubricated blends of pharmaceutical excipients. Powder Technology, 2010. 198(1): p. 101-107.

5. Pingali, K., et al., Evaluation of strain-induced hydrophobicity of pharmaceutical blends and its effect on drug release rate under multiple compression conditions. Drug Dev Ind Pharm, 2011. 37(4): p. 428-35.

499024 Particle Size Segregation in Granular Shear Flows

Abstract:

Gravity induced size-dependent segregation is a commonly observed, but usually undesired phenomenon in a variety of industrial processes. When a granular mixture with different size particles is sheared or vibrated, smaller particles have a tendency to fall into voids and segregate toward lower or interior portions of the flow. Inspired by recent work from our group on density-driven segregation, here we investigate the impact of different rheological characteristics on the size segregation of an "ideal solution" binary mixture, where a small number of intruders are immersed in an otherwise almost uniform bed. The Discrete Element Method is used to simulate the flow in a simple boundary-driven planar shear cell. Inside the cell is a mono-disperse particle packing that includes several smaller intruder particles which are used to measure size segregation for a variety of flow conditions: differing diameter ratios, shear rates, confining pressures and solid fractions. We find that in addition to the commonly used variables such as shear rate and solid fraction, pressure also has a profound but usually ignored effect on the segregation rate. Based on our simulation results, we propose a theoretical model linking size segregation with granular flow rheology.

487758 Quantification of Granular Size Segregation in a 3D Conical Bounded Heap: Theory, Simulations, and Experiments

Abstract:

Quantitative prediction of radial streamwise segregation in bounded conical heap flows of granular materials under a variety of flowrates and system sizes has yet to be demonstrated for problems at industry-scale, not least because of the difficulty in determining the kinematics below the free surface of an opaque material. We consider a continuum transport model for the local species concentration applied to the problem of particle segregation in an axisymmetric conical heap geometry. The continuum model is informed using mean-flow kinematics extracted from discrete element method (DEM) simulations with azimuthal periodic boundary conditions (to simulate an axisymmetric heap) for various flowrates, particle size ratios, and initial mixture concentrations. The DEM simulations are validated via comparison with experimental results for bidisperse streamwise segregation profiles and boundary kinematics obtained in a wedge-shaped conical flow apparatus with smooth walls. Although the measured streamwise segregation is qualitatively similar to previous results in quasi-2D bounded heaps at equivalent two-dimensional feed rates, key quantitative differences between the two geometries arise owing to differences in the mean velocity field and the local flowing layer depth, which varies non-trivially with radial distance away from the feed zone owing to the polar cylindrical geometry. To address this, we outline a method for incorporating the variable flowing layer thickness (whose scalings are provided by DEM) into a continuum advection-segregation-diffusion model, which is a necessary modification to the model to achieve quantitative agreement with measured segregation near the outer boundary wall. The continuum model also requires data for the species segregation velocity and diffusion coefficient, which depend on the local shear rate, particle concentration, and absolute particle diameters. Since the relations for the segregation velocity and diffusion coefficient are local, we use previous relations determined from quasi-2D bounded heap flows of bidisperse mixtures, which have also been shown to be applicable to other geometries. With the appropriate kinematic scalings incorporated, predictions from the continuum model show quantitative agreement with the steady-state radial streamwise segregation as measured in DEM simulations.

Funded by The Dow Chemical Company.

488652 Invited - Reactive Precipitation of Anhydrous Alkali Sulfide Nanocrystals with Concomitant Abatement of Hydrogen Sulfide and Co-Generation of Hydrogen

Abstract:

Anhydrous alkali sulfide (M_2S, M = Li and Na) nanocrystals (NCs) are important materials central to the development of next generation cathodes and solid state electrolytes for advanced batteries, but not commercially available at present. This work reports an innovative method to directly synthesize M_2S-NCs through alcohol-mediated reactions between alkali metals and hydrogen sulfide (H_2S). In the first step the alkali metal is complexed with alcohol in solution, forming metal alkoxide (ROM) and releasing hydrogen (H_2). Next, H_2S is bubbled through the ROM solution, where both chemicals are completely consumed to produce phase-pure M_2S-NC precipitates and regenerating alcohol that can be recycled. The M_2S-NC morphology may be tuned through choice of the alcohol and solvent. Both synthetic steps are thermodynamically favorable (â^†G_m^o < -100 kJ/mol), proceeding rapidly to completion at ambient temperature with ~100% atom efficiency. The net result, $H_2S + 2M \Rightarrow M_2S + H_2$, provides abatement of the hazardous industrial waste H_2S and delivers two value-added products that naturally phase separate for easy recovery. This scalable approach provides an energy-efficient and environmentally-benign solution to the production of nanostructured materials required in emerging battery technologies.

496490 Tunable Magnetic Core-Shell Nanoparticles: An Interplay between Composition, Size, Shape and Architecture

Abstract:

Over the past century, magnetic materials have played a key role in the development of the American economy through their incorporation in power plants, memory storage, and cancer treatments. Of particular interest is the creation of permanent magnet replacements, typically requiring expensive rare earth elements, for the application in energy generation (windmills) and energy conversion (electric vehicles). Typically, these permanent magnet replacements focus on iron based oxides due to their relative abundance in the earth's crust and natural magnetic moment. Unfortunately, these oxides are relatively weak magnets with dipoles which are highly susceptible to external magnetic fields, regardless of the composition. However, the saturation/remanent magnetism and coercivity can be engineered, producing magnets which are more robust to external fields (harder magnets), by increasing the shape anisotropy and controlling coupling across "hard"-"soft" magnetic interfaces. Specifically, highly anisotropic, core-shell iron oxide nanostructures are investigated to develop design criteria for engineering magnetic properties based on the relationship between material composition, structuring and architecture. The magnetic response of these nanostructures is known to be dependent on the crystallinity of the nanostructure, requiring careful synthesis to engineer the formation of defects within the crystal.

In this work, a modified thermal decomposition method is used in order to control the composition, morphology, architecture of Fe_3O_4 nanostructures with critical dimensions between 20-30 nm. Initially, composition of the nanostructures is determined through the synthesis of an M_xFe_{1-x} oleate (M= Co^{2+}, Ni^{2+}, $Fe^{2+/3+}$ or $Mn^{2+/3+}$) during the initial precursor synthesis, where the Co dopant increases coercivity and Ni/Mn dopants increases the remanent magnetization. These precursors are then used to synthesize spheres, cubes, plates, and pyramids based on the ratio of oleate to oleic acid precursor via a steric hindrance mechanism on the difference surface facets. Finally, core-shell nanostructures, with conformal and non-conformal coatings, are fabricated through a subsequent deposition on a magnetic core, in order to facilitate magnetic coupling across the interface, and non-magnetic structures, to overcome the ~20 nm critical dimension that results in agglomeration of the particles. Furthermore, this two-step process allows for the synthesis of high aspect ratio 1D and 2D magnetic structures which have not been previously reported. In addition to standard characterization techniques, the chemical and structural control will be highlighted using high resolution electron microscopy and specially resolved energy spectroscopy coupled with magnetic characterization to introduce design parameter for engineered materials.

496365 Effects of Alumina Incorporation By Particle Atomic Layer Deposition on Sintering, Microstructure, and Ionic Conductivity of Yttria-Stabilized Zirconia (8YSZ)

Abstract:

Yttria-stabilized cubic zirconia (YSZ) is the most-common electrolyte material for solid oxide fuel cells (SOFC) due to its reasonable oxygen-ion conductivity and chemical stability across a wide range of environments. To achieve suitable ionic conductivity, YSZ electrolytes must be near theoretical density; this necessitates high sintering temperatures, often exceeding 1450 °C. This high sintering temperature limits the viability of low-cost one-step sintering during manufacturing and thus increases the cost of SOFC fabrication. Previous researchers have successfully lowered the required YSZ sintering temperature through incorporation of aluminum oxide (Al_2O_3) particles with the YSZ. While this technique has proven to be successful in reducing the sintering temperature, the presence of Al_2O_3 in the electrolyte can reduce YSZ ionic conductivity. In this work, we report on the use of atomic layer deposition (ALD) to precisely and conformally coat individual YSZ particles with the desired amount of Al_2O_3. Constant-rate-of-heating (CRH) experiments were conducted using a horizontal push-rod dilatometer to extract the activation energy of sintering in the initial stage and also gain mechanistic insights into the active diffusion mechanisms as a function of ALD film thickness. Transmission electron microscopy (TEM) and scanning electron microscopy (SEM) were used to characterize the as-deposited thin Al_2O_3 films and the effects of the Al_2O_3 films on grain growth and microstructure during densification. Finally, electrochemical impedance spectroscopy was used to characterize the ionic conductivity of dense YSZ electrolytes.

488595 Atomic Layer Deposition Surface Functionalized Adsorbents for Adsorption of Metal Ions and Organic Pollutants

Abstract:

Ultrathin TiO_2 and Al_2O_3 films were deposited on silica gel particles and biochars by atomic layer deposition (ALD), respectively. TiO_2 coated silica gel particles were used as adsorbents for adsorption of metal and metalloid ions; their adsorption ability for a mixture of 19 trace elements of heavy metals and other toxic elements, including As(V), Se (IV), Be(II), Al(III), V(V), Cr(III), Mn(II), Co(II), Ni(II), Cu(II), Zn(II), Ba(II), Tl(I), Sb(III), Cd(II), Ag(I), Sr(II), Mo (VI), and Pb(II), from aqueous solutions was investigated. At pH 5, the 20 and 40 cycles of TiO_2 coated samples without heat treatment removed remarkable amount of As(V), Se(IV), V(V), Mo(VI), Pb(II), Sb(III), Ag(I), Cu (II), and Ba(II) from the solution simultaneously. The micron-sized adsorbent particles were separated easily from water due to their large particle size, making it practically suitable for trace contaminant remediation in water. Al_2O_3 deposited biochars were used for removal of organic pollutants and the adsorption capacities were evaluated by adsorption of methylene blue (MB). The data fit well with the Langmuir isotherm and the maximum adsorption capacities were found to be 26.8 and 35.0 mg/g at 25 °C for the uncoated biochar and 5 cycles of Al_2O_3 coated biochar, respectively. The improvement of adsorption capacity after Al_2O_3 ALD coating was due to its improved hydrophilia.

489091 Atomic Layer Deposition for Extended Surface Electrocatalyst Development

Abstract:

Platinum-nickel nanowires are a promising advanced catalyst material for the oxygen reduction reaction (ORR) in fuel cell applications. When synthesized by galvanic displacement of nickel nanowires (NiNWs), these extended surface electrocatalysts have demonstrated activity and durability superior to benchmark carbon-supported Pt catalysts. However, challenges in scale-up and reproducibility of this fabrication process have prompted investigation into atomic layer deposition (ALD) of Pt as a facile, scalable synthesis alternative for the production of Pt-NiNWs. Atomic layer deposition of Pt nanoparticles on NiNW substrate was performed in a packed bed reactor, using trimethyl(methylcyclopentadienyl)platinum(IV) ($MeCpPtMe_3$) as a metal source, and either H_2 or O_2 for ligand removal. Gases exiting the ALD reactor were monitored via an in-line mass spectrometer, which provided information on reaction mechanisms. Deposition of Pt was confirmed through inductively coupled plasma mass spectroscopy for a range of ALD cycles; oxidative ALD chemistry was found to yield greater Pt content at high cycle numbers. Transmission electron microscopy and elemental mapping techniques were also used to analyze the morphology and structure of ALD-modified NiNWs. The effects of post-processing treatments, such as H_2 annealing and acid leaching, were also studied. Electrochemical performance of ALD-synthesized catalysts in both half-cell and full-cell analyses was found to be comparable to that obtained with galvanic displacement catalysts, thereby demonstrating the efficacy of ALD as a fabrication method for extended thin film electrocatalyst structures.

499854 Ultrathin Hollow Graphene Oxide Membranes for Use As Nanoparticle Carriers for Energy and Biomedical Applications

Abstract:

We synthesize hollow spherical particles/membrane (*) sacks of graphene oxide loaded with nanoparticles to be used as nanoparticles carriers, through a new method based on emulsion precipitation and sublimation of the cores. We vary the synthesis parameters, such as shear rate, pH, and graphene oxide and oil concentration ratios. Our results show a concentration dependent membrane thickness that varies between 3 and 25 nm depending on the concentration, and their mean diameters vary between 500 nm and 70 μ m. In addition, polymeric nanoparticles are loaded inside the graphene oxide shells forming core–shell particles demonstrating that they can be used as carriers for nanoparticles. In our work we use a range of GO sheet sizes. To synthesize our larger Hollow Graphene Oxide Membrane (HGOM) particles we use larger graphene oxide sheets of approximately 4μm obtained with a bath sonication method and to synthesize the smaller micron sized HGOM particles, we use smaller GO sheets of the order of 0.52 μm obtained with a higher power probe sonication method. We have identified processing parameters, such as GO concentration, shear rate and pH that produce graphene oxide stabilized Pickering emulsions. When these emulsions are subsequently cooled they produced stable suspensions of a solid oil phase wrapped in graphene oxide membranes. Proper adjustment of the processing parameters can provide control over the size of the templated membranes and the thickness of the graphene oxide layer. When high shear rates and high concentrations of graphene oxide are used, particles of smaller diameters with thick membranes are produced. HGOM particles with diameters as small as 500– 3000 nm are produced for use as carriers for silicon and PTFE nanoparticles, respectively. By decreasing the concentration of graphene oxide, membranes as thin as 3 nm can be produced. Plotting the various processing parameters and the resulting particle characteristic results in three distinctive regions, which appear to be characteristic to this process. Region I consists of particles with nearly constant GO membrane thickness and varying particle diameter. Region II consists of a transition region in which the GO concentration aff ects both membrane thickness and particle size. Region III consists of particles with constant diameter as the GO concentration is increased for a fixed shear rate. Decreasing the pH assists in forming multiple layers of graphene oxide, since lower pH values favor the interactions between individual graphene oxide sheets and these interactions are more favorable than the interactions with the aqueous phase. We found that pH conditions which resulted in zeta potentials of approximately −20 to −30 mV provided very stable emulsions. It is worth mentioning that stable emulsions using graphene oxide and naphthalene or other aromatic compounds can also be obtained without further decreasing the pH of the emulsion. However, by adding extra acid, membranes become more hydrophobic, and hence, the amount of GO that migrates to the water phase is minimized. Thus, in this way GO sheets tend to be more attracted to the oil phase and stack to each other. Our particles are characterized via laser diffraction, zeta potential, FE-SEM, TEM, BET, and AFM. Potential applications of this work include applications that benefit from core– shell structures and nanoparticle carriers, including drug formulation, catalysis, and electrochemical applications.

(*) Langmuir March 2017. Kurt B. Smith, M. Silvina Tomassone, DOI: 10.1021/acs.langmuir.6b04583

Application of an efficient discrete particle model to simulate an industrial FCC regenerator

Liqiang Lu* and Sofiane Benyahia

National Energy Technology Laboratory, Morgantown, West Virginia 26507, United States

* Corresponding author: Liqiang.Lu@netl.doe.gov

In past decades, the continuum approach was the only practical technique to simulate large-scale fluidized bed reactors because discrete approaches suffer from the cost of tracking huge numbers of particles and their collisions[1,. This study significantly improved the computation speed of discrete particle methods in two steps: First, the time-driven hard-sphere (TDHS) algorithm with a larger time-step is proposed allowing a speedup of 20-60 times; second, the number of tracked particles is reduced by adopting the coarse-graining technique gaining an additional 2-3 orders of magnitude speedup of the simulations. A new velocity correction term was introduced and validated in TDHS to solve the over-packing issue in dense granular flow. The TDHS was then coupled with the coarse-graining technique to simulate the heat transfer and chemical reaction mechanisms in an industrial FCC regenerator in a reasonable time with little computational resources. The simulation results compared well with available industrial data and proved that this new approach can be used for efficient and reliable simulations of industrial-scale fluidized bed systems.

References:

1. Lu, L., Liu, X., Li, T., Wang, L., Ge, W., Benyahia, S., 2017. Assessing the capability of continuum and discrete particle methods to simulate gas-solids flow using DNS predictions as a benchmark. Powder Technology 321, 301-309.
2. Lu, L.; Gopalan, B.; Benyahia, S., 2017. Assessment of different discrete particle methods ability to predict gas-particle flow in a small-scale fluidized bed. Industrial & Engineering Chemistry Research, *56*, 7865–7876
3. Lu, L., Benyahia, S., Li, T., 2017. An efficient and reliable predictive method for fluidized bed simulation. AIChE Journal 63, 5320-5334.
4. Lu, L.; Konan, A.; Benyahia, S., 2017. Influence of grid resolution, parcel size and drag models on bubbling fluidized bed simulation. Chemical Engineering Journal, *326*, 627-639.
5. Lu, L.; Morris, A.; Li, T.; Benyahia, S., 2017. Extension of a coarse grained particle method to simulate heat transfer in fluidized beds. *Int. J. Heat Mass Transfer*, *111*, 723-735.
6. Lu, L., Yoo, K., Benyahia, S., 2016. Coarse-Grained-Particle Method for Simulation of Liquid–Solids Reacting Flows. Industrial & Engineering Chemistry Research 55, 10477-10491.
7. Lu, L., Xu, J., Ge, W., Gao, G., Jiang, Y., Zhao, M., Liu, X., Li, J., 2016. Computer virtual experiment on fluidized beds using a coarse-grained discrete particle method—EMMS-DPM. Chemical Engineering Science 155, 314-337.
8. Lu, L., Xu, J., Ge, W., Yue, Y., Liu, X., Li, J., 2014. EMMS-based discrete particle method (EMMS–DPM) for simulation of gas–solid flows. Chemical Engineering Science 120, 67-87.

487860 Validated Hydrodynamic CFD Model for Catalytic Fast Pyrolysis

Abstract:

Catalytic fast pyrolysis (CFP) is a promising technology for converting lignocellulosic biomass into renewable transportation fuels and chemicals. In CFP, small particles of biomass are injected into a circulating fluidized bed (CFB) of hot catalyst particles. Inside the reactor, the biomass particles are rapidly heated and pyrolyzed, and the vapors are catalytically deoxygenated. As in Fluidized Catalytic Cracking (FCC), coke combustion in the regenerator supplies the heat required by the endothermic process. The advantages of CFP over competing thermochemical biofuels processes are twofold. First, a single particle is both heat carrier and catalyst, which greatly simplifies the solids handling and reactor engineering, in turn reducing capital requirements and operational complexity. Second, the resulting bio-oil is much lower in oxygen than typical pyrolysis oil, and therefore, much more stable. When combined with a hydrogen addition upgrading step, the overall yield and economics of the CFP process are superior to its competitors.

Inaeris Technologies (formerly KiOR) is a world leader in the development and commercialization of the CFP technology. Reactor design is a crucial element of the Inaeris technology. Depending on the design concept employed, it is possible to have multiple hydrodynamic regimes in the same reactor, ranging from moving bubbling bed, to dense phase riser, to dilute phase riser (fast fluidization). Furthermore, simultaneous feed of solid biomass and catalyst particles, and evolution of pyrolysis gases, complicates the hydrodynamics analysis. Both of these factors make it difficult to use standard CFB scale-up correlations and methods. As a result, Inaeris decided in late 2014 to start a program to develop computational fluid dynamics (CFD) models capable of supporting scale-up from pilot scale. Other future uses such as optimization, troubleshooting, and ultimately, blank sheet reactor design were also contemplated as a justification for this effort.

In early 2015, Inaeris selected the Barracuda Virtual Reactor® software package, and embarked on a collaboration with CPFD Software, for two reasons. The first was CPFD's experience in rapid model development; the software's incorporation of the MP-PIC numerical method was the second.

After two years, 1000+ cold flow CFB tests, and 500+ simulations, the collaboration has produced a single CFD model which has been experimentally validated for bubbling, dilute phase and dense phase regimes across a wide range of reactor configurations [1-4]. Equally important is that the model accurately predicts hydrodynamics for catalyst-only circulation, as well as those for catalyst and biomass co-feeding. This is important, because somewhere inside the reactor, the hydrodynamics transition from catalyst plus biomass to catalyst only (with some residual char particles). Thus the model must be able to accurately predict the extreme cases.

In this presentation, details of the model will be presented, and application of the model to reactor scale-up will be illustrated. Further recommendations for model refinement, as well as guidelines for general CFD model development for CFP, will also be given.

References

[1] B. Adkins, N. Kapur, J. Pendergrass, J. Parker, P. Blaser, KiOR Update: Incorporating Barracuda in our CFP Development Process, Proceedings of the Inaugural Barracuda Virtual Reactor Users' Conference, Santa Ana Pueblo, New Mexico, 2015.

[2] B. Adkins, N. Kapur, T. Dudley, S. Webb, P. Blaser, Experimental validation of CFD hydrodynamic models for catalytic fast pyrolysis (CFP), Fluidization XV, May 22-27, 2016, Quebec, Canada.

[3] N. Kapur, B. Adkins, T. Dudley, S. Webb, P. Blaser, "Cold Flow Validation of Catalyst-Biomass Hydrodynamics in Catalytic Fast Pyrolysis", 2016 AIChE Annual Meeting, San Francisco, California, November 2016.

[4] B. Adkins, N. Kapur, T. Dudley, S. Webb, P. Blaser, "Experimental Validation of CFD Hydrodynamic Models for Catalytic Fast Pyrolysis", Powder Technology, published online, December 2016.

European Research Council

Sine-squared scaling of drag coefficient for different non-spherical particles

Sathish K. P. Sanjeevi, Johan T. Padding

Delft University of Technology, Netherlands

Motivation

- Renewable energy is the need of the future - **Biomass**.
- Particles **simplified as spheres** traditionally.
- **Orientation** effects, lift, torque ?
- Large parameter space - **simplified models required** for Euler-Lagrangian simulations of non-spherical particles.

Fluidization of biomass-like spherocylindrical particles

Numerical method

- **Lattice Boltzmann method** with multi-relaxation time (MRT) [1].
- MPI parallelized. Linear scaling up to **300,000 cores.**
- **Interpolated bounce back**[2] scheme applied for particle surface.

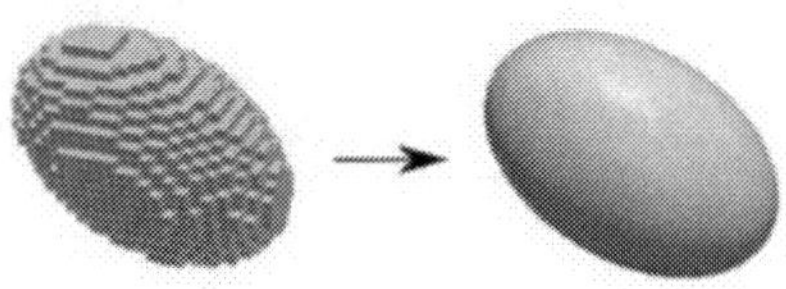

[1] d'Humières, D. *et al.*, 2002. Phil. Trans. R. Soc. A, 360(1792), pp.437-451.
[2] Lallemand P, Luo LS. (2003) *J Comput Phys.* 184(2):406-21.

Simulation setup

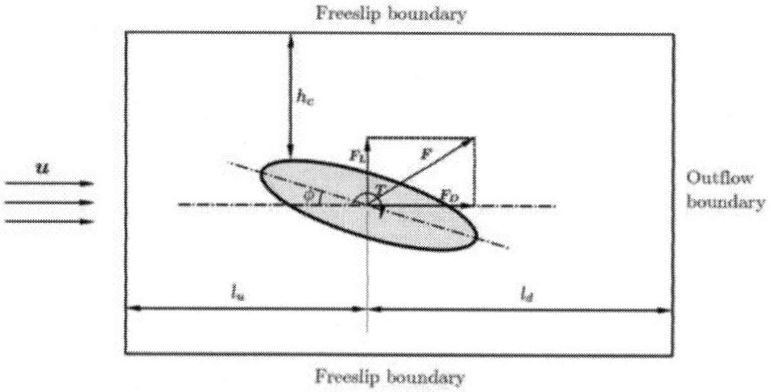

Parameter of interest

Drag coefficient, $C_D = \frac{|\vec{F_D}|}{\frac{1}{2}\rho|\vec{u}|^2\frac{\pi}{4}D_p^2}$

D_p = Diameter D of volume equivalent sphere

Range

$0.1 \leqslant Re \leqslant 2000$

$0^\circ \leqslant \phi \leqslant 90^\circ$

An interesting phenomenon …

$$C_{D,\phi} = C_{D,\phi=0^\circ} + (C_{D,\phi=90^\circ} - C_{D,\phi=0^\circ})\sin^2\phi$$

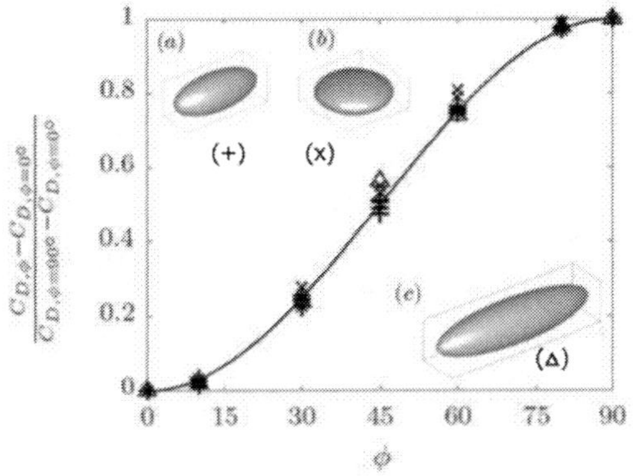

- 3 different particles tested. Data from **Re=0.1 up to 2000**.
- Observed even at **complex, unsteady flows** of Re=2000.
- **Take home message**: two simulations are sufficient: $\phi = 0^\circ$ and $\phi = 90^\circ$

Sanjeevi, S. K. P., & Padding, J. T. (2017). J. Fluid Mech., **820**, R1.

TUDelft Challenge the future 5

Understanding the phenomenon …

- Sine-squared drag behavior is observed at **creeping flow.**
- This is due to the **linearity** of the velocity and pressure fields:

$$u_\phi = u_{\phi=0^\circ}\cos\phi + u_{\phi=90^\circ}\sin\phi,$$
$$p_\phi - p_\infty = (p_{\phi=0^\circ} - p_\infty)\cos\phi + (p_{\phi=90^\circ} - p_\infty)\sin\phi.$$

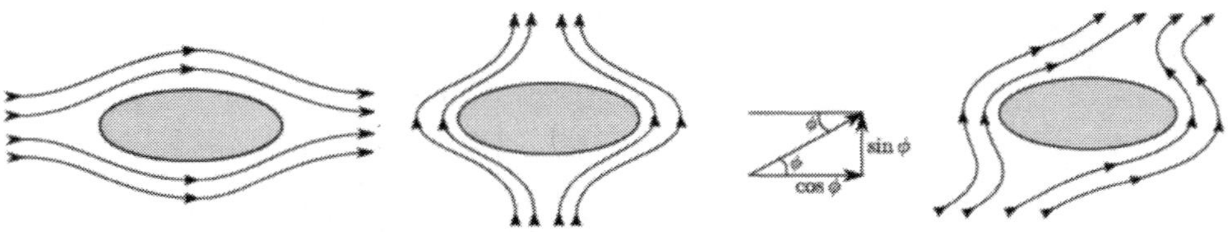

- Above equations **do not** hold at **high Re** due to flow nonlinearities.
- We test for **linearity** sufficiently **close to the particle.**

TUDelft Challenge the future 6

Test of linearity

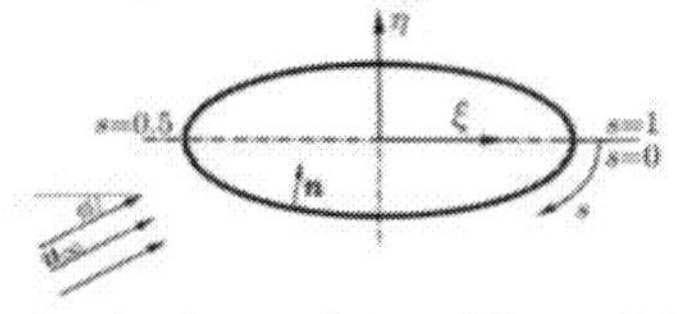

Local reference frame of the particle

Velocity field linearity:

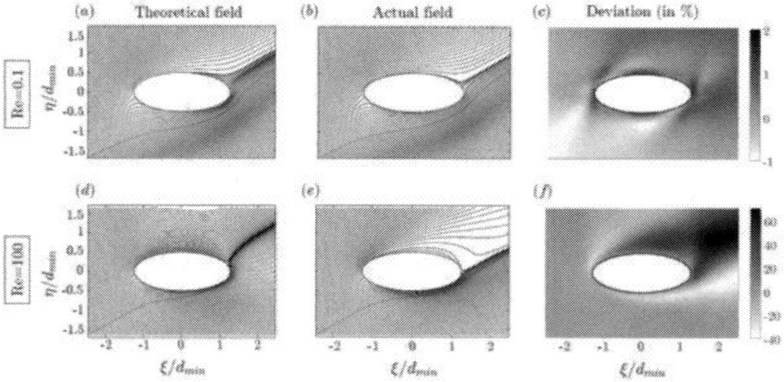

$$u_\phi = u_{\phi=0^\circ}\cos\phi + u_{\phi=90^\circ}\sin\phi$$

Pressure linearity:

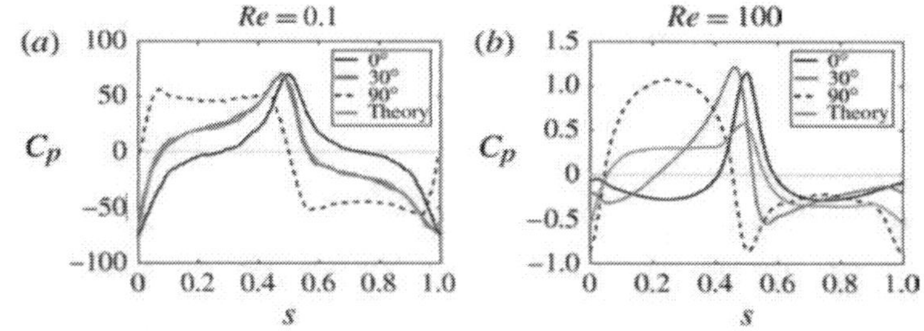

It is **not due to identical theory** observed in Stokes flow.

Drag components:

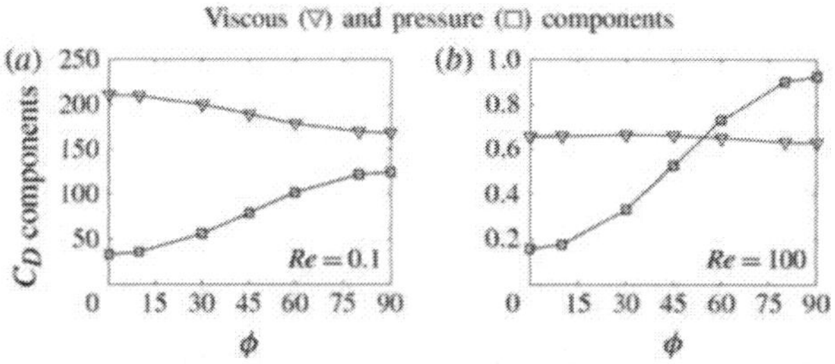

At high Re, angle dependence is purely due to **pressure component only**.

Sine-squared behavior at high Re

For high Re, we **propose** the pressure coefficient takes the form:

$C_p = -m + (1+m)(\hat{\boldsymbol{u}} \cdot \boldsymbol{n})^2$

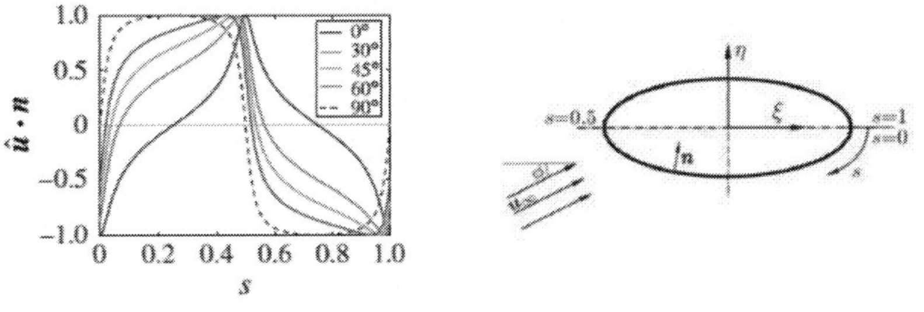

Proposed form is inspired from the **inviscid flow** around sphere:

$C_p = 1 - c\sin^2\theta$, with $c = 9/4$

Sanjeevi, S. K. P., & Padding, J. T. (2017). J. Fluid Mech., **820**, R1.

Comparison between theory and simulations

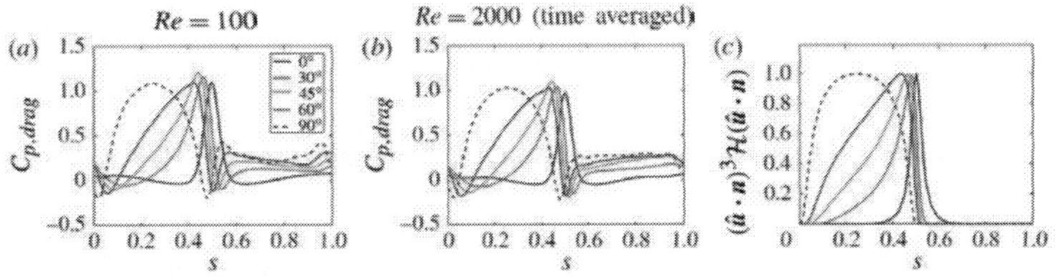

$C_p = (p - p_\infty)/(1/2\rho|\boldsymbol{u}_\infty|^2)$

$C_{p,drag} = C_p \hat{\boldsymbol{u}} \cdot \boldsymbol{n}.$

$$\mathcal{H}(x) = \begin{cases} 1 & \text{if } x > 0, \\ 0 & \text{otherwise.} \end{cases}$$

Comparison between theory and simulations

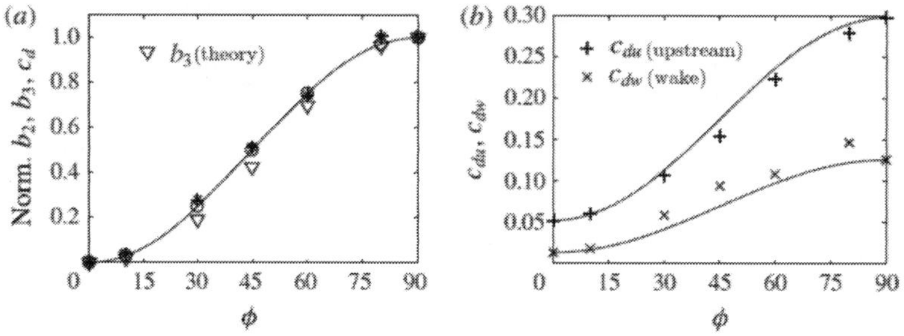

$$b_3 = \int_0^1 (\hat{u}\cdot n)^3 \mathcal{H}(\hat{u}\cdot n)\, \mathrm{d}s \qquad c_{du} = \int_0^1 C_{p,drag} \mathcal{H}(\hat{u}\cdot n)\, \mathrm{d}s$$

- The proposed theory **agrees well** with the simulation results.

Particle shapes

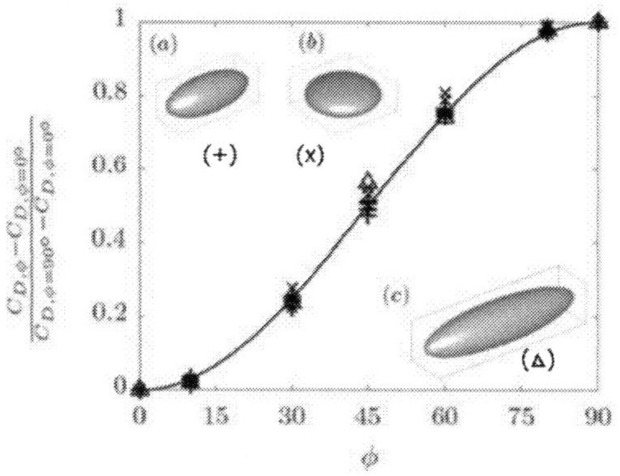

$$C_{D,\phi} = C_{D,\phi=0^\circ} + (C_{D,\phi=90^\circ} - C_{D,\phi=0^\circ}) \sin^2\phi$$

Applicable for

- low and high aspect ratio prolate (needle-like) ellipsoids.
- low aspect ratio oblate (disc-like) ellipsoids.

Sanjeevi, S. K. P., & Padding, J. T. (2017). J. Fluid Mech., **820**, R1.

Limitations and regime map

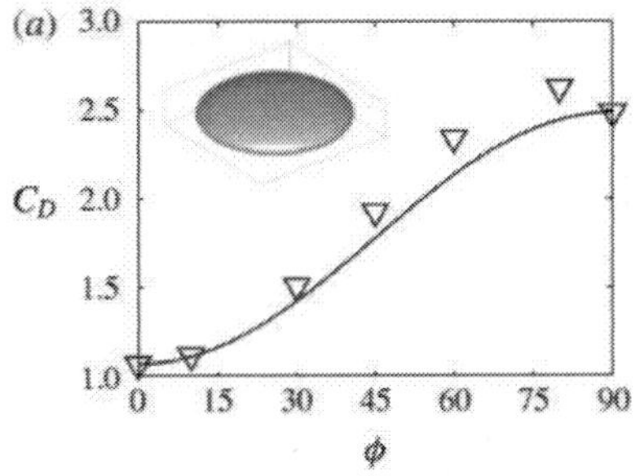

Disc of aspect ratio 4:1 at Re=100

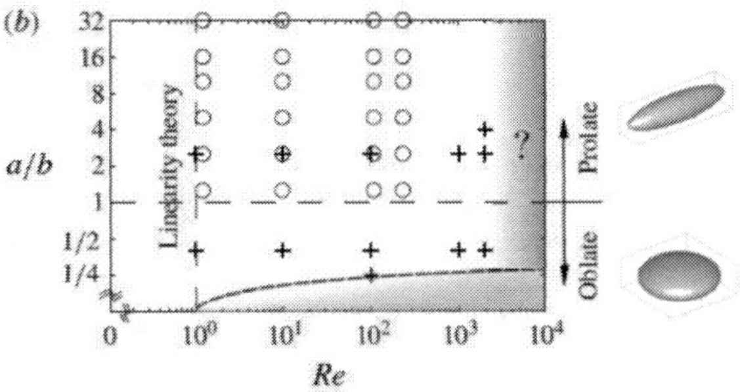

Possible valid regime (unshaded)

- Sine-squared drag law not applicable for **flatter discs**.
- Re=2000 **limitation** is from flow solver.
- **Unknown** for Re>2000.

Lift coefficients

- Theoretical lift equation at Stokes limit is

$$C_{L,\phi} = (C_{D,\phi=90^\circ} - C_{D,\phi=0^\circ}) \sin\phi \cos\phi$$

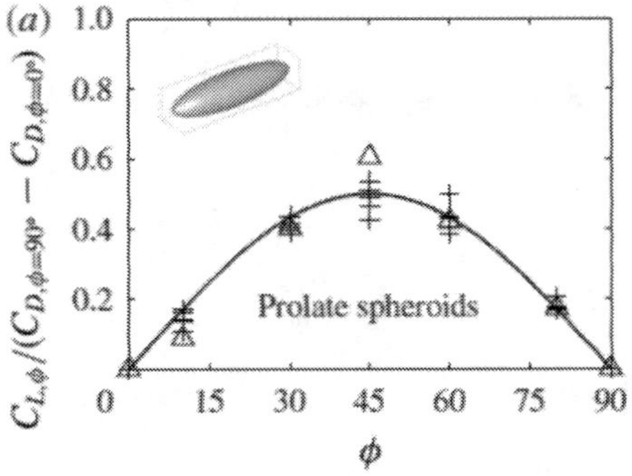

Prolate spheroids of aspect ratio 2.5 (+) up to Re=2000 and aspect ratio 4 (△) at Re=2000.

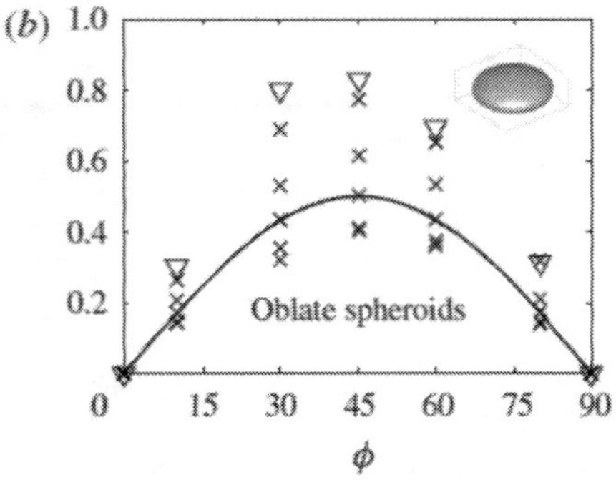

Oblate spheroids of aspect ratio 2.5 (+) up to Re=2000 and aspect ratio 4 (▽) at Re=100.

Conclusions[1]

- **Reason** for the sine-squared behavior at high Re explained.
- Applicability of **sine-squared** scaling.
- **Take home message** - Just two simulations are required: $\phi = 0^\circ$ and $\phi = 90^\circ$
- Very useful for **CFD-DEM** community!

[1] Sanjeevi, S. K. P., & Padding, J. T. (2017). J. Fluid Mech., **820**, R1.

Multi-scale modeling of non-spherical particles

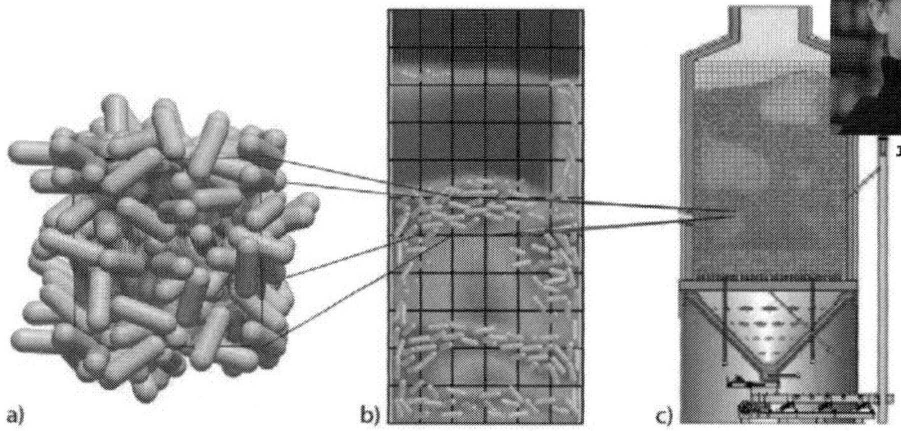

Direct numerical simulations (DNS) | **Euler-Lagrangian: CFD-DEM** | **Multi-particle collision (MPC)**

Sathish Sanjeevi, Ahad Zarghami, Vinay Mahajan, Ivan Mema, Barry Fitzgerald, Vikrant Verma

Acknowledgements

Financial support

European Research Council

Computational support

SURF SARA

(Dutch national HPC agency)

Questions?

492167 Predicting and Improving FCC Reactor Cyclone Performance Using CFD Techniques

Abstract:

TechnipFMCÂ Process Technology uses computational fluid dynamics (CFD) as an FCC design and troubleshooting tool. CFD provides qualitative as well as quantitative information which is used to visualize and understand gas/catalyst flow behavior. TechnipFMC is actively using CFD to identify unit mal-performance and evaluate hardware and operational changes required to optimize FCC units.

This paper highlights the application of CFD modeling for troubleshooting FCC reactor cyclone operation. The paper discusses the effect of inlet configuration on cyclone performance. Barracuda CFD software is used to model the performance of a riser â€“ cyclone system operating with high catalyst losses. The cyclone performance is then compared with cold flow study as well as with commercial data and CFD results from a similar unit operating at higher efficiencies. The results show that the riser exit configuration design can cause flow pulsation and turbulence at the cyclone inlet which affects cyclone pressure drop and catalyst collection efficiency. The paper also discusses the challenges faced in modeling gas solid flow in industrial reactor cyclones.

CFD-DEM modeling on exascale computer architectures

M. Syamlal and J. Musser

National Energy Technology Laboratory

A. Almgren and J. Bell

Lawrence Berkeley National Laboratory

C. Hrenya and T. Hauser

University of Colorado, Boulder

Session "Industrial Application of Computational and Numerical Approaches to Particle Flow I," AIChE Annual Meeting, October 29 - November 3, 2017, Minneapolis, MN

EXASCALE COMPUTING PROJECT

U.S. DEPARTMENT OF ENERGY | Office of Science

Acknowledgments

This research was supported by the Exascale Computing Project (http://www.exascaleproject.org), a joint project of the U.S. Department of Energy's Office of Science and National Nuclear Security Administration, responsible for delivering a capable exascale ecosystem, including software, applications, and hardware technology, to support the nation's exascale computing imperative.

Project Number: 1.2.1.21

EXASCALE COMPUTING PROJECT

www.ExascaleProject.org

U.S. DEPARTMENT OF ENERGY | Office of Science

The Exascale Computing Project (ECP)

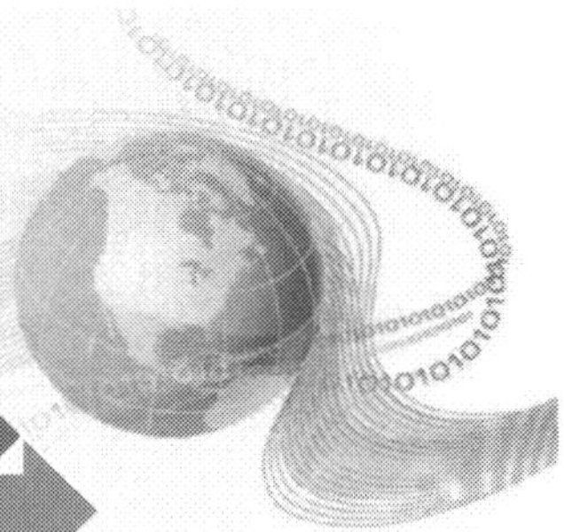

Collaboration

2 US Department of Energy organizations
- Office of Science
- National Nuclear Security Administration

Execution

800 researchers (22 laboratory and agency partners; 39 universities) engaged in:
- 66 software projects
- 25 science application projects
- 5 co-design centers

Goal

Drive pre-exascale science, application development, hardware and software R&D to ensure that the US has a **capable exascale ecosystem** in 2021

ECP EXASCALE COMPUTING PROJECT

MFIX-Exa: Exascale simulation for the design of industrial-scale chemical reactors

Goal: Develop an efficient high-fidelity multiphase flow modeling capability to aid in the design of industrial-scale chemical reactors

Simulation with high-fidelity, physics-based models is essential to scaling up from lab → pilot → commercial scale reactors
- Reduction in cost
- Reduction in time to deployment
- Risk mitigation at large scales

Proposed increase in fidelity will aid in the development of CO_2 capture technology (supported by DOE-FE) as well as unlock the ability to simulate a host of relevant problems in energy, chemical processing and pharmaceutical industries

Lab-scale testing of a novel CO_2 capture method at NETL

Petra Nova, world's largest post-combustion CO_2 capture plant, began operation in January 2017

NATIONAL ENERGY TECHNOLOGY LABORATORY MFIX Exa

ECP EXASCALE COMPUTING PROJECT

MFIX-Exa will be based on MFIX-DEM

Gas Phase – Navier-Stokes like equations

$$\frac{\partial}{\partial t}(\varepsilon_g \rho_g) + \frac{\partial}{\partial x_j}(\varepsilon_g \rho_g U_{gj}) = 0$$

$$\frac{\partial}{\partial t}(\varepsilon_g \rho_g U_{gi}) + \frac{\partial}{\partial x_j}(\varepsilon_g \rho_g U_{gj} U_{gi}) = -\varepsilon_g \frac{\partial P_g}{\partial x_i} + \frac{\partial \tau_{gij}}{\partial x_j} + f_{gi} + \varepsilon_g \rho_g g_i$$

Particles – Newton's law

$$\frac{dx_{pi}}{dt} = u_{pi}$$

$$m_p \frac{du_{pi}}{dt} = m_p g_i + f_{pi} + m_p A_{coll}$$

$$I_{ij} \frac{d\omega_{pj}}{dt} = T_{pi}$$

MFIX DEM

Advantages CFD-DEM over TFM

- No numerical diffusion in particle phase
- Does not require (uncertain) constitutive relations for the particle phase
- Able to describe slowly shearing or stagnant regions
- Able to track the evolution of particle-scale properties: size, density, chemical conversion ...

Garg R, Galvin J, Li T, Pannala S. Documentation of open-source MFIX–DEM software for gas-solids flows, From URL https://mfix.netl.doe.gov/documentation/dem_doc_2012-1.pdf. (2012)

5 Exascale Computing Project

NETL National Energy Technology Laboratory · Boulder · MFIX Exa · ECP Exascale Computing Project

MFIX-Exa challenge problem

Simulate 1 MWe Chemical Looping Reactor with CFD-DEM

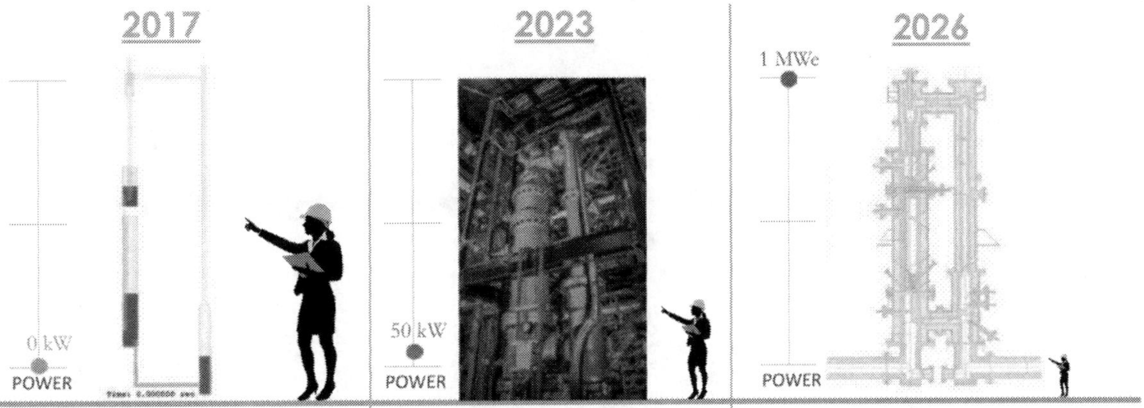

Particle Count: 60×10^6
Time to Solution: 600 days

Particle Count: 5×10^9
Time to Solution: 0.5 days

Particle Count: 100×10^9
Time to Solution: 2 days

Time-to-solution is estimated for 5 minutes of real time in all cases; the 2023/2026 values are guestimates.

6 Exascale Computing Project

NETL National Energy Technology Laboratory · Boulder · MFIX Exa · ECP Exascale Computing Project

Achieving the desired performance in MFIX-Exa

The 10 Year Challenge Problem

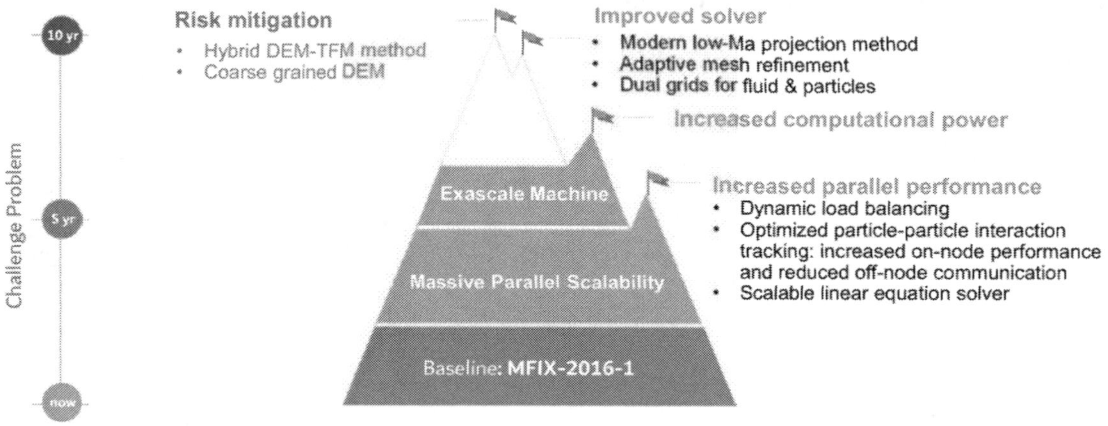

From Giga to Exa, via Tera & Peta*

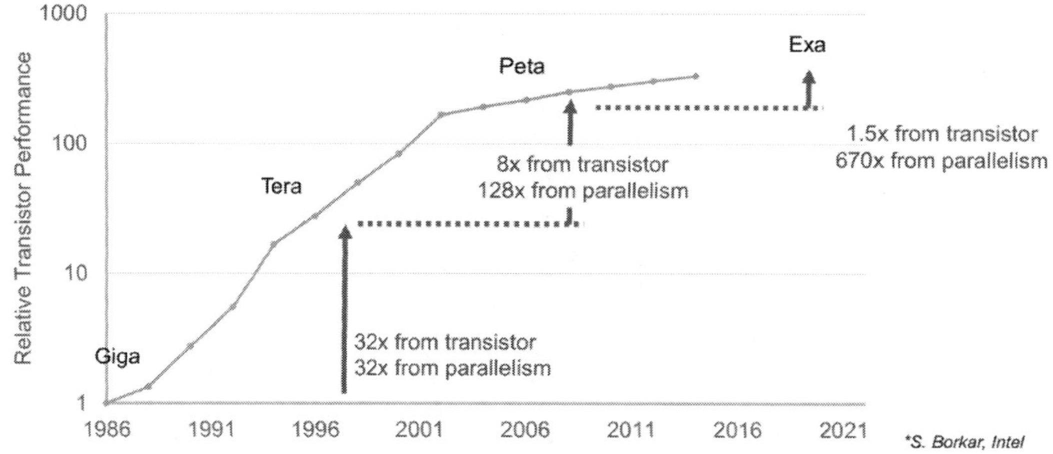

*S. Borkar, Intel

Performance from parallelism

Exascale requires rethinking of algorithms and software

Expected hardware characteristics

1. Increased cores/node parallelism
   - Reduced memory/core and memory bandwidth
2. Reducing power consumption is a paramount concern

New algorithms and software required

- Reduce data movement, the main source of power consumption
- Exploit vastly more parallelism, contending with less memory and memory bandwidth
- Work on heterogeneous compute nodes
- Support resilience
- Manage, analyze, and visualize large data sets

Functionality	Energy (pJ)	Performance (ps)
Core/FP Unit	50	125
Register bank	10	250
Cache/SRAM	100	2000
Memory/DRAM	1000	100,000

Source: T. Hoefler. Energy-aware Software Development for Massive-Scale Systems. Keynote at EnA-HPC, 2011, Hamburg, Germany

1. Exascale Programming Challenges, DOE-ASCR report 2011; 2. Shalf J, Dosanjh S, Morrison J. Exascale Computing Technology Challenges. 9th International Conference on High Performance Computing for Computational Science, VECPAR 2010, Berkeley, CA, USA, in June 2010

MFIX-Exa brings together three teams and two codes

- 60+ years of experience in multiphase modeling and MFIX (NETL and CU)
- 60+ years of experience in large-scale, multiscale multiphysics applications (LBNL)
- 90+ years of experience in high performance computing

MFIX Exa

MFIX + AMReX

- 30+ years of development
- 12 developers at NETL
- 4,000+ registered users
- 175+ downloads per month
- 200+ citations per year
- Applied for reactor design and troubleshooting in fossil, bio, nuclear, and solar energy; chemicals industry; and nuclear waste treatment

- Block-structured AMR software framework supported by ECP Co-Design Center
- Supports multiple DOE codes: accelerator modeling, astrophysics, combustion, cosmology, and subsurface
- Long development history

First version of MFIX-Exa developed and verified

Many verification cases

Nightly-regression-test results available at
https://ccse.lbl.gov/pub/RegressionTesting/MFIX/

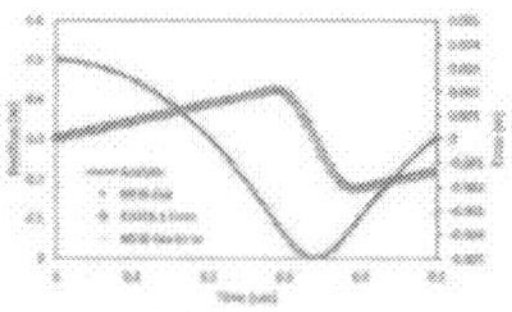

Freely falling particle with wall collision

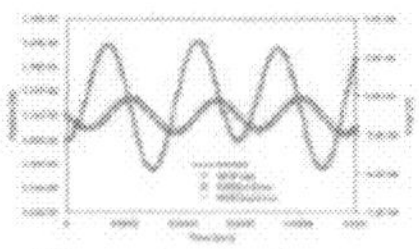

Two stacked compressed particles

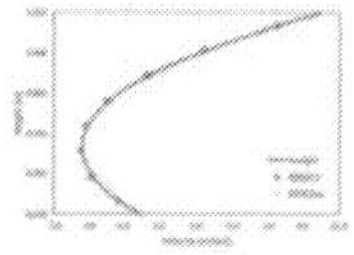

Couette flow in a channel

Four benchmark cases that mimic sections of a CLR

Fluidized Bed

Settling Bed

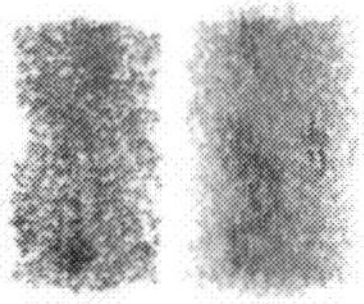

Riser Flow

Homogeneous Cooling System

Preliminary performance analysis conducted

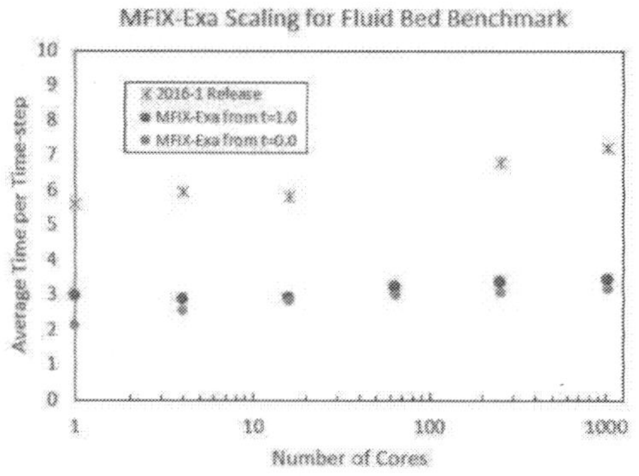

2X reduction in time-to-solution observed in a weak scaling study of MFIX-Exa and MFIX-2016-1 on Cori-KNL (run for 50ms)

MFIX-Exa released with hybrid parallelism and dynamic load balancing

- Take full advantage of many-core architectures through Hybrid parallelism (MPI + OpenMP)
- Minimize run time through Dynamic load balancing

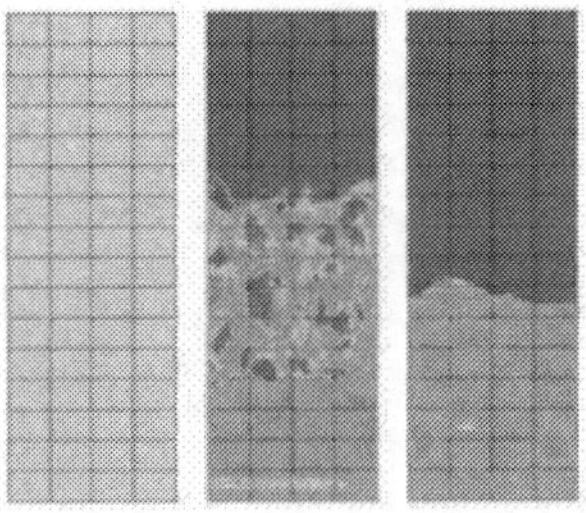

Based on number of grid cells

Two load balancing strategies

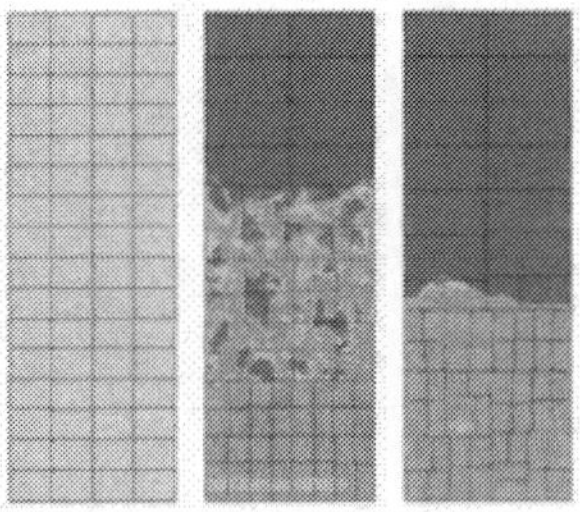

Based on number of particles

MFIX-Exa Status and Development Plans

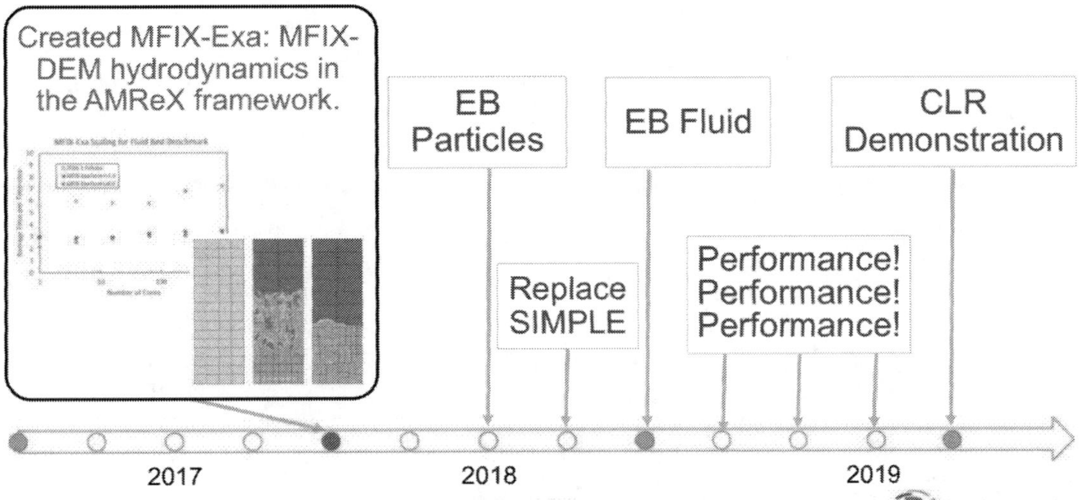

Thank You!

For more information
https://mfix.netl.doe.gov/

Madhava Syamlal, PI, MFIX-Exa project
madhava.syamlal@netl.doe.gov

490243 Hybrid Simulations of Iron Ore Reduction in Fluidized Beds

Abstract:

Fluidized bed and moving bed reactors are one of the most important technologies in several branches of process industry. Especially, it is known since decades that iron can be reduced rapidly and efficiently from iron carrier materials using such. The primary energy sources and reducing agents are natural gas, coal, coke, pulverized coal, which are finally released as CO_2 and in a lesser extent as H_2O to the environment. Iron reduction consumes about 70% of the energy during steelmaking therefore offering potential in energy and CO_2 savings. Due to the limited accessibility for measurements, simulation methods have become one of the most important tools for optimizing the iron making processes. While the filtered two-fluid model (Schneiderbauer & Pirker 2014) would be a good candidate to attack the simulation of large-scale multi-phase processes it lacks from a proper representation of the particle size distribution and the related physical phenomena. This, in turn, gives rise to particle-based approaches, such as the coupling between CFD and DEM methods, which can easily handle particle segregation, particle growth and particle mixing. Furthermore, chemical reactions can be evaluated per particle and it is not required to transfer these reactions to a continuum representation. However, CFD-DEM approaches require an appropriate coarse-graining to considerably reduce their computational demands.

We, therefore, present a generalization of the Lagrangian-Eulerian hybrid model for the numerical assessment of reacting poly-disperse gas-solid flows (Schneiderbauer et al. 2016) to fluidized beds used for iron ore reduction. The main idea of such a modeling strategy is to use a combination of a Lagrangian discrete phase model (DPM) and a coarse-grained two-fluid model (TFM) to take advantage of the benefits of those two different formulations. On the one hand, the DPM model unveils additional information such as the local particle size distribution, which is not covered by TFM. On the other hand, the TFM solution deflects the DPM trajectories due to the inter-particle stresses. Moreover, sub-grid drag corrections are applied to account for the impact of the small unresolved scales on the gas-solid drag force. This hybrid approach further enables the efficient evaluation of the gas-solid phase reduction of iron ore at a particle level using DPM. The predictive capability and numerical efficiency of this reactive hybrid modeling approach is demonstrated in the case of a lab-scale fluidized bed. The results show that the model is able to correctly predict fractional reduction of the iron ore. The results further give a closer insight about the temperatures and reaction gas consumption due to the reduction process. Finally, the scale-up to an industrial scale fluidized bed is shown.

REFERENCES

Schneiderbauer, S. et al., 2016. A Lagrangian-Eulerian Hybrid Model for the Simulation of Industrial-scale Gas-solid Cyclones. *Powder Technology*, 304, pp.229--240.

Schneiderbauer, S. & Pirker, S., 2014. Filtered and heterogeneity based sub-grid modifications for gas-solid drag and solids stresses in bubbling fluidized beds. *AIChE Journal*, 60(3), pp.839â€"854. Available at: http://onlinelibrary.wiley.com/doi/10.1002/aic.14321/abstract.

Verification study of anisotropic filtered Two Fluid Model closures

Jan Hendrik Cloete[1], Schalk Cloete[2], Stefan Radl[3], & Shahriar Amini[1,2]

[1]Norwegian University of Science and Technology, Trondheim, Norway

[2]SINTEF Materials and Chemistry, Trondheim, Norway

[3]Graz University of Technology, Graz, Austria

Abstract

This study presents a comparison of isotropic and anisotropic filtered Two Fluid Model (fTFM) closures for drag and mesoscale solids stresses. Anisotropic modelling of drag is important to correctly predict the drag correction in directions not aligned with gravity, while anisotropic stress modelling predicts the deviatoric component of the mesoscale stress tensor with much higher accuracy than closures based on the conventionally employed Boussinesq approximation. Comparisons to a highly computationally expensive resolved simulation showed that the anisotropic stress model had the largest positive effect on model accuracy. Clear improvements relative to the isotropic model were observed for all flow variables, including reaction rates. Anisotropic drag modelling had a smaller, albeit still significant effect. In general, the good performance of the newly proposed anisotropic fTFM closures observed in this study serves as a clear recommendation for future study in this direction.

Introduction

Studying industrial scale fluidized beds by means of Computational Fluid Dynamics remains challenging due to the computational demand of simulating the behavior of the large number of particles contained in these processes. A promising method for overcoming this challenge is the filtered Two Fluid Model (Igci et al., 2008) or fTFM. In this approach, the solids phase is firstly assumed to be continuous, as in the Two Fluid Model (TFM) closed by the Kinetic Theory of Granular Flow. Secondly, sub-grid closures are used to model the effects of mesoscale structures that are not resolved on the computational grid. Mesoscale structures in fluidized beds take the form of gas bubbles and solids clusters and play a crucial role in determining the macro scale flow behavior by affecting the overall drag between the phases and the momentum dispersion in the system. fTFMs therefore allow the mesoscale structures to be accounted for without having to resolve these structures on a fine grid, which remains computationally infeasible for industrial scale simulations.

Over the past decade, several groups have contributed different closures (Cloete et al., 2017c; Igci et al., 2008; Igci and Sundaresan, 2011; Milioli et al., 2013; Ozel et al., 2013; Parmentier et al., 2012; Sarkar et al., 2016; Schneiderbauer, 2017; Schneiderbauer and Pirker, 2014). However, most of these studies assumed the effects of the mesoscale scale structures to be isotropic, and have therefore derived closures that are directionally independent and are expressed as functions of directionally independent variables. Anisotropy can be expected to be important in fluidized beds due to the critical influence of gravity on fluidization behavior, which is a result of the large density difference of the phases. Indeed, recent work has shown the anisotropy of the sub-grid drag corrections (Cloete et al., 2017c).

For this reason, this study will assess the importance of incorporating anisotropic effects in fTFM closures. This is done by comparing the performance of isotropic and anisotropic closures for the two most important hydrodynamic fTFM closures, the filtered drag and the solids mesoscale stresses. The performance of the closures is evaluated by comparing coarse grid simulations results using fTFM models to the results of resolved TFM simulations. The conclusions from this study will therefore reveal the importance of considering anisotropy in future fTFM development.

Simulations

Filtered closures

This section outlines the different fTFM closures used in this study. The detailed equations for the closures evaluated in this study are given in the appendix.

Filtered momentum equation

As outlined in a previous study (Cloete et al., 2017a), the filtered conservation equations can be obtained by performing a spatial averaging (or filtering) operation on the governing equations and then separating variables by using phase-weighted averages and fluctuating variables. For modeling the hydrodynamics of fluidized beds, all the terms requiring closure can be identified in the following filtered solids momentum equation:

$$\frac{\partial}{\partial t}\left(\rho_s \overline{\alpha_s} \widetilde{\vec{v}_s}\right) + \nabla \cdot \left(\rho_s \overline{\alpha_s} \widetilde{\vec{v}_s} \widetilde{\vec{v}_s}\right) = -\overline{\alpha_s} \nabla \bar{p} - \nabla \overline{p_s}$$

$$-\nabla \cdot \left(\rho_s \overline{\alpha_s \vec{v}_s{}' \vec{v}_s{}'}\right) + \nabla \cdot \overline{\bar{\bar{\tau}}_s} + \overline{\alpha_s} \rho_s \vec{g} + \overline{K_{gs}\left(\vec{v}_g - \vec{v}_s\right)} - \overline{\alpha_s{}' \nabla p'} \qquad \textbf{Equation 1}$$

At large filter sizes that are relevant to industrial scaled simulations, the filtered kinetic theory stresses (terms 2 and 4 on the right-hand side) can be neglected. However, the filtered frictional pressure has been found to be significant and is therefore modelled by the closure given in an earlier study (Cloete et al., 2017b). The gas mesoscale stresses (not shown in Equation 1) can be neglected due to the gas density being much smaller than the solids density. As a result, the hydrodynamic closures requiring further discussion are those for the solids mesoscale stresses (third term on the right of Equation 1), the filtered drag force (sixth term on the right) and the pressure gradient fluctuation contribution (seventh term on the right).

Solids mesoscale stress closures

Isotropic model (S-)

The isotropic model closes the mesoscale solids stresses using the Boussinesq approximation, similar to what is commonly done in fTFM literature (Milioli et al., 2013; Sarkar et al., 2016; Schneiderbauer, 2017). In this approach, the mean of the normal stresses is modelled as a filtered solids pressure (Equation 17) and the deviatoric stresses using a filtered solids viscosity (Equation 18), as shown in Equation 2.

$$\rho_s \overline{\alpha_s u_{s,i}{}' u_{s,j}{}'} \approx p_{s,fil}\delta_{ij} - \overline{\alpha_s}\mu_{s,fil}\left(\frac{\partial \widetilde{u_i}}{\partial x_j} + \frac{\partial \widetilde{u_j}}{\partial x_i} - \frac{2}{3}\frac{\partial \widetilde{u_k}}{\partial x_k}\delta_{ij}\right) \qquad \textbf{Equation 2}$$

Both of these values are closed as a function of the filter size, the filtered solids volume fraction and the filtered deviatoric shear rate magnitude (Equation 19 and Equation 20). However, it has been found that this approach is unable to correctly predict the deviatoric stresses, therefore the interest in an anisotropic closure for the solids mesoscale stresses.

Anisotropic model (S+)

The anisotropic model closes each component of the solids mesoscale stress tensor individually (Equation 21 and Equation 22). This is done based on the observation that a filtered variance quantity in the form of $\overline{\alpha X'Y'}$ can be modelled by a function of the form $\bar{\alpha}(\frac{d\tilde{X}}{dx}\frac{d\tilde{Y}}{dx} + \frac{d\tilde{X}}{dy}\frac{d\tilde{Y}}{dy})$, termed the gradient product marker (GPM). The anisotropic model therefore replaces the isotropic filtered deviatoric shear rate magnitude, used as independent variable in the isotropic model, with the anisotropic GPM (Equation 23), which contains the velocity gradients relevant to the stress component being considered. This approach predicts the solids mesoscale stresses in resolved simulations much more accurately than the Boussinesq approach based models.

Interphase momentum exchange closures

Isotropic model (D-)

In the isotropic model (Cloete et al., 2017c), the filtered drag and the pressure gradient fluctuation contribution are treated together as follows:

$$\overline{K_{gs}(\vec{v}_g - \vec{v}_s)} - \overline{\alpha_s' \nabla p'} = C K_{gs,hom}(\widetilde{\vec{v}_g} - \widetilde{\vec{v}_s}) \qquad \textbf{Equation 3}$$

Here, $K_{gs,hom}$ is the interphase momentum exchange coefficient evaluated at the filtered solids volume fraction and the filtered slip velocity magnitude. The drag correction factor, C, must be closed. This closure is expressed as a function of the filter size, the filtered solids volume fraction and the scaled filtered slip velocity magnitude (Equation 24). The filtered slip velocity magnitude is scaled by the steady state sedimentation velocity of a homogenous suspension (Equation 7 to Equation 16). It has been found that this scaling provides better statistics for closure fitting and a simpler closure expression (Cloete et al., 2017c).

Similar to other works in literature (Milioli et al., 2013; Sarkar et al., 2016), the drag correction factor is closed based on the direction aligned with gravity, assuming that this is the dominant direction for the drag. The same drag correction factor is then applied in the other directions. This approach has been shown to substantially overpredict the drag in the lateral directions (Cloete et al., 2017c).

Anisotropic model (D+)

The anisotropic model closes the filtered drag and pressure gradient fluctuation contributions separately. The filtered drag is closed as follows:

$$\overline{K_{gs}(\vec{v}_g - \vec{v}_s)} = C K_{gs,hom}(\widetilde{\vec{v}_g} - \widetilde{\vec{v}_s} - \overrightarrow{v_d}) \qquad \textbf{Equation 4}$$

This approach closes a drift velocity, $\overrightarrow{v_d}$, similar to the work from the INPT group (Ozel et al., 2013; Parmentier et al., 2012). However, in contrast to these prior works, the non-linearity correction factor, C, is used to account for the effect of assumptions used to derive the drift velocity form of the filtered drag. It has been found that the effect of C, although much smaller than that of the drift velocity, is significant at larger filter sizes.

The drift velocity, the non-linearity correction factor and the pressure gradient fluctuation contribution are all closed as functions of the filter size and the filtered solids volume fraction. In addition, the drift velocity for a specific direction (Equation 25 to Equation 31) is a function of the scaled filtered slip velocity in the direction considered and the GPM for the drift velocity, which is a function of the filtered gas velocity and volume fraction gradients (Equation 30). Also, the non-linearity correction factor (Equation 32) uses the scaled filtered slip velocity magnitude as an additional independent variable, whereas the pressure gradient fluctuation contribution (Equation 33 to Equation 35) uses the scaled filtered slip velocity in the direction considered.

The anisotropic model accurately predicts the drag in the lateral directions, when compared to resolved simulation data. Additionally, the anisotropic model solves three different closure equations for the drift velocity, the non-linearity correction factor and the pressure gradient fluctuation contribution, whereas these three effects are combined in the drag correction factor in the isotropic model. Also, for the drift velocity, which is the most important of the three closures, an additional independent variable is considered compared to the isotropic model. As a result of these added complexities, the anisotropic model explains more of the variance in the resolved simulation data than the isotropic model, even for the direction aligned with gravity.

Reactive closures

Considering an isothermal, first-order, solids catalyzed reaction of reactant A to product B, the filtered species equation can be obtained for the reactant as follows:

$$\frac{\partial}{\partial t}\left(\rho_g \overline{\alpha}_g \widetilde{X}_A\right) + \nabla \cdot \left(\rho_g \overline{\alpha}_g \widetilde{X}_A \widetilde{\vec{v}}_g\right) = \nabla \cdot \left(D\overline{\alpha_g \nabla X_A}\right) - \nabla \cdot \left(\rho_g \overline{\alpha_g X_A{}' \vec{v}_g{}'}\right) - k_A \rho_g \overline{\alpha_s X_A} \qquad \textbf{Equation 5}$$

The filtered microscopic diffusion is assumed to be negligible (first term on the right) (Loezos and Sundaresan, 2002) and the mesoscale species dispersion (second term on the right) is modelled using the closure by Agrawal et al. (2013). It was previously shown (Cloete et al., 2017b) that for this case the effect of the mesoscale species dispersion is small.

Finally, the filtered reaction rate (third equation on the right) is closed as follows:

$$k_A \rho_g \overline{\alpha_s X_A} = R k_A \rho_g \overline{\alpha_s} \widetilde{X_A} \quad \textbf{Equation 6}$$

The reaction rate correction factor, R, is closed as a function of the filter size, the filtered solids volume fraction and the scaled filtered slip velocity magnitude as in an earlier work (Cloete et al., 2017b).

Simulation setup

Simulations are performed for spherical particles with a diameter (d_s) of $75\mu m$ and a density of $1500 \frac{kg}{m^3}$. Coarse grid simulations using fTFM closures are performed using grid sizes of 20 mm ($267d_s$), 40 mm ($533d_s$) and 80 mm ($1067d_s$). A resolved TFM simulation, used a benchmark to assess the performance of fTFM closures, is performed on a grid of 0.884 mm ($11.8d_s$).

The simulations are performed in 2D to ensure that the resolved simulation used for the detailed verification in this study is computationally feasible for the large domain considered. Even considering this simplification, the resolved simulation required approximately 3 months of computational time on 128 cores to obtain sufficiently averaged data. The fTFM closures were also derived from 2D simulations, thus allowing a direct comparison.

The geometry shown in Figure 2 is simulated in this study. A rectangular region of 0.96m by 1.6m is considered, with an inlet at the bottom and periodic boundaries at the sides. Periodic boundaries are selected to eliminate uncertainties arising from wall effects in this verification, since the closure models evaluated in this study were derived from data collected from periodic simulations. To prevent numerical instabilities due to backflow, walls slope at $45°$ from the top of the rectangular region towards an outlet with a width of 0.08m. The solids boundary condition at the walls are specified as free-slip to prevent backflow into the rectangular region and thereby minimize the effect of the walls on flow in this region.

A mean solids flux of $150 \frac{kg}{m^2 s}$ is chosen at the inlet and the average superficial gas velocity is chosen to operate at the center of the bubbling regime according to Bi and Grace (1995). Additionally, non-uniform boundary conditions, shown in Figure 1, are specified at the inlet to allow faster cluster formation in the absence of walls. The gas and solid velocities are set equal at the inlet and are specified to be half the mean superficial gas velocity required at the sides and then increase linearly towards the center. The solids volume fraction is set to half of the mean solids volume fraction required at the center and increases linearly towards the sides. The flow at the inlet therefore changes from fast and dilute at the center to denser and slower at the sides of the domain. Additionally, the reactant is injected with a mass fraction of 1 at the center of the inlet and decreases linearly towards 0 at the sides.

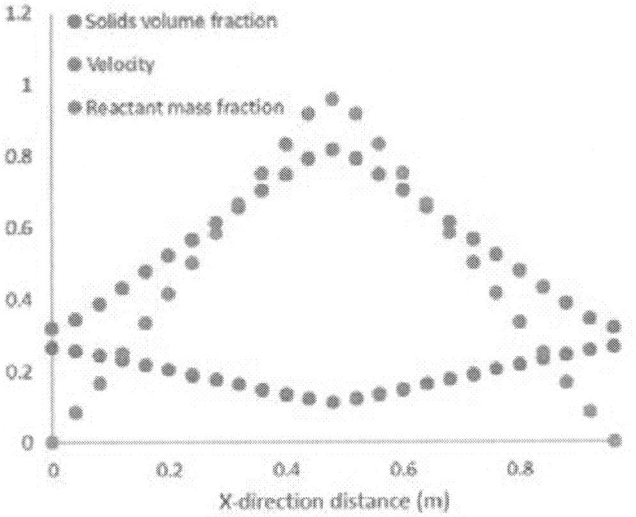

Figure 1: Non-uniform inlet boundary conditions.

Results and discussion

In this study, the four possible combinations of the anisotropic and isotropic closures for the filtered interphase momentum exchange and the solids mesoscale stresses are evaluated at three different grid sizes. Time-averaged vertical profiles of relevant quantities are then compared to the results of the resolved simulation to assess the performance of the different closures.

Inlet effect

The non-uniform inlet conditions shown in Figure 1 was specified to introduce some mean gradients into the flow, in the absence of walls for most of the domain, thus providing more interesting verification data. However, this boundary condition also introduced some uncertainty due to the region of uniform intermediate volume fraction that can be observed at the bottom of the domain in Figure 2. In filtered simulations, a region of intermediate volume fraction is generally associated with significant clustering, leading to significant corrections for drag, stresses and reaction rates. As a result, the filtered simulations generally predicted significant coarse-grid corrections in the uniform inlet region where no correction should actually be imposed.

This is clearly observed in the slip velocity prediction along vertical lines in the center and at the side of the domain shown in Figure 3. It is shown that the match between the resolved and coarse grid simulations is generally very good, except in the lower region on the center line. This is the region of intermediate volume fraction where the models predict a correction, while no correction should be imposed. Due to this uncertainty in the central region of the domain, the rest of the paper will present results on the side periodic boundaries where clustering takes place along the entire height of the domain.

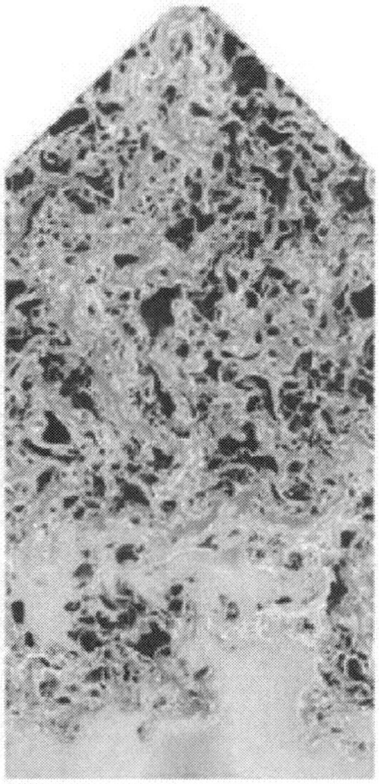

Figure 2 – Contour plots of the instantaneous solids volume fraction from the resolved TFM simulations.

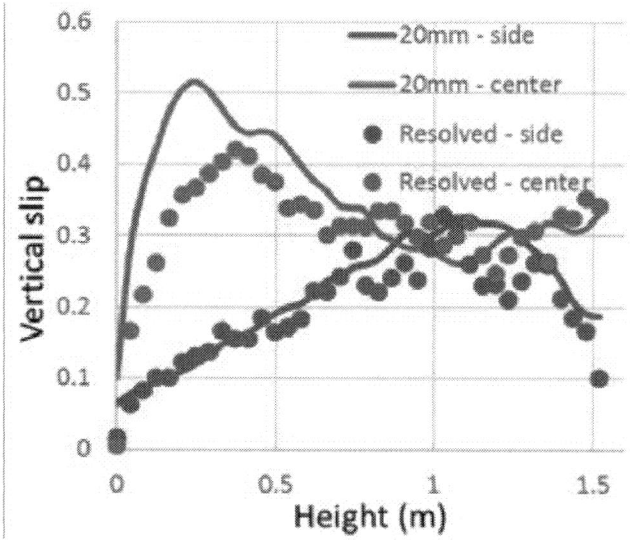

Figure 3 – Comparison of time-averaged vertical profiles of the vertical slip velocity along lines at the center and at the side of the domain. Coarse grid simulations are performed on a 20 mm grid and using anisotropic closures for both the drag and the stresses.

It should be noted that fTFM closures in literature are generally expected to perform poorly in such cases with homogenous regions at intermediate volume fractions, since all the proposed closures make use of algebraic equations. Solving transport equation for the closure quantities would, however, allow accurate modelling of such cases on coarse grids. Such models have been proposed (Schneiderbauer, 2017), but were subsequently simplified to algebraic relations before application, in order to eliminate the complexity of solving additional transport equations. However, in practical applications clustering general occurs rapidly at the boundaries and such homogenous regions at intermediate volume fractions do not occur. As a result, this weakness of algebraic fTFM closures is not expected to cause significant errors in most practical applications.

Solids volume fraction

The model predictions of time-averaged solids volume fraction along a vertical line on the side periodic boundary of the domain is shown in Figure 4. Clearly, all models return reasonable predictions of this quantity, but the cases with anisotropic stress modelling generally produce a closer match to the resolved simulation data. Anisotropic stress modelling also appears to be important for good grid independence behavior, especially when transitioning from the 40 mm to the 80 mm grid. The anisotropic drag model also appears to outperform the isotropic drag model, except for the coarsest grid.

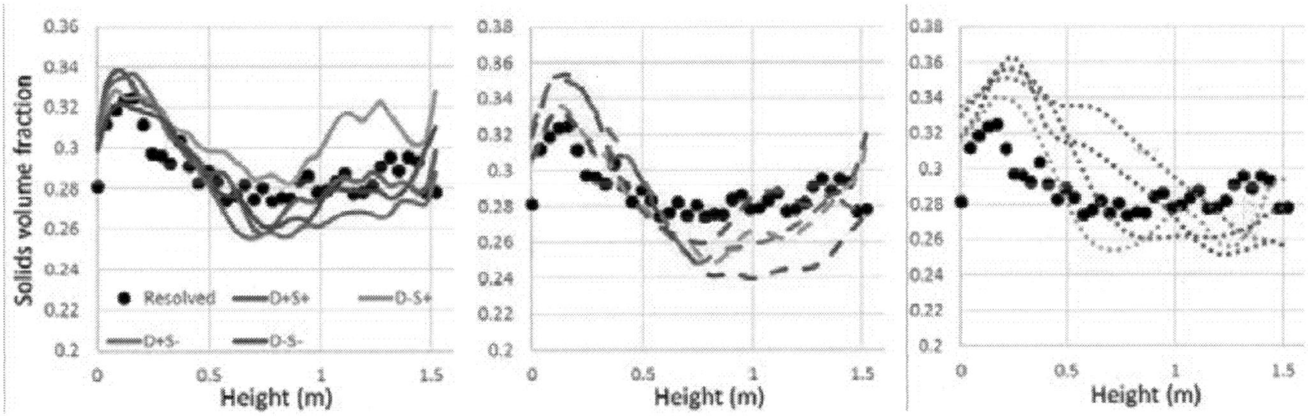

Figure 4 - Time-averaged vertical profiles of the solids volume fraction along the side boundary. From left to right the coarse grid simulations are performed on grids of 20 mm, 40 mm and 80 mm.

Vertical direction slip velocity

The insights from the solids volume fraction profiles in Figure 4 are also reflected in the slip velocity predictions plotted on Figure 5. Once again, the anisotropic stress model appears to be important for ensuring good predictions, especially on coarser grids. In this case, no clear improvement from the anisotropic drag model can be observed when used in combination with the anisotropic stress model. However, in combination with the isotropic stress model, the anisotropic drag model performs better on the two smaller grid sizes.

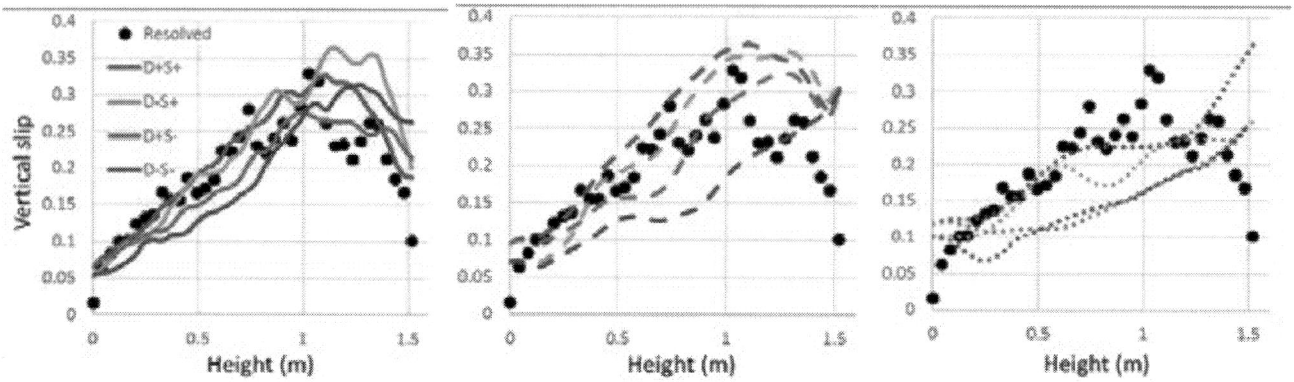

Figure 5 - Time-averaged vertical profiles of the vertical slip velocity along the side boundary. From left to right the coarse grid simulations are performed on grids of 20 mm, 40 mm and 80 mm.

Volume fraction variance

As may be expected, Figure 6 shows that all coarse grid simulations predict a significantly lower solids volume fraction variance than the resolved simulations, especially as the grid is coarsened. Figure 7 offers a graphical illustration of this effect. However, it is clearly shown that the anisotropic stress model significantly increases the degree of solids volume fraction variance resolved on the coarse grids. The anisotropic drag model also resolved greater flow details on the coarser grids.

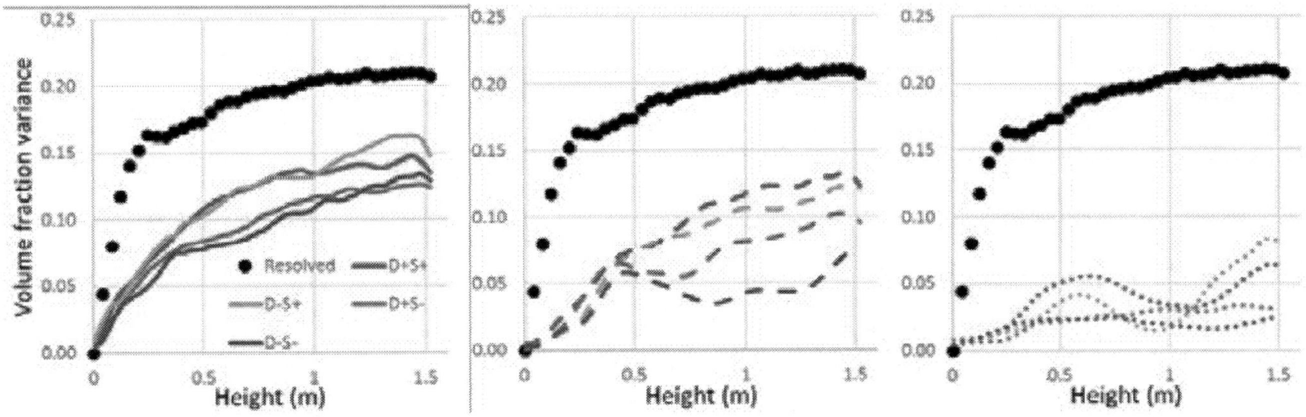

Figure 6 - The vertical profiles of the solids volume fraction variance along the side boundary. From left to right the coarse grid simulations are performed on grids of 20 mm, 40 mm and 80 mm.

The improved performance of the anisotropic stress closures can be explained based on the structure resolution. From Figure 7 it can be seen that the isotropic stress models predict less distinct macro-scale 'clusters' and 'bubbles'. More gas flow will then tend to pass through these macro-scale clusters, whereas more of the gas flow will tend to slip around the more resolved macro-scale clusters predicted by the anisotropic mesoscale stress closure. As a result, the vertical drag is increased when using the isotropic mesoscale solids stress closure, leading to underpredictions in the vertical slip (Figure 5) and the solids volume fraction (Figure 4, with the exception of the coarsest grid).

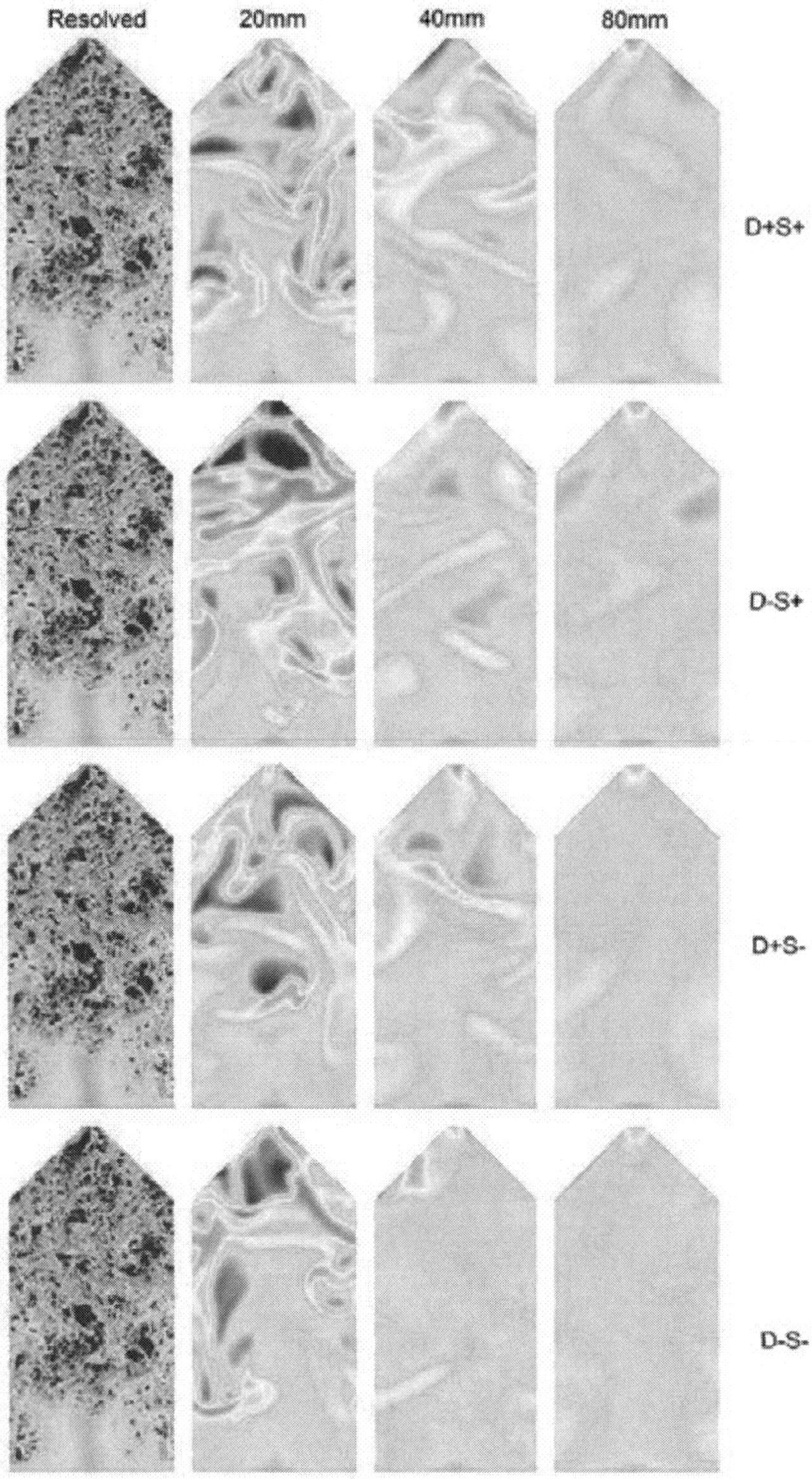

Figure 7: Comparison of the instantaneous volume fraction contours for the resolved simulations and for the different closure combinations at different coarse grid sizes.

Conversion rate

Finally, the performance of the filtered reaction rate closure is displayed in Figure 8. Once more, the anisotropic stress model appears to the critical for attaining accurate results on the coarser meshes. This result illustrates the importance of accurate resolution of the hydrodynamic flow details when completing reactive fluidized bed simulations. In this case, the effect of the anisotropic drag model was insignificant.

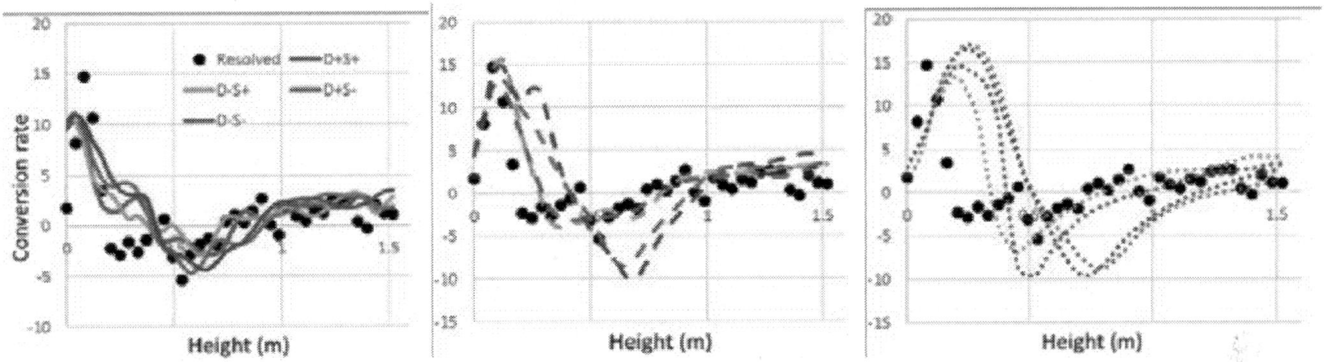

Figure 8 - The time-averaged vertical profiles of the conversion rate ($\frac{d(-log(X_A))}{dy}$) along the side boundary. From left to right the coarse grid simulations are performed on grids of 20 mm, 40 mm and 80 mm.

Conclusions

This study clearly illustrated the importance of accounting for anisotropy in the filtered modelling of fluidized bed reactors. The inclusion of anisotropic stress modelling was found to be particularly influential. The Boussinesq approximation that is commonly employed in filtered modelling of fluidized beds tends to overpredict momentum dispersion by modelling the deviatoric stresses with a filtered viscosity. This misrepresentation causes significant mismatches in all flow variables when comparing to resolved simulation data.

Anisotropic drag modelling was significant in some cases, but generally had a smaller effect than anisotropic stress modelling. This is against the convention that drag modelling is generally much more important than mesoscale stress modelling in filtered TFM simulations. Although drag remains the most influential closure, improvements in mesoscale stress modelling can still lead to significant improvements in model accuracy.

In general, the newly proposed anisotropic closures produced an excellent fit to the resolved simulation data. It is also notable that this good fit was achieved by setting the filter size equal to the cell size in the coarse grid simulations. Higher filter sizes to cell size ratios are sometimes employed, presumably as a tuning factor to attain reasonable matches with experiments. Based on these promising results, further research into anisotropic fTFM closures is recommended.

Acknowledgements

The authors would like to express their gratitude for the financial support from the European Commission under the NanoSim grant (project number: 604656), as well as for the computational resources provided at NTNU by UNINETT Sigma2 AS, https://www.sigma2.no, as part of the national infrastructure for computational science in Norway.

Nomenclature

Main Symbol definitions:

A	Area (m^2)
d	Particle diameter (m)
$\vec{g}$	Gravity vector (m/s^2)
$\overline{\overline{I}}$	Identity matrix
K_{sg}	Interphase momentum exchange coefficient $(kg/m^3 s)$
p	Pressure (Pa)
$\overline{\overline{S}}$	Deviatoric shear rate tensor $(1/s)$
t	Time (s)
v_t	Particle terminal velocity (m/s)
α	Volume fraction
μ	Viscosity $(kg/(m.s))$
ρ	Density (kg/m^3)
$\overline{\overline{\tau}}$	Stress tensor (Pa)
$\vec{\upsilon}$	Velocity vector (m/s)
υ_d	Drift velocity (m/s)
Δ_f	Filter size (m)

M Gradient product marker

Sub- and superscript definitions:

fil Filtered value

g Gas

i, j Coordinate indexes

s Solid

Operator definitions

$\bar{x}$ Volume average

$\tilde{x}$ Phase-weighted volume average

x' Fluctuation from mean

$\hat{x}$ Non-dimensionalized value

x^* Scaled value

$\vec{x}$ Vector quantity

$|x|$ Absolute value

$\|x\|$ Magnitude of value

Appendix

All closure models used in this study are a function of the filter size, which is set to be equal to the length of the grid cells used in the coarse grid simulations, therefore in 2D: $\Delta_f = A_{cell}^{1/2}$. The filter size is non-dimensionalized by the particle relaxation length, $\widehat{\Delta_f} = \frac{\Delta_f}{v_t^2/g}$. Additionally, the closures in this study uses a scaled filter size, where $\Delta_f^* = \min(\widehat{\Delta_f} - \widehat{\Delta_f}_{min}, 0)$. By setting $\widehat{\Delta_f}_{min}$ to the dimensionless grid size at which the resolved simulations for model derivation were performed, $\widehat{\Delta_f}_{min} = 0.1285$, this ensures that no sub-grid corrections are predicted if the fTFM closures are used on grid sizes equal to or smaller than those used in the model derivation.

Where the filtered slip velocity is used as independent variable, it is non-dimensionalized using the steady state sedimentation velocity, which is defined as the slip velocity where the drag force balances the gravitational force for a specific volume fraction.

$$\widetilde{v_{slip}}^* = \frac{\widetilde{v_{slip}}}{\bar{v}_{ss\_slip}}$$ **Equation 7**

The steady state sedimentation velocity ($\bar{v}_{ss\_slip}$) is approximated from the Huilin-Gidaspow drag law (Huilin and Gidaspow, 2003) as follows:

$$\bar{v}_{ss_{slip},WY} = \frac{\mu_g \overline{\mathrm{Re}}_{ss_{slip},WY}}{\rho_g d_s} \quad \textbf{Equation 8}$$

$$\overline{\mathrm{Re}}_{ss\_slip,WY} \approx \frac{A}{\bar{\alpha}_g (1 + 0.2296 A^{0.5329})^{0.7642}} \quad \textbf{Equation 9}$$

$$A = \frac{\bar{\alpha}_g^{3.65} g (\rho_s - \rho_g) d_s^3 \rho_g}{18 \mu_g^2} \quad \textbf{Equation 10}$$

$$\bar{v}_{ss\_slip,Ergun} = \frac{-b + \sqrt{b^2 - 4ac}}{2a} \quad \textbf{Equation 11}$$

$$a = 1.75 \frac{\rho_g \bar{\alpha}_s}{d_s} \quad \textbf{Equation 12}$$

$$b = 150 \frac{\bar{\alpha}_s^2 \mu_g}{\bar{\alpha}_g d_s^2} \quad \textbf{Equation 13}$$

$$c = -(\rho_s - \rho_g) \bar{\alpha}_s g \quad \textbf{Equation 14}$$

$$\psi = \frac{1}{2} + \frac{\arctan\left(262.5(\bar{\alpha}_s - 0.2)\right)}{\pi} \quad \textbf{Equation 15}$$

$$\bar{v}_{ss\_slip} = \psi \bar{v}_{ss\_slip,Ergun} + (1 - \psi) \bar{v}_{ss\_slip,WY} \quad \textbf{Equation 16}$$

Appendix A: Isotropic solids mesoscale stress model

The following closures for the filtered solids pressure and the filtered solids viscosity are used in the Boussinesq approximation in Equation 2:

$$\frac{p_{s,fil}}{\rho_s v_t^2} = \frac{2}{\pi} x_1 \overline{\alpha_s}^{x_2} \mathrm{atan}(x_3 \Delta_f^{*\,x_4} \|\hat{S}_{s,fil}\|^{x_5} (x_6 - \overline{\alpha_s})) \Delta_f^{*\,x_7}$$
$$\times \|\hat{S}_{s,fil}\|^{x_8 + x_9 \Delta_f^{*\,x_{10}}} + \frac{2}{\pi} x_{11} \overline{\alpha_s}^{x_{12}} \mathrm{atan}(x_{13}(x_{14} - \overline{\alpha_s})) \Delta_f^{*\,x_{15}} \quad \textbf{Equation 17}$$

$x_1 = 0.774$, $x_2 = 1.72$, $x_3 = 0.403$, $x_4 = 0.610$, $x_5 = 1.19$, $x_6 = 0.684$, $x_7 = 1.57$, $x_8 = 1.00$, $x_9 = 0.331$, $x_{10} = -0.103$, $x_{11} = 0.123$, $x_{12} = 0.621$ $x_{13} = 2.89$, $x_{14} = 0.591$ and $x_{15} = 1.05$

$$\frac{\mu_{s,fil}}{\rho_s v_t^3/g} = \frac{2}{\pi} x_1 \overline{\alpha_s}^{x_2} \mathrm{atan}(x_3 \Delta_f^{*\,x_4} \|\hat{S}_{s,fil}\|^{x_5}(x_6 - \overline{\alpha_s}))\Delta_f^{*\,x_7}$$
$$\times \|\hat{S}_{s,fil}\|^{x_8 + x_9 \Delta_f^{*\,x_{10}}} + \frac{2}{\pi} x_{11} \overline{\alpha_s}^{x_{12}} \mathrm{atan}(x_{13}(x_{14} - \overline{\alpha_s}))\Delta_f^{*\,x_{15}} \|\hat{S}_{s,fil}\|^{-1} \qquad \textbf{Equation 18}$$

$x_1 = 0.350$, $x_2 = 0.545$, $x_3 = 2.43$, $x_4 = 0.141$, $x_5 = 0.772$, $x_6 = 0.624$, $x_7 = 1.83$, $x_8 = 0.400$, $x_9 = 0.348$, $x_{10} = -0.0905$, $x_{11} = 0.130$, $x_{12} = -0.498$, $x_{13} = 3.58$, $x_{14} = 0.618$ and $x_{15} = 0.968$.

The dimensionless filtered deviatoric shear rate magnitude is defined as:

$$\|\hat{S}_{s,fil}\| = \frac{\sqrt{2\bar{\bar{S}}_{s,fil} : \bar{\bar{S}}_{s,fil}}}{g/v_t} \qquad \textbf{Equation 19}$$

$$\bar{\bar{S}}_{s,fil} = \frac{1}{2}(\nabla \vec{v}_s + \nabla \vec{v}_s^T) - \frac{1}{3}\nabla \cdot \vec{v}_s \bar{\bar{I}} \qquad \textbf{Equation 20}$$

Appendix B: Anisotropic solids mesoscale stress model

The different components of the mesoscale stress tensor are described as follows:

The non-dimensionalized normal stresses:

$$\frac{\rho_s \overline{\alpha_s u_{s,i}' u_{s,i}'}}{\rho_s v_t^2} = \left(\frac{2}{\pi}\right)^2 x_1 \mathrm{atan}\left(x_2 \Delta_f^{*\,x_3} \widehat{M}_{ii}^{\,x_4} (\bar{\alpha}_s + x_5)\right)$$
$$\times \mathrm{atan}\left(x_6 \Delta_f^{*\,x_7} \widehat{M}_{ii}^{\,x_8} (x_9 - \bar{\alpha}_s)\right) \Delta_f^{*\,x_{10}} \widehat{M}_{ii}^{\,x_{11} + x_{12}\Delta_f^{*\,x_{13}}} \qquad \textbf{Equation 21}$$
$$+ \left(\frac{2}{\pi}\right)^2 x_{14} \mathrm{atan}(x_{15}\bar{\alpha}_s) \mathrm{atan}(x_{16}(x_{17} - \bar{\alpha}_s))\Delta_f^{*\,x_{18}}$$

$x_1 = 0.550$, $x_2 = 2.91$, $x_3 = 0.102$, $x_4 = -0.515$, $x_5 = 0.0358$, $x_6 = 3.04$, $x_7 = 0.960$, $x_8 = 0.240$, $x_9 = 0.691$, $x_{10} = 1.75$, $x_{11} = 0.80$, $x_{12} = 0.608$, $x_{13} = -0.219$, $x_{14} = 53.2$, $x_{15} = 2.49$, $x_{16} = 0.00719$, $x_{17} = 0.604$ and $x_{18} = 1.08$

The non-dimensionalized shear stresses:

$$\frac{\rho_s \overline{\alpha_s u_{s,i}' u_{s,j}'}}{\rho_s v_t^2} = \left(\frac{2}{\pi}\right)^2 x_1 \text{atan}\left(x_2 \Delta_f^{*x_3} \widehat{M}_{ij}^{x_4} (\bar{\alpha}_s + x_5)\right)$$

$$\times \text{atan}\left(x_6 \Delta_f^{*x_7} \widehat{M}_{ij}^{x_8} (x_9 - \bar{\alpha}_s)\right) \Delta_f^{*x_{10}} \widehat{M}_{ij}^{x_{11} + x_{12} \Delta_f^{*x_{13}}}$$

Equation 22

$x_1 = 0.630,\ x_2 = 1.76,\ x_3 = 0.0844,\ x_4 = -0.394,\ x_5 = 0.0767,\ x_6 = 4.89,\ x_7 = 0.514,\ x_8 = 0.157,\ x_9 = 0.685,\ x_{10} = 1.79,\ x_{11} = 0.80,\ x_{12} = 0.528$ and $x_{13} = -0.167.$

The non-dimensionalized gradient product marker for the mesoscale solids stresses is defined as follows:

$$\widehat{M}_{ij} = \bar{\alpha}_s \left(\frac{d\widetilde{v_{s_i}}}{dx}\frac{d\widetilde{v_{s_j}}}{dx} + \frac{d\widetilde{v_{s_i}}}{dy}\frac{d\widetilde{v_{s_j}}}{dy}\right)\frac{v_t^2}{g^2}$$

Equation 23

Appendix C: Isotropic drag model

$$-\log(C) = \text{atan}(x_1 \Delta_f^* \bar{\alpha}_s)\text{atan}\left(x_2 \Delta_f^* max(\bar{\alpha}_{\max} - \bar{\alpha}_s, 0)\right)\text{atan}(x_3 \Delta_f^*)\left(\frac{2}{\pi}\right)^3$$

$$\times \left(x_4 \log\left\|\widetilde{v_{slip}}^*\right\| + x_5 \Delta_f^{*x_6} + x_7\left(\log\left\|\widetilde{v_{slip}}^*\right\|\right)^2 \left(1 - \frac{\text{atan}(x_8 \Delta_f^*)}{\pi/2}\right)\right)$$

Equation 24

$x_1 = 36.59,\ x_2 = 22.63,\ x_3 = 1.676,\ x_4 = 0.8350,\ x_5 = 0.1399,\ x_6 = 0.1881,\ x_7 = 1.329,\ x_8 = 3.280$ and $\bar{\alpha}_{max} = 0.5511.$

Appendix D: Anisotropic drag model

The drift velocity in direction i can be calculated as follows:

$$v_{d,i}^* = \frac{v_{d,i}}{\bar{v}_{ss\_slip}}$$

Equation 25

$$v_{d,i}^* = k_1 + \tilde{v}_{slip,i}^*(1 - 10^{-k_2}) + k_3$$

Equation 26

$$k_1 = x_1\left(\frac{2}{\pi}\right)(\overline{\alpha_s}^{x_2})\ \text{atan}\left(x_3 \max(x_4 - \overline{\alpha_s}, 0)\right)$$

Equation 27

$k_1 = 0$ if i is in the direction perpendicular to gravity

$$k_2 = \left(\frac{2}{\pi}\right)^2 \operatorname{atan}(x_5\overline{\alpha_s}) \operatorname{atan}\left(x_3 \max(x_4 - \overline{\alpha_s}, 0)\right)\left(x_6\left(\frac{2}{\pi}\right)\operatorname{atan}\left(x_7\overline{\alpha_s}^{x_8}\left|\tilde{v}^*_{slip,i}\right|\right) log\left|\tilde{v}^*_{slip,i}\right| + x_9\overline{\alpha_s}^{x_{10}}\right) \quad \textbf{Equation 28}$$

$$k_3 = \left(\frac{2}{\pi}\right)^3 \operatorname{atan}(x_{11}\overline{\alpha_s}) \operatorname{atan}\left(x_{12} \max(x_4 - \overline{\alpha_s}, 0)\right) atan\left(x_{13}\mathrm{M}_{drift,i}\right)\left\{x_{14} + \left|\tilde{v}^*_{slip,i}\right|^{x_{15}}\left[\left(x_{16}\left|\overline{\alpha_s} - \frac{x_4}{2}\right|^{x_{17}} + x_{18}\right) + \min\left(sign\left(\tilde{v}^*_{slip,i}\mathrm{M}_{drift,i}\right), 0\right) x_{19}\left|\overline{\alpha_s} - \frac{x_4}{2}\right|^{x_{17}}\right]\right\} \quad \textbf{Equation 29}$$

$x_1 = -1.74\left(\frac{2}{\pi}\right) atan\left(1.71\Delta^*_{fil}\right)$, $x_2 = 1.54$, $x_3 = 21.8\left(\frac{2}{\pi}\right) atan\left(0.708\Delta^*_{fil}\right)$, $x_4 = 0.558$, $x_5 = 45.6{\Delta^*_{fil}}^{2.10}$, $x_6 = 0.308$, $x_7 = 6.35\Delta^*_{fil}$, $x_8 = 1.22$, $x_9 = 2.09\left(\frac{2}{\pi}\right) atan\left(0.331\Delta^*_{fil}\right)$, $x_{10} = 0.248$, $x_{11} = 4030$, $x_{12} = 194$, $x_{13} = 0.0742{\Delta^*_{fil}}^{1.89}$, $x_{14} = \frac{1}{3.53\left(\frac{2}{\pi}\right)atan\left(0.303\Delta^*_{fil}\right)}$, $x_{15} = \frac{1}{1+4.69\left(\frac{2}{\pi}\right)atan\left(0.0604\Delta^*_{fil}\right)}$, $x_{16} = 9.18$, $x_{17} = 2.84$, $x_{18} = \frac{1}{2.24\Delta^*_{fil}}$ and $x_{19} = \frac{1}{0.0742\left(\frac{2}{\pi}\right)atan\left(0.419\Delta^*_{fil}\right)}$.

Here the GPM, $\mathrm{M}_{drift,i}$, is defined as:

$$\mathrm{M}_{drift,i} = \frac{\left(\frac{d\overline{\alpha_g}}{dx}\frac{d\overline{v}_{g_i}}{dx} + \frac{d\overline{\alpha_g}}{dy}\frac{d\overline{v}_{g_i}}{dy}\right)}{\overline{\alpha_s}\,\overline{\alpha_g}\overline{v}_{ss\_slip}}\frac{v_t^4}{g^2} \quad \textbf{Equation 30}$$

$$\overline{u_g} = \tilde{u}_g - \bar{\alpha}_s v_d \quad \textbf{Equation 31}$$

The non-linearity correction factor is closed as follows:

$$C = 1 + \left\{\left[\operatorname{atan}(x_1\overline{\alpha_s}) + x_2 \operatorname{atan}\left(x_3\max(\overline{\alpha_s} - x_4, 0)\right)\right] x_5 + \left[\operatorname{atan}\left(x_6\left\|\widetilde{v_{slip}}^*\right\|^{x_7}\overline{\alpha_s}\right) + \operatorname{atan}\left(x_8\left\|\widetilde{v_{slip}}^*\right\|^{x_7}\max(\overline{\alpha_s} - x_4, 0)\right)\right] x_9\left\|\widetilde{v_{slip}}^*\right\|^{x_{10}}\right\} \quad \textbf{Equation 32}$$

$$\times\left(\frac{2}{\pi}\right)^2 \text{atan}\left(x_{11}\max(\overline{\alpha_s}-x_{12},0)\right)$$

$x_1 = 23.5, x_2 = -1.29, x_3 = 7.73 - 2.17\left(\frac{2}{\pi}\right)\text{atan}(0.194\Delta^*_{fil}), x_4 = 0.154,$

$x_5 = 2.04\left(\frac{2}{\pi}\right)\text{atan}(0.710\Delta^*_{fil}), x_6 = 135, x_7 = -0.451, x_8 = -43.6,$

$x_9 = -0.547\left(\frac{2}{\pi}\right)\text{atan}(0.825\Delta^*_{fil}), x_{10} = 0.390, x_{11} = 13.0$ and $x_{12} = 0.570.$

Finally, the dimensionless pressure gradient fluctuation contribution is closed as follows:

$$-\frac{\overline{\alpha_s'\nabla p'}}{\rho_g g} = k_1 + k_2$$ **Equation 33**

$$k_1 = x_1\left(\frac{2}{\pi}\right)\overline{\alpha_s}^{x_2}\, atan\left(x_3 max(\overline{\alpha_s}-x_4,0)^{x_5}\right)$$ **Equation 34**

$k_1 = 0$ if i is in a direction perpendicular to gravity

$$k_2 = x_1 x_6\left(\frac{2}{\pi}\right)^3 atan(x_7\overline{\alpha_s})\, atan\left(x_8\, max(x_4-\overline{\alpha_s},0)\right) atan\left(x_9\tilde{v}^*_{slip,i}\right)$$ **Equation 35**

$x_1 = 989\left(\frac{2}{\pi}\right)\text{atan}\left(0.660\Delta^*_{fil}\right), x_2 = 1.66, x_3 = 6.91, x_4 = 0.570, x_5 = 1.79,$

$x_6 = 0.879 - 0.805\left(\frac{2}{\pi}\right)\text{atan}\left(0.752\Delta^*_{fil}\right), x_7 = 2.71, x_8 = 2.64$ and $x_9 = 0.430.$

References

Agrawal, K., Holloway, W., Milioli, C.C., Milioli, F.E., Sundaresan, S., 2013. Filtered models for scalar transport in gas–particle flows. Chemical Engineering Science 95, 291-300.

Bi, H.T., Grace, J.R., 1995. Flow regime diagrams for gas-solid fluidization and upward transport. International Journal of Multiphase Flow 21, 1229-1236.

Cloete, J.H., Cloete, S., Municchi, F., Radl, S., Amini, S., 2017a. The sensitivity of filtered Two Fluid Model to the underlying resolved simulation setup. Powder Technology.

Cloete, J.H., Cloete, S., Radl, S., Amini, S., 2017b. Verification of filtered Two Fluid Models for reactive gas-solid flows, CFD 2017, Trondheim, Norway.

Cloete, J.H., Cloete, S., Radl, S., Municchi, F., Amini, S., 2017c. Marker selection for improving drag closures used in filtered Two Fluid Models. AIChE Journal Under review.

Huilin, L., Gidaspow, D., 2003. Hydrodynamics of binary fluidization in a riser: CFD simulation using two granular temperatures. Chemical Engineering Science 58, 3777-3792.

Igci, Y., Andrews, A.T., Sundaresan, S., Pannala, S., O'Brien, T., 2008. Filtered two-fluid models for fluidized gas-particle suspensions. AIChE Journal 54, 1431-1448.

Igci, Y., Sundaresan, S., 2011. Constitutive Models for Filtered Two-Fluid Models of Fluidized Gas–Particle Flows. Industrial & Engineering Chemistry Research 50, 13190-13201.

Loezos, P.N., Sundaresan, S., 2002. The role of meso-scale structures on dispersion in gas–particle flows, Circulating Fluidized Beds VII, pp. 427-434.

Milioli, C.C., Milioli, F.E., Holloway, W., Agrawal, K., Sundaresan, S., 2013. Filtered two-fluid models of fluidized gas-particle flows: New constitutive relations. AIChE Journal 59, 3265-3275.

Ozel, A., Fede, P., Simonin, O., 2013. Development of filtered Euler–Euler two-phase model for circulating fluidised bed: High resolution simulation, formulation and a priori analyses. International Journal of Multiphase Flow 55, 43-63.

Parmentier, J.-F., Simonin, O., Delsart, O., 2012. A functional subgrid drift velocity model for filtered drag prediction in dense fluidized bed. AIChE Journal 58, 1084-1098.

Sarkar, A., Milioli, F.E., Ozarkar, S., Li, T., Sun, X., Sundaresan, S., 2016. Filtered sub-grid constitutive models for fluidized gas-particle flows constructed from 3-D simulations. Chemical Engineering Science 152, 443-456.

Schneiderbauer, S., 2017. A spatially-averaged two-fluid model for dense large-scale gas-solid flows. AIChE Journal 63, 3544-3562.

Schneiderbauer, S., Pirker, S., 2014. Filtered and heterogeneity-based subgrid modifications for gas–solid drag and solid stresses in bubbling fluidized beds. AIChE Journal 60, 839-854.

491791 Particle Motion in Vertical Bladed Mixers

Abstract:

Vertical axis mixers are common particle-particle mixers in industry. Particle motion in vertical axis mixers is similar to other equipment utilized in industry, including high shear granulators in pharmaceutical applications and filter dryers in food processing. The motion of individual particles within industrial equipment is subject to competing mechanisms of mixing including diffusion, convection, and shear mixing. In spite of the importance of particle mixing in industry, understanding the mixing fundamentals is still challenging. For example, mixing processes can lead to particle segregation under certain operating conditions, and in some instances, mixing and segregation occur simultaneously but in different regions.

In order to study the mixing process it is imperative to study the motion of individual particles under different mixing mechanisms arising from blade movement. In this study, experiments were performed in a transparent 15.2 cm ID acrylic cylinder fitted with two 45° blades and filled with particles of varying diameter and density. Tagged particles were loaded in different positions and image analysis techniques were employed to follow the trajectories of the tagged particles. Observations were obtained for tagged particle initial location at the free surface as well as along the transparent walls. Particle Tracking Velocimetry (PTV) techniques were used to obtain the particle position, orientation, and velocity. The effects of particle size, density, initial location, and impeller speed on the trajectory and final location of the tagged particles are reported.

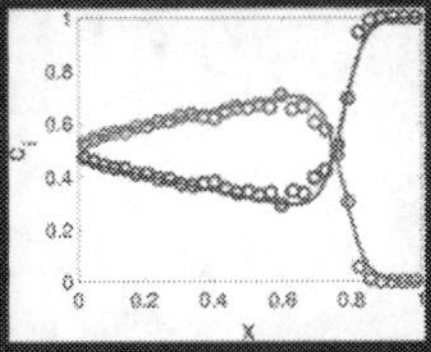

ESTABLISHING CONTINUUM MODEL SEGREGATION PARAMETERS FOR PRACTICAL PARTICLE MIXTURES

Alexander M. Fry[1], Vidyapati[2], John P. Hecht[2], Paul B. Umbanhowar[1], Julio M. Ottino[1], Richard M. Lueptow[1]

[1] Northwestern University
[2] Procter & Gamble, Global Engineering

AIChE Annual Meeting
November 2, 2017

P&G Northwestern | McCORMICK SCHOOL OF ENGINEERING

Segregation in gravity driven granular flow

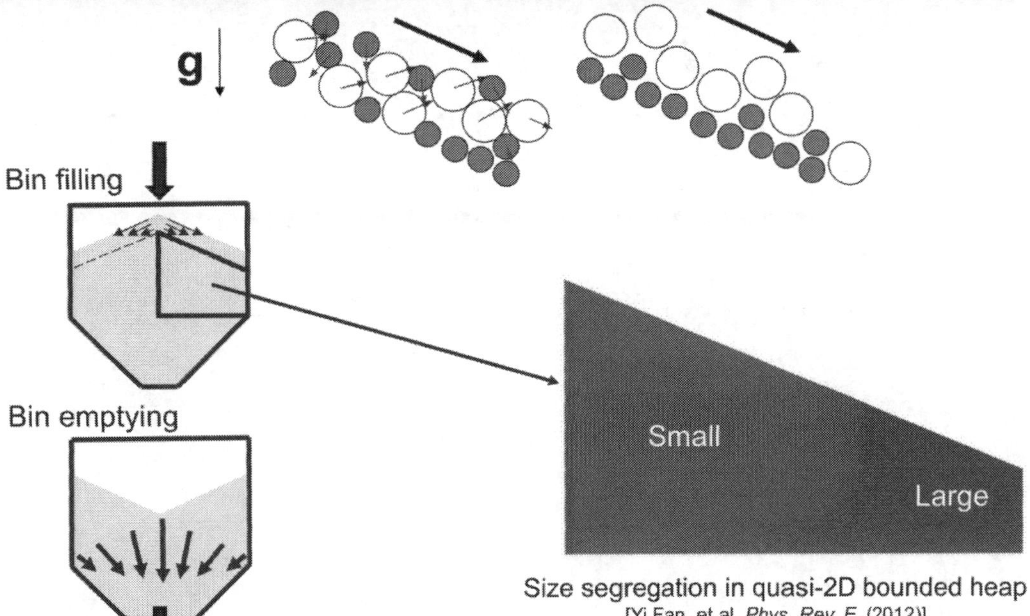

Size segregation in quasi-2D bounded heap
[Yi Fan, et al. *Phys. Rev. E.* (2012)]

Segregation-Advection-Diffusion Equation

$$\frac{\partial c_i}{\partial t} + \underbrace{\frac{\partial (uc_i)}{\partial x} + \frac{\partial (wc_i)}{\partial z}}_{\text{advection}} + \underbrace{\frac{\partial (w_{p,i}c_i)}{\partial z}}_{\text{segregation}} - \underbrace{\left[\frac{\partial}{\partial x}\left(D\frac{\partial c_i}{\partial x}\right) + \frac{\partial}{\partial z}\left(D\frac{\partial c_i}{\partial z}\right)\right]}_{\text{diffusion}} = 0$$

Transport equation for quasi-2D bounded heap

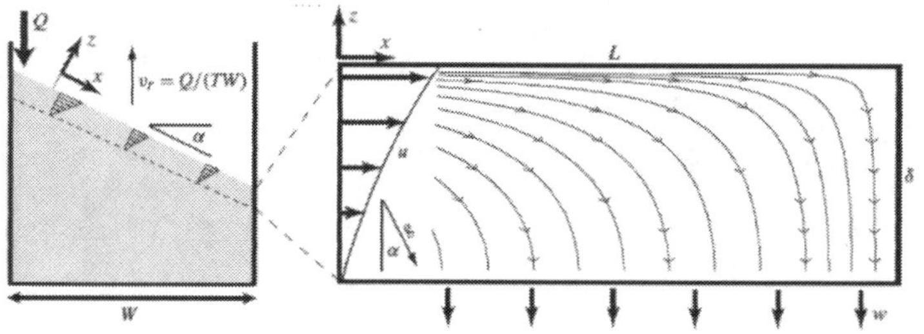

Quasi-2D bounded heap flow

Yi Fan, et al. *J. Fluid Mech.* (2014)

Segregation-Advection-Diffusion Equation

$$\frac{\partial c_i}{\partial t} + \underbrace{\frac{\partial (uc_i)}{\partial x} + \frac{\partial (wc_i)}{\partial z}}_{\text{advection}} + \underbrace{\frac{\partial (w_{p,i}c_i)}{\partial z}}_{\text{segregation}} - \underbrace{\left[\frac{\partial}{\partial x}\left(D\frac{\partial c_i}{\partial x}\right) + \frac{\partial}{\partial z}\left(D\frac{\partial c_i}{\partial z}\right)\right]}_{\text{diffusion}} = 0$$

Transport equation for quasi-2D bounded heap

Velocity profile

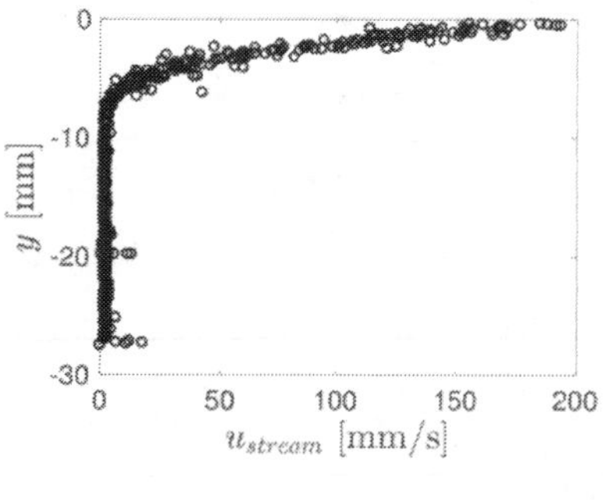

Diffusion coefficient

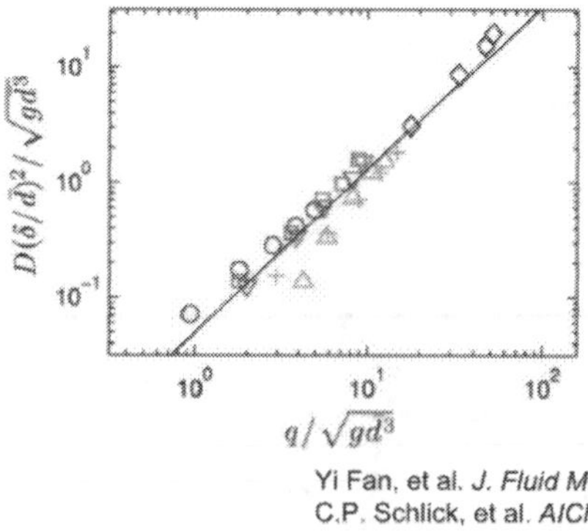

Yi Fan, et al. *J. Fluid Mech.* (2014)
C.P. Schlick, et al. *AIChE J.* (2015)

Segregation-Advection-Diffusion Equation

$$\frac{\partial c_i}{\partial t} + \underbrace{\frac{\partial (u c_i)}{\partial x} + \frac{\partial (w c_i)}{\partial z}}_{\text{advection}} + \underbrace{\frac{\partial (w_{p,i} c_i)}{\partial z}}_{\text{segregation}} - \underbrace{\left[\frac{\partial}{\partial x}\left(D\frac{\partial c_i}{\partial x}\right) + \frac{\partial}{\partial z}\left(D\frac{\partial c_i}{\partial z}\right)\right]}_{\text{diffusion}} = 0$$

Transport equation for quasi-2D bounded heap

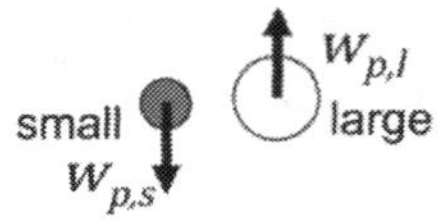

$$w_{p,i} = S\dot{\gamma}(1 - c_i)$$

Yi Fan, et al. *J. Fluid Mech.* (2014)
Based on Savage & Lun, *J. Fluid Mech.* (1988)

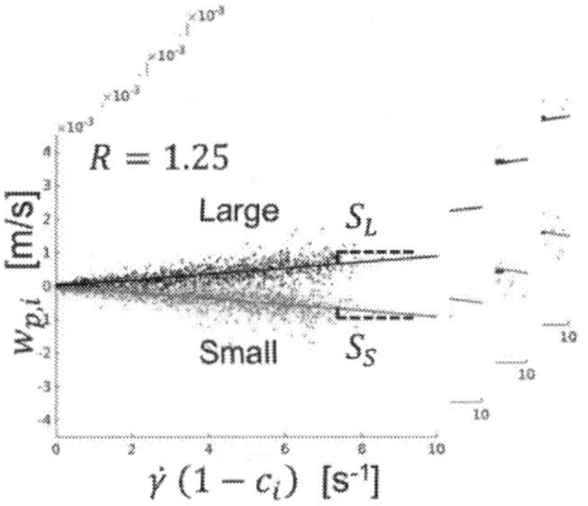

Yi Fan, et al. *J. Fluid Mech.* (2014)

Segregation-Advection-Diffusion Equation

$$\frac{\partial c_i}{\partial t} + \underbrace{\frac{\partial (u c_i)}{\partial x} + \frac{\partial (w c_i)}{\partial z}}_{\text{advection}} + \underbrace{\frac{\partial (w_{p,i} c_i)}{\partial z}}_{\text{segregation}} - \underbrace{\left[\frac{\partial}{\partial x}\left(D\frac{\partial c_i}{\partial x}\right) + \frac{\partial}{\partial z}\left(D\frac{\partial c_i}{\partial z}\right)\right]}_{\text{diffusion}} = 0$$

Transport equation for quasi-2D bounded heap

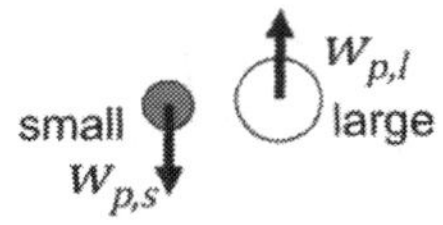

$$w_{p,i} = S\dot{\gamma}(1 - c_i)$$

Yi Fan, et al. *J. Fluid Mech.* (2014)
Based on Savage & Lun, *J. Fluid Mech.* (1988)

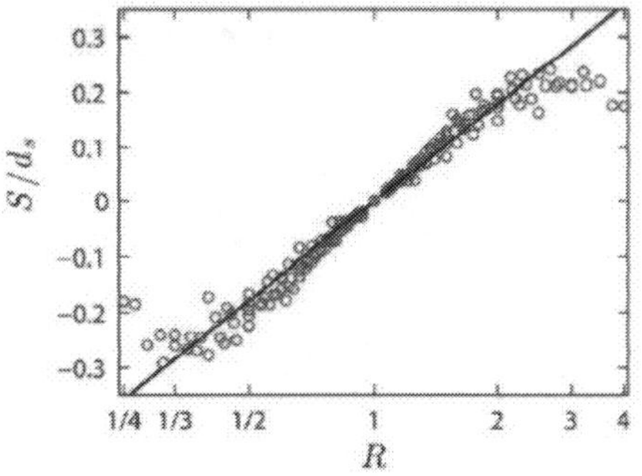

Yi Fan, et al. *J. Fluid Mech.* (2014)
C.P. Schlick et al. *AIChE J.* (2015)

Two methods to generate segregation model parameters

- Local method

DEM simulations

↓

Parameter measurement (S, D, δ)

↓

$$\frac{\partial c_i}{\partial t} + \frac{\partial (u c_i)}{\partial x} + \frac{\partial (w c_i)}{\partial z} + S\frac{\partial}{\partial z}[\dot{\gamma}(1-c_i)c_i] - \left[\frac{\partial}{\partial x}\left(D\frac{\partial c_i}{\partial x}\right) + \frac{\partial}{\partial z}\left(D\frac{\partial c_i}{\partial z}\right)\right] = 0$$

↓

Confirm via small scale experiment

- Global method

Large scale process simulation

↑

Parameter estimation (S, D, δ)

↑

$$\frac{\partial c_i}{\partial t} + \frac{\partial (u c_i)}{\partial x} + \frac{\partial (w c_i)}{\partial z} + S\frac{\partial}{\partial z}[\dot{\gamma}(1-c_i)c_i] - \left[\frac{\partial}{\partial x}\left(D\frac{\partial c_i}{\partial x}\right) + \frac{\partial}{\partial z}\left(D\frac{\partial c_i}{\partial z}\right)\right] = 0$$

↑

Small scale experiment

But, in industrial granular flows the segregation parameters are unknown.

Measure steady state concentration profiles

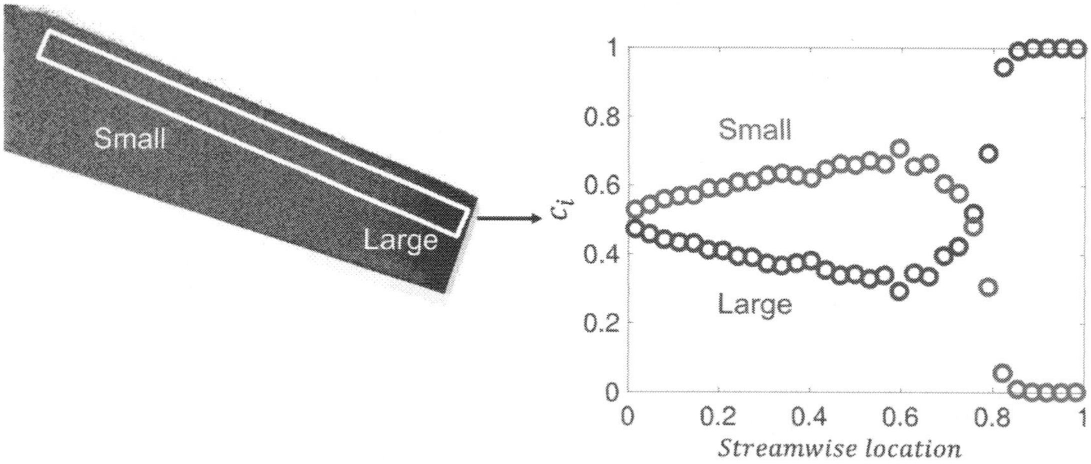

Parameter estimation method

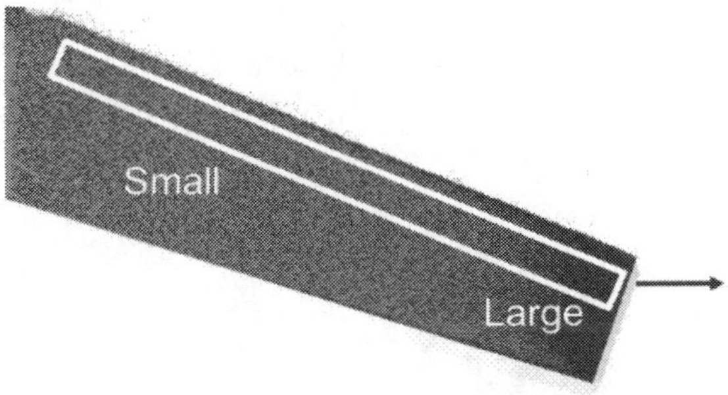

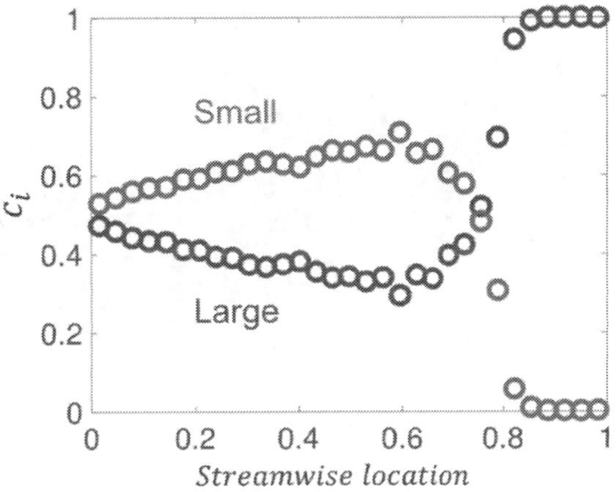

- Minimize error between model and experimental data by adjusting S

$$\frac{\partial c_i}{\partial t} + \frac{\partial (uc_i)}{\partial x} + \frac{\partial (wc_i)}{\partial z} + S\frac{\partial}{\partial z}[\dot{\gamma}(1-c_i)c_i] - \left[\frac{\partial}{\partial x}\left(D\frac{\partial c_i}{\partial x}\right) + \frac{\partial}{\partial z}\left(D\frac{\partial c_i}{\partial z}\right)\right] = 0$$

Parameter estimation method

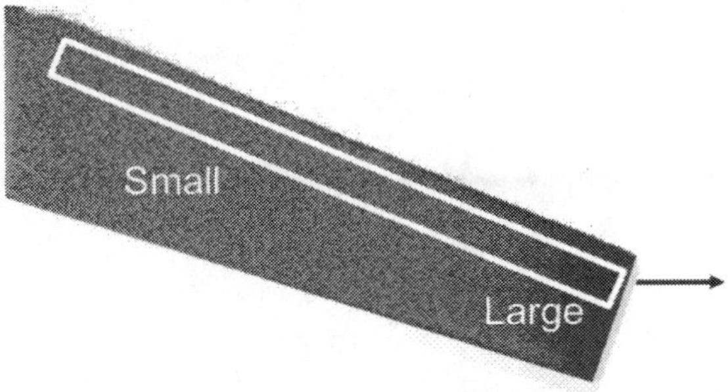

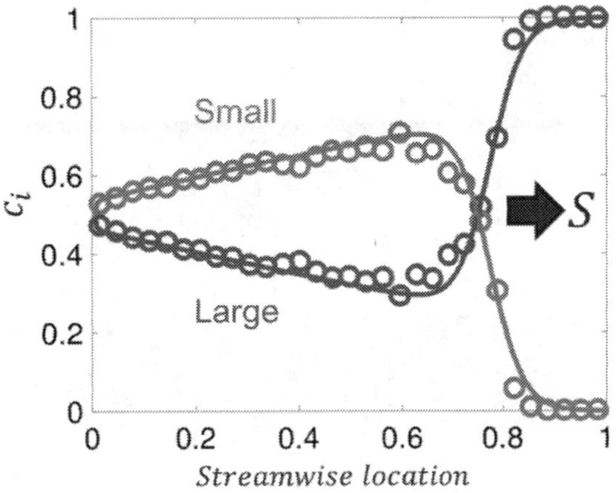

- Minimize error between model and experimental data by adjusting S

$$\frac{\partial c_i}{\partial t} + \frac{\partial (uc_i)}{\partial x} + \frac{\partial (wc_i)}{\partial z} + S\frac{\partial}{\partial z}[\dot{\gamma}(1-c_i)c_i] - \left[\frac{\partial}{\partial x}\left(D\frac{\partial c_i}{\partial x}\right) + \frac{\partial}{\partial z}\left(D\frac{\partial c_i}{\partial z}\right)\right] = 0$$

Validate parameter estimation method with DEM data

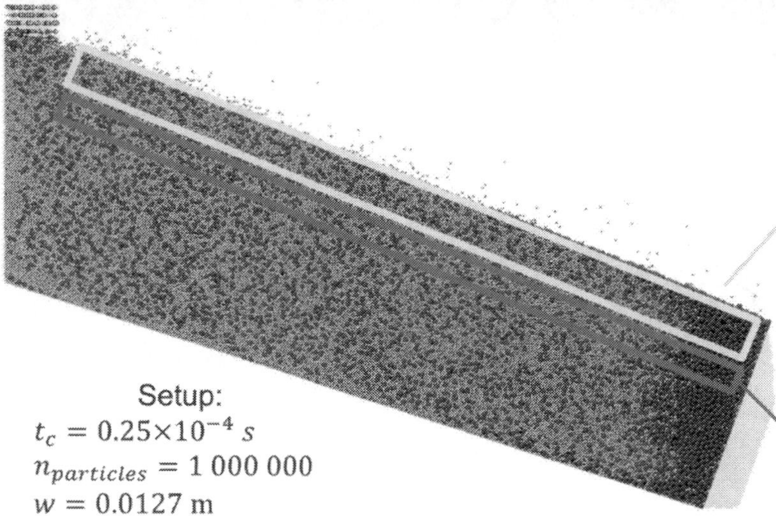

Local: Measure S using kinematic data from DEM

Global: Back-calculate S using concentration field from DEM

Setup:
$t_c = 0.25 \times 10^{-4}\ s$
$n_{particles} = 1\ 000\ 000$
$w = 0.0127$ m
$l = 0.62$ m
$R = 1 - 2$
$\bar{d} \approx 1.5$ mm
$\rho = 1450\ kg/m^3$

Validation routine

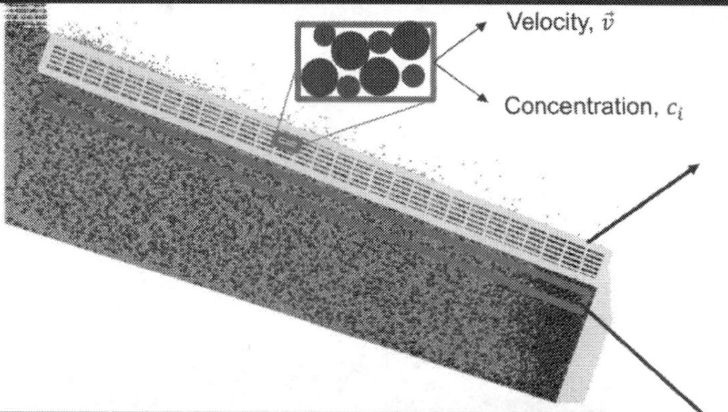

Local: Measure S

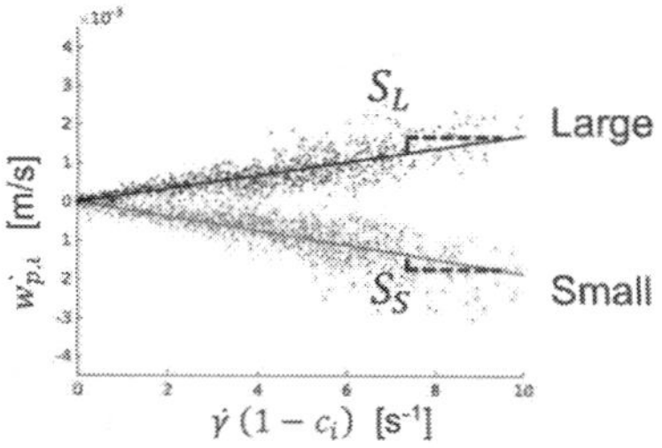

Global: Fit S

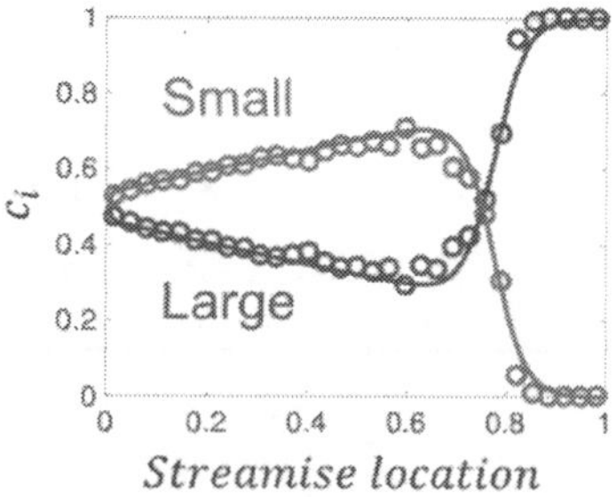

Size Ratio R	Local S [mm]	Global S [mm]	Relative error
1.25	0.091	0.095	5%
1.5	0.144	0.124	13%
1.75	0.171	0.148	13%
2	0.1975	0.1965	<1%

Experimental setup

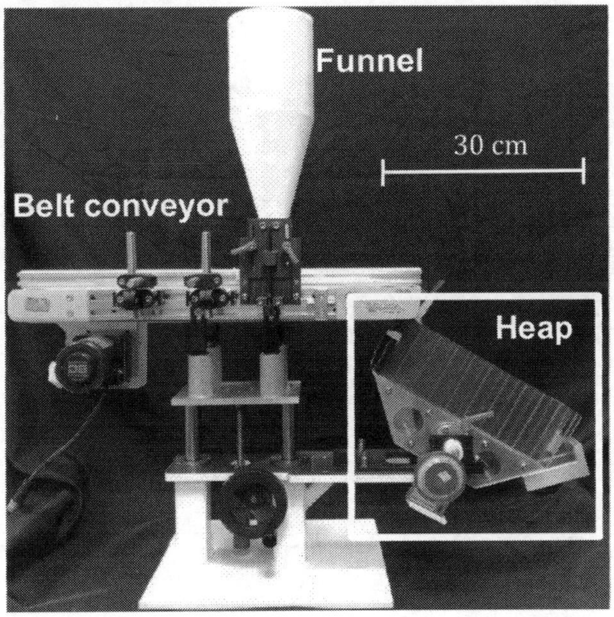

a) Fill heap at known feed rate q

b) Insert three-piece sample cards

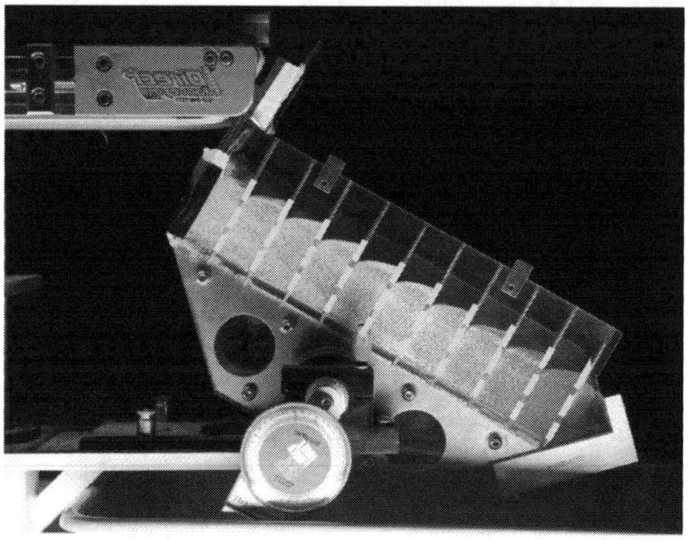

c) Collect middle sample (steady-state)

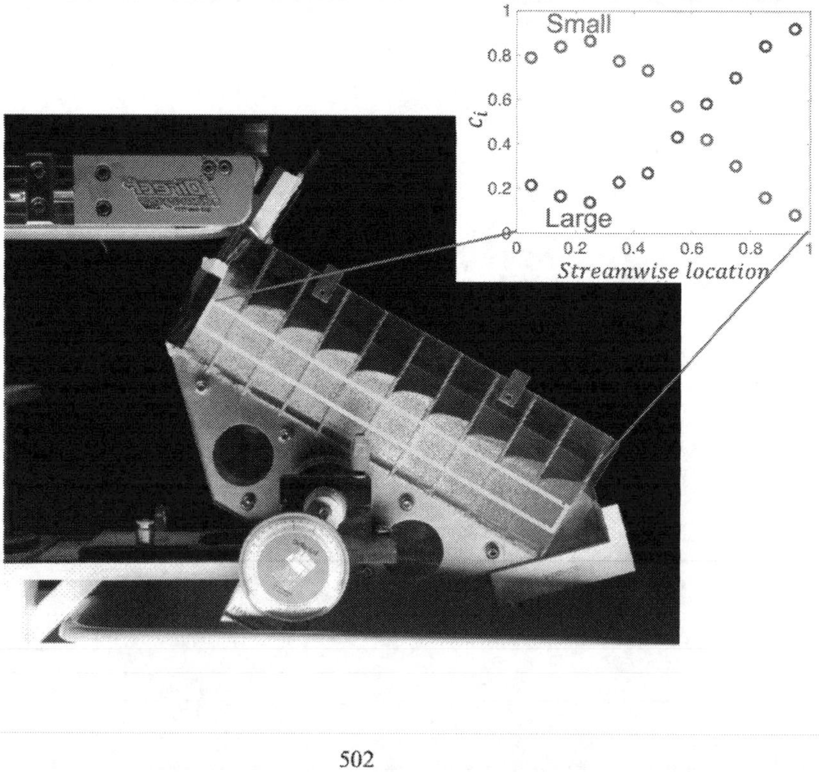

Initial study using Cellets

- Various size ratios $1.4 < R < 5.5$
- Species mean particle diameters $150\ \mu m$ to $1200\ \mu m$

$d \approx 280\ \mu m$

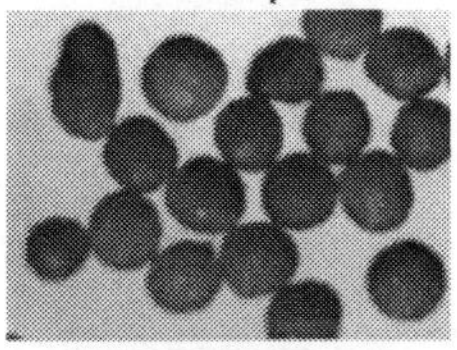

$d \approx 425\ \mu m$

Cellets particles

$R = 1.5$

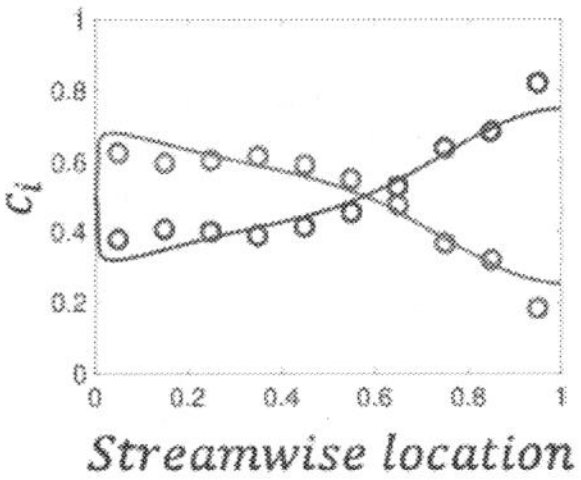

Best fit

$S = 0.055$ mm

Initial study using Cellets

- Various size ratios $1.4 < R < 5.5$
- Species mean particle diameters $150\ \mu m$ to $1200\ \mu m$

$d \approx 280\ \mu m$

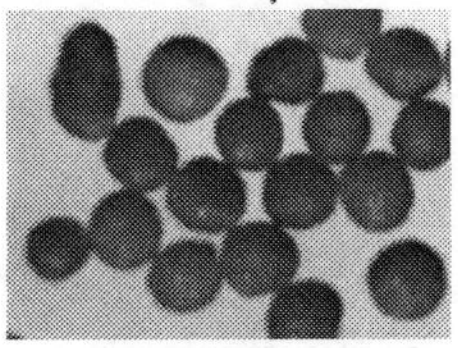

$d \approx 825\ \mu m$

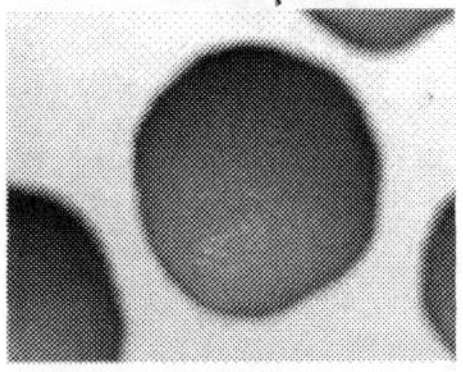

Cellets particles

$R = 2.9$

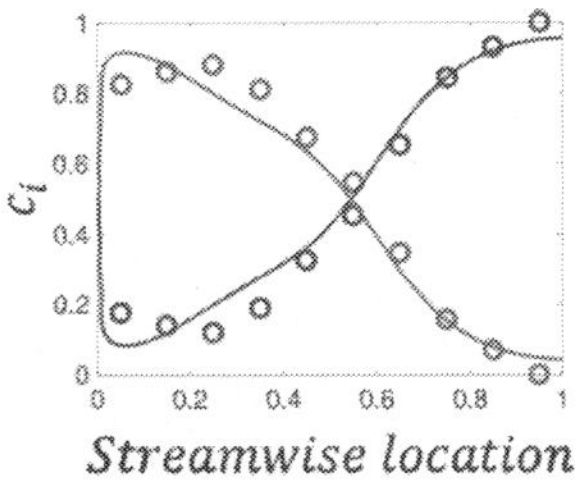

Best fit

$S = 0.11$ mm

Initial study using Cellets

- Various size ratios $1.4 < R < 5.5$
- Species mean particle diameters $150\ \mu m$ to $1200\ \mu m$

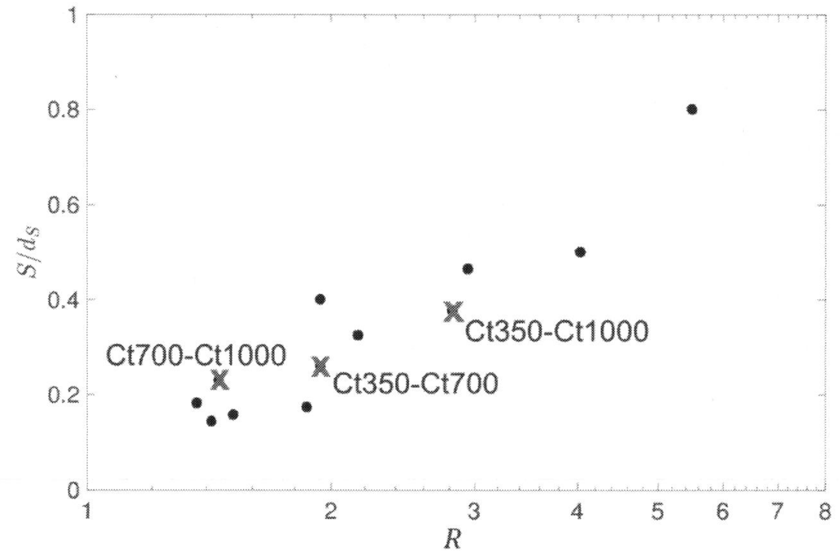

Will bidisperse parameters predict tridisperse segregation?

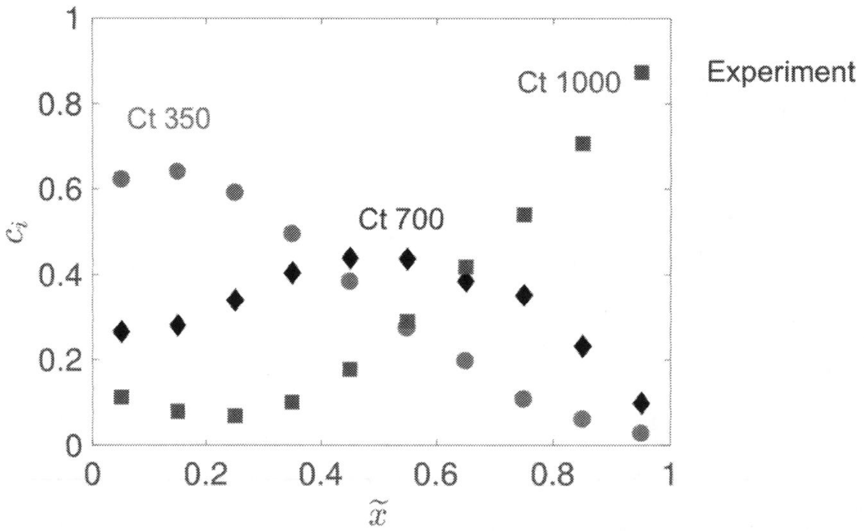

Will bidisperse parameters predict tridisperse segregation?

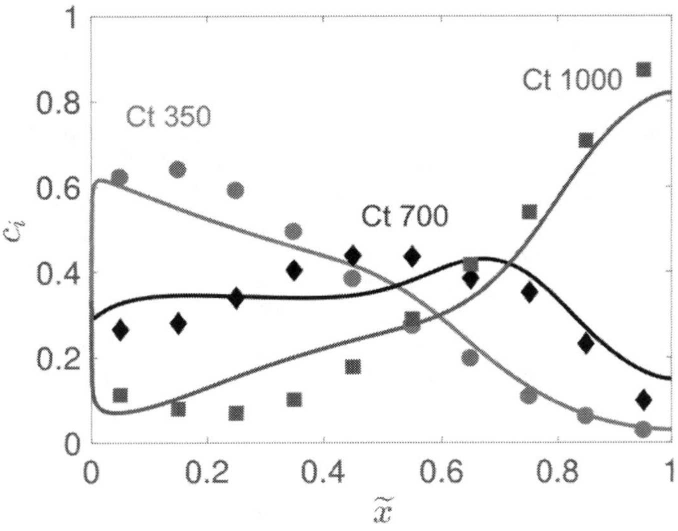

Experiment & Model prediction

Summary & Next Steps

- Segregation parameter from global measurement
- Works across broad size range ($1.4 < R < 5.5$)
- Tested in tridisperse case
- NEXT: Scale-up, incorporate in bin filling/emptying code

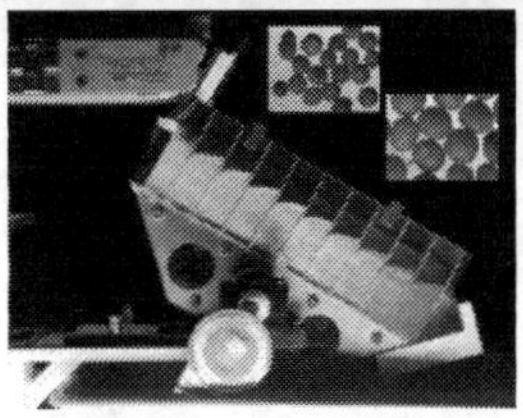

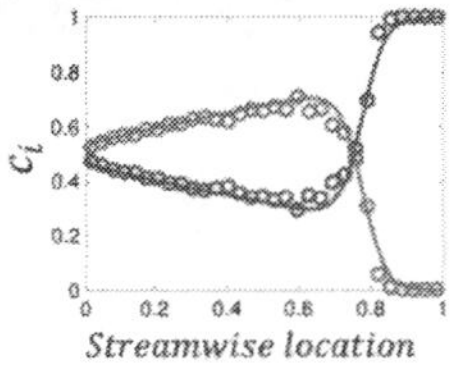

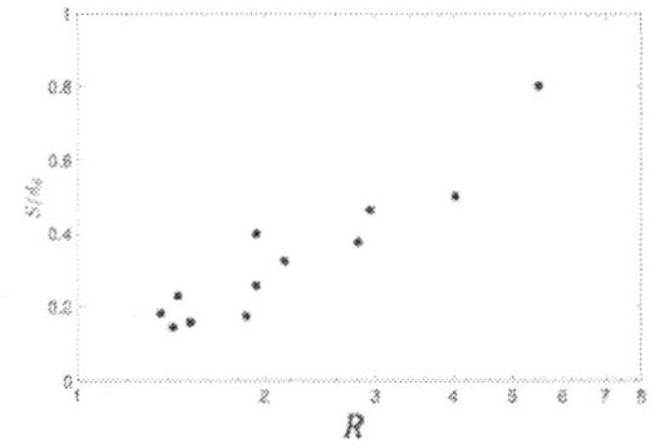

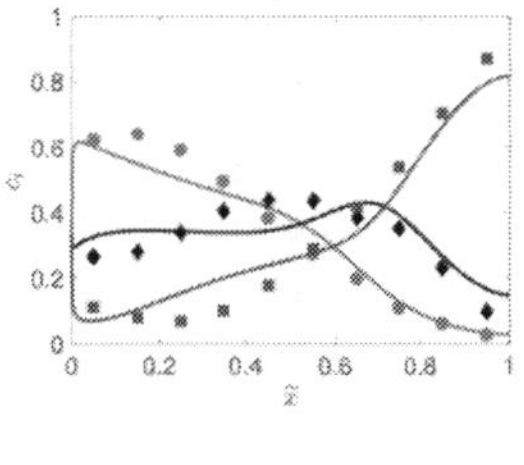

492740 Segregation of Spherical and Non-Spherical Particles in DEM Simulations

Abstract:

Â Most studies of segregation in flowing granular materials use spherical particles to evaluate the effects of particle size and density ratios, as well as particle surface roughness. However, in practical applications particles are generally non-spherical. Here we describe a discrete element method (DEM) simulation approach for spherical and non-spherical particles which are described mathematically as super-ellipsoids and, for contact resolution purposes, treated as potential particles in that the function that describes the particle also is used to determine if a point is on the interior, exterior, or surface of the particle. The contact resolution algorithm for non-spherical particles is more complex than for spherical particles. Screening methods limit the computationally expensive contact resolution algorithm to contacting particles, and the simulation is validated through comparison with previous results for â€œtrue geometryâ€ models of cylindrical particles. We simulate bidisperse mixtures of non-spherical particles in quasi-2D bounded heaps and measure the segregation velocity as a function of local shear rate, species concentration, and shape difference. Shape-driven segregation velocities are compared to size-driven segregation velocities, which may eventually allow for shape differences for non-spherical particle mixtures to be mapped to equivalent size differences in spherical particle mixtures.

Partially funded by the Dow Chemical Company.

495261 Experimental Investigation of Residence Time Distribution of Free-Flowing Particles in a Lab-Scale Cylindrical Silo

Abstract:

The effect of hopper half-angle and wall condition on residence time distribution (RTD) of 2 mm free flowing particles (mustard seed) in an 80 mm cylindrical silo has been investigated. Two wall conditions, rough and smooth, and hopper half-angles of 20, 24, 30, 45, 60; and 90° (flat bottom) were used; the discharge opening diameter was 14 mm in all experiments. Tracer experiments were carried out using samples of the test material dyed to be optically distinct; red and blue dyes were used as these had negligible impact of the flow properties of the particles. In each experiment, material exiting the silo was collected in the compartments of a rotary table sampler turning at a constant rate. In all experiments the relative proportions, concentrations, of red and blue particles were determined from photographs of all the particles in each sample. Classical analysis of the response to a step change, here a change in the colour of the test material in the silo, was then used to extract RTD from the relative proportions of red and blue particles in the silo discharge; the timeline was established from the angular speed of the rotary sampler. The effect of increased wall friction was significant for hoppers with a small half-angle; for the larger half-angles residence time distribution changed only minimally when the silo walls were roughened.

493105 Effect of Material Properties on the Mass Hold up Dynamics and Residence Time Distribution in Continuous Powder Blenders

Abstract:

Pharmaceutical research development has focused in characterizing blender performance using a formulation based approach. This approach consists on studying blender performance for a specific blend (e.g., mixture of excipients, lubricant, and API) and a desired process flow rate [1-4]. Although efficient for the early stages of continuous manufacturing, this approach does not provide process developers with sufficient information to ensure process performance on instances where process throughput and/or formulation fractions – due to dosing changes – vary from the ones measured during characterization. The formulation based approach, although effective for the specific product used for development, does not provide a vast amount of information that can be used to design new processes for new formulations. For these reasons, we performed an experiment to develop a correlation between incoming material properties and the mixing behavior in a continuous tubular blender. As a result, we developed a model to predict blender performance metrics such as mass hold up and mixing behavior using process parameters and material properties as model inputs.

The new paradigm focuses on characterizing the blender using pharmaceutically relevant materials that span a large range of material properties. The objective is not to develop a specific blend of material, but to understand how do blender process outputs (e.g., mass hold up, residence time, and flow rate out) are affected when material properties change. To establish the largest range of material properties, we evaluated our in-house material property database containing twenty-five (25) different materials using a statistical method for orthogonal transformation, known as Principal Component Analysis. Out the twenty-five (25) possible materials available to us, eight (8) were found to have a unique set of material properties. All 8 materials were run individually through a GEA continuous convective blender. Each material was fed individually to reduce the effects of changing material properties due to mixing. The effect of impeller speed and total mass flow rate were investigated along with the effect of material properties. The study was performed using a 3x2x8 experimental design for the blade speed, mass flow rate, and materials, respectively; yielding a total of 48 experiments. In each experiment, the mass flow rate out of blender was dynamically monitored over time by placing a weight scale at the blender outlet. Using the information from the feeders coupled with the blender and the outgoing flow rate, the mass hold up was computed dynamically. Mass hold up transitions between impeller speeds and mass flow rates were also investigated at all conditions for all eight materials. From the results provided, we developed a model to calculate mass hold up dynamically as a function of material properties.

Besides the mass hold up inside of the blender, we also investigated the effect of material properties on the mixing behavior of the unit. Residence time distribution (RTD) has been shown to be a powerful tool for understanding and characterizing the mixing behavior inside of continuous powder blenders [5-7]. RTD is defined as the distribution of time a material stays inside a unit operation in a continuous flow system [8]. In the context of continuous blenders, RTD enables the understanding of material flow profiles inside the blender (i.e., mixing and dispersion), which can be used to determine the blender's ability to filter incoming noise from unit operations upstream. RTDs are also an integral piece in the development of control strategies for a specific unit operation and operations in series. The RTD profile of the blender was investigated for the eight materials by using a pulse experiments. In each pulse experiment, a known amount of tracer was introduced at the process's steady state while monitoring the tracer's concentration at the blender output over time. The effects of impeller speed, blade configuration, and mass flowrate on the RTD of continuous blenders have been examined for a variety of convective blending systems [7, 9]. Nevertheless, the role of material properties on RTD parameters, remains poorly understood.

The mixing behavior in the GEA blender was characterized by using a modified version of the Tank-in-Series RTD model. The modification of the Tanks-in-Series model included a delay time associated with the transport of powder inside of the convective blender. For each of the eight (8) material, a pulse of tracer material was introduced at each of the 3x2 experimental conditions – blade speed and flow rate – and the concentration of the tracer was measured dynamically at the blender's outlet. For each of the tracer concentration profiles, time-constant parameters for the delay Tanks-in-Series RTD model were regressed. The RTD model constants fit were the number of tanks, mean residence time, and delay time. A correlation was developed between the model parameters, material properties, process parameters, and operating conditions. Each of those regressed constants were then correlated with the material properties, leading to a two (2) level model that allows to predict the residence time distribution of material in the system based on its properties. This material property correlation and empirical modeling strategies can enable savings in time and material in future process and product development, as well as improve our understanding of how material properties affect residence time in continuous tubular blenders. Tracer properties and its effect on RTD measurements are also briefly discussed.

References

1. Berthiaux, H., K. Marikh, and C. Gatumel, *Continuous mixing of powder mixtures with pharmaceutical process constraints.* Chemical Engineering and Processing, 2008. **47**(12): p. 2315-2322.

2. Bridgwater, J., *Mixing of powders and granular materials by mechanical means—A perspective.* Particuology, 2012. **10**(4): p. 397-427.

3. Koller, D.M., et al., *Continuous quantitative monitoring of powder mixing dynamics by near-infrared spectroscopy.* Powder Technology, 2011. **205**(1-3): p. 87-96.

4. Portillo, P.M., Ierapetritou, M.G., Muzzio, F.J., *Effects of rotation rate, mixing angle, and cohesion in two continuous mixers—A statistical approachwder.* Powder Technology, 2009. **194**: p. 217-227.

5. Danckwerts, P.V., *Continuous flow systems: Distribution of residence times.* Chemical Engineering Science, 1953. **2**(1): p. 1-13.

6. Florian-Algarin, M. and R. Méndez, *Blend uniformity and powder phenomena inside the continuous tumble mixer using DEM simulations.* AIChE Journal, 2015. **61**(3): p. 792-801.

7. Vanarase, A.U., J.G. Osorio, and F.J. Muzzio, *Effects of powder flow properties and shear environment on the performance of continuous mixing of pharmaceutical powders.* Powder Technology, 2013. **246**: p. 63-72.

8. Gao, Y., et al., *Characterizing continuous powder mixing using residence time distribution.* Chemical Engineering Science, 2011. **66**(3): p. 417-425.

9. Vanarase, A.U. and F.J. Muzzio, *Effect of operating conditions and design parameters in a continuous powder mixer.* Powder Technology, 2011. **208**(1): p. 26-36.

497702 Solids Mixing Studies in Fluidized Beds Using Fluorescent Tracer Technique

Abstract:

A fluorescent solid tracer technique has been used to study various properties such as solids mixing time, solids residence distribution time and the effects of internals on mixing in fluidized beds. Studying these parameters is of great importance in controlling the back-mixing of solids or verifying that solids segregate in a fluidized bed. The technique may also be used to compare the flow patterns of different solids, understand the hydrodynamics associated with different internals at different operating conditions, and to optimize the size and the shape of a reactor. In these sets of experiments, FCC catalyst was coated with a suitable UV dye in a 12-in (30-cm)-diameter spray column using a suitable solvent to dissolve the dye. The dyed catalyst was then injected into a 6-in (15-cm)- diameter Plexiglas column fluidized at superficial gas velocities of 1 ft/s (0.31 m/s) and 0.25 ft/s (0.08 m/s) to analyze mixing of FCC equilibrium catalyst. Fluorescent probes, which were inserted at different locations in the bed, captured the mixing phenomena of the catalyst in the bed. A photodiode converted an optical signal to an electrical signal and the data were recorded using a data acquisition system and was analyzed to study the mixing behavior of particulate solids in a small scale fluidized bed.

487547 Continuum Modeling of Segregation for Polydisperse Granular Materials in Bounded Heap Flow

Abstract:

Mixing and segregation of polydisperse granular materials remain challenging problems in many industrial processes. However, most studies have focused either on bidisperse materials, which are not representative of most real mixtures, or on polydisperse materials in an idealized periodic chute geometry using discrete element method (DEM) simulations. In this research, we develop and implement a continuum-based model that captures the effects of segregation, diffusion, and advection on polydisperse granular flow in a quasi-two-dimensional bounded heap. Two dimensionless parameters (obtained from experiments or DEM simulations) characterize the model: the Peclet number, *Pe*, which is defined as the ratio of the advection rate to the diffusion rate, and Î›, which is defined as the ratio of the segregation rate to the advection rate. As *Pe* increases with fixed Î›, advection dominates segregation and diffusion. As Î›Â increases with fixed *Pe*, segregation dominates diffusion and advection. Unlike other segregation models that require fitting parameters for each comparison between model and DEM simulations, the parameters in our model can be obtained from just one or two additional DEM simulations. Quantitative agreement of continuum model results with DEM simulations over a wide range of flow conditions and particle distributions supports our modeling approach.
Funded by NSF Grant # CBET-1511450.

488587 "Core–Shell" Nanostructured Supported Size-Selective Catalysts Prepared By Molecular Layer Deposition

Abstract:

Heterogeneous catalysts are widely used due to many advantages, including the ease of separating the catalysts by filtration after reaction. However, heterogeneous catalysts usually cannot selectively convert specific molecules in a reactant mixture to only desired products. Size-selective catalysis is an important concept for improving the selectivity of heterogeneous catalysts. In most previous studies, unsupported and nonporous substrates supported metallic nanoparticles were encapsulated in relatively thick porous structures to realize the size-selectivity of heterogeneous catalysts by the size discrimination effect of the porous shell. However, the catalytic activity of the catalysts greatly decreased due to mass transfer limitations brought by relatively thick shells and the contact areas between the porous shells and active sites. In our study, we deposited an ultra-thin porous oxide shell on high surface area substrate-supported Pt nanoparticle catalysts, using molecular layer deposition (MLD), followed by oxidation to remove the organic components in hybrid organic/inorganic MLD films. The encapsulated catalyst showed very high selectivity in catalyzing hydrogenation reactions of n-hexene over cis-cyclooctene. However, there was a decrease of the catalytic activity due to contact areas between the porous shell and metallic nanoparticles. To reduce the contact areas, we introduced gaps between the Pt nanoparticles and porous shells using a sacrifical layer of self-assembled monolayers (SAMs) on the Pt nanoparticles, before the deposition of a MLD layer. This novel nanostructured catalyst showed much higher activity, as compared to the catalyst that had been directly coated with the MLD layer.

500149 Novel ALD-Formed Cobalt/Alumina Nanostructures Active for Fischer-Tropsch Synthesis

Abstract:

Cobalt metal was deposited onto alumina (Al_2O_3) supports using atomic layer deposition (ALD) to create dispersed catalysts for Fischer-Tropsch Synthesis (FTS). We show that the phase of the Al_2O_3 support greatly influences the FTS activity of the resulting catalyst. ALD literature predicts that Co deposited on Al_2O_3 with low cycle numbers should only produce Co nanoparticles less than 2 nm in diameter, which fall into the well-characterized non-active region of nanoparticle sizes for FTS. Our FTS experiments confirm that this inactivity is indeed the case for Co deposited on amorphous Al_2O_3. However, Co deposited on calcined Î³-Al_2O_3 is in fact active, suggesting an alternative ALD deposition regime not previously recognized. Aberration corrected STEM/EELS imaging shows that in addition to nanoparticle deposition, cobalt has also deposited on Î³-Al_2O_3 as semi-continuous planar layers likely less than 3 monolayers thick, which must be responsible for the catalystâ€™s FTS activity. Further, it was found that the number of ALD cycles greatly alters the activity of these catalysts, with 4 and 8 cycle catalysts being much more active than the 1 cycle catalyst. These data suggest Co requires the presence of crystalline Al_2O_3 surfaces in order to achieve a thin semi-continuous planar growth regime. Finally, density functional theory (DFT) calculations were carried out to understand the thermodynamics of Co small particle vs. planar layer (monolayer and bilayer) growth on Al_2O_3 and explain the activity shown by these catalysts.

490326 Controlled Production of Nanostructured Noble-Metal Catalysts Using Atomic Layer Deposition

Abstract:

Controlled production of nanostructured noble-metal catalysts using atomic layer deposition

J. Ruud van Ommen, Fabio Grillo, Hao Van Bui, Jacob A Moulijn, Michiel T Kreutzer

Already an established technique for the growth of conformal thin films, atomic layer deposition (ALD) is also an attractive technique for the controlled growth of nanoparticles (NPs). As such, it can play an important role in the synthesis of nanostructured catalysts. Compared to conventional synthesis routes, ALD boasts a number of advantages: atomic-level control over the amount of material being deposited, conformality, scalability to high-surface-area substrates, and its being solvent-free. Yet, bringing the unparalleled precision of ALD of thin films to ALD of NPs is not straightforward. The precision of ALD relies, in fact, on cyclic repetitions of self-saturating surface reactions that lead to the deposition of less than a monolayer per cycle. Hence, if the growth proceeds in a layer-by-layer fashion, as is the case with ALD of thin films, varying the number of cycles translates into tuning the film thickness with atomic-level precision. However, if the as-deposited atoms form into NPs upon deposition, a growth per cycle of less than a monolayer does not necessarily enable atomic-level control over the NP size. This is because the NP morphology (e.g., size, shape and number density) is dictated by atomistic processes other than "ALD reactions" such as surface diffusion and aggregation of atoms and NPs, and atom attachment to and from NPs. Understanding the role of such kinetic processes during the ALD of NPs is therefore crucial to the development of ALD routes for the synthesis of NPs with a well-defined morphology and thus functionality.

Here, we present an atomistic model of thermal ALD of Pt and Pd NPs on oxides nanoparticles and graphene nanoplatelets. In particular, we study the effect of temperature, number of cycles, coreactant partial pressure and exposure time on the evolution of the NP size distribution and metal loading. Atomistic modelling of our experiments shows that the NPs grow mostly due to Smoluchowski aggregation, that is, NP diffusion and coalescence, rather than through single atom processes such as precursor chemisorption, single atom diffusion and attachment, and Ostwald ripening [1]. While the metal loading can be precisely controlled over a wide range of temperatures, atomic-level precision over the NP size is retained only at low deposition temperatures ($T \leq 100$ °C) when growth by atom attachment/deposition becomes the dominant process. Furthermore, we show that the coreactant partial pressure and exposure time are far more important parameters for the NP size than the number of cycles. We will demonstrate the favourable effect of optimizing the synthesis conditions on the catalytic performance of the obtained material.

[1] Grillo, F., et al. "Understanding and Controlling the Aggregative Growth of Platinum Nanoparticles in Atomic Layer Deposition: An Avenue to Size Selection." The Journal of Physical Chemistry Letters 8.5 (2017): 975.

492271 Controlled Deposition of Metal Oxide Layers By Non-Hydrolytic Sol-Gel for Tailored Acid Sites Generation and Beyond

Abstract:

Bulk materials are omnipresent in catalysis due to their ease of production, but they often suffer from low selectivity and stability due to the lack of precise design of their surface functionalities. We present a new technique for the controlled deposition of metal oxides layers, where precise tuning of material surface functionalities yields optimal performances for various applications.

The method is based on non-hydrolytic sol-gel chemistry using mixtures of metal alkoxides and metal halides. For instance, addition of a 1/3 mol/mol mixture of $Ti(iPrO)_4$ and $TiCl_4$ to a toluene suspension of Cu_2O spheres led to the formation of a thin conformal TiO_2 overcoat. Thickness can be adjusted by the number of deposited layers with a growth rate of 0.5 nm/cycle. Furthermore, this method allows deposition of TiO_2 layers inside the pores of SBA-15 while maintaining 2D hexagonal structure and high porosity. The mixed oxide created at the interface generates medium strength Lewis acid sites which catalyze 1-phenylethanol dehydration to styrene. High Lewis acid sites density with medium acid strength was obtained with 6 TiO_2 layers which lead to the highest yields of styrene and the highest catalyst stability. In contrast, a zeolitic catalyst often used for this reaction (HZSM-5) deactivates due to polystyrene formation on strong Lewis acid sites. The technique has also been employed to overcoat supported metal particle yielding metal-acid bifunctional catalysts, where overcoat porosity can be adjusted by ageing in basic media. Finally, an analogous method was developed for deposition of alumina, yielding strong acid sites.

The easy implementation and tunable porosity of the present technique contrasts with the heavy instrumentation and limited structural control associated with gas phase methods such as atomic layer deposition or chemical vapor deposition on powder substrates. We anticipate that the method reported can be used for the rational design of catalytic materials used in a wide range of applications.

498798 Enhanced Hydrothermal Stability of Phosphonate-Coated Al_2O_3 Supported Catalysts

Abstract:

Gamma phase alumina (γ-Al_2O_3) is a common catalyst support because its high abundance, mechanical and chemical stability, and high surface area, however, γ-Al_2O_3 is unstable in the presence of high temperature/high pressure steam used in aqueous phase reforming. Under these conditions, the support undergoes a phase transition to boehmite (AlO(OH)) accompanied with a significant loss in the support's surface area as the pore structure deteriorates. Consequently, catalytic performance is greatly diminished as active metal clusters become less accessible or detach from support.

To prevent catalyst degradation, hydrophobic coatings can be deposited onto the Al_2O
increased hydrothermal stability. Specifically, we chose to coat catalysts with phosphonic acids (PAs) since these capping agents selectively bind to the oxide support and form robust bonds with the surface. Additionally, the physical properties of catalysts can be further tuned by changing the phosphonate's tail group.

In this work, we discuss recent results that have shown the deposition of alkyl phosphonic acids on alumina and supported catalysts drastically improves their hydrothermal stability. We examine how the alkyl tail length (or hydrophobicity) affects the suppression of boehmite formation and improves the retention of both metal and support surface areas. The beneficial roles of both phosphonate head and tail groups will be assessed as well as their impact on catalytic performance.

498271 Catalytic Activity of Porphyrin-Supported Iron Oxide Clusters for Methane Oxidation

Abstract:

During the last decade, natural gas production in the United States has significantly increased every year, and it is predicted to increase even more rapidly during the coming years. This has provided new incentives for selectively transforming natural gas components and their derivatives into more valuable chemicals through energy efficient and environmentally friendly processes. However, the current route for converting methane into more useful chemicals is through a complex, multi-step industrial process that requires multiple reactors, separations, and high temperatures and pressures[1]. For this reason, there is a great interest in designing catalysts that accomplish this transformation directly while avoiding deeper oxidation to carbon dioxide.

Methane monooxygenase (MMO), an enzyme found in biological systems, is capable of oxidizing methane to methanol using O2 at mild conditions. MMO has two different forms: (1) particulate MMO (pMMO), which is postulated to have a dicopper-oxo active site, and (2) soluble MMO (sMMO), which contains a diiron site.[2] The identification of different iron and copper sites in these enzymes has inspired the synthesis of catalysts that mimic their activity towards the partial oxidation of methane.[3]

Using density functional theory, we have studied the growth and catalytic activity of iron oxide nanoclusters that can be grown via atomic layer deposition (ALD) on a porphyrinic substrate and mimic the structure of the active site in the enzyme for the oxidation of methane to methanol. The porphyrin-supported iron oxide active site consists of an $Fe^{IV}_2(\mu\text{-}O)_2(OH)_2$ diamond core supported by a bridging $Al^{III}(\mu\text{-}O)_2(OH)$ moiety. After a series of ALD cycles to deposit the Al atom and the Fe^{II} sites on the porphyrin substrate, the bridging oxygen atoms that initiate the catalytic cycle can be incorporated via O_2 or N_2O activation. Preliminary results show that N_2O activation is more favorable, but may lead to the formation of sites that contain only one bridging oxygen and two Fe^{III} sites. Two mechanisms for the oxidation of methane using the $Fe^{IV}_2(\mu\text{-}O)_2(OH)_2$ active site and N_2O as the oxidant are proposed: (1) a concerted mechanism in which the C-H bond is cleaved heterolytically and (2) a rebound mechanism in which a bridging oxygen or a terminal oxo group abstracts a hydrogen atom from methane. Our preliminary computational results show that the rebound mechanisms are more likely to occur due to lower activation barriers of about 70 kJ/mol. We also find activation barriers between 100-120 kJ/mol for N_2O activation and CH_3OH desorption. To further understand the effect of cluster composition on the activity for the various oxygen sites in the di-iron cluster, we replaced the aluminum atom and one of the iron atoms with other transition metals, calculated the main reaction barriers, and performed population analysis studies. Using these computational tools, we aim to understand the ways in which we can tune the activity of the proposed catalysts for the selective oxidation of methane.

References

1. Crabtree, R. H. *Chem. Rev.* 95, pp 987–1007 (1995).

2. Lippard, S. J. and Feig, A. L. *Chem. Rev.* 94, pp 759–805 (1994).

3. Wulfers, M. J., Teketel, S., Ipek, B., and Lobo, R. F. *Chem. Commun.* 51, pp 4447-4450 (2015).

FACULTY OF ENGINEERING AND ARCHITECTURE

AICHE ANNUAL MEETING 2017 – MINNEAPOLIS. NOVEMBER 2 - 2017.

INVESTIGATION OF PARTICLE CLUSTER MORPHOLOGY IN A DOWNFLOW REACTIVE SYSTEM VIA LARGE EDDY SIMULATIONS

Noel Gómez, Laurien A. Vandewalle, Pieter A. Reyniers, David J. Van Cauwenberge, Alejandro Molina, Rodney O. Fox, Kevin M. Van Geem, Guy B. Marin.

TO MOVE IN GROUPS IS ALWAYS EASIER…

For **birds…**

… lift is increased, and both wingbeat and heart rate are reduced to a half compared with a lonely and sad bird. *[Nature magazine]*

For **cyclists…**

… energy savings up to 40% because drag is reduced *[specialists]* … and you get huge motivation from your colleagues *[a friend]*

GHENT UNIVERSITY LCT UNIVERSIDAD NACIONAL DE COLOMBIA SEDE MEDELLÍN

… BUT IT IS NOT ALWAYS GOOD

Clusters are groups or agglomerations of particles that increase mass transfer limitations

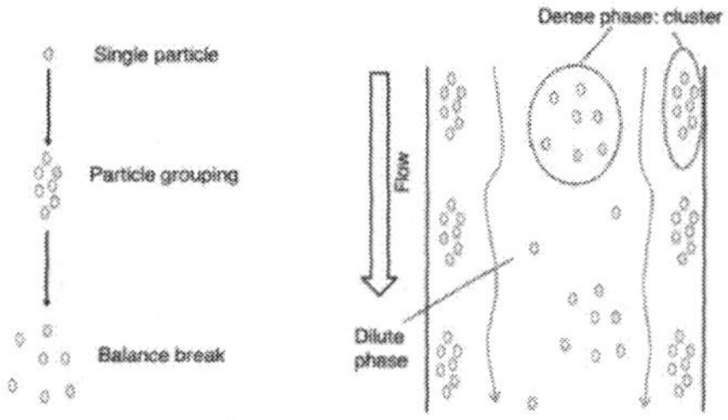

Graphical representation of clusters in a downer

- Clusters cause up to 15% reduction in the rates of reaction. Capecelatro et al., 2015.
- Most of experimental studies about clusters were carried out under cold flow conditions.
- No experimental data available about clusters under reactive flow conditions

The study of **Clusters**, particularly, under reactive conditions is challenging

GHENT UNIVERSITY LCT UNIVERSIDAD NACIONAL DE COLOMBIA SEDE MEDELLÍN

J. Capecelatro, et al *Chem. Eng. Sci.*, 2015.

INVESTIGATION ON PARTICLE CLUSTERING IN DOWNERS

- Determine size and spatial distribution of clusters in a downer reactor during the process of fluid catalytic cracking via large eddy simulations.

GHENT UNIVERSITY LCT UNIVERSIDAD NACIONAL DE COLOMBIA SEDE MEDELLÍN

5

OUTLINE

- Fluid catalytic cracking
- Simulations
  - Reactive flow validation
    - LES-DEM simulation
    - Comparison: 1D vs LES-DEM
  - Validation of the methodology to detect clusters
  - Analysis of cluster characteristics in the reactive flow downer
  - Modification of the 1D model
- Conclusions

GHENT UNIVERSITY LCT UNIVERSIDAD NACIONAL DE COLOMBIA SEDE MEDELLÍN

6

FLUID CATALYTIC CRACKING (FCC) OF GASOIL AND DOWNER REACTORS

FCC is a process for producing high-market-value products with low molecular weight (gasoline, propylene, and ethylene) from heavy H-C molecules.

Downers are used as an alternative to risers in gas-solid processes, such as Fluid Catalytic Cracking (FCC), because they provide a better particle distribution that prevents back-mixing and under-or over-cracking.

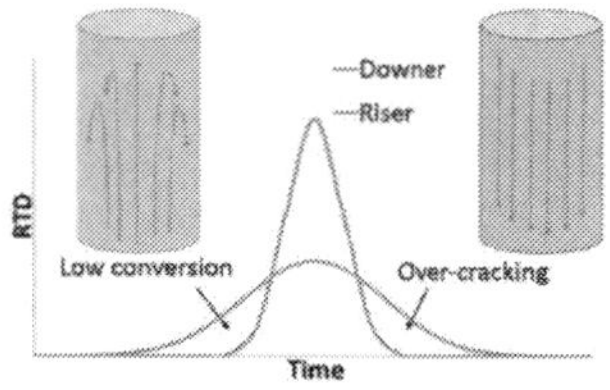

Residence time distribution comparison for downers and risers. Adapted from Lambert et al., 2014.

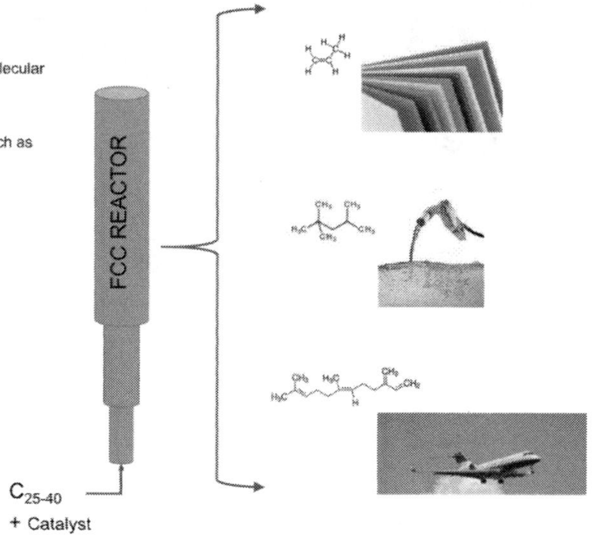

7

SIMULATIONS

A one-meter cylindrical downer was chosen as canonical geometry to carry out the simulations.

- Experimental data available
  - A. Corma, C., et al. *Appl. Catal. A Gen.* 2002.
  - S. Krol, et al.," *Powder Technol.*, Mar. 2000.

Four-way coupled simulations

- Smagorinsky subgrid scale model for the gas phase
- EMMS drag model
- DEM for particle-particle collisions
- Four-lump kinetic model

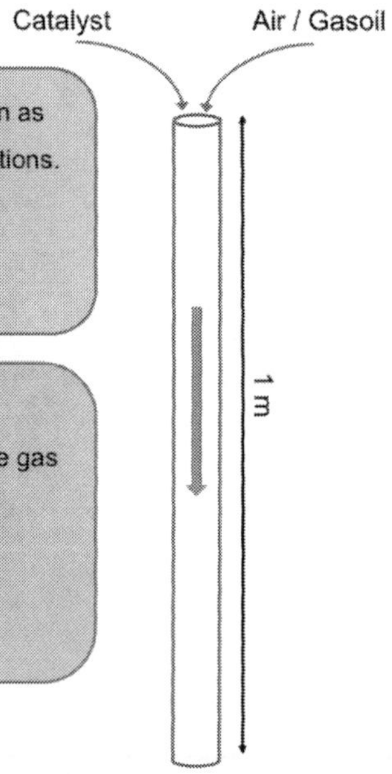

Mesh

Parameter	Value
Length	1 m
Diameter	0.0133 m
Grid	~7 MM cells
y+	~0.5

Gas

Parameter	Value
Inlet velocity	1 m/s
Inlet pressure	1 bar
Inlet temperature	823 K
Time-step	2e-5 s

Catalyst

Parameter	Value
Mean particle diameter	83 μm
Particle density	1525 kg/m^3
Particle loading	~5 MM
Mean particle volume fraction	0.01

8

VALIDATION: REACTIVE-FLOW DOWNER

The four-way coupled DEM-LES was validated by comparing outlet mass fractions with experimental data from Corma et al., 2002

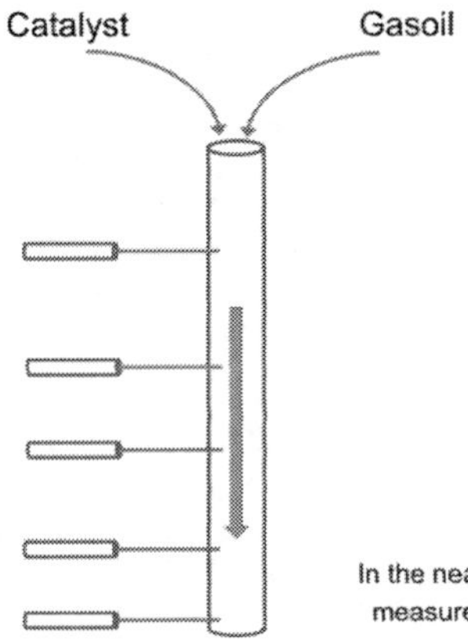

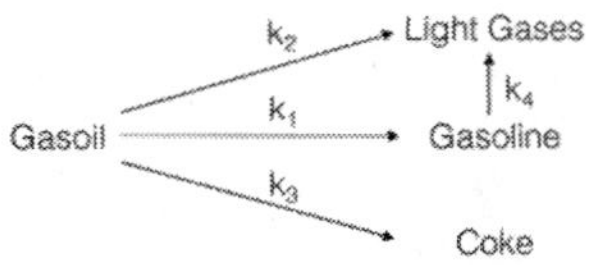

Simulation results were in good agreement with experimental data at the outlet

In the near future, we will be able to do in-situ measurements of the chemistry using laser techniques.
Oral session: 279d

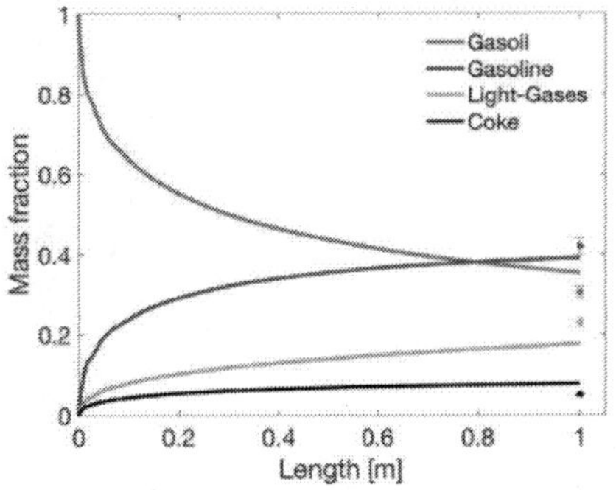

Symbols: Experimental data from MicroDowner developed by Corma et al.
Catalyst to Oil ratio (w/w): 10. Temperature of reaction: 823 K

A. Corma, C., et al. *Appl. Catal. A Gen.* 2002. 9

COMPARISON WITH 1D SIMULATIONS

Several 1D simulations were carried out to predict the performance of the downer reactor under the same operating conditions and the kinetic model that were employed in the 3D LES-DEM.

The 1D model includes:

- Mass and momentum balances for the gas phase.
- Momentum balance for the particle phase (2-way coupled)

Advantage of 1D

Simulation	Wall-clock time
1D simulation (steady state)	0.15 s
3D LES-DEM (2 s of flow time)*	~ 1.5 months

*Simulation was carried out using 112 cores in parallel in the Flemish supercomputer Tier-1 (Belgium).

Drawback of 1D

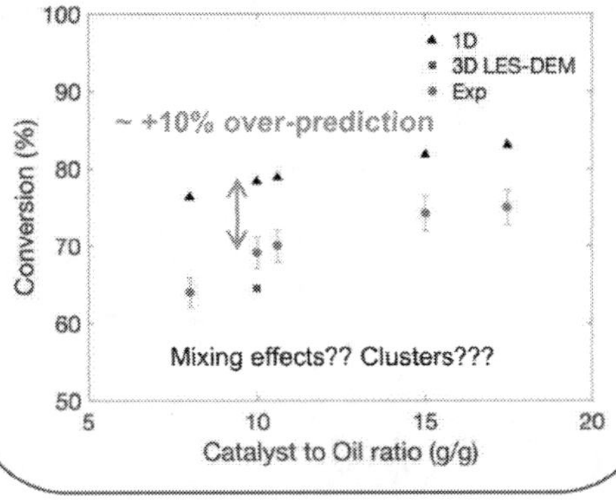

10

VALIDATION: METHODOLOGY FOR CLUSTER DETECTION

Validation was carried by identifying clusters in the bottom of the cold-flow downer (last 0.1 m) using the location of particles and the distance between them, and comparing the results (size of clusters) with experimental data from Krol et al., 2000.

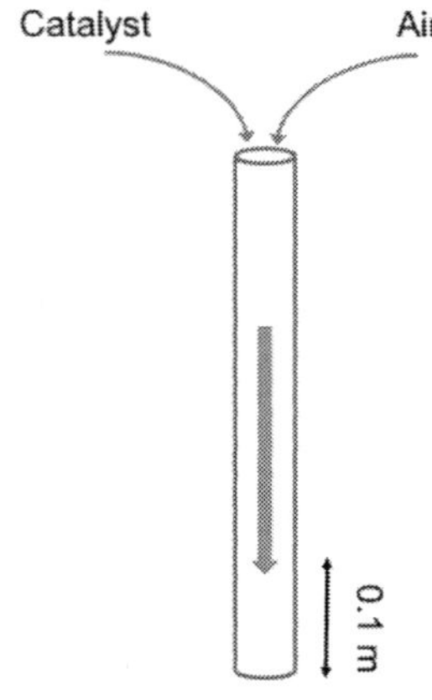

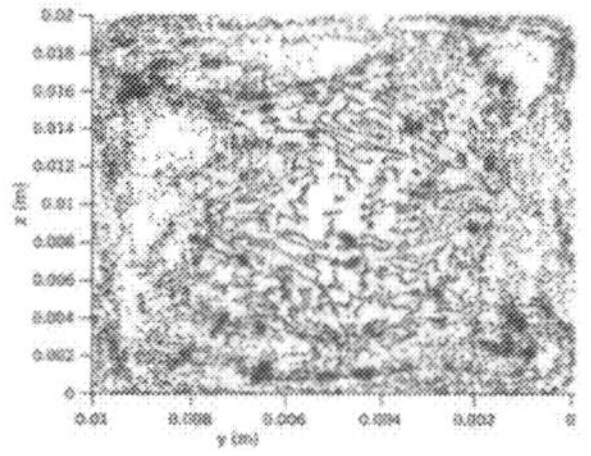

The maximum distance between 2 particles in a clusters is controlled by the particle volume fraction

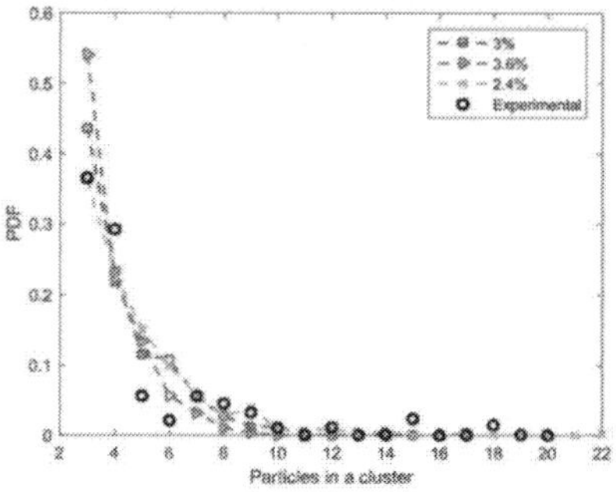

- The results were validated when using a threshold of 2.4% for particle volume fraction.
- Agglomerations of particles with 2.4% of particle volume fraction or above were considered clusters.
- Most of clusters are formed by 3 particles.

S. Krol, et al.," *Powder Technol.*, Mar. 2000.

GHENT UNIVERSITY LCT UNIVERSIDAD NACIONAL DE COLOMBIA SEDE MEDELLÍN

11

REACTIVE FLOW DOWNER: CLUSTER SIZE DISTRIBUTION

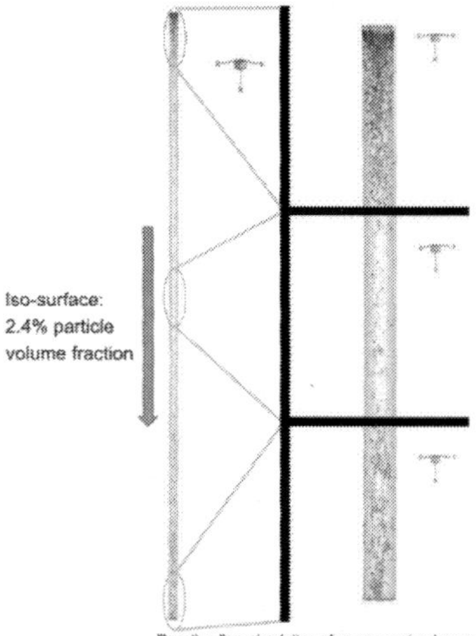

Reactive flow simulation of an one-meter downer

Qualitatively, clusters tend to be smaller at the entrance zone of the reactor; however, particles group each other towards the bottom of the downer and create larger clusters

The tendency was confirmed by using the methodology for identifying clusters.

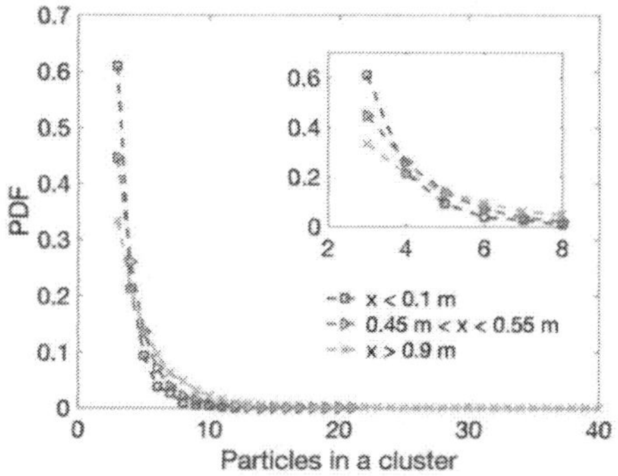

GHENT UNIVERSITY LCT UNIVERSIDAD NACIONAL DE COLOMBIA SEDE MEDELLÍN

12

CLUSTERS CHARACTERISTICS: FREQUENCY (f_c) AND MEAN LIFETIME (θ_c)

The identification of clusters allowed the calculation of other characteristics such as frequency and mean lifetime.

The frequency, defined as the number of clusters that appear in a certain zone per second, decreases along the reactor, forming fewer but larger clusters.

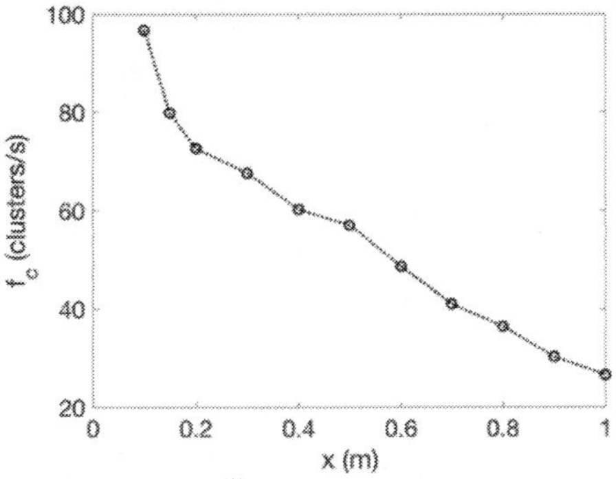

The mean lifetime, which is the average time that a cluster remains with a particle volume fraction above the threshold, increases. This means long-lasting clusters.

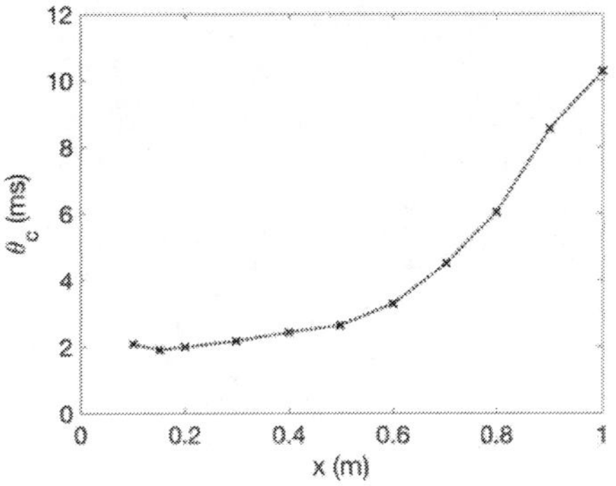

GHENT UNIVERSITY LCT UNIVERSIDAD NACIONAL DE COLOMBIA SEDE MEDELLÍN

13

CLUSTERS CHARACTERISTICS: PREDICTION

- Two empirical expressions to predict frequency and mean lifetime of clusters were proposed.
- The parameters a, b and c were optimized using 100 points obtained from the 3D DEM-LES

$$\zeta = j \cdot Re_p^a \cdot St^b + c$$

Frequency and mean lifetime of clusters respond to the hydrodynamic conditions of the system. Such conditions can be described using dimensionless numbers such as:

$$Re_p = \frac{\varepsilon_g \rho_g d_p |\tilde{u}_g - v_p|}{\mu_g}$$

$$St = \frac{\rho_p d_p^2 \tilde{u}_g}{18 \mu_g d_h}$$

Frequency (f_c)

Parameter	Value
a	0.79 ± 0.04
b	2.09 ± 0.06
c	14.44 ± 3.89
j	1

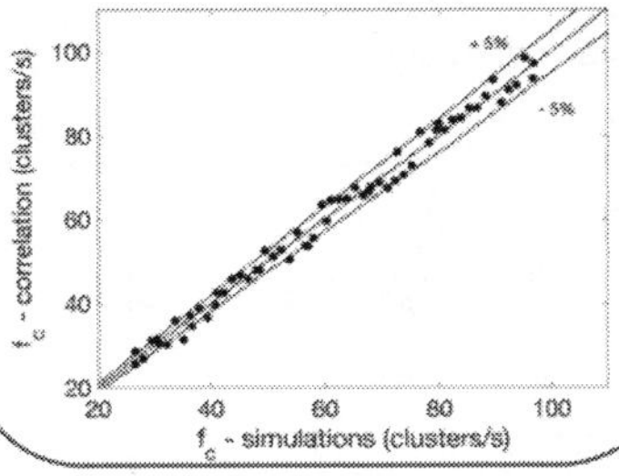

Mean lifetime (θ_c)

Parameter	Value
a	0.36 ± 0.004
b	1.75 ± 0.03
c	26.01 ± 1.09
j	-1

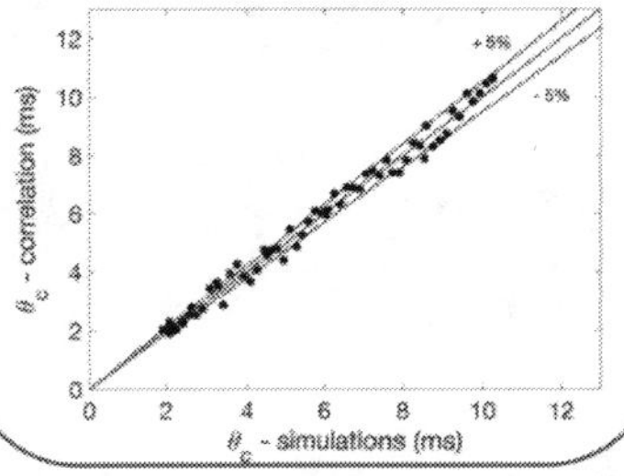

GHENT UNIVERSITY LCT UNIVERSIDAD NACIONAL DE COLOMBIA SEDE MEDELLÍN

14

IMPLEMENTATION OF THE EMPIRICAL CORRELATIONS

The mass balances in the 1D model were modified by including the terms of frequency and mean lifetime of clusters. Such terms multiply the rates of reaction (r_j), "punishing" them due to the presence of clusters.

$$r_j^* = (1 - f_c \cdot \theta_c) \cdot r_j$$

- $f_c \cdot \theta_c$: Dimensionless factor that represents the fraction of time during which clusters exist.
- r_j: Rate of reaction j
- r_j^*: Modified rate of reaction j

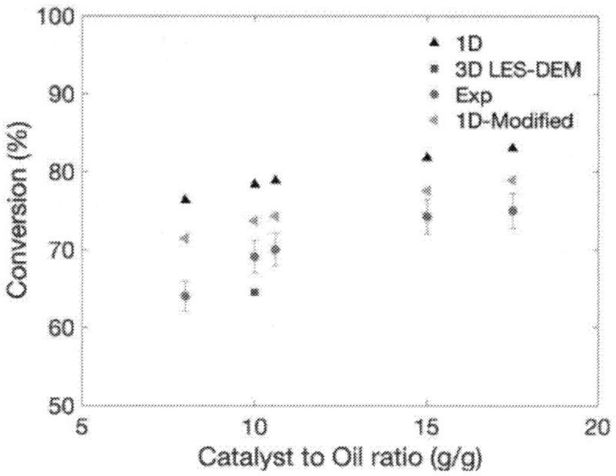

- Predictions of the 1D model improved by including the modification on the rates of reaction.
- Next step: Model will be enhanced by including the cluster-size distribution

GHENT UNIVERSITY LCT UNIVERSIDAD NACIONAL DE COLOMBIA SEDE MEDELLÍN

15

CONCLUSIONS AND FINAL REMARKS

- The particle clustering phenomenon was successfully predicted and characterized in different zones in downer reactors using LES.
- 1D simulations over-predicted conversion of gasoil by approximately 10% at different CTO ratios, presumably, due to the presence of cluster.
- DEM-LES was more accurate when predicting gasoil conversion; however, the prediction was lower than experimental data.
- Most of clusters that were identified have 3 particles; however, the probability of finding larger clusters at the bottom of the downer is higher.
- Two empirical correlations were proposed to predict frequency and mean lifetime of clusters in a downer reactor during the catalytic cracking of gasoil.
- Better predictions of gasoil conversion were achieved when using the correlations in the 1D approach.

GHENT UNIVERSITY LCT UNIVERSIDAD NACIONAL DE COLOMBIA SEDE MEDELLÍN

16

ACKNOWLEDGEMENTS

17

BIBLIOGRAPHY

[1] J. Capecelatro, P. Pepiot, and O. Desjardins, "Numerical investigation and modeling of reacting gas-solid flows in the presence of clusters," *Chem. Eng. Sci.*, vol. 122, pp. 403–415, 2015.

[3] K. Tuzla, a. K. Sharma, J. C. Chen, T. Schiewe, K. E. Wirth, and O. Molerus, "Transient dynamics of solid concentration in downer fluidized bed," *Powder Technol.*, vol. 100, no. 2–3, pp. 166–172, 1998.

[4] D. J. Van Cauwenberge, C. M. Schietekat, J. Floré, K. M. Van Geem, and G. B. Marin, "CFD-based design of 3D pyrolysis reactors : RANS vs. LES," *Chem. Eng. J.*, vol. 282, pp. 66–76, 2015.

[5] S. B. Kuang, K. Li, R. P. Zou, R. H. Pan, and A. B. Yu, "Application of periodic boundary conditions to CFD-DEM simulation of gas-solid flow in pneumatic conveying," *Chem. Eng. Sci.*, vol. 93, pp. 214–228, 2013.

[6] G. Mallouppas and B. van Wachem, "Large Eddy Simulations of turbulent particle-laden channel flow," *Int. J. Multiph. Flow*, vol. 54, no. 3, pp. 65–75, 2013.

18

QUESTIONS?

GHENT UNIVERSITY LCT UNIVERSIDAD NACIONAL DE COLOMBIA SEDE MEDELLÍN

19

FACULTY OF ENGINEERING AND ARCHITECTURE

Noel Gomez
PhD Candidate

E nagomezm@unal.edu.co / Noel.GomezMendoza@ugent.be
M +57 317 851 13 26

GHENT UNIVERSITY LCT UNIVERSIDAD NACIONAL DE COLOMBIA SEDE MEDELLÍN

502416 Numerical Simulation of an Industrial Gas-Solid Flow By the Coarse Grain Model

Abstract:

The Discrete Element Method (DEM) is often employed in various granular and multi-phase flows in industries [1]. Although the DEM becomes an established approach, there is a critical problem from a viewpoint of limit of number of the calculated particles. When the DEM simulation is performed on a single PC, number of the calculated particles becomes at most millions. On the other hand, over billion particles are generally required in industrial powder systems. In order to solve this problem, the authors' group developed a scaling law model of the DEM, which is referred as coarse grain model [2]. In previous studies, the authors' group proved the applicability of the coarse grain model in fluidized beds [3,4] and a pneumatic conveying system [2]. In this study, the coarse grain model was applied to solid mixing in a spouted bed. In the calculations, DEM-CFD method was employed to simulate the gas and solid phases in the spouted bed. When the mixing state was examined, Lacey's mixing index was used in the current study. In this study, 4 cases of the verification tests were performed to show the adequacy of the coarse grain model, namely, original particle system, coarse grain model systems (coarse grain ration was 2.0 and 3.0) and simply triple size particle system (without the coarse grain model). The macroscopic properties such as pressure drop and velocity distribution were in good agreement between the original particle system and coarse grain systems, though double and triple sized particles were used in the coarse grain model systems. As far as the mixing state was considered, it also well agreed in the original and coarse grain model systems. Consequently, applicability of the coarse grain model of the solid mixing is shown through this study.

Acknowledgement

This study was financially supported by the Initiatives for Atomic Energy Basic and Generic Strategic Research by the Ministry of Education, Culture, Sports, Science, and Technology of Japan.

References

[1] M. Sakai, "How should the discrete element method be applied in industrial systems?: A review," KONA Powder and Particle Journal, 33, 169–178 (2016)

[2] M. Sakai and S. Koshizuka, "Large-scale discrete element modeling in pneumatic conveying," Chem. Eng. Sci., 64, 533–539 (2009)

[3] M. Sakai, H. Takahashi, C.C. Pain, J-P Latham, J. Xiang, "Study on a large-scale discrete element model for fine particles in a fluidized bed," Adv. Powder Technol., 23, 673–681 (2012)

[4] M. Sakai, M. Abe, Y. Shigeto, S. Mizutani, H. Takahashi, A. Vire, J.R. Percival, J. Xiang, C.C. Pain, "Verification and validation of a coarse grain model of the DEM in a bubbling fluidized bed," Chem. Eng. J.,

244, 33-43 (2014)

487275 Dynamic Multiscale Method for Gas-Solid Flow

Abstract:

Multiscale simulation of gas-solid flow, including direct numerical simulations, discrete particle methods and continuum models, is popular in fluidization community. Especially, there is a rapid development of the so-called correlative multiscale method (Li and Kwauk, 2003), where lower level models are used to obtain the constitutive relationships needed in higher level models (van der Hoef et al., 2008; Tenneti and Subramaniam, 2014). For example, direct numerical simulations have been extensively used to obtain the drag correlations used in two-fluid models and discrete particle methods. However, one-way and offline coupling features of correlative multiscale method make itself not the best way to mimic the two-way and dynamic coupling nature of gas-solid flow. In this study, dynamic multiscale method is proposed to couple different simulation methods in a two-way and online coupling manner (Chen and Wang, 2015; Chen et al., 2016; Chen and Wang, 2017). Two-fluid model is dynamically coupled with discrete particle method to simulate gas-solid flow in a channel. This study proofs the possibility of dynamic multiscale method for simulating gas-solid flow.

Chen X, Wang J. 2015. Hybrid Discrete-continuum Model for Granular Flow. Procedia Engineering 102, 661-667.

Chen X, Wang J. 2017. Dynamic multiscale method for gas-solid flow via spatiotemporal coupling of two-fluid model and discrete particle model. AIChE Journal, DOI:10.1002/aic.15723.

Chen X, Wang J, Li J. 2016. Multiscale modeling of rapid granular flow with a hybrid discrete-continuum method. Powder Technology 304, 177-185.

Li J, Kwauk M. 2003. Exploring complex systems in chemical engineering: the multi-scale methodology. Chemical Engineering Science 58, 521-535.

Tenneti S, Subramaniam S. 2014. Particle-Resolved Direct Numerical Simulation for Gas-Solid Flow Model Development. Annual Review of Fluid Mechanics 46, 199-230.

van der Hoef MA, van Sint Annaland M, Deen NG, Kuipers JAM. 2008. Numerical simulation of dense gas-solid fluidized beds: A multiscale modeling strategy. Annual Review of Fluid Mechanics 40, 47-70.

493081 Applying Data-Driven Dimension Reduction Techniques to Constitutive Model Formulation for Gas-Particle Flows

Abstract:

Introduction: Gas-particle flows in fluidized beds and circulating fluidized beds are inherently unstable and manifest structures on a wide range of length and time scales. Researchers have sought to develop filtered models to simulate large flow structures without having to resolve fine-scale flow structures, which are computationally very expensive to resolve [1,2]. Highly resolved Euler-Euler (two-fluid model) and Euler-Lagrange simulations (such as CFD-DEM) of gas-particles flows are being used to generate the computational data to infer the level of complexity of the filtered models that one must employ and associated constitutive equations [2,3,4]. Systematic filtering of the results from highly resolved flow simulations yields a large dataset, which form the basis for the sub-filter (commonly referred to as sub-grid) constitutive models. Formulation of these constitutive models is hampered by the fact that the optimal set of filtered variables and sub-grid correlations one should employ to formulate the constitutive model is not known. In literature studies, plausible sets of variables to employ in constitutive models are identified through a trial and error procedure [2,3,4]; while such approaches have led to significant progress, their further improvements are prohibitively tedious. In the present study, we have employed machine learning techniques, in particular data-driven dimension reduction, to identify optimal set of filtered variables and sub-grid correlations.

Methodology: A nonlinear dimension reduction technique, namely diffusion map [5], was found to provide fast and accurate identification of the intrinsic dimension of datasets. The diffusion map algorithm exploits a low dimensional geometric embedding of a high dimensional dataset. Starting from a dataset organized in the form of an $n_{samples} \times n_{variables}$ matrix, the diffusion map technique reduces the dimension to $n_{samples} \times m$, where m is less than or equal to $n_{variables}$. If and when m is smaller than $n_{variables}$, the diffusion map technique allows us to identify the smaller set of variables we need for the modeling process; further analysis of the low dimensional diffusion map coordinates can provide us with insight into their correlation to the original variables. This dimension reduction process can simplify constitutive model formulation process and make further model fitting methods such as neural networks computationally affordable.

Results: A dataset was first obtained by filtering the results from highly resolved Euler-Lagrange simulations; the dataset takes the form of a collection of filtered variables, sub-filter scale correlations and the drag correction for which we seek a model. Diffusion map analysis of the dataset revealed the intrinsic dimension to be two, and helped establish a new two-dimensional representation of the dataset. This finding implies that the drag correction can be parametrized by two independent (sub-filter scale) correlations, and points the way to further refinement of sub-filter scale constitutive models beyond what has been achieved in the literature.

Summary: Data-driven dimension reduction techniques have shed more light on constitutive model formulation for filtered models for gas-particle flows.

References:

[1] Agrawal, K., Loezos, P. N., Syamlal, M., & Sundaresan, S. (2001). The role of meso-scale structures in rapid gas–solid flows. *Journal of Fluid Mechanics, 445*, 151-185.

[2] Ozel, A., Fede, P., & Simonin, O. (2013). Development of filtered Euler–Euler two-phase model for circulating fluidised bed: high resolution simulation, formulation and a priori analyses. *International Journal of Multiphase Flow, 55*, 43-63.

[3] Igci, Y., Andrews, A. T., Sundaresan, S., Pannala, S., & O'Brien, T. (2008). Filtered two-fluid models for fluidized gas-particle suspensions. *AIChE Journal*, *54*(6), 1431-1448.

[4] Ozel, A., Kolehmainen, J., Radl, S., & Sundaresan, S. (2016). Fluid and particle coarsening of drag force for discrete-parcel approach. *Chemical Engineering Science*, *155*, 258-267.

[5] Lafon, S. and Lee, A.B. (2006). Diffusion maps and coarse-graining: a unified framework for dimensionality reduction, graph partitioning, and data set parameterization. *IEEE Transactions on Pattern Analysis and Machine Intelligence*, 28(9), 1393-1403.

Spatially-Averaged Models for Dense Gas-Solid Flows

JOHANNES KEPLER UNIVERSITÄT LINZ

Simon Schneiderbauer

AIChE Annual Meeting 2017
Minneapolis, November 2nd, 2017

CD Laboratory for Multi-Scale Modelling of Multiphase Processes
Department of Particulate Flow Modelling
Johannes Kepler University | Linz | Austria

CIT – SA-TFM
Problem description

Single phase fluid — riser — Bubbling fluidized bed → **Mass loading**

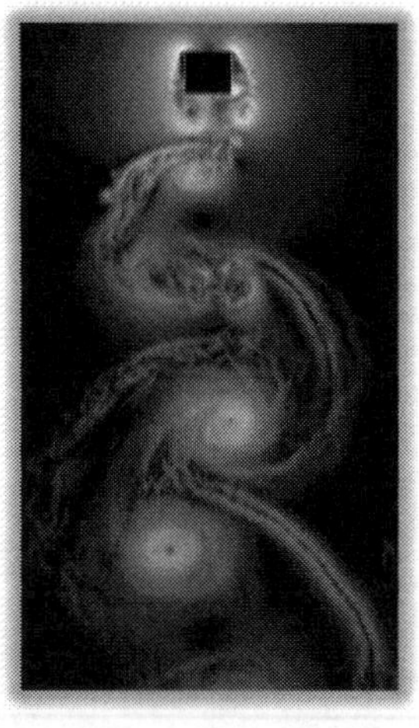

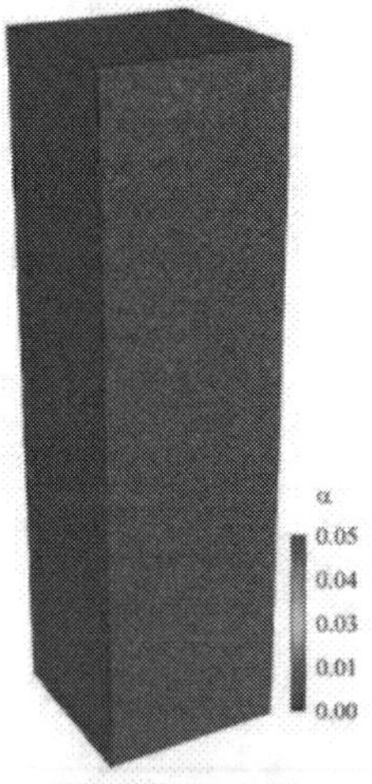

Capecelatro et al. 2015, JFM

At higher mass loading, i.e. in collision dominated flows, the situation is not such apparent

Mol. Fluc. Of particles
↕
Solid turbulence
↕
gas turbulence

Schneiderbauer et al., 2014, AIChE J

CIT – SA-TFM
Modelling large-scale gas-solid flows

Industrial scale

W_g = 0.21m/s
d_s = 150μm
ρ_s = 2500kg/m³

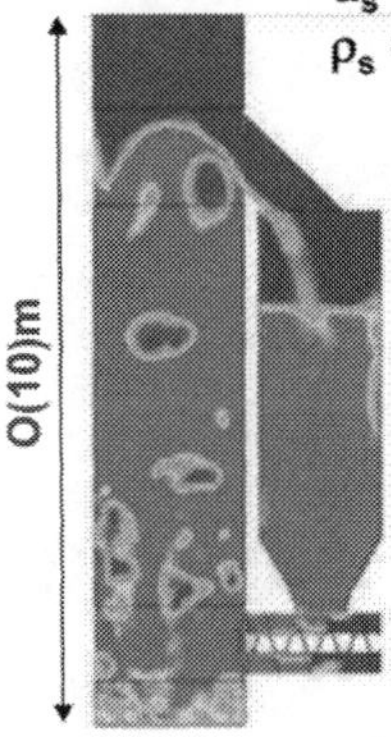

> O(10⁶) cells
> 4 weeks for 10s on 32 CPUs

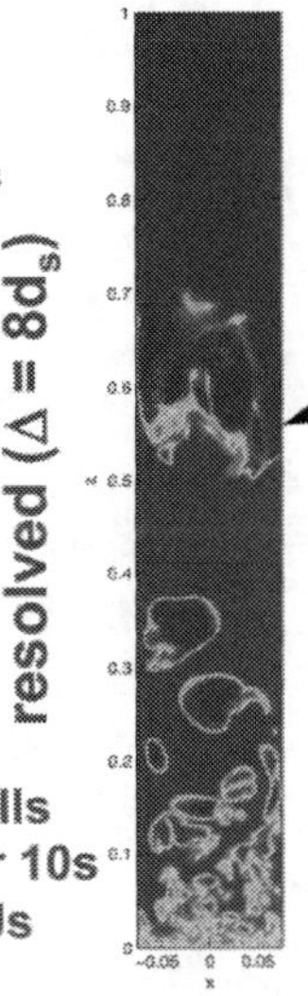

Considerable overestimation of bed expansion (80%)

Speedup > O(10²)

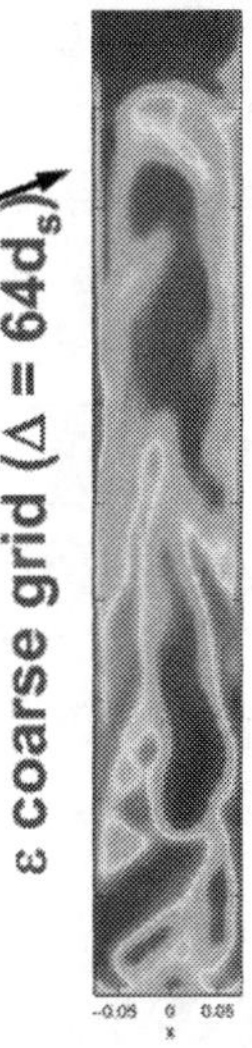

< O(10⁴) cells
< 1 day for 10s on 8 CPUs

CIT – SA-TFM
SA-TFM equations (for solid phase)

Filter:

$$\bar{g}(\boldsymbol{x},t) = \iiint G_\Delta(\boldsymbol{x}-\boldsymbol{y})g(\boldsymbol{y},t)\,\mathrm{d}\boldsymbol{y}$$

with $$\iiint G_\Delta(\boldsymbol{x}-\boldsymbol{y})\,\mathrm{d}\boldsymbol{y} = 1$$

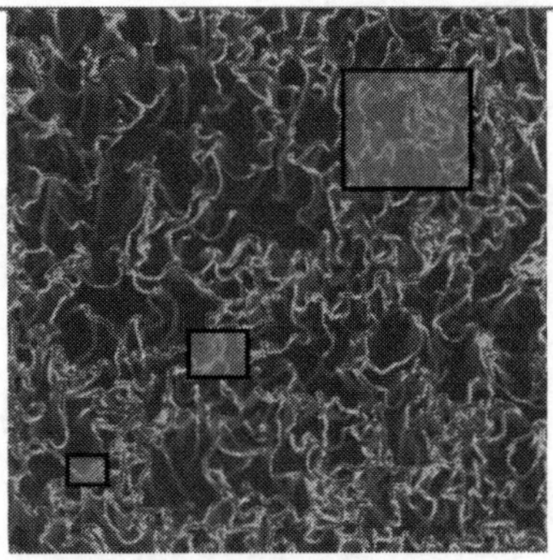

Milioli, C.C. et al., 2012. *Constitutive model for the fluid-particle drag coefficient in filtered two-fluid models for gas-particle flows. Aiche Meeting*

Similar to RANS/LES we define:

$$\phi' = \phi - \bar{\phi}$$

$$\boldsymbol{u}'' = \boldsymbol{u} - \widetilde{\boldsymbol{u}}$$

where $\widetilde{\boldsymbol{u}} = \dfrac{\overline{\phi \boldsymbol{u}}}{\bar{\phi}}$ **(Favre average)**

CIT – SA-TFM
Filtered balance equations for solid phase

Applying the filter to the balance equations for the solid phase yields[1]:

$$\frac{\partial \overline{\phi}}{\partial t} + \frac{\partial}{\partial x_k}\left(\overline{\phi}\widetilde{u}_k\right) = 0$$

$$\frac{\partial \overline{\phi}\widetilde{u}_i}{\partial t} + \frac{\partial}{\partial x_k}\left(\overline{\phi}\widetilde{u}_i\widetilde{u}_k\right) = \frac{1}{\rho_s}\left(\overline{f}_{si} + \overline{f^{\mathrm{R}}_{si}}\right)$$

Averaged RHS of momentum equation → filtered drag and stresses

Reynolds-stress like contribution stemming from the unresolved vel. fluctuations

1) Schneiderbauer, S., 2017. A spatially-averaged two-fluid model for dense large-scale gas-solid flows. *AIChE Journal.*

CIT – SA-TFM
Filtered balance equations for solid phase

Filtered forces on solid phase:

$$\overline{f}_{si} = -\overline{\overline{\phi}\frac{\partial \overline{p}}{\partial x_i}} + \frac{\partial}{\partial x_j}\overline{\Sigma^{\mathrm{kc}}_{ij}} + \overline{\beta(v_i - u_i)} + \overline{\phi}\rho g_i$$

Filtered molecular stresses

Filtered gas-solid drag

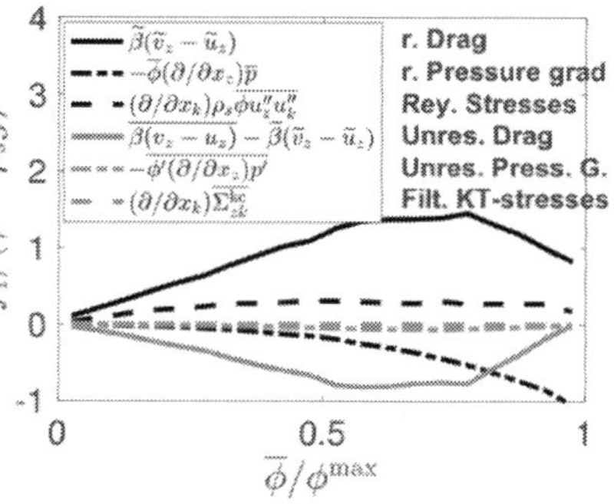

"Reynolds" stresses:

$$\overline{f^{\mathrm{R}}_{si}} = \frac{\partial}{\partial x_k}\rho_s\left(\overline{\phi}\widetilde{u}_i\widetilde{u}_k - \overline{\phi u_i u_k}\right) \equiv -\frac{\partial}{\partial x_k}\rho_s\overline{\phi u''_i u''_k}$$

CIT– SA-TFM
New approach for filtered drag force

State-of-the-art filtered models:[1-6]

$$\overline{\beta(v_i - u_i)} \approx \left(1 - H(\overline{\phi}, u_{gs}, \Delta_f)\right)\widetilde{\beta}(\widetilde{v}_i - \widetilde{u}_i)$$

H was derived from curve fitting.

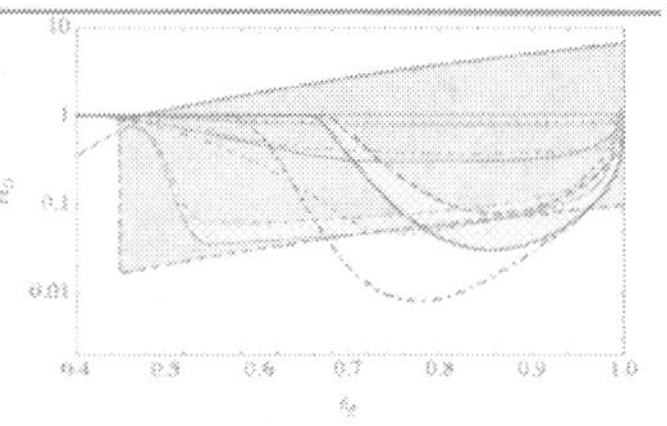

NEW: Taylor expansion of microscopic drag coefficient (neglecting HOT)

$$\beta \approx \beta(1-\overline{\phi}, \widetilde{\boldsymbol{v}}, \widetilde{\boldsymbol{u}}) + \left(\frac{\partial \beta}{\partial v_j}\right)_{1-\overline{\phi},\widetilde{\boldsymbol{v}},\widetilde{\boldsymbol{u}}} v_j'' + \left(\frac{\partial \beta}{\partial u_j}\right)_{1-\overline{\phi},\widetilde{\boldsymbol{v}},\widetilde{\boldsymbol{u}}} u_j'' - \left(\frac{\partial \beta}{\partial (1-\phi)}\right)_{1-\overline{\phi},\widetilde{\boldsymbol{v}},\widetilde{\boldsymbol{u}}} \phi'$$

yields functional dep. of constitutive relation for filtered drag.

1. Sarkar, A. et al., 2016. Filtered sub-grid constitutive models for fluidized gas-particle flows constructed from 3-D simulations. *Chemical Engineering Science*, 152, pp.443–456.
2. Ozel, A., Fede, P. & Simonin, O., 2013. *International Journal of Multiphase Flow*, 55, pp.43–63.
3. Milioli, C.C. et al., 2013. Filtered two-fluid models of fluidized gas-particle flows: new constitutive relations. *AIChE Journal*, 59(9), pp.3265–3275.
4. Igci, Y. et al., 2008. Filtered Two-Fluid Models for Fluidized Gas-Particle Suspensions. *AIChE Journal*, 54(6), pp.1431–1448.
5. Parmentier, J.-F., Simonin, O. & Delsart, O., 2012. A functional subgrid drift velocity model for filtered drag prediction in dense fluidized bed. *AIChE Journal*, 58(4), pp.1084–1098.
6. Schneiderbauer, S. & Pirker, S., 2014. *AIChE Journal*, 60(3), pp.839–854.

CIT– SA-TFM
Filtered drag force

Substituting the series expansion and averaging yields

$$\begin{aligned}\overline{\beta(v_i - u_i)} \approx\ & \widetilde{\beta}(\widetilde{v}_i - \widetilde{u}_i) + \widetilde{\beta}\sqrt{\frac{2}{3}\overline{\phi'^2}}\left(\frac{\xi_{\phi g}\sqrt{k_g}}{1-\overline{\phi}} + \frac{\xi_{\phi s}\sqrt{k_s}}{\overline{\phi}}\right)\\ & + \left(\frac{\partial \beta}{\partial v}\right)_{1-\overline{\phi},\widetilde{\boldsymbol{v}},\widetilde{\boldsymbol{u}}}\left(\frac{2}{3}\left(k_g - \xi_{gs}\sqrt{k_g k_s}\right) + \frac{\xi_{\phi g}}{1-\overline{\phi}}\sqrt{\frac{2}{3}k_g\overline{\phi'^2}}(\widetilde{v}_i - \widetilde{u}_i)\right)\\ & + \left(\frac{\partial \beta}{\partial u}\right)_{1-\overline{\phi},\widetilde{\boldsymbol{v}},\widetilde{\boldsymbol{u}}}\left(\frac{2}{3}\left(\xi_{gs}\sqrt{k_g k_s} - k_s\right) - \frac{\xi_{\phi s}}{\overline{\phi}}\sqrt{\frac{2}{3}k_s\overline{\phi'^2}}(\widetilde{v}_i - \widetilde{u}_i)\right)\\ & - \left(\frac{\partial \beta}{\partial (1-\phi)}\right)_{1-\overline{\phi},\widetilde{\boldsymbol{v}},\widetilde{\boldsymbol{u}}}\sqrt{\frac{2}{3}\overline{\phi'^2}}\left(\xi_{\phi g}\sqrt{k_g} - \xi_{\phi s}\sqrt{k_s}\right)\end{aligned}$$

However, requires closures for TKE (k_s, k_g) and bulk density fluctuations

CIT– SA-TFM
Filtered drag force

Budget analysis:

$$\overline{\beta(v_i - u_i)} \approx$$

$$\approx \widetilde{\beta}(\widetilde{v}_i - \widetilde{u}_i) + \widetilde{\beta}\sqrt{\frac{2}{3}\overline{\phi'^2}}\left(\frac{\xi_{\phi g}\sqrt{k_g}}{1-\overline{\phi}} + \frac{\xi_{\phi s}\sqrt{k_s}}{\overline{\phi}}\right)$$

$$+ \left(\frac{\partial\beta}{\partial v}\right)_{1-\overline{\phi},\widetilde{\mathbf{v}},\widetilde{\mathbf{u}}}\left(\frac{2}{3}\left(k_g - \xi_{gs}\sqrt{k_g k_s}\right) + \frac{\xi_{\phi g}}{1-\overline{\phi}}\sqrt{\frac{2}{3}k_g\overline{\phi'^2}}(\widetilde{v}_i - \widetilde{u}_i)\right)$$

$$+ \left(\frac{\partial\beta}{\partial u}\right)_{1-\overline{\phi},\mathbf{v},\mathbf{u}}\left(\frac{2}{3}\left(\xi_{gs}\sqrt{k_g k_s} - k_s\right) - \frac{\xi_{\phi s}}{\overline{\phi}}\sqrt{\frac{2}{3}k_s\overline{\phi'^2}}(\widetilde{v}_i - \widetilde{u}_i)\right)$$

$$- \left(\frac{\partial\beta}{\partial(1-\phi)}\right)_{1-\overline{\phi},\widetilde{\mathbf{v}},\widetilde{\mathbf{u}}}\sqrt{\frac{2}{3}\overline{\phi'^2}}\left(\xi_{\phi g}\sqrt{k_g} - \xi_{\phi s}\sqrt{k_s}\right)$$

$$H = \frac{\left(\frac{\partial\beta}{\partial(1-\phi)}\right)_{1-\overline{\phi},\widetilde{\mathbf{v}},\widetilde{\mathbf{u}}}\sqrt{2\overline{\phi'^2}}\left(\xi_{\phi g}\sqrt{k_g} - \xi_{\phi s}\sqrt{k_s}\right)}{\widetilde{\beta}(\widetilde{v}_i - \widetilde{u}_i)}$$

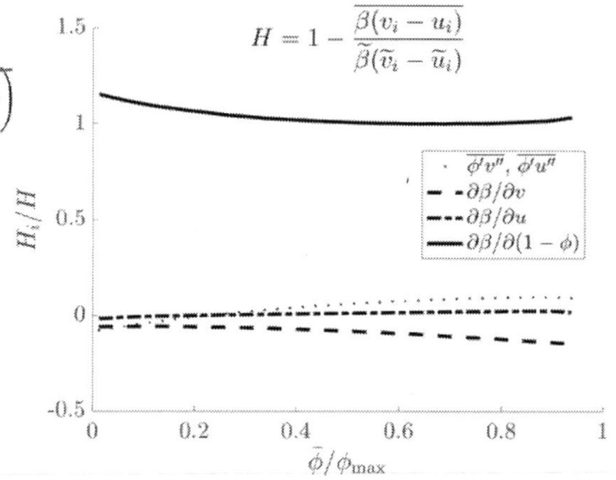

Schneiderbauer, AIChE Journal, 2017

CIT – SA-TFM
Closure model for TKEs

Transport equation for TKEs

Contribution from interfacial work

$$\frac{\partial\overline{\phi}k_s}{\partial t} + \frac{\partial}{\partial x_j}\left(\overline{\phi}k_s\widetilde{u}_j\right) = \underbrace{-\overline{\phi u_i'' u_j''}\frac{\partial\widetilde{u}_i}{\partial x_j} + \frac{\widetilde{\beta}_\phi}{\rho_s}\left(\overline{\phi u_i'' v_i''} - \overline{\phi u_i'' u_i''}\right)}_{\text{production}} - \underbrace{\left(\overline{\phi}\epsilon - \frac{1}{\rho_s}p_\phi^{\text{kc}}\overline{\phi\frac{\partial u_i''}{\partial x_j}}\right)}_{\text{dissipation}}$$

$$\underbrace{-\frac{\partial}{\partial x_j}\left(\frac{1}{2}\overline{\phi u_i'' u_i'' u_j''} + \frac{1}{\rho}\overline{u_i''\left(\Sigma_{ij}^{\text{kc}}\right)'}\right)}_{\text{diffusion}} - \frac{1}{\rho_s}\overline{\phi u_i''\frac{\partial p'}{\partial x_i}}.$$

In addition to the classical production by mean shear (**SIT**), the transport equation for TKEs contains an additional production term due interfacial work. In turn, this drag production mechanism gives rise to a separate class of multiphase turbulence, which is referred to as cluster-induced turbulence (**CIT**).[1]

1. Capecelatro, J., Desjardins, O. & Fox, R.O., 2015. On fluid-particle dynamics in fully developed cluster-induced turbulence. *Journal of Fluid Mechanics*, 780, pp.578–635.

CIT – SA-TFM
Closure model for fluctuations of volume frac.

Transport equation for variance of solids volume fraction (VSV)

$$\frac{\partial \overline{\phi'^2}}{\partial t} + \frac{\partial}{\partial x_k}\left(\overline{\phi'^2}\widetilde{u}_k\right) = \underbrace{-\overline{\phi'^2}\frac{\partial \widetilde{u}_k}{\partial x_k} - 2\overline{\phi' u_k''}\frac{\partial \overline{\phi}}{\partial x_k}}_{\text{div. \& gradient production}}$$

div. & gradient production

$$-2\overline{\phi'\phi\frac{\partial u_k''}{\partial x_k}} + \overline{\phi'^2\frac{\partial u_k''}{\partial x_k}} - \frac{\partial}{\partial x_k}\overline{\phi'^2 u_k''}$$

Attenuation / dissipation — **Diffusion**

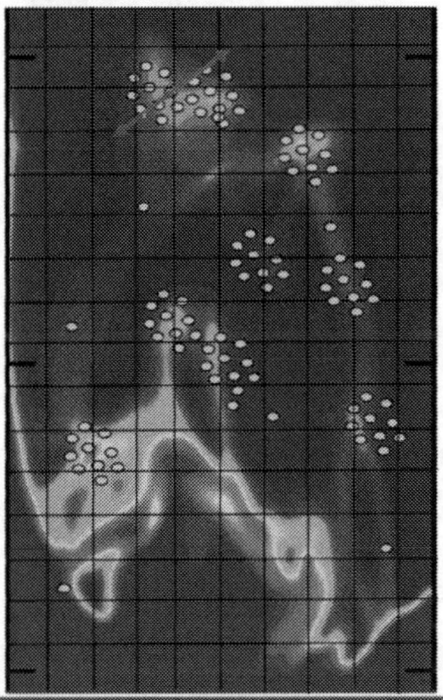

SA-TFM
Closure models

TKEs and VSV are dominated by local production and dissipation → transport negligible[1]

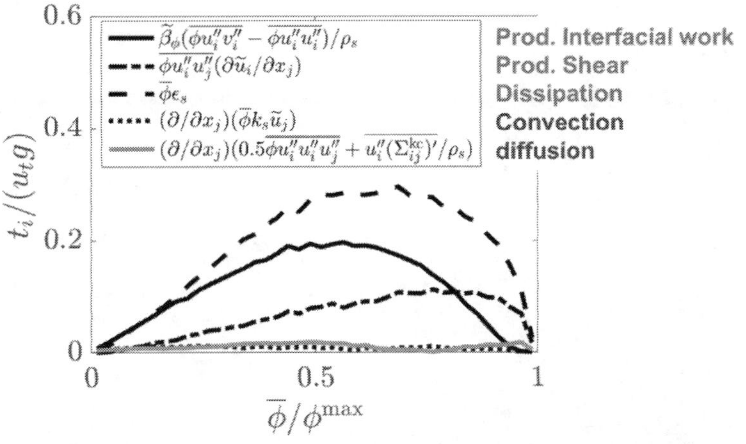

SA-TFM
Closure models

Thus, we obtain

$$k_s = \frac{1}{C_{\epsilon s}^2}\left(-\frac{\widetilde{\beta}_\phi l_{ms}}{\rho_s} + \sqrt{\left(\frac{\widetilde{\beta}_\phi l_{ms}}{\rho_s}\right)^2 + 2\left(l_{ms}^2 \widetilde{S}_{s,ij}\widetilde{S}_{s,ij} + \frac{\xi_{gs}\widetilde{\beta}_\phi l_{ms} k_g^{1/2}}{\rho_s}\right)}\right)^2$$

$$k_g = \frac{1}{C_{\epsilon g}^2}\left(-\frac{\overline{\phi}\widetilde{\beta}_\phi l_{mg}}{(1-\overline{\phi})\rho_g} + \sqrt{\left(\frac{\overline{\phi}\widetilde{\beta}_\phi l_{mg}}{(1-\overline{\phi})\rho_g}\right)^2 + 2\left(l_{mg}^2 \widetilde{S}_{g,ij}\widetilde{S}_{g,ij} + \frac{\xi_{gs}\overline{\phi}\widetilde{\beta}_\phi l_{mg} k_s^{1/2}}{(1-\overline{\phi})\rho_g}\right)}\right)^2$$

$$\overline{\phi'^2} = \frac{8}{3}\frac{\xi_{\phi s}^2 k_s \left(\frac{\partial\overline{\phi}}{\partial x_k}\right)^2}{\left(\frac{\partial\widetilde{u}_k}{\partial x_k} + C_{\phi s}C_{\epsilon s}\frac{k_s^{1/2}}{l_{ms}}\right)^2}.$$

CIT – SA-TFM
Validation of SA-TFM

Zhu, H., Zhu, J., Li, G. & Li, F. Detailed measurements of flow structure inside a dense gas-solids fluidized bed. Powder Technol. 180, 339–349 (2008).

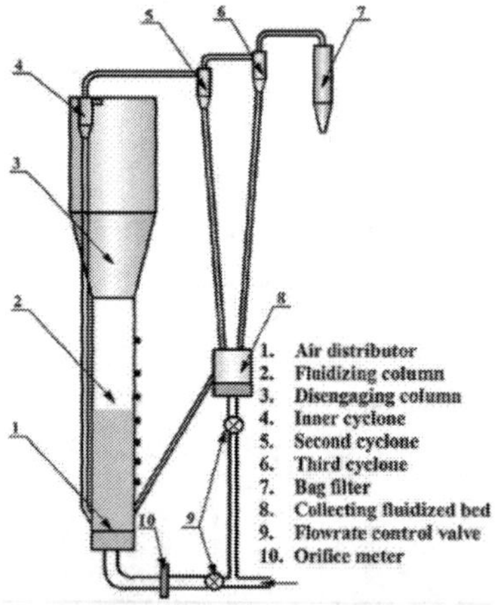

65 μm FCC particles
i.d. 0.267 m

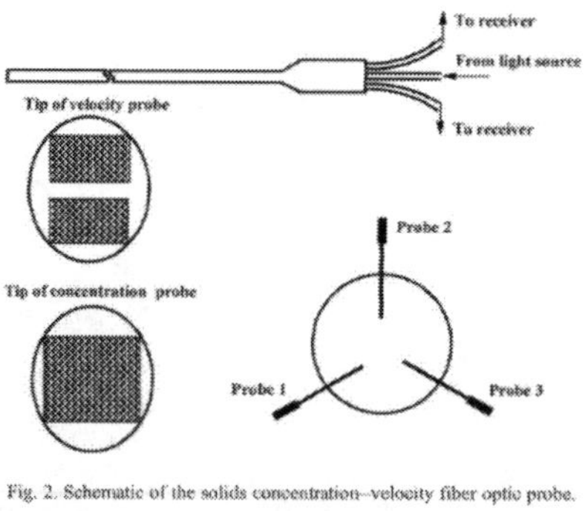

Fig. 2. Schematic of the solids concentration–velocity fiber optic probe.

CIT – SA-TFM & SA-DDPM
Validation: Zhu et al. 0.4 m/s

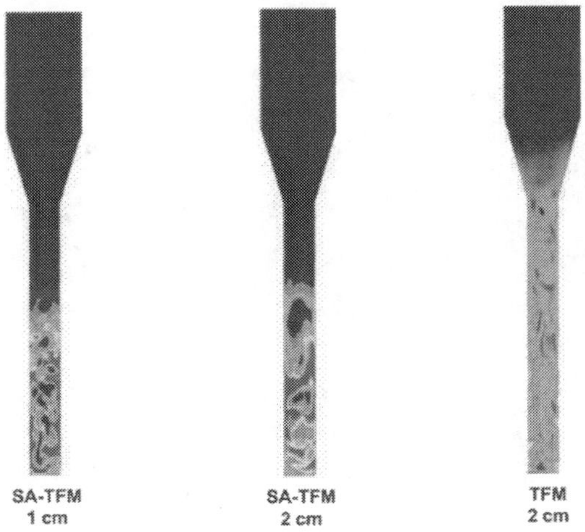

CIT – SA-TFM
Validation of SA-TFM: Zhu et al. 0.4m/s

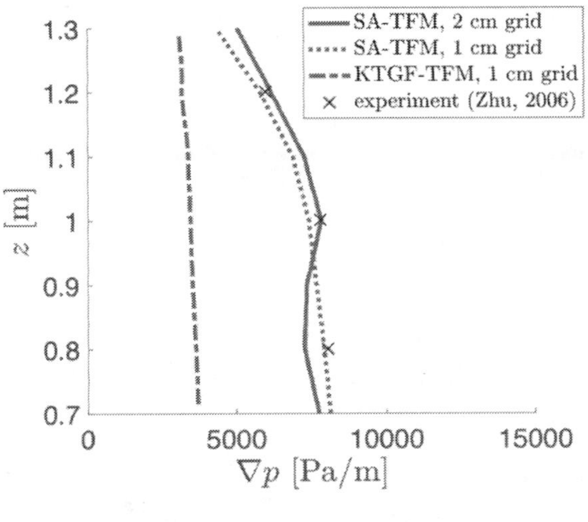

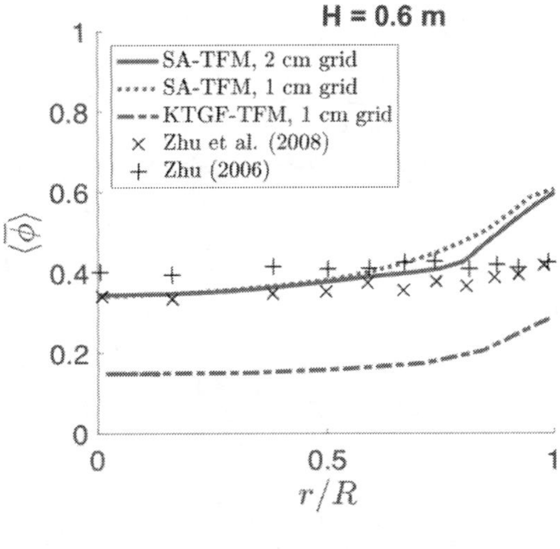

CIT – SA-TFM
Validation of SA-TFM

NETL challenge problem III

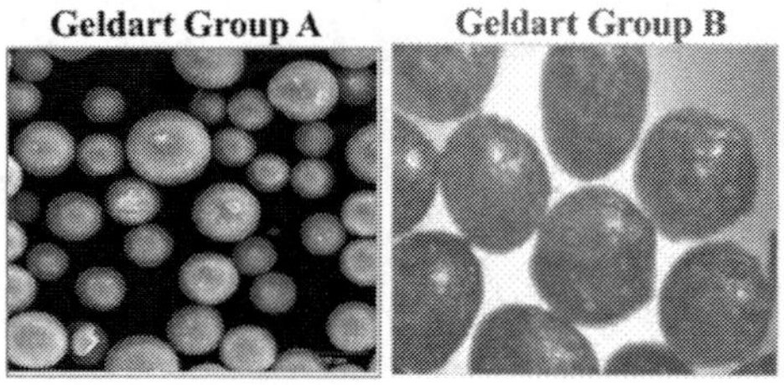

Statistical Experimental Design

Group A: CASE 1 Dilute
CASE 2 Core Annular

Group B: CASE 3 Dense Transport (near FFB)
CASE 4 Core Annular
CASE 5 Dense Suspension Upflow

Material	Case	U_{gi} (m/sec)	M_s (kg/sec)	F_R (SCMs)	F_{SL} (SCMs)	RH (%)	T (°C)	P_b (kPa)
Group A	1	5.14	1.44	0.683	0.001	0.14	20.5	182
	2	5.14	9.26	0.682	0.002	0.14	20.5	167
Group B	3	5.71	5.54	0.476	0.025	48	23	100
	4	7.58	7.03	0.599	0.025	49	23	102
	5	7.58	14.00	0.640	0.029	46	23	105

JKU | CD-Laboratory for Multi-scale Modelling of Multiphase Processes

CIT – SA-TFM
Validation of SA-TFM: CFB riser case 2

Pressure gradient:

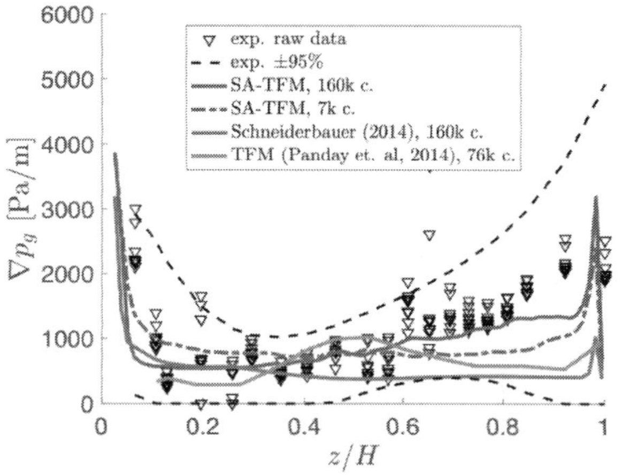

Pressure drop:
Exp: 18.3 kPa
Sim: 16.5 kPa

JKU | CD-Laboratory for Multi-scale Modelling of Multiphase Processes

CIT – SA-TFM
Validation of SA-TFM: CFB riser case 2

Velocity profiles:

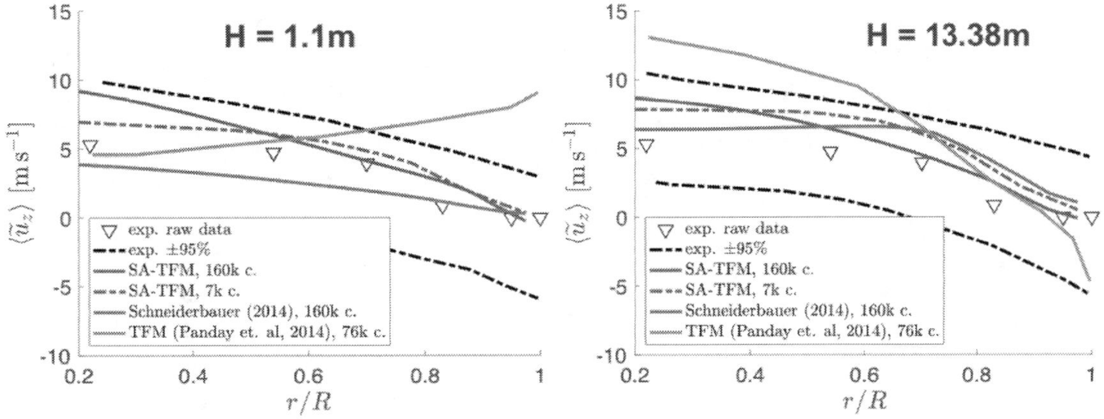

JKU | CD-Laboratory for Multi-scale Modelling of Multiphase Processes

CIT – SA-TFM
Validation of SA-TFM: CFB riser case 5

Pressure gradient:

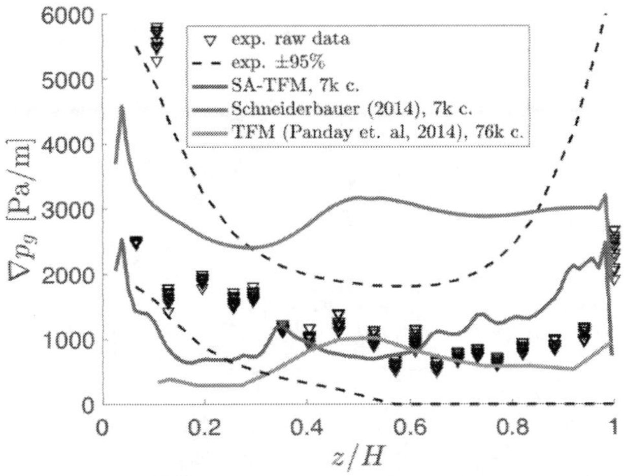

Pressure drop:
Exp: 20.6 kPa
Sim: 17 kPa

JKU | CD-Laboratory for Multi-scale Modelling of Multiphase Processes

CIT – SA-TFM
Validation of SA-TFM: CFB riser case 5

H = 6.23 m

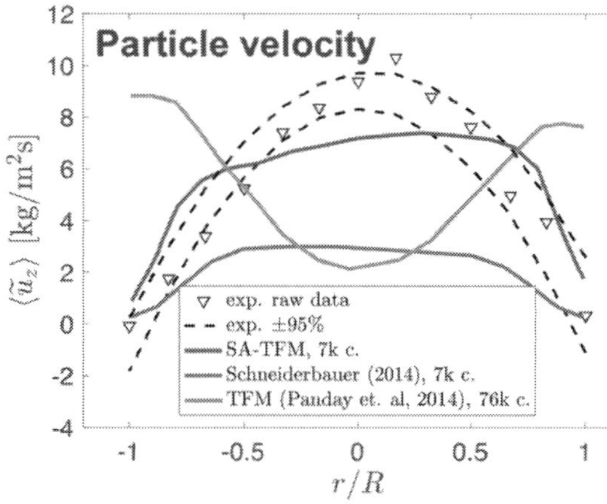

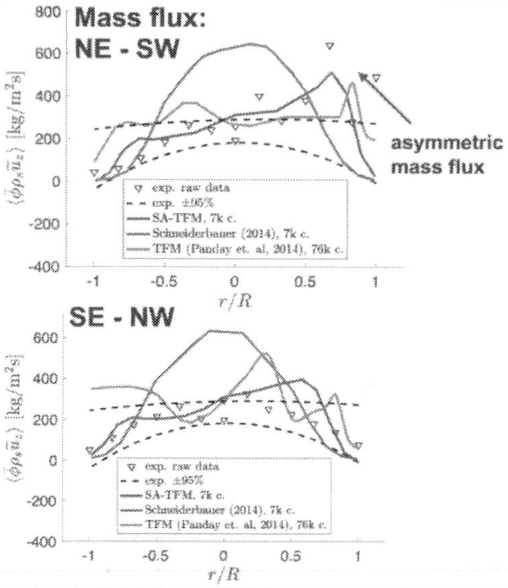

CIT – SA-TFM
Validation of SA-TFM: CFB riser case 5

H = 8.88 m

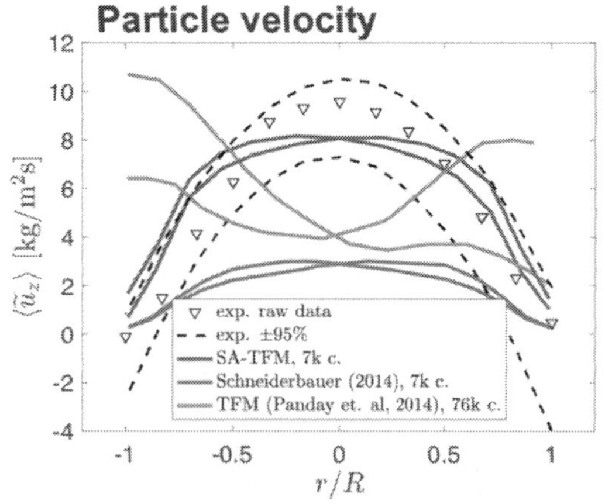

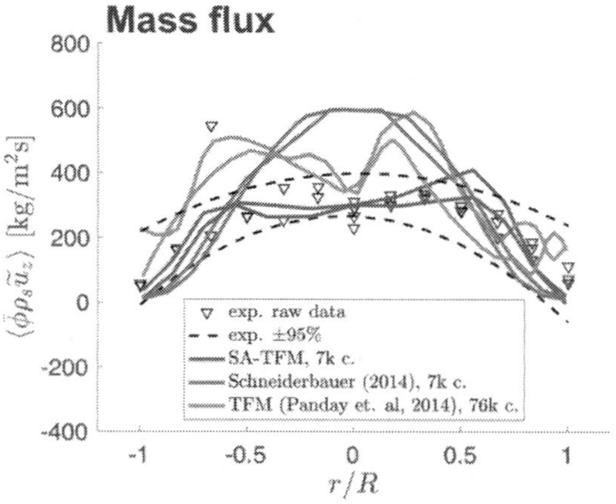

CIT – SA-DDPM
Sub-grid modifications for parcel based appr.

Radl, S. & Sundaresan, S. (2014).

Fine grid/grained simulations

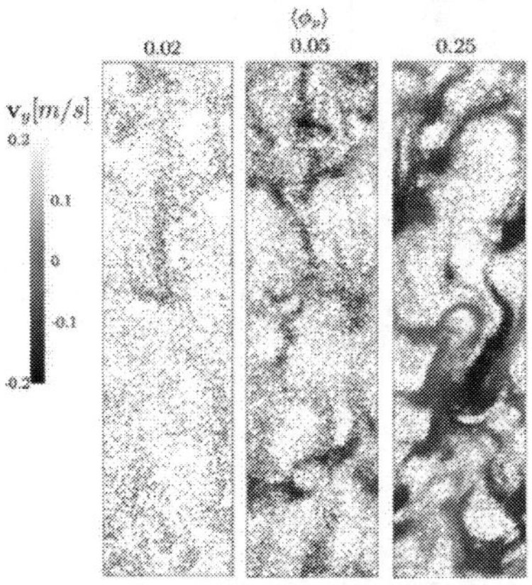

Fluid coarsening

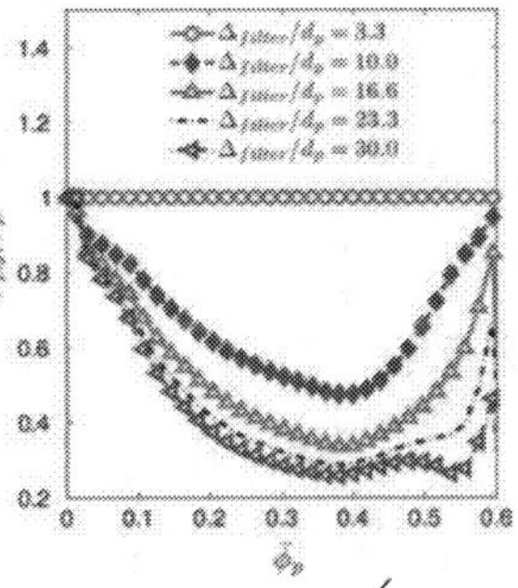

Particle coarsening

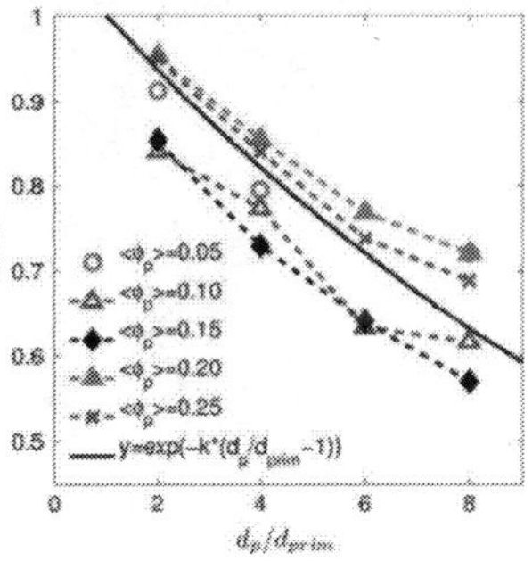

Curve fitting

$$\frac{\overline{\beta}_p}{\beta_p} = \left(1 - f\left(\frac{\Delta_{filter}}{L_c}, \overline{\phi}_p\right) h(\overline{\phi}_p)\right) c_{corr}(\alpha, \overline{\phi}_p).$$

JKU | CD-Laboratory for Multi-scale Modelling of Multiphase Processes

CIT – SA-DDPM
Sub-grid modifications for parcel based appr.

Fluid coarsening:

Variance of fluid velocity and of solids volume fraction are lost

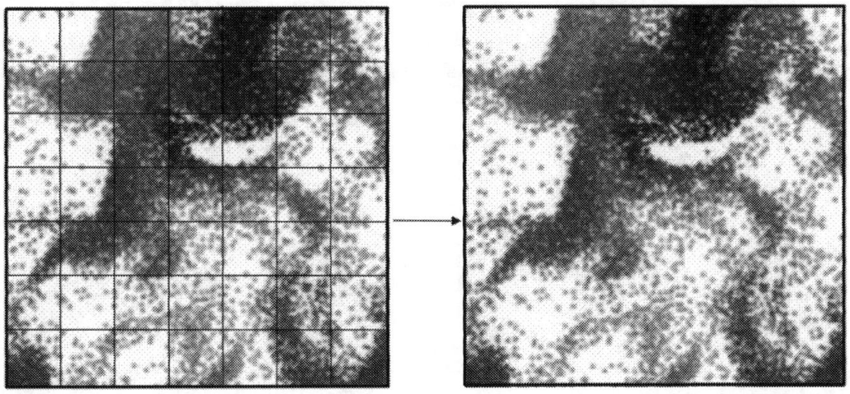

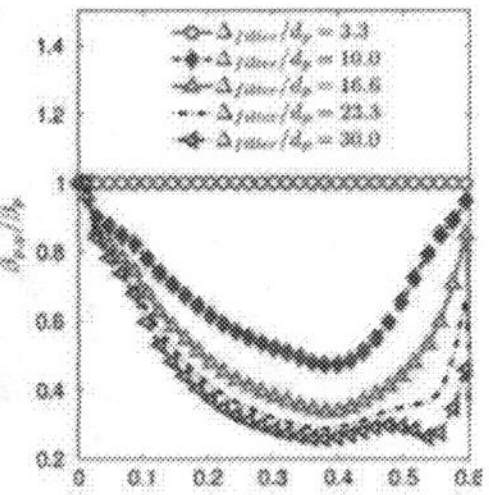

These sub-grid heterogeneities can be quantified by

$$k_g, \ \overline{\phi'^2}$$

JKU | CD-Laboratory for Multi-scale Modelling of Multiphase Processes

CIT – SA-DDPM
Sub-grid modifications for parcel based appr.

Particle coarsening:

Variance of solid velocity is lost

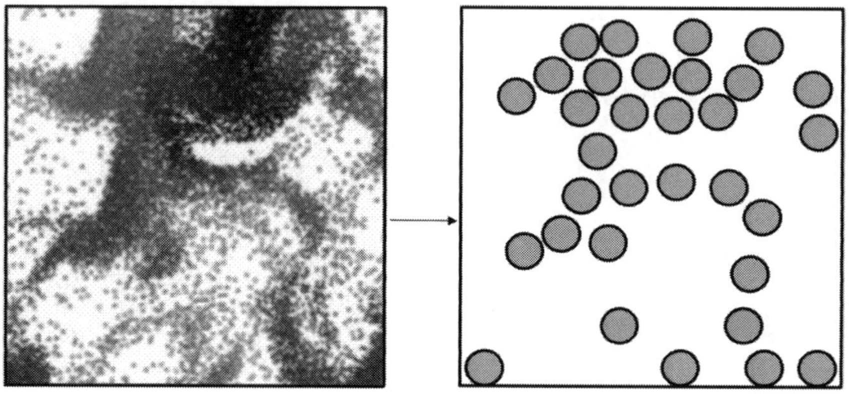

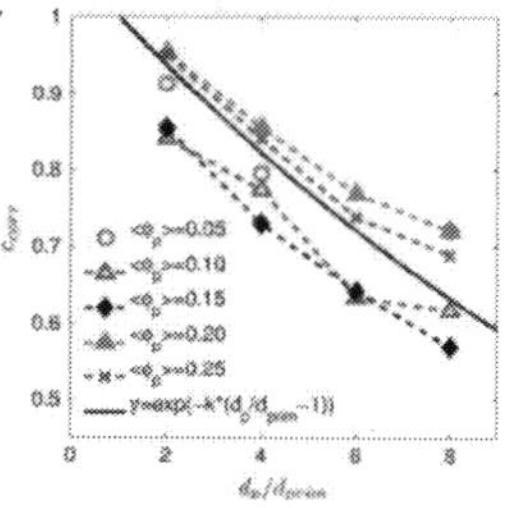

This sub-grid heterogeneity can be quantified by

$$k_s$$

CIT – SA-TFM & SA-DDPM
Validation: Zhu et al. 0.4 m/s

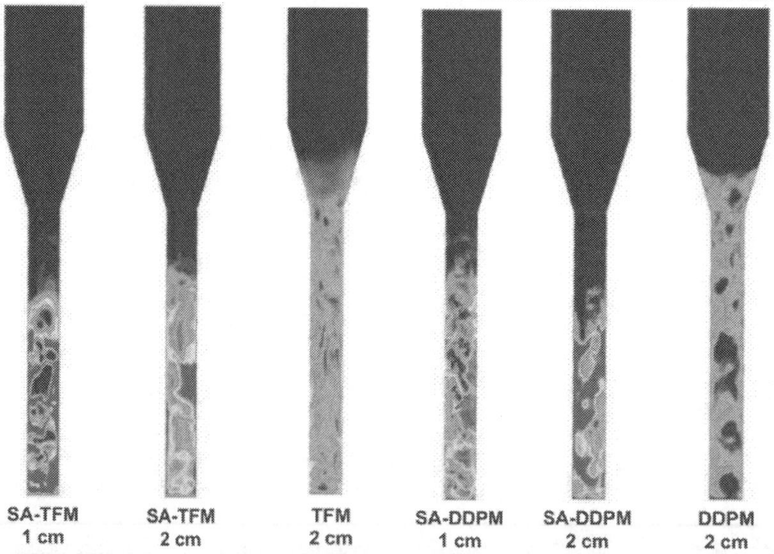

CIT – SA-TFM & SA-DDPM
Validation: Zhu et al. 0.4 m/s

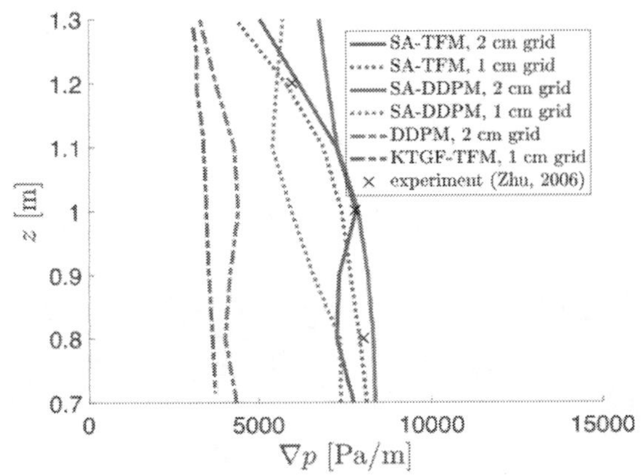

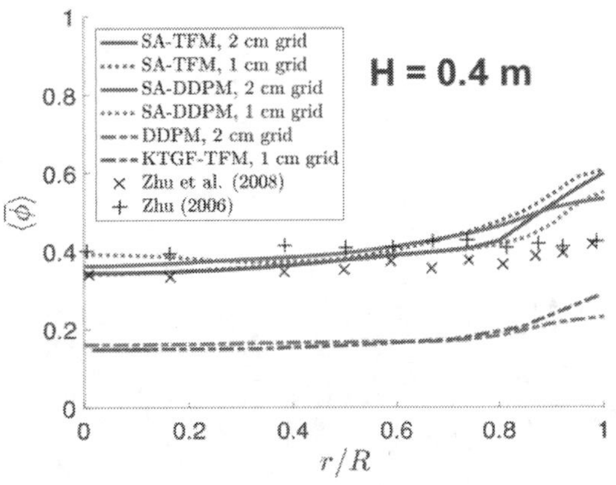

CIT – SA-TFM
Conclusions

1) The filtered drag force shows a considerable dependence on the variance of the solids volume fraction and the TKEs of the gas and solid phase.
2) At least in dense gas-solid flows the filtered large scale shear viscosity can be approximated by a mixing length model.
3) In regions of low shear rates the TKEs are determined by the interfacial work, while in highly sheared regions the phase interactions are getting negligible.
4) The presented approaches for the filtered drag force, Reynolds stresses, TKEs and the variance of the solids volume fraction show fairly good agreement with the fine grid data obtained for considerably different particle properties and fluidization velocities.
5) The presented SA-TFM approach provides fairly good estimates of the hydrodynamics of bubbling fluidized beds and risers using coarse grids.

JOHANNES KEPLER
UNIVERSITÄT LINZ

Thank you for your attention !

Simon Schneiderbauer

Christian Doppler Laboratory for Multiscale Modeling of Multiphase Processes
Johannes Kepler University | Linz | Austria

494528 CFD Analysis for Boundary Effect of Particle Flow in Pipeline Manifold

Abstract:

CFD Analysis for Boundary Effect of Particle Flow in Pipeline Manifold

Mohanrao Rampure

SABIC Technology & Innovation, SPADC Center, Riyadh (KSA) 11551

rampuremr@sabic.com

Gaseous feed stocks in Petrochemical plants has limits on impurities such as; undesired hydrocarbon gas/condensate components, and solid particles to ensure the performance and reliability of downstream unit operation. Pipelines contains iron oxides corrosion/erosion materials, and hydrocarbon condensate droplets carried by feedstock pipelines to plants systems that can be detrimental to the continued safe and reliable plant operation and business profitability. Solid particles and condensate droplets in gaseous feedstock streams tend to separates and re-distribute unevenly from main flows when feedstock is branched through distribution networks piping systems and headers. Design, sizing, and layout of feedstock receiving and filtration/cleaning systems becomes important to effectively capture these impurities and protect downstream plant operation. In this study, a multiphase solid/gas flow assessment in and industrial scale natural gas feedstock header is considered. This main header is branched into three smaller pipes of different sizes. Pipe branches travel in various size diameters based on the different plant capacity downstream. 3D CFD model is developed to determine the multiphase flow distribution between the branches along with a parametric study to assess the flow sensitivity to increased solid particles loading. Figure 1 below shows the geometry and velocity vector plot for the gas phase. The boundary conditions and CFD geometry to be considered plays an important role determining the particulate distribution at the outlet manifold. Particles such as iron oxides and oil droplets are studied to analyze the particles flow mal-distribution exiting the main header. The particles are solved in the Lagrangian frame of reference while the gas in Eulerian frame.

Numerical Simulations of PSRI Cold-Flow FCC Stripper Experiment with Subway Grating Baffles

Session 751: Industrial Application of Computational and Numerical Approaches to Particle Flow II

Sam Clark, CPFD Software, LLC, sam.clark@cpfd-software.com
Peter Blaser, CPFD Software, LLC, peter.blaser@cpfd-software.com
Allan Issangya, Particulate Solid Research, Inc. (PSRI), allan.issangya@psri.org
Ray Cocco, Particulate Solid Research, Inc. (PSRI), ray.cocco@psri.org

Table of Contents

Abstract

Industrial fluid catalytic cracking units (FCCUs) employ fluidized bed strippers to recover hydrocarbons carried away with the downward flowing spent catalyst particles from the reactor, preventing these valuable hydrocarbons from entering and combusting in the regenerator. In the stripper, steam (the "gas") flows upward, counter-current to the catalyst flow direction, stripping the hydrocarbon vapors from and around the catalyst particles before the particles enter the regenerator. Effective stripper operation requires intimate contact between the gas and particle phases. Commercial stripper designs often use internals, such as subway grating trays, chevrons (also called sheds), disk-and-donut baffles, structured packing or other configurations, to promote better mixing of gas and particles and improve stripping efficiency.

Numerical simulations were performed of a PSRI cold-flow FCC stripper experiment with subway grating baffles. The experimental unit had an internal diameter of 3 feet (0.9 m), and a height of 25 feet (7.62 m). The subway grating baffles within the vessel were intended to promote more uniform fluidization of particles by preventing the formation of large bubbles as well as preventing gas bypassing in the stripper. The commercial computational fluid dynamics (CFD) software Barracuda Virtual Reactor®, which implements the multi-phase particle-in-cell (MP-PIC) numerical method, was used to perform the simulations. Simulation results were compared with experimental measurements of apparent bed density over a wide range of solids flux conditions, at superficial gas velocities typical of industrial FCC stripper operation. The simulation results were in good agreement with the experimental bed density measurements, and matched the experimentally observed trends of bed density variation as a function of gas superficial velocity and catalyst mass flux through the stripper vessel.

Experiment

Full details of the experimental work for the system under consideration are given in PSRI's *Research Report No. 88* (Issangya et al, 2014), which is available to PSRI member companies. Permission has been granted by PSRI to provide sufficient details from *Research Report No. 88* to describe the geometry and operating conditions of the experiment used for creating the numerical models which are the main subject of the current work.

Physical Geometry

A schematic diagram of the experimental system is shown in Figure 1. The system consisted of a 3 ft (0.9 m) diameter fluidized bed with subway grating baffles; a 14 in (0.36 m) diameter standpipe below the fluidized bed; two 12 in (0.3 m) diameter risers lifting particles back to the top of the system; and a two-stage cyclone system to direct particles from the risers into the fluidized bed. The circulation rate of solids through the system was controlled by two pneumatically operated slide valves positioned before each riser. The gas distributor for the 3 ft (0.9 m) diameter stripper column was a 24 in diameter ring sparger, photographs of which are shown in Figure 2.

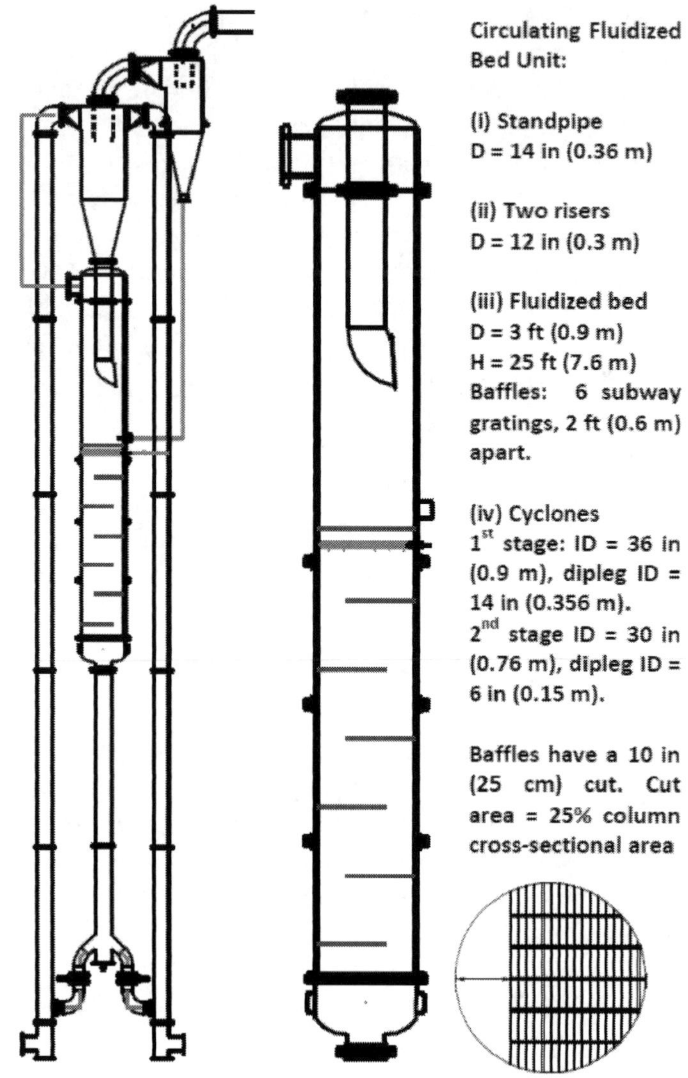

Figure 1. 3-ft (0.9-m)-diameter fluidized bed stripping unit equipped with subway grating internals.

Figure 1: Experimental unit geometry. Reproduction of Figure 1 from Research Report No. 88 (Issangya et al, 2014)

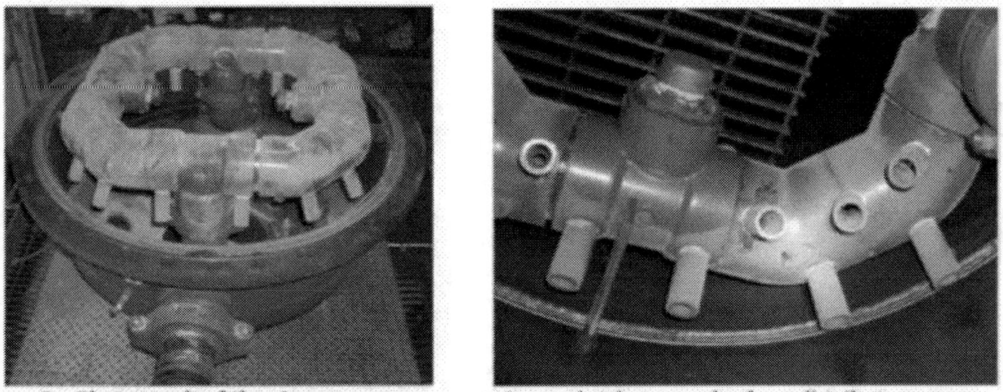

Figure 6a. Photograph of the ring sparger gas distributor installed in the 3-ft (0.9-m)-diameter stripper.

Figure 6b. Photograph of gas distributor as seen from below

Figure 2: Photographs of fluidized bed ring sparger. Reproduction of Figure 6 from Research Report No. 88 (Issangya et al, 2014)

Particle Properties

FCC catalyst particles with an 8% fines content were used in the experimental system. The particle size distribution (PSD) of these particles is shown in Figure 3, and particle properties are listed in Table 1.

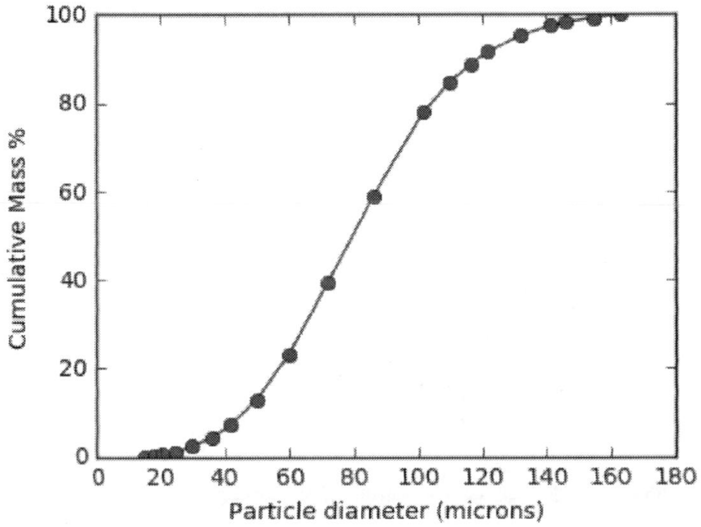

Figure 3: Particle size distribution of FCC catalyst particles

Table 1: FCC catalyst particle properties

Parameter	Value
Particle density	93 lb/ft^3 (1490 kg/m^3)
Sauter mean diameter	70.3 microns

Experimental Results

Over a range of operating conditions, PSRI reported the apparent bed density in the baffled section of the fluidized bed stripper, based on the pressure drop over an 11.2 ft (3.42 m) section of the bed. The apparent bed density is calculated from the measured pressure drop by the formula:

$$\rho_{bed} = \frac{\Delta P}{gL} \tag{1}$$

where:

ρ_{bed} is the apparent bed density (kg/m^3)

ΔP is the measured pressure drop (Pa)

g is the gravitational constant, 9.81 m/s^2

L is the length over which the pressure drop is measured (m)

The experimental results are shown in Figure 4. The apparent bed density was found to be inversely proportional to the superficial gas velocity, and also inversely proportional to the solids flux.

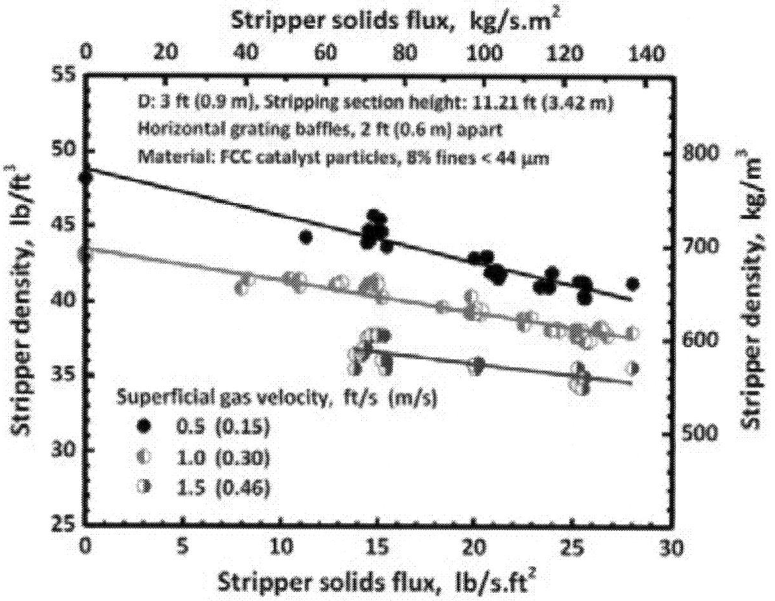

Figure 21. Stripper density as a function of solids flux for three superficial gas velocities in the 3-ft (0.9-m)-diameter fluidized bed stripper with subway gratings using 8% fines FCC catalyst particles.

Figure 4: Experimental results of apparent bed density over the baffled section of the fluidized bed stripper. Reproduction of Figure 21 from Research Report No. 88 (Issangya et al, 2014)

Simulations

Computational models of the stripper section of the experimental unit were set up and run using Barracuda Virtual Reactor version 17.1. Barracuda Virtual Reactor is a commercial multiphase CFD code based on the multiphase particle-in-cell (MP-PIC) numerical method (Snider, et al, 1998; Snider, 2001). It has been used to simulate fluidized bed systems such as FCC regenerators (Singh, et al, 2017), coal gasifiers (Snider, et al, 2011), circulating fluidized bed (CFB) combustors (Blaser, et al, 2012), polysilicon reactors (Parker, 2011), polymerization processes (Kashyap, 2016), and chemical looping combustors (Parker, 2014).

Multiple simulations of the PSRI 3 ft (0.9 m) diameter stripper with subway grating baffles were run, corresponding to various gas superficial velocities and various solids fluxes. The purpose was to simulate a wide range of operating conditions and compare with the experimental data reported by PSRI, enabling model validation and providing confidence in the use of simulation for scale-up and troubleshooting.

Domain and Boundary Conditions

A 3-dimensional CAD model of the fluidized bed stripper section was created and used in the simulations. The standpipe, risers, and external cyclones were not considered as part of the computational domain. A computational grid consisting of 727,000 real cells was used to resolve the geometry. Virtual Reactor™ uses cut cells to represent walls and internal structures. In the current model, the cyclone dipleg, fluidized bed ring sparger, helium pipe sparger (used in stripping efficiency tests) and subway grating baffles were captured by the computational grid.

Figure 5 shows the boundary conditions (BCs) used in simulations with non-zero particle flux through the stripper section; cases with zero particle flux used the same BCs, except that they did not include the bottom particle withdrawal flow BC. In all cases, particles were allowed to exit the system through the top pressure BC if they were entrained with the gas flow. All particles entrained from the system were returned via a BC Connector through the cyclone dipleg flow BC. This particle return was assumed to be 100% efficient, in order to maintain a constant particle mass in the simulation.

In cases with non-zero solids flux, a particle withdrawal flow BC was defined and the solids mass flux was specified. All particles withdrawn at this flow BC were returned via a second BC Connector to the cyclone dipleg flow BC. Again, this return was assumed to be 100% efficient, in order to keep the particle mass in the system constant. The particle withdrawal rates used for the flux values (Gs) considered in this work are listed in Table 2. Interstitial air was allowed to exit with the particles at the withdrawal flow BC, mimicking standpipe behavior.

Table 2: Particle mass flow rates, for solids mass flux values considered

Gs ($lb/ft^2/s$)	Gs ($kg/m^2/s$)	Particle mass flow rate (kg/s)
0	0	0
15	73.4	48
20	97.9	64
30	146.8	96

Fluidizing air was supplied via 28 discrete injection BCs placed near the fluidized bed ring sparger. The inner injection BC locations were moved inward from the physical nozzle locations (toward the center of the vessel), approximating a 4-inch jet penetration length. The total mass flow rate of air through these injection BCs was specified to match the superficial gas velocity used in the experimental system (either 0.5, 1.0, or 1.5 ft/s, depending on the condition being simulated). The air entering the system through the ring sparger was assumed to be at a temperature of 299.8 K, and a pressure of 140 kPa, resulting in a gas density of 1.63 kg/m^3. The resulting gas mass flow rates used for the superficial velocities considered in the current work are listed in Table 3.

Table 3: Gas mass flow rates from air ring, for superficial gas velocities considered

Ug (ft/s)	Ug (m/s)	Gas mass flow rate (kg/s)
0.5	0.15	0.163
1.0	0.30	0.326
1.5	0.46	0.489

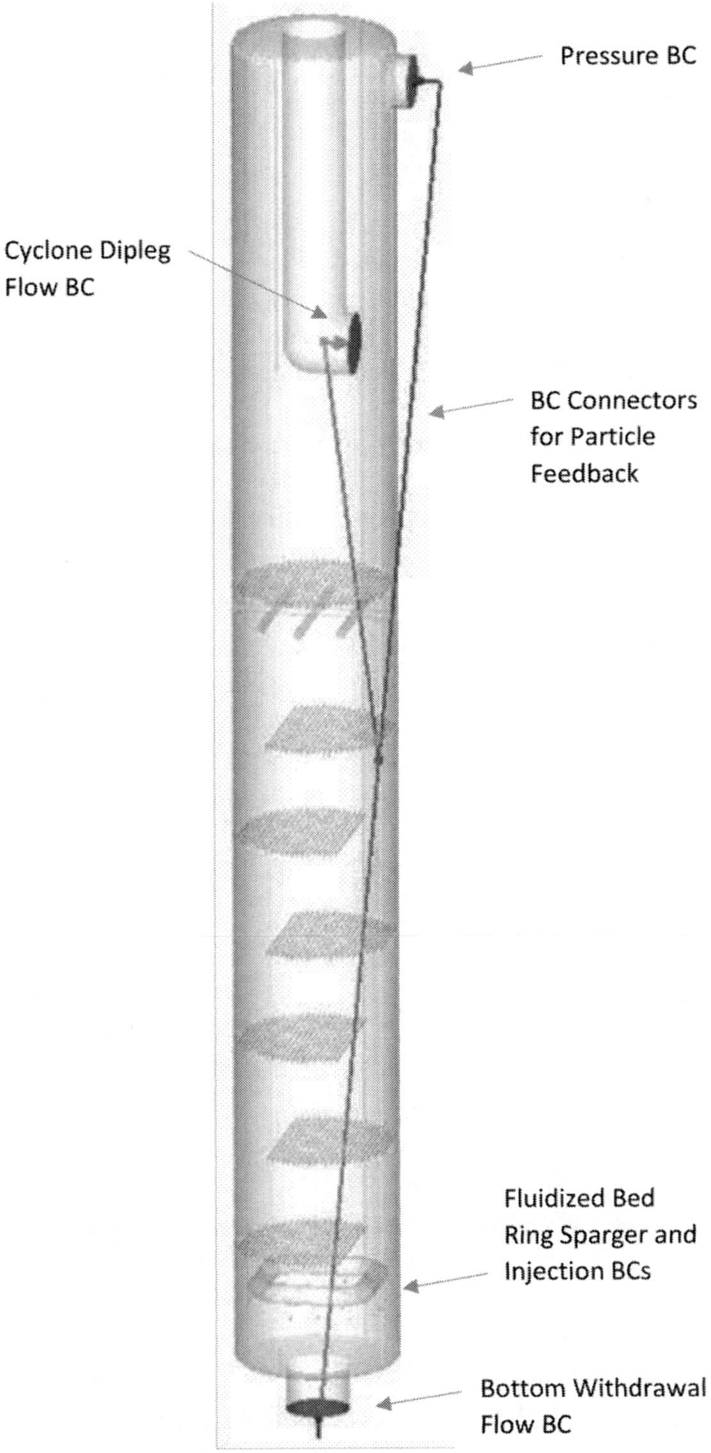

Figure 5: Computational model boundary conditions

Model Parameters

Simulations of fluid and particle dynamics are based on a combination of fundamental physics concepts, such as conservation laws, supplemented with empirical models. In this simulation, as in many multiphase simulations of fluidized bed systems, one of the most influential empirical models was the drag used to couple the particle and fluid phases. So-called "drag laws" are not physical laws, but rather models used by necessity to describe the forces exchanged between the particle and fluid phases.

The drag model used in this work was a blend of the drag model proposed by Ergun (1949) and that proposed by Wen and Yu (1966), with Sauter-mean diameter used as a basis for the drag calculation. A multiplier of 0.7 was applied to the Wen and Yu portion of the drag model, which reduced the drag force applied to particles when they were not near the close-pack particle volume fraction. This multiplier was chosen based on a calibration process in which various multiplier values were explored and compared with available test data. While calibration is standard practice for fluid-particle simulations, it should be noted that the model calibration was performed for a single operating condition, with all subsequent simulations using the identical empirical and numerical model settings.

Simulation Results

A visual representation of the simulation results for Ug = 1 ft/s, and various solids flux values, is shown in Figure 6. This is a half-slice view of the system at its center plane. Quantitatively, the Virtual Reactor simulation results are compared with experimental data for superficial gas velocities of 0.5 ft/s, 1.0 ft/s, and 1.5 ft/s in Figure 7, Figure 8, and Figure 9, respectively. In general, the simulation results compared very well with the experimental data. The inverse relationship between solids flux and apparent bed density was captured, as was the inverse relationship between gas superficial velocity and apparent bed density.

The largest deviation between simulation results and experimental data was observed in the solids flux range of 15 to 20 $lb/ft^2/s$ for the lowest superficial velocity (0.5 ft/s), as shown in Figure 7. These are cases where the amount of gas withdrawn at the bottom exit BC was approaching the amount of gas being introduced by the fluidizing ring sparger. However, even in the low gas flow case, the trend was captured by the simulation results.

Discussion

Scale-up and troubleshooting of fluidized bed processes remains challenging due to non-linear process scalability combined with the limited data available at full scale, industrial facilities. Large scale test facilities enable the exploration of concepts and designs with strong industrial applicability. However, the large scale tests also provide data useful for calibration of numerical and empirical models used in CFD simulation.

In this work, the model calibration was performed for a single operating condition (Gs = 0 $lb/s/ft^2$, and Ug = 1.0 ft/s) before being verified to capture the experimentally-observed behavior at ten other combinations of gas and solids flux. This provides a modeler confidence in the applicability of the model to capture the general fluidization trends, rather than precise agreement for a single case. For cases where large scale test data is inaccessible, model calibration can be performed using available operational data. Ideally a computational model can be calibrated and verified using fundamental test data as well as operational data, but this work demonstrates the power of a single operational point to inform model settings with a significant range of applicability.

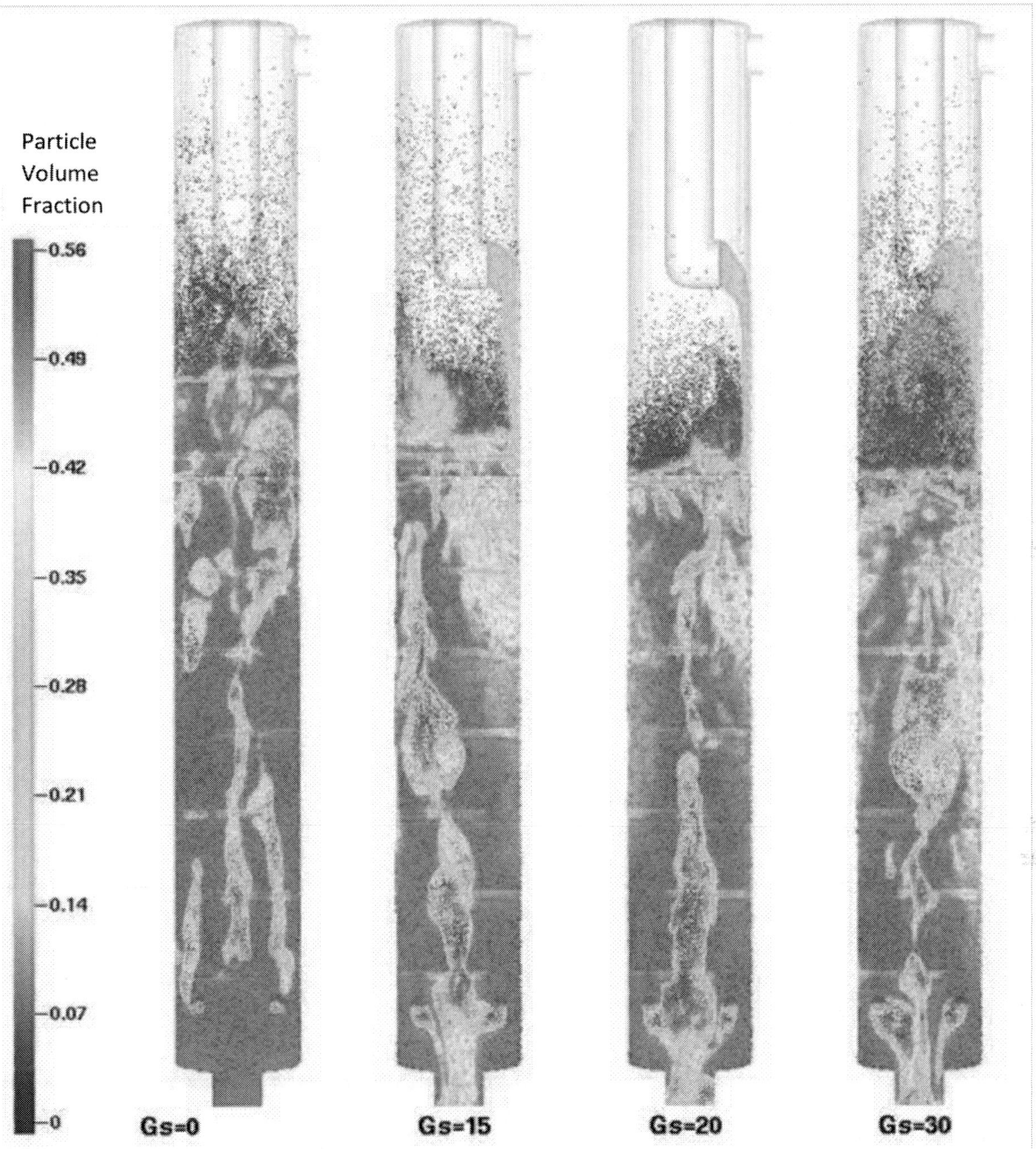

Figure 6: Half-slice view of particles colored by volume fraction, for simulations with Ug = 1 ft/s

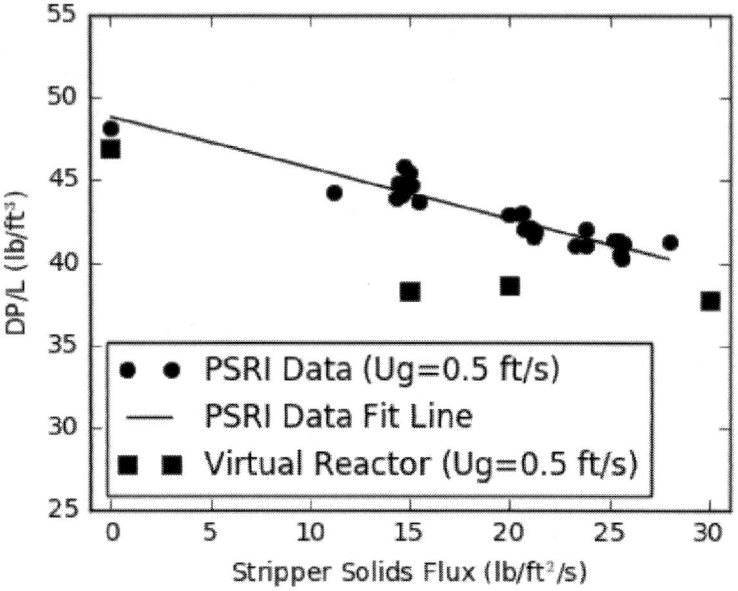

Figure 7: Simulation results compared with experimental data for Ug = 0.5 ft/s

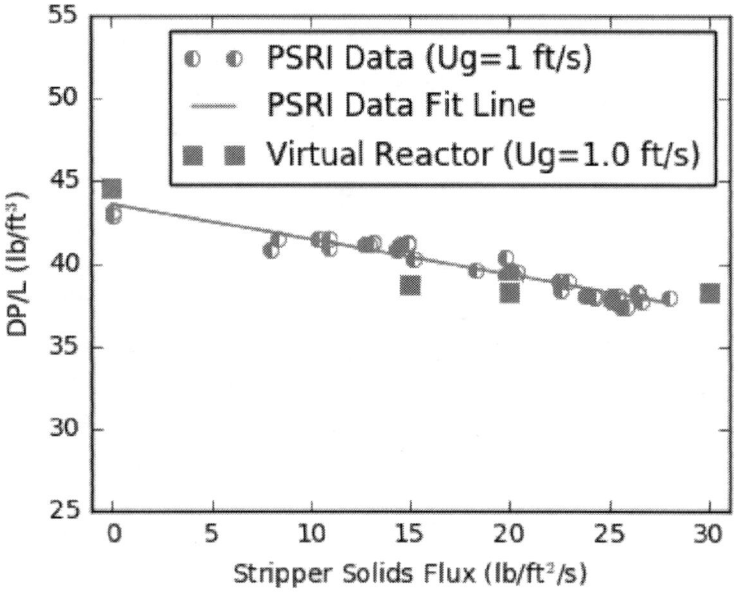

Figure 8: Simulation results compared with experimental data for Ug = 1.0 ft/s

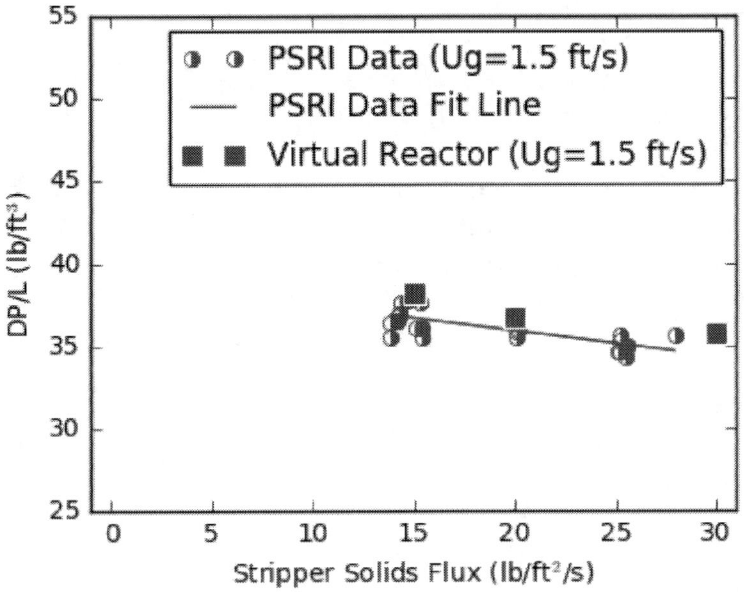

Figure 9: Simulation results compared with experimental data for Ug = 1.5 ft/s

Conclusions

Numerical simulations have been performed of a PSRI cold-flow FCC stripper experiment with subway grating baffles. The experimental unit had an internal diameter of 3 feet, and a height of 25 feet. The subway grating baffles within the vessel were intended to promote more uniform fluidization of particles by preventing the formation of large bubbles as well as preventing gas bypassing in the stripper. The commercial computational fluid dynamics (CFD) software Barracuda Virtual Reactor, which implements the multi-phase particle-in-cell (MP-PIC) numerical method, was used to perform the simulations. Simulation results were compared with experimental measurements of apparent bed density over a wide range of solids flux conditions, at superficial gas velocities typical of industrial FCC stripper operation. The simulation results were in good agreement with the experimental bed density measurements, and matched the experimentally observed trends of bed density variation as a function of gas superficial velocity and catalyst mass flux through the stripper vessel.

References

1. Issangya, A. S., Karri, S. B. R., and Cocco, R. A., 2014. "PSRI Research Report No. 88: Performance of Subway Grating Baffles in a 3-ft (0.9-m)-Diameter Fluidized Bed Stripper Using FCC Catalyst Particles." Particle Solid Research, Inc. Full report available to PSRI member companies only. Figures and data referenced in the current work have been used with permission from PSRI.
2. Snider, D.M., P.J. O'Rourke, and M.J. Andrews. 1998. "Sediment Flow in Inclined Vessels Calculated Using a Multiphase Particle-in-Cell Model for Dense Particle Flows." *International Journal of Multiphase Flow* 24 (8): 1359–82. doi:10.1016/S0301-9322(98)00030-5.

3. Snider, D.M. 2001. "An Incompressible Three-Dimensional Multiphase Particle-in-Cell Model for Dense Particle Flows." *Journal of Computational Physics* 170 (2): 523–49. doi:10.1006/jcph.2001.6747.
4. Singh, Raj, and Eusebius Gbordzoe. 2017. "Modeling FCC Spent Catalyst Regeneration with Computational Fluid Dynamics." *Powder Technology* 316 (July): 560–68. doi:10.1016/j.powtec.2016.10.058.
5. Snider, Dale M., Samuel M. Clark, and Peter J. O'Rourke. 2011. "Eulerian–Lagrangian Method for Three-Dimensional Thermal Reacting Flow with Application to Coal Gasifiers." *Chemical Engineering Science* 66 (6): 1285–95. doi:10.1016/j.ces.2010.12.042.
6. Blaser, Peter J., and Giorgio Corina. 2012. "Validation and Application of Computational Modeling to Reduce Erosion in a Circulating Fluidized Bed Boiler." *International Journal of Chemical Reactor Engineering* 10 (1). doi:10.1515/1542-6580.3001.
7. Parker, James M. 2011. "Validation of CFD Model for Polysilicon Deposition and Production of Silicon Fines in a Silane Deposition FBR." *International Journal of Chemical Reactor Engineering* 9 (1). doi:10.1515/1542-6580.2549.
8. Mayank Kashyap, "Application of multiphase flow CFD in the gas phase polymerization process" in "Fluidization XV", Jamal Chaouki, Ecole Polytechnique de Montreal, Canada Franco Berruti, Wewstern University, Canada Xiaotao Bi, UBC, Canada Ray Cocco, PSRI Inc. USA Eds, ECI Symposium Series, (2016). http://dc.engconfintl.org/fluidization_xv/65
9. Parker, James M. 2014. "CFD Model for the Simulation of Chemical Looping Combustion." *Powder Technology* 265 (October): 47–53. doi:10.1016/j.powtec.2014.01.027.
10. Ergun, S. 1949. "Fluid flow through packed columns." *Chemical Engineering Progress, 48:89,*
11. Wen, C.Y., and Yu, Y.H. 1966. "Mechanics of fluidization." *Chemical Engineering Progress Symposium*, pages 100–111.

502888 Silica Nanoparticles as Wettability Modifier and Mobility Control Agent in Enhanced Oil Recovery

Abstract:

Oil is recovered from a reservoir in three different stages. In the primary stage, oil is recovered due to natural pressure of the reservoir. Only a small fraction (5-10%) of total oil in place is recovered in the primary stage. In secondary stage, reservoir pressure is maintained by injection of external fluids such as water. Only 15% to 25% of original oil in place in an oil reservoir can be recovered after successive use of primary recovery and secondary recovery methods. To recover the remaining oil, tertiary oil recovery or enhanced oil recovery (EOR) methods are used. Chemical EOR is one of the most important techniques in which oil recovery is enhanced by certain chemicals such as surfactants, polymers and/or alkalis by improving the fluid properties. Water soluble polymers improve the oil/water mobility by enhancing the viscosity of displacement fluid (water). Surfactants are injected to change the wettability of rock and to lower the interfacial tension between water and oil. Recently, nanoparticles have also attracted the attention due to their ability to change the wettability of rock, to lower the interfacial tension and to improve the rheological properties of water soluble polymers.

The effect of silica nanoparticles on carbonate rock wettability, interfacial tension and on rheological properties of partially hydrolyzed polyacrylamide was investigated. Rheological properties were assessed using Discovery hybrid rheometer while interfacial tension and the contact angle were measured using a tensiometer. The addition of silica nanoparticles significantly improves the rheological properties of polyacrylamide. However, the improvement in rheological properties is concentration dependent and an optimum concentration is required. Detail results will be discussed in the presentation.

488461 Hierarchical Porous Reduced Graphene Oxide As High-Performance Anode for Li-Ion Batteries

Abstract:

Lithium-ion batteries (LIBs) are modern energy storage devices that have shown substantial promise in solving the global energy demand. Among possible electrode materials, graphene has emerged as one of the foremost candidates, due to their unique material properties. A scalable and cost-effective approach for fabricating graphene-based electrode materials involve the use of graphene oxide (GO) obtained via direct exfoliation of graphite. However, due to the van der Waals interactions and high aspect ratio of GO sheets, GO sheets can easily restack and form a bulk graphite-like structure during conventional fabrication process. The restacking of graphene sheets result in significant loss of usable Li-insertion sites, and consequently, a decrease in specific capacity and rate capability of the electrode. Therefore, it is necessary to develop synthesis strategies that preserve graphene interlayer space and Li-insertion sites. In this work, with the assistance of micro oil-gel as a template, we report the design of hierarchical macro/meso-porous reduced graphene oxide (RGO) materials as anodes in LIBs. Macropores with diameter ~200nm could serve as â€œhighwayâ€ for Li ion transport, while mesoporous structure along the macropore walls enhance the overall surface area (~276 m^2/g) of RGO anodes and further shorten the overall Li ion transport length. The resulting RGO anode material shows a high reversible capacity (~720 mAh/g at 0.2C) and high rate capability (~160 mAh/g at 20C). Furthermore, the process to fabricate such hierarchical structures is novel, facile and scalable, and opens up a promising strategy to develop enhanced electrodes for lithium ion batteries.

485307 Invited - Synthesis, Fabrication, and Characterization of Thick Lithium-Ion Battery Electrodes

Abstract:

While materials chemistry limits the maximum electrochemical energy stored in a battery cell, the design and processing of the battery electrode can have significant impact on the power and energy density at the battery cell level. Generally, thicker battery electrodes will result in higher energy density at the cell level; however, fabricating thick electrodes in some cases results in processing challenges because of mechanical limitations of the electrode structure. In addition, thick electrodes result in greater path lengths for the electrons and ions that must traverse the electrode structure curing charge and discharge and thus resistances can result that make practical rates of charge and discharge a challenge. This talk will introduce a method developed in our group to synthesize sub-micrometer lithium-ion battery particles that are fabricated into thick electrode architectures. These particles can be processed into architectures with both nanoscale and macroscale porosity that reduce the transport limitations within the cell architecture. In addition, we will describe methods to characterize these thick electrodes which provide insights into the tortuosity of the electrodes and the process-property dependence of these battery materials.

494007 Flower-Structured 1T Metallic Phase MoS_2 As Electrode Materials for Solar Cells

Abstract:

Molybdenum disulfide (MoS_2) layered nanosheets on a fluorine-doped tin oxide (FTO) glass substrate were prepared as a platinum (Pt)-free counter electrode (CE) for dye-sensitized solar cells (DSSCs). It was found that the DSSC with 1T phase MoS_2 as counter electrode exhibited a much higher efficiency than that with 2H phase MoS_2 as counter electrode. The excellent electrocatalytic activity of the 1T phase MoS_2 counter electrode was further demonstrated by cyclic voltammograms, in which the 1T phase MoS_2 showed higher peak current density and lower peak-to-peak separation positions than 2H phase MoS_2.

499843 Scalable Self-Assembly of Nanoparticle Antireflection Coatings

Abstract:

Antireflection coatings on transparent substrates (e.g., glass) are important components for a large number of optical and optoelectronic devices, such as displays, lenses, and photovoltaic panels. Traditional quarter-wavelength antireflection coatings produced by vacuum deposition suffer from high operating and equipment costs, limited material selection, low throughput, and small coating areas. Although some solution processing technologies have been developed to reduce the manufacturing cost and improve the production throughput, many of these techniques involve multiple steps, are limited to single-sided coatings on planar substrates, are not very reproducible over large areas, and/or are not inherently parallel for industry-scale manufacturing. Here we report a simple, inherently parallel, and scalable bottom-up approach for fabricating nanoparticle antireflection coatings on large glass substrates. Negatively charged silica nanoparticles are electrostatically adsorbed onto a surface-functionalized glass substrate with positive surface charges to form a monolayer nanoparticle coating. This innovative technology enables the simultaneous coating of multiple 5-in.-sized glass substrates with high and reproducible qualities. Importantly, we combine experiments with theoretical simulations to address a basic question for monolayer nanoparticle-based antireflection coatings - what is the optimal nanoparticle surface area coverage for best antireflection performance? Surprisingly, the theoretical optimal nanoparticle surface area coverage is found to be ~60% instead of 74% for a close-packed colloidal array. We have demonstrated that this optimal nanoparticle coverage is readily achievable by the current electrostatics-assisted bottom-up technology.

483681 Hybrid Nanocoatings of Graphene-ZnO-Binder on Steel Surfaces to Reduce Friction and Wear Under High Load Conditions

Abstract:

A novel tertiary graphene-zinc oxide-PVDF composite film is developed on the steel surfaces as a solid-state lubricant for friction and wear reduction under extreme load conditions. Superior tribological performance was achieved under ambient conditions using a ball-on-disk tribometer with contact pressures up to 1.02 GPa and sliding distances up to 450 m. The graphene-rich lubricant demonstrates substantial friction and wear reduction (ca. 90 %) compared to unlubricated sliding. The exceptional tribological performance of graphene-rich composite is ascribed to the synergistic adhesion effect from zinc oxide on steel surface. Malleable zinc adheres graphene to the contact interface, maintaining improved tribological performance under high contact pressure. The stability and flexibility of this new hybrid adhesive coating suggest excellent potential as a dry lubricant for high load-bearing applications.

496417 Laying to Rest the Magnetically Dead Layer in Magnetic Nanoparticles

Abstract:

Decades of research focused on size and shape control of iron oxide nanoparticles have led to methods of synthesis that afford excellent control over physical size and shape, but comparatively poor control over magnetic properties. Synthesis methods based on thermal decomposition of organometallic precursors yield particles with mixed iron oxide phases resulting in poor magnetic properties. One feature of this is the existence of a relatively thick (several nm) "magnetically dead layer" calculated from the difference between the physical and magnetic diameters of the nanoparticles. Guided by the hypothesis that traditional thermal decomposition syntheses are carried out under conditions where oxygen is the limiting reactant, we identified conditions for the safe stoichiometric addition of molecular oxygen as a reactive species.[1] Synthesis of nanoparticles by popular routes, such as the heating-up and extended LaMer methods, with and without controlled oxygen addition into the reactor clearly shows that particles synthesized with controlled oxygen addition have magnetic properties approaching those of bulk magnetite and have thin (< 1 nm) magnetically dead layers, whereas particles synthesized in the absence of oxygen have relatively thick (up to 5 nm) magnetically dead layers, particularly for particle physical diameters above 20 nm. Furthermore, high resolution electron microscopy suggests that particles synthesized with controlled addition of oxygen consist of uniform single crystals with minimal defects. The improved magnetic properties and thin magnetically dead layers lead to improvements in functional particle properties relevant for biomedical applications.

References:

1. Mythreyi Unni, Amanda Uhl, Shehaab Savliwala, Benjamin Savitzky, Roham Dhavalikar, Nicolas Garraud, David Arnold, Lena Kourkoutis, Jennifer Andrew, and Carlos Rinaldi, "Thermal decomposition synthesis of iron oxide nanoparticles with diminished magnetic dead layer by controlled addition of oxygen." *ACS Nano*, **11**(2):2284-2303, 2017. [doi: 10.1021/acsnano.7b00609]

497648 *In vitro* Study of a Drug Delivery Systems Constituted By Levan and 5-Fluorouracil with Different Cancer Cell Lines with an Overexpression in GLUT5

Abstract:

Levan is a fructose polymer that shows a great potential for biomedical applications due to its amphiphilic character and its affinity for a fructose transporter (GLUT5). This transporter is overexpressed in some cancer types, such as liver, colon or breast. However, drug delivery systems constituted by levan and a pharmaceutical have not been tested *in vitro* with the previous cancer cell lines. Based on the previous facts, we developed a drug delivery system levan and 5-fluorouracil and we tested *in vitro* its effect in different cancer cell lines with an overexpression in GLUT5 (liver, colon and breast cells). We obtained nanoparticles levan-5-fluorouracil of about 200-400 nm with a Z-potential value of -25 mV and *in vitro* results showed that levan can be used as a carrier for this purpose, decreasing the cell viability. Besides, levan without the drug also decreases the cell viability, showing a greater effect for colon cancer cell lines (cell line with the higher GLUT5 overexpression). According to the results, this biodegradable polymer can be used without a drug to treating cancer cell lines with overexpressed GLUT5.

488361 Modulating Antimicrobial Activity and Mammalian Cell Biocompatibility with Glycosylated Miktoarm Star Polymers

Abstract:

The development of novel reagents and antibiotics for combating multi-drug resistance bacteria has been receiving significant attention in recent years. The use of synthetic biomacromolecules/nanoparticles prepared via controlled polymerization techniques is regarded as one of the most promising approaches to overcome such challenge because of the ability to precisely engineer the chemical functionality, and hence the biological properties of the nanomaterial (Lam et al., 2016). Several research groups have reported on the synthesis of antimicrobial polymers to combat the rise of antibiotic-resistant pathogens. However, the majority of these materials exhibit poor selectivity (Hancock and Sahl, 2006). Even though they are effective in killing bacteria, they are inherently harmful towards healthy mammalian cells, which hinder their potential for clinical applications.

In here, we describe the fabrication of new antimicrobial star polymers (< 30 nm in size) consisting of mixtures of polylysine and glycopolymer arms that not only possess good antimicrobial efficacy towards Gram-positive bacteria, including methicillin-resistant Staphylococcus aureus (MRSA) and vancomycin-resistant Enterococci (VRE) (with MIC values as low as 16 Î¼g mL-1), but are also non-hemolytic (HC50 > 10000 Î¼g mL-1) and exhibit excellent mammalian cell biocompatibility. This new class of antimicrobial star polymers is prepared via a strategic combination of modern synthetic polymer chemistry protocols that entails click chemistry, reversible addition fragmentation chain transfer (RAFT) polymerization, N-carboxyanhydride ring-opening polymerization (NCA-ROP), and macromolecular assembly/cross-linking. The antimicrobial activity and mammalian cell biocompatibility of the star nanoparticles can be tuned efficiently depending on the molar ratio of polylysine to glycopolymers arms and the type of sugars employed (e.g., glucosamine, glucose and galactose). The technology described herein thus represents an innovative and effective approach in fighting deadly infectious diseases.

492031 Electrospinning Polymer Nanomedicines Extends Shelf-Life and Size Stability

Abstract:

Polymeric nanoparticle drug formulations can increase drug bioavailability and lower toxicity when compared to the free drug. However, nanoparticles require drying to improve long-term size stability and shelf-life for clinical use e.g. chemotherapy. To maintain threptic efficacy, drying techniques must avoid irreversible aggregation, redispersion should not require high-energy mixing e.g. sonication, and the redispersed solution must have appropriate osmolarity for intravenous administration. Current drying techniques, freeze-drying and spray-drying, require high concentrations of cyroprotectants (sugars) to avoid irreversible aggregation but the resulting dispersions can be hypertonic. We propose to dry nanoparticles via electrospinning with water-soluble polymers to decrease the osmolality of the redispersed nanoparticles while maintaining nanoparticle size. In electrospinning, a dissolved polymer is extruded, as an electric field is applied, and the polymer jet is elongated into nanofibers that are collected in a nonwoven pattern. Flash Nanoprecipitation (FNP) a bottom-up, polymer directed self-assembly method, was used to create polymer nanoparticles because it is a rapid and scalable method for encapsulating drugs into nanoparticles with high drug loading and narrow size distribution. Polyvinyl alcohol (PVA) was used for electrospinning as proof-of-concept. The polymer nanoparticles were added to dissolved polymer solution and electrospun. The nanoparticles were redispersed in aqueous solution using hand-mixing, a low energy method. Nanoparticle to nanofiber diameter (NP:NF) ratio was found to influenced nanoparticle size stability. Higher NP:NF ratio improve size stability after redispersion up to no change in size. A liquid core (vitamin E) nanoparticle was shown to have a 2-fold lower change in particle size after drying compared to solid core (polystyrene) nanoparticle. We attribute the difference due to liquid particle deformation under the shear stresses of fiber formation. Nanoparticle loading of 3.2% mass of nanoparticles to mass of polymer was achieved with no change in nanoparticle size upon reconstitution. Additionally, nanoparticle size upon reconstitution was not altered by storage time of at least 5 months at room temperature.

499770 Rapid and Facile PET-Activation of Preformed Phthalocyanine-Nanoparticles for Imaging Applications

Abstract:

Early detection of tumors is key to the success of clinical intervention and patient outcomes. Thus, there is strong demand to develop diagnostic technologies that can detect Alzheimer's, infections, and cancers at earlier stages when the disease is most manageable. Positron emission tomography (PET) is currently among the most sensitive imaging modalities for cancer diagnostics but require radionucleotide localization to targeted areas. Nanoparticle (NP) contrast agents can drastically increase local concentrations of radionuclides, lowering the detection limit and increasing the sensitively of PET imaging to enable detection of small and early-stage cancers that would otherwise be overlooked. However, previous methods to load NPs with PET radionuclides for *in vivo* imaging require harsh loading conditions or NP surface modifications that alter or diminish targeting abilities. We here present the technology for encapsulating phthalocyanine dyes inside water dispersible and polyethylene coated (PEG) NPs, using the self-assembly technique Flash Nanoprecipitation. Under the very mild conditions at pH6 and 37 C, these phthalocyanine NPs rapidly and spontaneously chelate metals, and can act as sinks for PET radionuclides, such as ^{64}Cu, to produce PET-active NPs **(Fig. 1)**. By measuring the absorbance shifts of the encapsulated dye over time, we developed a reaction rate model that can be used to predict chelation rates of non-radioactive 'cold' metals with given reaction parameters. This model was then verified with radioactive 'hot' metals, where chelation efficiency could be assessed directly by measuring NP radioactivity. We demonstrated that, within three hours, 762 NPs could chelate radioactive ^{64}Cu with 98% efficiency, and retained > 90% activity even after incubation with ethylenediaminetetraacetic acid for 12 hours **(Fig. 2)**. This result matched perfectly with the previous reaction rate model and demonstrates the use of dye absorbance shifts to accurately predict and optimize chelation reaction conditions, reducing the number of radioactive procedures required. This development, optimization, and verification of a rapid and facile radionuclide labeling method promises to increase sensitivity and expand applications of NP-based PET imaging.

498127 The Influence of Fluid Dynamics on Nanomaterial Delivery Efficiency: Elucidating the Roles of Particle Size and Cell Model

Abstract:

Colloidal silver nanoparticles (AgNPs) are being increasingly utilized in biomedical applications, such as drug/gene delivery and bio-imaging techniques. However, the effectiveness of these procedures are highly dependent upon sustained, strong interactions between AgNPs and the surrounding environment; referred to as the nano-bio interface. Therefore, prior to the development of effective NP-based therapeutics, a reliable and accurate means of assessing NP delivery must be established. Both *in vitro* and *in vivo* methodologies are currently being utilized for evaluation, however, limited correlation exists between these models due to their fundamental differences. One way to overcome this limitation is through the development of enhanced *in vitro* environments that retain the advantages of cellular systems but more accurately mimic true physiology. For example, the introduction of fluid dynamics into a cellular system can impact particokinetics, the composition of the nano-bio interface, and the subsequent cellular response.

In the work presented here, a dynamic *in vitro* environment was successfully utilized to characterize the AgNP deposition efficiency. Dynamic flow was generated through the use of a peristaltic pump, operating at a flowrate to produce a linear tube velocity of 0.2 cm/s: equivalent to know capillary rates. To better understand how dynamic flow impacted deposition, two experimental cell models were utilized; an adherent lung epithelial model (A549) and a suspension monocyte model (U937). Additionally, as bio-transport mechanisms are a function of particle size we included two experimental, citrate-coated AgNPs â€“ 5 and 50 nm.

The AgNPs were introduced into both cell models and incubated for 24 hours under either static or dynamic conditions. After this exposure, the AgNP deposition was evaluated and found to vary as a function of flow environment, cell model, and primary particle size. As expected, dynamic flow significantly decreased the delivered dose of AgNPs to the adherent lung cells; with the 5 nm AgNPs experiencing the greatest drop in deposition. However, AgNP delivery was increased within a dynamic environment for the monocytes, due to increased nano-cellular interactions. While an increase in deposition was seen for both the 5 and 50 nm particles, the 5 nm AgNPs were associated with a greater increase, roughly 20%. For both cell models, the subsequent cytotoxicity, stress, and inflammatory responses correlated to delivered NP dosages. Specifically, reactive oxygen species (ROS) and p53 levels were used to assess cellular stress and the secretion of pro-inflammatory cytokines were the selected marker for inflammatory activation. This work highlights the need for NP deposition, safety, and efficacy evaluations to be carried out in a physiologically relevant system, as environmental variables are capable of influencing NM properties, behaviors, and downstream bioeffects.

498449 Discrete Element Modeling to Predict Pharmaceutical Powder Flow at Different Humidity Conditions

Abstract:

Pharmaceutical powder flow can alter significantly based on the exposed humidity conditions. Lack of predictive models to understand the flow variability at different humidity conditions may undermine its subsequent impediments. A Discrete Element Model (DEM) has been developed to predict the effects of humidity on powder flow, altering the cohesive forces based on granular bond numbers in simple hopper geometries. Experiments similar to the simulations are further performed for different pharmaceutical excipients at different humidity conditions. To predict cohesive properties for powders, those did not flow under gravity; vertical vibration has been introduced in both experiments and simulations. The equivalent DEM bond numbers are in good accordance with the experimental results. Pharmaceutical powders with lower moisture sorption show more variability in flow properties in comparison to cohesive powders having higher moisture contents. The equivalent DEM parameters suggest the same and are being implemented to predictpowder flow variability in different geometries and humidity conditions.

491765 Long-Term Coating Predictions for in a Wurster Fluidized-Bed Coater with Using a Combined CFD-DEM and Markov-Chain Monte Carlo Approach

Abstract:

Coating of particles, in particular, microspheres, using the Wurster fluid-bed coater is a common unit operation in the pharmaceutical industry. Such coatings are typically used to layer active drugs or other ingredients that control drug release, and also for taste/odor masking and enteric barrier coatings that prevent dissolution/disintegration. To ensure control over the drug-release rate and to achieve a consistent inter-particle drug loading (potency), a high degree of coating uniformity is essential. However, relating the coating uniformity of the beads to the process parameters (such as fluidization air-flow rate, atomizing air-flow rate, Wurster-gap height, and bead size) is not trivial.

Experimental methods to characterize bead-flow dynamics inside Wurster coaters are difficult for large systems; moreover, sampling techniques are insufficient to quantify the coating uniformity of millions of beads. In such situations, coupled CFD-DEM simulations (computational-fluid dynamics coupled with discrete-element method) can model the complex particle and airflow in a fluidized bed. However, due to the high computational costs (several days per process-second for an industrial system), simulating several hours of a coating process using CFD-DEM is infeasible. Therefore, a multi-scale model is required to predict the long-term coating uniformity, based on inputs from the CFD-DEM simulations. The Markov-chain Monte-Carlo-based compartment model is one such approach to develop a reduced-order model.

The compartment model subdivides the coater into non-overlapping regions, i.e., compartments. The exchange of particles between adjoining compartments is modeled as stochastic process described by a probability matrix *P*, where P_{ij} describes the probability of a particle moving from compartment *i* to compartment *j* during some small compartment-model time step Î"t_{CM}. This probability matrix *P* is calculated from the detailed CFD-DEM particle trajectories; therefore, the exact positions of the particles are not tracked in the compartment model. Changes in the process parameters affect the flow behavior of the beads which is reflected as changes in the probability-matrix entries P_{ij}.

Certain compartments near the spray nozzle are designated as the spray zone; every particle that visits these spray compartments gains some coating mass, based on the spray rate, particle size, and the compartment-model time stepÂ Î"t_{CM}. The probability matrix *P* is then used to update the positions of the particles in various compartments. By tracking the flow of a large, representative sample of beads, the evolution of the coating mass distribution can be tracked with time - these predictions can be generated in minutes for hours of real process time using the reduced-order Markov-chain Monte Carlo method.

Results from the compartment model for a given set of process conditions were successfully validated against

experimental results of particle diameters obtained using Sympatec QicPic and coating thickness from optical-coherence tomography (OCT). After validation, a virtual design of experiments (DoE) study with varying process conditions has been performed. The results show that a larger Wurster-gap height reduces the variation of the coating mass and thickness but, owing to the initial size distribution of the uncoated beads, the variation in the coating potency increases. Depending on the application, it may be more desirable to optimize the coating potency (to maintain a constant coating-mass fraction for coatings that contain active drug), or optimize the coating thickness (e.g., for protective coatings where a predefined barrier thickness is needed). The present Markov-chain Monte Carlo approach, in combination with CFD-DEM simulations, can be used to design the optimal process parameters for both applications.

496208 Microfluidization as an Enabling Technology for Solubility Enhancement

Abstract:

Microfluidization as an enabling technology for solubility enhancement

Iris Duarte, Clara Sa Couto, Tiago Porfirio, João Vicente and Marcio Temtem

R&D Drug Product Development, Hovione Farmaciência S.A, Sete Casas, 2674-506 Loures, Portugal

INTRODUCTION

Emerging drugs are essentially poorly water-soluble and such limited aqueous solubility has been one of the major hurdles in the development of oral-dosage forms, as poor solubility typically hinders oral bioavailability. This challenge has been addressed with different formulation strategies and enabling technologies, including: complexation with cyclodextrines (CD), production of amorphous solid dispersions (ASD) or particle-size reduction, such as the production of nanosuspensions and/or nanocrystals [1].

This work describes a technique which encompasses a microfluidization followed by a spray drying step, to produce nano- to microparticles of crystalline drug-alone, cyclodextrins complexes or amorphous solid dispersions. Three different case-studies will be detailed in this work. The manufacturing process is scalable to industrial scales and can operate either in batch or continuous manufacturing mode as benchmarked in the current work.

The microfluidization step is defined as a dynamic high-pressure process where two liquid streams, solution or suspension, pass through micro-channels toward an impingement area through which the fluids flow and interact [2]. The microfluidization step renders in either a liquid solution or liquid suspension that is typically followed by a spray drying step to remove the solvents and produce the dried particulate system. A schematic illustration of the process train is shown in Figure 1. This process was successfully demonstrated using fluticasone propionate as drug alone, itraconazole (ITZ) / Eudragit L-100 amorphous solid dispersions and ITZ / cyclodextrins inclusion complexes.

Case study I: production of micro / nanocrystals using microfluidization: Fluticasone propionate was dissolved in acetone. As anti-solvent, a mass of deionized water corresponding to 10 times that of the solvent was used. The two solutions were fed and mixed using PureNano^{TM} in a single passage. The particle size of the isolated product was characterized through scanning electron microscopy, SEM (Figure 2). Roughly, particles with nano-scale were obtained [3].

Case study II: complexation of a poorly water soluble API using cyclodextrines: ITZ was dissolved in DCM and SBE-ß-CD dissolved in water. The two solutions were fed and mixed using PureNano^{TM} in a single passage. Two different trials at different temperatures were done, a filtration step was done prior to the spray dryer to remove the

API not complexed. 3% was the maximum drug load obtained in the dried product [4].

Case study III: co-precipitation of nano-solid dispersions: ITZ and Eudragit L-100, in a mass proportion 1:9 and 40:60, were dissolved in a mixture of ethanol / acetone in a volume proportion of 1:1. Acidified cold water (pH=2) was used as anti-solvent with a ratio solvent:anti-solvent of 1:10 w/w. The results indicated that amorphization of ITZ was successful when using the present method (Figure 4) [5].

CONCLUSION

This work demonstrates that the technique herein described, consisting in a microfluidization step followed by an isolation step via spray drying is able to produce drug-alone nano-crystals, cyclodextrins complexes and amorphous solid dispersions. The process is scalable to industrial scales.

Figure 1 - Experimental set-up for the batch and continuous controlled precipitation process.

Figure 2 - SEM of spray dried particles of fluticasone propionate.

Figure 3 - i) X-ray diffractograms of ITZ solid dispersion at two different ratios and its crystalline form and ii) SEM of spray dried dispersion of ITZ.

[1] Perrie Y., Rades T. "Pharmaceutics - Drug Delivery and Targeting", Pharmaceutical Press, 2010;

[2] Rakesh P. Patel, Ashok H Baria, Nikunjana A Patel, An overwiew of size reduction technologies in the field of pharmaceutical manufacturing, Asian Journal of Pharmaceutics October-December (2008) 216-220.

[3] Fonseca, T., Duarte, I., Vicente, J., Temtem, M. "Continuous production of particles", Patent No. WO2016156841 A1, 2016.

[4] Lisboa, H., Santos, F., Vicente, J., Temtem, M. "Continuous complexation of active pharmaceutical ingredients", Portuguese Patent No. 109117, 2016.

[5] Temtem, M., Pereira, R., Vicente, J., Duarte, I., Gaspar, F., Duarte, I. "A method of preparing amorphous solids dispersions in submicron range by co-precipitation", Patent No. WO2016016665 A1, 2016.

491353 Droplet-Templated Anti-Solvent Spherical Crystallization of Hydrophilic and Hydrophobic Drugs with an *in situÂ* Formed Binder

Abstract:

Conventional pharmaceutical crystallization processes in batch reactors often lead to broad crystal size distributions with frequent batch-to-batch variations. These crystals subsequently undergo several secondary formulation processes for crystal and dosage uniformity. To consolidate these processes, spherical crystallization is a promising method, which combines crystallization, milling, and granulation into a single unit operation. Droplet-templated spherical crystallization methods have been developed recently to achieve highly monodisperse crystal microspheres with droplet microfluidics and evaporation techniques. We present here a novel droplet-templated anti-solvent spherical crystallization method applicable to both hydrophilic and hydrophobic drugs. In both cases, an alginate hydrogel binder forms in situ simultaneously, effectively encapsulating the drug crystals into highly monodisperse spheres. We study in detail the crystallization and gelation processes with mass transfer modelling, and the characterization of the obtained alginate/drug spheres. While glycine and carbamazepine are used as model hydrophilic and hydrophobic drugs, this method can be generalized to other drugs as well. In particular, this method offers additional benefits such as minimal thermal impact, fast crystallization rates, high drug-binder loading ratios, and high selectivity towards metastable polymorphs.

Tailoring CSD & Shape of Ibuprofen to Improve Flow Properties in Batch & Continuous Crystallization

Yunliang He, Brian Glennon, Yuan Gao
SSPC, School of chemical & bioprocess engineering, University College Dublin
Anne-Marie Healy, Zelalem Worku
SSPC, School of pharmacy & pharmaceutical sciences, Trinity College Dublin

❖ The significance of powder flow properties in pharmaceutical industry

In pharmaceutical industry, most API and excipients are manufactured or processed in the form of powders.
Flow properties are critical for handling, storing and formulation of powders

borderline flow properties for high speed tableting

flow factor: 6.74(±0.27) and bulk density and tap density is 0.353 and 0.476 g/ml respectively.

❖ The effect of size distribution & shape on flow properties

Common belief

Big particles show good flow properties, particles >300 microns are free-flowing
Plates and needles are always poor-flowing

❖ Tailor crystal size distribution and shape of ibuprofen

micronization
Granulation
Batch Crystallization
Continuous Crystallization

Is it possible to prepare ibuprofen for direct compression from crystallization?

SSPC
sfi
www.sspc.ie INNOVATION THROUGH COLLABORATION

Batch & continuous crystallization of ibuprofen

❖ Fast crystallization to prepare fine crystals

10~30g ibuprofen/100g 50%m Ethanol-Water or 20~30g ibuprofen/100ml hexane was heated above 50℃ to dissolve and crash out by quick cooling to 5 ℃(in 20 mins).

❖ Slow crystallization to prepare big crystals

10~30g ibuprofen/100g 50%m Ethanol-Water or 25g ibuprofen/100ml hexane was heated above 50C to dissolve and then cool to about 40 cool slowly (2~6 ℃/h) from about 40C to about 5C

❖ Continuous crystallization to prepare narrow distribution big crystals

MSMPR(mixed suspension mixed product removal)
Feeding solution: 15g per 100g 50%m ethanol-water
Feeding rate:1~2ml/s
Working volume: 80ml
Residence time:40~80min
Generally, after 6~8 residence time, crystallization will reach steady state

Control of crystal size based on metastable zone

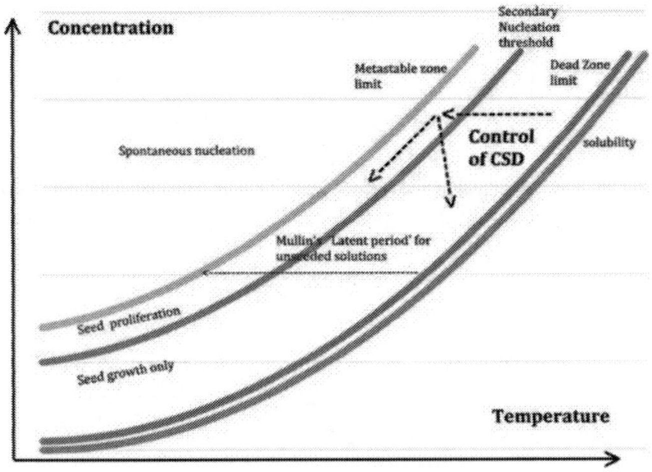

Primary and secondary metastable zone

Threlfall and Coles, Cryst. Eng. Comm. 2016, 18, 369

For batch crystallization

If crystallization was restricted in secondary metastable zone, there will be only crystal growth and no nucleation. The counts and average size R could be controlled by seed counts N.

$$R = \sqrt[3]{\frac{mass}{density * N * shape\ factor}}$$

For continuous crystallization

Crystallization must be operated in primary metastable zone. Continuous secondary nucleation will provide enough seeds to maintain steady state during continuous suspension removal. Crystal size is controlled by supersaturation and residence time.

Control of crystal size in batch crystallization

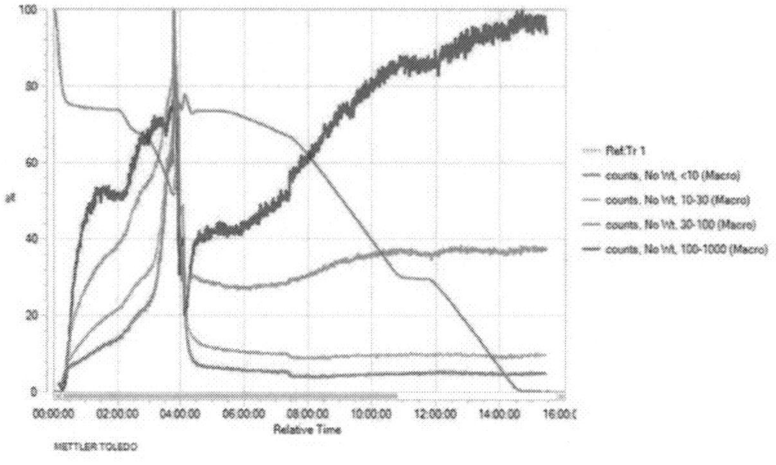

Failed Control of Crystal Size in Batch Crystallization

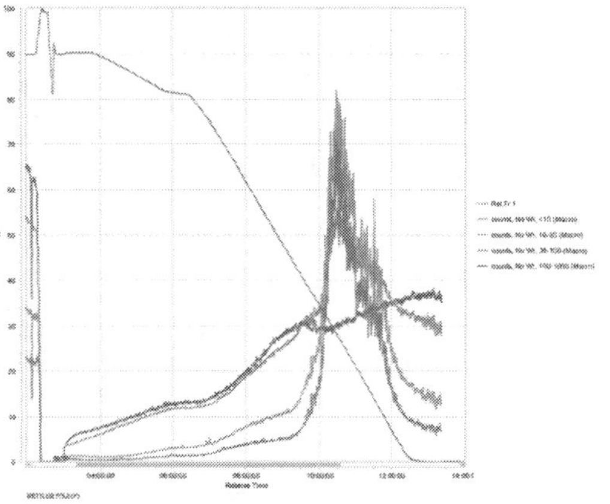

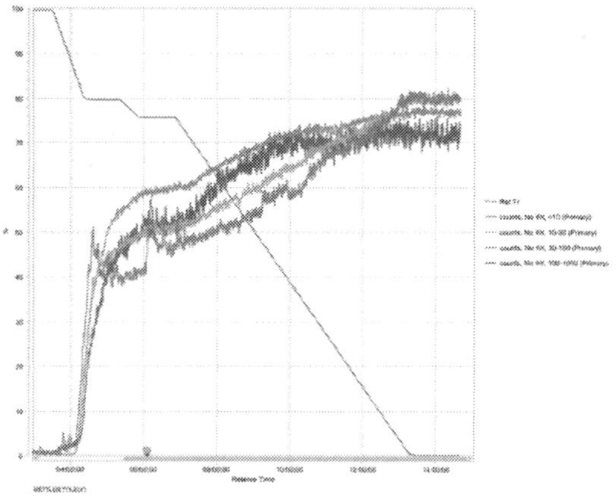

Control of crystal size in continuous crystallization

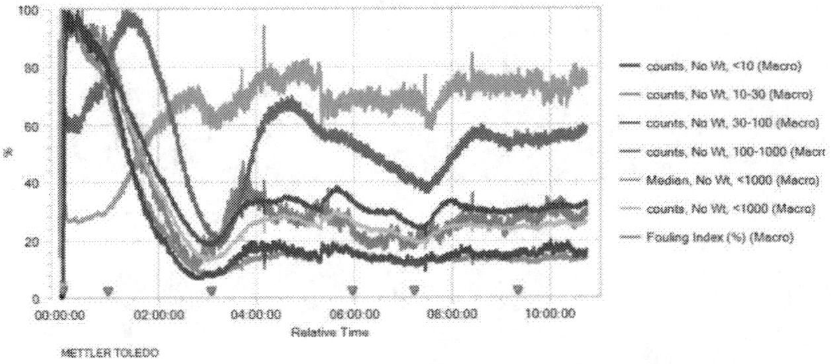

Crystal growth mode and control of shape

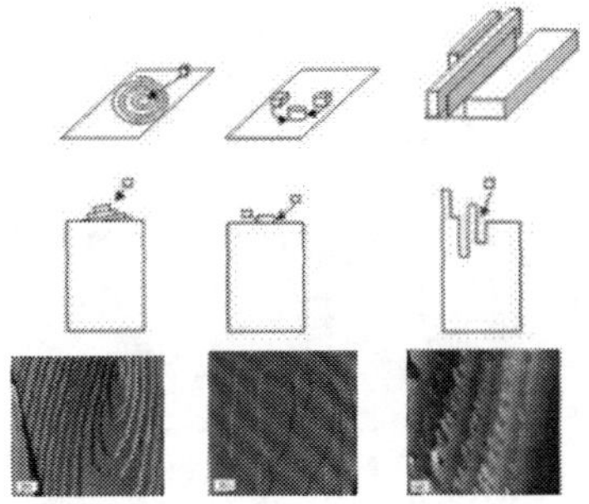

Dislocation **2D nucleation** **Continuous**

Y. Matsuzuki et al. J. Cryst. Growth 242(1-2):192

Factors on crystal shape
- ◆Intrinsic crystal anisotropy
- ◆Interfacial energy
- ◆Supersaturation

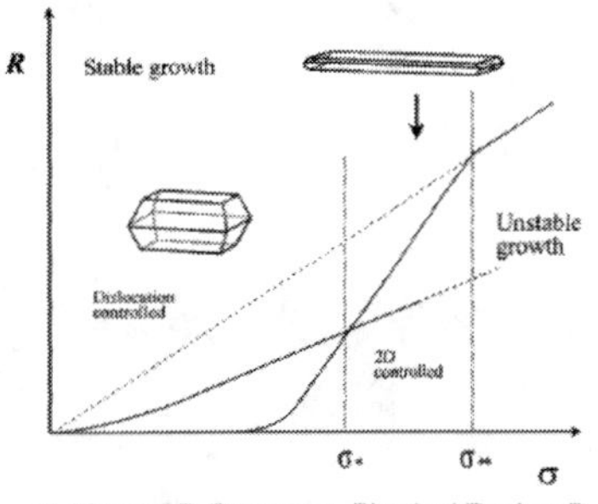

X. Liu and P. Bennema, Physical Review B 53(5):2314

Ibuprofen Crystal Shape From Controlled Crystallization

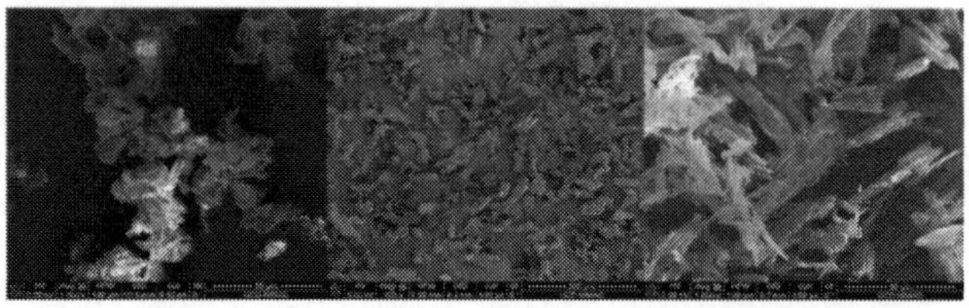

fast batch product from E-W(left & middle) and hexane(right)

continuous product from E-W

slow batch product from E-W

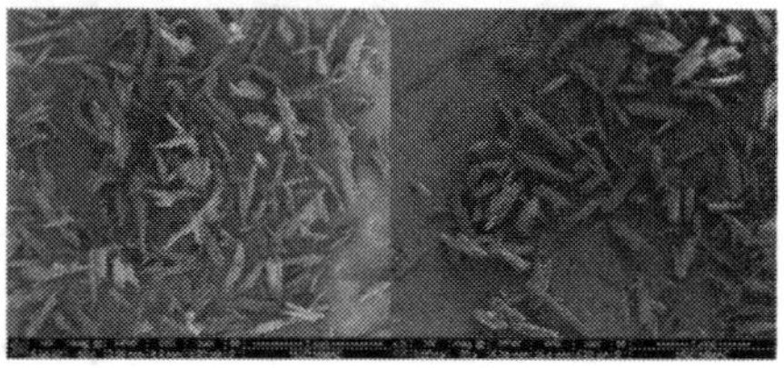

slow batch product from hexane

Comparison of CSD from batch & continuous crystallization

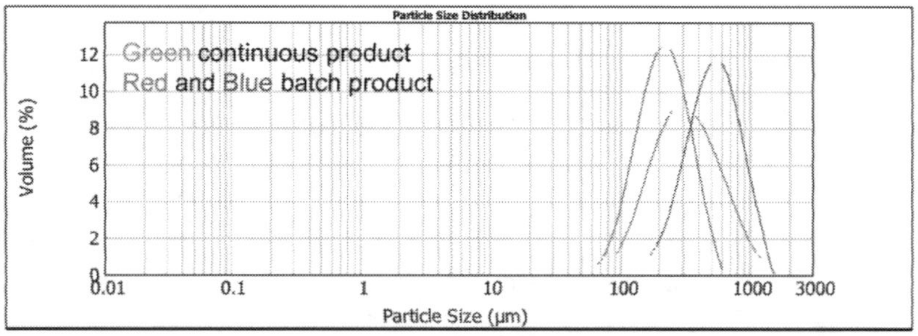

sample		Span	D [4, 3]	D [3, 2]	d (0.1)	d (0.5)	d (0.9)
continuous	Green	1.26	229.85	169.06	110.23	212.59	378.74
slow batch 1	red	1.88	364.50	163.45	123.72	303.34	693.37
slow batch 2	blue	1.40	542.40	277.64	218.59	506.72	929.39

ibuprofen crystal size distribution

Table 1 size distribution of batch and continuous crystallization product in ethanol-water

sample	Span	D [4, 3]	D [3, 2]	d (0.1)	d (0.5)	d (0.9)
Continuous1	1.42	373.07	259.46	156.62	341.15	639.69
Continuous2	1.44	275.20	194.24	119.19	247.18	475.34
Continuous3	1.32	200.34	143.47	90.89	184.74	335.40
Continuous4	1.39	252.26	166.16	108.05	231.41	430.38
Continuous5	1.18	215.49	164.80	109.80	200.61	346.29
Continuous6	1.26	229.85	169.06	110.23	212.59	378.74
Fast batch1	8.25	511.40	17.38	11.57	164.65	1370.03
Fast batch2	2.41	117.41	59.04	32.45	88.51	245.56
Fast batch3	1.57	79.73	49.34	31.46	69.95	141.52
Slow batch1	1.40	542.40	277.64	218.59	506.72	929.39
Slow batch2	1.88	364.50	163.45	123.72	303.34	693.37
Slow batch3	1.22	326.10	171.05	145.53	315.02	530.69
Slow batch4	1.18	160.11	114.33	77.85	140.03	243.60
Slow batch5	1.53	178.78	112.27	74.00	159.14	317.84

Table 2 Size distribution of product from n-hexane

sample	Span	D [4, 3]	D [3, 2]	d (0.1)	d (0.5)	d (0.9)
Fast batch	**3.29**	**44.07**	**14.40**	**7.19**	**18.31**	**67.43**
slow batch1	**3.57**	**354.63**	**105.74**	**67.86**	**219.16**	**849.73**
Slow batch2	**2.43**	**193.91**	**100.92**	**59.74**	**144.71**	**410.69**
Slow batch3	**2.06**	**288.17**	**134.67**	**85.25**	**236.76**	**573.11**

FT4 Powder Rheometer

FT4

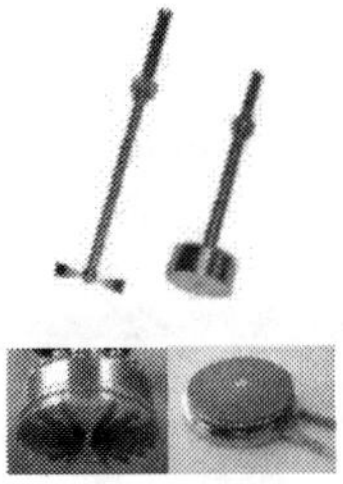
Accessories

Shear Cell Test

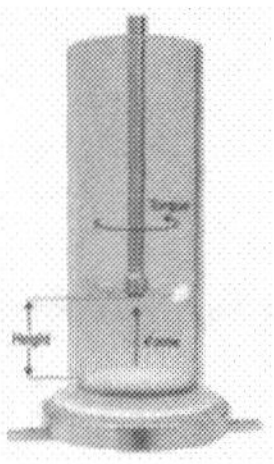
Stability &VFR

Shear cell test

FF vs D[3,2] and D[4,3]

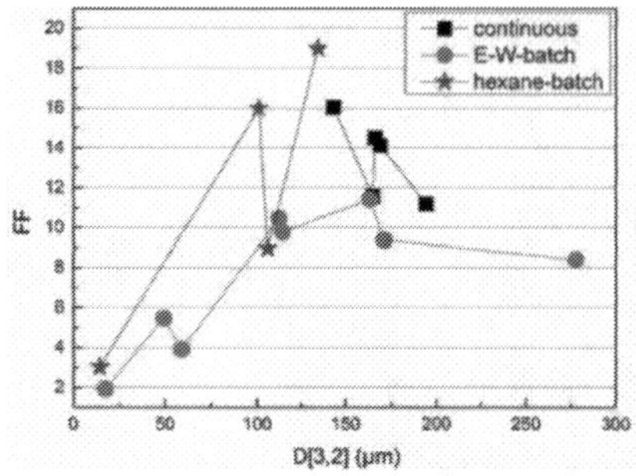

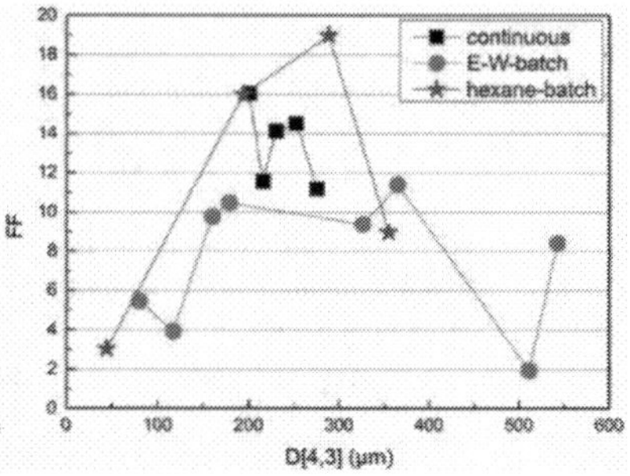

Type of flow	Flow Function Value
Free-flowing	10 < FF
Easy-flowing	4 < FF < 10
Cohesive	2 < FF < 4
Very cohesive and non-flowing	FF < 2

cohesion vs D[3,2] and D[4,3]

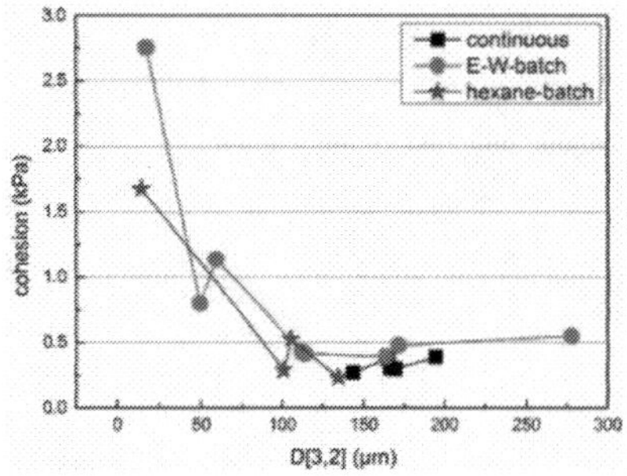

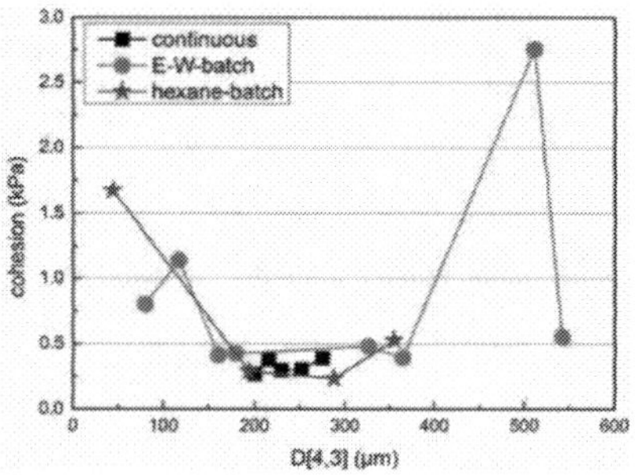

Density in shear cell vs D[3,2] and D[4,3]

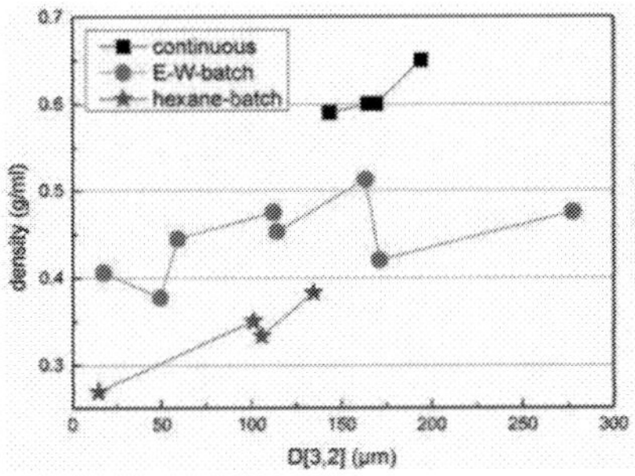

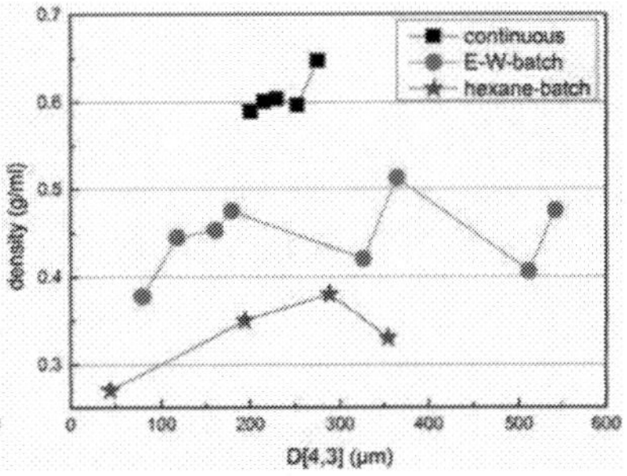

Stability & VFR test

BFE vs D[3,2] and D[4,3]

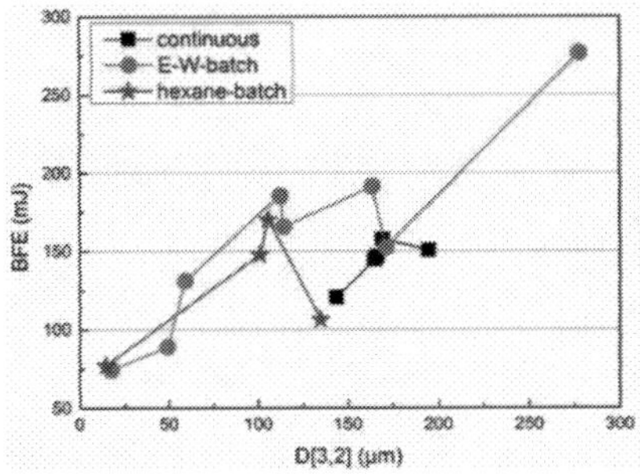

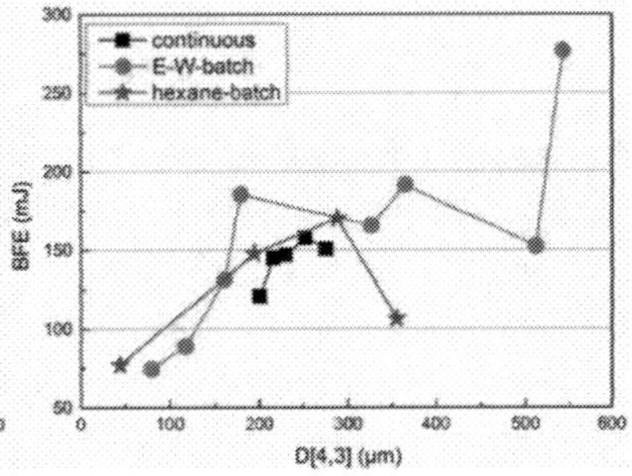

SE vs D[3,2] and D[4,3]

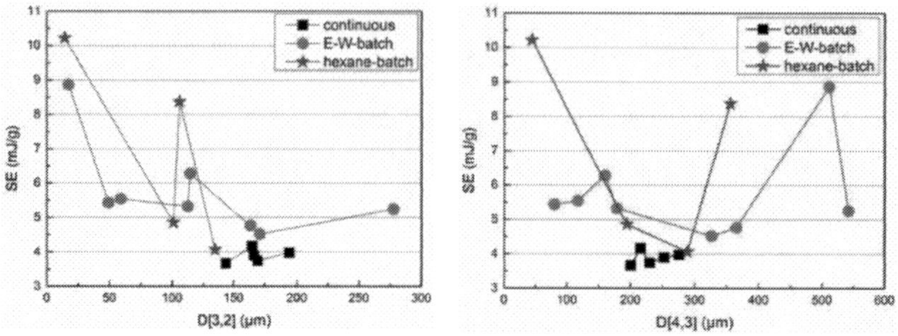

CBD vs D[3,2] and D[4,3]

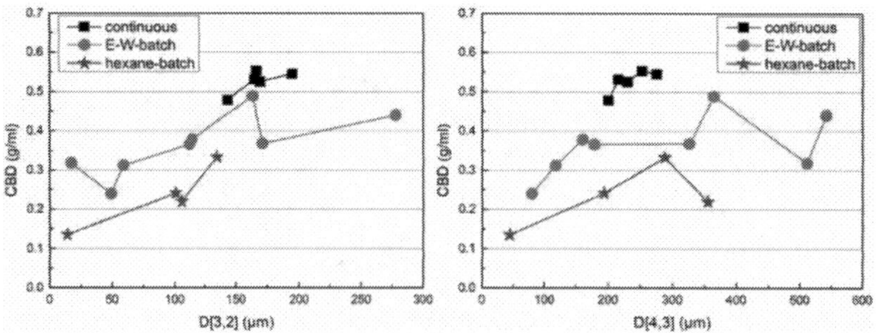

Conclusion

- Continuous products show equant shape, free-flowing, high density and small variance. Continuous crystallization is promising in preparing free-flow powders.
- Flow properties are better correlated with D[3,2] than D[4,3]
- Flow function is mainly determined by size while density is more influenced by shape than size. Rod crystals could be free-flowing by increasing size above 100 μm.
- Flowability(FF) doesn't simply increase with size. Conglomerate and shape also play critical roles in flowability.

Further study

- Investigate the direct compression and tableting behavior of tailor-made ibuprofen
- Set up a continuous manufacture process from crystallization to tableting

Thanks for your attention!

Any Questions?

www.sspc.ie INNOVATION THROUGH COLLABORATION

497920 Formation and Dissolution Characteristics of Paracetamol Granules

Abstract:

The ability of active pharmaceutical ingredients (APIs) to affectively influence the targeted region is of great importance. Therefore, there is a need to offer lasting solutions and improve upon the current methods of formulation. This research analyzes physical and dissolution characteristics to determine the bonding tendencies of paracetamol granules formed using the drop method. Paracetamol is a common API and is considered a Category III drug. Initially, a series of experiments were conducted to form similarly sized granules. The process involved creating a paracetamol powder bed and dropping liquid binder onto the bed. Granules were separated from the powder bed using sieve trays. The cohesiveness of the granules was tested by dropping the granules onto a flat surface from various heights. The weights of each granule was measured before and after dropping to determine a percent attrition. The final investigation involved the ability of the granule to dissolve over time. This testing method involved the use of a USP Apparatus 3 to prevent coning seen in Apparatus 2. Further testing was conducted using UV Vis. Various binders were considered for this research including natural and synthetic polymers. Results show that using 8 drops of binder to form granules will deliver close to the normal dosage (325 â€“ 500 mg) of the API. Synthetic binders produced more cohesive granules during the drop testing. Granules formed using polyvinylpyrrolidone (PVP) binder had the smallest percent attrition (0.1%) when dropped from a height of 24 inches. The rate of dissolution differed for each material examined; however, polyethylene glycol had a very consistent dissolution rate and reached a percent dissolution of 85% in one hour.

492697 Size Reduction through Flash Nanoprecipitation to Improve Solubility, Dissolution, and Bioavailability of Clofazimine

Abstract:

Clofazimine is a poorly soluble riminophenazine antibiotic (logP= 7.66) approved by Food and Drug Administration (FDA) with good safety record. Thus, a reliable, efficacious, and fast-release formulation of clofazimine is highly desired. To improve the drug bioavailability and dissolution rate, clofazimine was encapsulated in nanoparticles prepared with three different surface stabilizers, hypromellose acetate succinate (HPMCAS), lecithin, and zein, through the Flash Nanoprecipitation (FNP) process with high encapsulation efficiency. To convert liquid colloidal suspensions to powders, lyophilization and spray drying, were applied to solidify nanoparticles. Dissolution tests were carried out to evaluate release kinetics of clofazimine nanoparticles prepared by different surface coatings, comparing to raw clofazimine powder and its currently market product Lamprene®. Remarkably improved dissolution rate and supersaturation levels were observed with all clofazimine nanoparticles despite different release kinetics. Additionally, calorimetry technique was explored to study the crystallinity of clofazimine in nanoparticles. The results strongly suggest that the new clofazimine nanoparticles prepared with affordable materials can be used as viable cryptosporidiosis therapeutics.

AUTHOR INDEX

AUTHOR INDEX

AUTHOR INDEX

AUTHOR INDEX

AUTHOR INDEX

AUTHOR INDEX

AUTHOR INDEX

AUTHOR INDEX